JOURNAL OF CHROMATOGRAPHY LIBRARY — volume 51A

chromatography, 5th edition

fundamentals and applications of chromatography and related differential migration methods

part A: fundamentals and techniques

JOURNAL OF CHROMATOGRAPHY LIBRARY — volume 51A

chromatography, 5th edition

fundamentals and applications of chromatography and related differential migration methods

part A: fundamentals and techniques

edited by

E. Heftmann

P.O. Box 928, Orinda, CA 94563, U.S.A.

ELSEVIER
Amsterdam — Oxford — New York — Tokyo 1992

ELSEVIER SCIENCE PUBLISHERS B.V.
Sara Burgerhartstraat 25
P.O. Box 211, 1000 AE Amsterdam, The Netherlands

Distributors for the United States and Canada:

ELSEVIER SCIENCE PUBLISHING COMPANY INC.
655 Avenue of the Americas
New York, NY 10010, U.S.A.

Library of Congress Cataloging-in-Publication Data

Chromatography : fundamentals and applications of chromatography and related differential migration methods / edited by E. Heftmann. -- 5th ed.
p. cm. -- (Journal of chromatography library ; v. 51A-B)
Includes bibliographical references and indexes.
Contents: pt. A. Fundamentals and techniques -- pt. B. Applications.
ISBN 0-444-88404-1 (set : alk. paper). -- ISBN 0-444-88236-7 (pt. A : alk. paper). -- ISBN 0-444-88237-5 (pt. B : alk. paper)
1. Chromatographic analysis. I. Heftmann, Erich. II. Series.
QD79.C4C485 1992
543'.089--dc20 91-35963
CIP

ISBN 0-444-88236-7 (Part A)
ISBN 0-444-88404-1 (set)

This book is printed on acid-free paper.

Printed in The Netherlands

TO BIBI

CONTENTS

List of Authors

Dr. Hugo A. H. Billiet, Laboratorium voor Analytische Scheikunde, Technische Hogeschool Delft, De Vries van Heystplantsoen 2, NL-2628 RZ Delft, The Netherlands (Part B)

Dr. Phyllis R. Brown, Department of Chemistry, University of Rhode Island, Kingston, RI 02881, USA (Part B)

Dr. Shirley C. Churms, Department of Chemistry, University of Cape Town, Private Bag, Rondebosch 7700, South Africa (Part B)

Dr. Karen M. Gooding, SynChrom, Inc., P.O. Box 310, Lafayette, IN 47902-0310, USA (Part B)

Dr. Paul R. Haddad, Department of Analytical Chemistry, University of New South Wales, P.O. Box 1, Kensington, N.S.W. 2033, Australia (Part B)

Dr. Lars Hagel, Pharmacia LKB Biotechnology, S-751 82 Uppsala, Sweden (Part A)

Dr. Jeffrey B. Harborne, Department of Botany, University of Reading, Reading RG6 2AS, Great Britain (Part B)

Dr. F. Xavier de las Heras, Petroleum Geochemistry Group, Energy Center Building, School of Geology & Geophysics, 100 East Boyd Street, University of Oklahoma, Norman, OK 73019, USA (Part B)

Dr. Robert S. Hodges, Department of Biochemistry, University of Alberta, Edmonton, Alta. T6G 2H7, Canada (Part B)

Dr. Yoichiro Ito, Laboratory of Technical Development, National Heart, Lung, and Blood Institute, National Institutes of Health, 9000 Rockville Pike, Bethesda, MD 20892, USA (Part A)

Dr. Karl Jacob, Institut für Klinische Chemie am Klinikum Grosshadern der Ludwig-Maximilians-Universität München, Postfach 701260, D-8000 München 70, Germany (Part B)

Dr. Josef Janča, Institute of Analytical Chemistry, Czechoslovak Academy of Sciences, Leninova 82, CS-611 42 Brno, Czechoslovakia (Part A)

Dr. Nan-In Jang, Zenith Laboratories, 140 Legrand Ave., Northvale, NJ 07647, USA (Part B)

Dr. Jan-Christer Janson, Pharmacia LKB Biotechnology AB, S-751 82 Uppsala, Sweden (Part A)

Dr. F. W. Karasek, Department of Chemistry, University of Waterloo, Waterloo, Ont. N2L 3G1, Canada (Part B)

Dr. Arnis Kuksis, Banting & Best Department of Medical Research, University of Toronto, Toronto, Ont. M5G 1L6, Canada (Part B)

Dr. Karel Macek, Institute of Physiology, Czechoslovak Academy of Sciences, Budejovicka 1083, CS-142 20 Praha 4, Czechoslovakia (Part B)

Dr. Jan Macek, Research Institute of Rheumatic Diseases, Na slupi 4, CS-128 50 Praha 2, Czechoslovakia (Part B)

Dr. Colin T. Mant, Department of Biochemistry, University of Alberta, Edmonton, Alta. T6G 2H7, Canada (Part B)

Dr. Thomas H. Mourey, Eastman Kodak Research Laboratories, Rochester, NY 14650, USA (Part B)

Dr. K. P. Naikwadi, Department of Chemistry, University of Waterloo, Waterloo, Ont. N2L 3G1, Canada (Part B)

Dr. Szabolcs Nyiredy, Gyógynövény Kutató Intézet, József A. u. 68, H-2011 Budakalász, Hungary (Part A)

Dr. Emilios Patsalides, School of Chemistry, University of Sydney, Sydney, N.S.W. 2006, Australia (Part B)

Dr. Terry M. Phillips, Immunochemistry Laboratory, Ross Hall, George Washington University Medical Center, 2300 Eye Street N.W., Washington, DC 20037, USA (Part A)

Dr. R. P. Philp, Petroleum Geochemistry Group, Energy Center Building, School of Geology and Geophysics, 100 East Boyd Street, University of Oklahoma, Norman, OK 73019, USA (Part B)

Dr. Colin F. Poole, Department of Chemistry, Wayne State University, Detroit, MI 48202, USA (Part A)

Dr. Salwa K. Poole, Department of Chemistry, Wayne State University, Detroit, MI 48202, USA (Part A)

Dr. Hans Poppe, Laboratorium voor Analytische Scheikunde, Universiteit van Amsterdam, Nieuwe Achtergracht 166, NL-1018 WV Amsterdam, The Netherlands (Part A)

Dr. Fred E. Regnier, Department of Biochemistry and Chemistry, Purdue University, West Lafayette, IN 47907, USA (Part B)

Dr. Tadeus Reichstein, Institut für Organische Chemie, Universität Basel, St. Johanns-Ring 19, CH-4056 Basel, Switzerland (Part A)

Dr. Pier Giorgio Righetti, Dip. di Scienze e Tecnologie Biomediche, Sez. Chimica Organica e Biochimica, Università di Milano, Via Celoria 2, I-20133 Milano, Italy (Part A)

Dr. Peter J. Schoenmakers, Philips Natuurkundig Laboratorium, Postbus 8000, NL-5600 JA Eidhoven, The Netherlands (Part A)

Dr. T. C. Schunk, Eastman Kodak Research Laboratories, Rochester, NY 14650, USA (Part B)

Dr. Joseph Sherma, Department of Chemistry, Lafayette College, Easton, PA 18042, USA (Part A)

Dr. Lloyd R. Snyder, LC Resources Inc., 26 Silverwood Ct., Orinda, CA 94563, USA (Part A)

Dr. Louis G. M. Uunk, Philips Natuurkundig Laboratorium, Postbus 8000, NL-5600 JA Eidhoven, The Netherlands (Part A)

Dr. Harold F. Walton, Cooperative Institute for Research in Environmental Sciences, University of Colorado, Campus Box 449, Boulder, CO 80309, USA (Part A)

Dr. Nian En Zou, Department of Biochemistry, University of Alberta, Edmonton, Alta. T6G 2H7, Canada (Part B)

List of Abbreviations

A

A	ampere
Å	ångström = 10^{-8} cm
AA	acetic acid
AAS	atomic absorption spectrometry
AASP	Advanced Automated Sample Processor
Ab	antibody
ABA	abscisic acid
AC	alternating current
Ac	acetyl
acac	acetylacetonate
ACM	acryloylmorpholine
ADAM	1-aminoadamantane
Ade	adenine
Ado	adenosine
ADP	adenosine 5′-diphosphate
AFID	alkali flame-ionization detector
AFS	atomic fluorescence spectrometry
Ag	antigen
ag	attogram = 10^{-18} g
Ala	alanine
AMD	automated multiple development
amol	attomol = 10^{-18} mol
AMP	adenosine 5′-monophosphate
AMPS	2-acrylamido-2-methylpropanesulfonic acid
AOAC	Association of Official Analytical Chemists
APIM	N-(4-anilinophenyl)isomaleimide
APIP	N-(4-anilinophenyl)isophthalimide
aq.	aqueous
Arg	arginine
Asp	aspartic acid
ASTM	American Society for Testing and Materials
Asn	asparagine
atm	atmosphere = 1 bar = 760 torr = 14.7 psi = 10^5 Pa
ATP	adenosine 5′-triphosphate
AUFS	absorbance units full scale
AZT	3′-azido-2′,3′-dideoxythymidine

B

BAC	*N,N′*-bisacrylylcystamine
BAMPITC	4-(*t*-butyloxycarbonylaminomethyl)phenyl isothiocyanate

BAP	bisacrylylpiperazine
bar	atmosphere = ca. 14.7 psi
BAW	*n*-butanol/acetic acid/water (4:1:5)
BHT	2,6-di-*t*-butyl-*p*-cresol (butylated hydroxytoluene)
Bis	*N,N'*-methylene bisacrylamide
BN chamber	Brenner/Niederwieser chamber
bp	base pair
br	boiling range
Bu	butyl
BuHCl	*n*-butanol/2 *N* HCl (1:1)
BuSA	butanesulfonic acid

C

C	centigrade, celsius
C_4	leukotriene C_4
CA	carrier ampholyte
ca.	circa
CAD	collision-activated dissociation
cAMP	adenosine 3',5'-cyclic monophosphate
CCC	countercurrent chromatography
CCD	chemical composition distribution; countercurrent distribution
CCK	cholecystokinin
cCMP	cytidine 3',5'-cyclic monophosphate
4CDD	tetrachlorodibenzo-*p*-dioxin
5CDD	pentachlorodibenzo-*p*-dioxin
6CDD	hexachlorodibenzo-*p*-dioxin
7CDD	heptachlorodibenzo-*p*-dioxin
8CDD	octachlorodibenzo-*p*-dioxin
CDP	cytidine 5'-diphosphate
CE	capillary electrophoresis; cholesteryl ester
CEC	cation-exchange chromatography
CER	ceramide
CFFF	concentration field-flow fractionation
CGE	capillary gel electrophoresis
cGMP	guanosine 3',5'-cyclic monophosphate
CHFE	contour-clamped homogeneous field electrophoresis
CI	chemical ionization
CIE	crossed immunoelectrophoresis
cIMP	inosine 3',5'-cyclic monophosphate
CL	cardiolipin
CLC	column liquid chromatography
CLSA	closed-loop stripping analysis
CM	carboxymethyl
cm	centimeter
CMC	critical micelle concentration; 1-cyclohexyl-3-(2-morpholinoethyl)carbodiimide metho-*p*-toluenesulfonate
CMP	cytidine 5'-monophosphate
conc.	concentrated
cP	centipoise
CPC	centrifugal planar (partition) chromatography; coil planet centrifuge
CPCl	cetylpyridinium chloride
CsA	cephalosporin A

CTAB	cetyltrimethylammonium bromide
Ctd	cytidine
CTP	cytidine 5′-triphosphate
cUMP	Uridine 3′,5′-cyclic monophosphate
CV	coefficient of variation
Cys	cysteine
Cyt	cytosine
CZE	capillary zone electrophoresis

D

D	dalton
2-D	2-dimensional
D_4	leukotriene D_4
DABITC	4′-*N,N*-dimethylamino-4-azobenzene isothiocyanate
DABS	4′-*N,N*-dimethylamino-4-azobenzenesulfonyl
DABTH	dimethylaminoazobenzene thiohydantoin
dAdo	2′-deoxyadenosine
dADP	2′-deoxyadenosine 5′-diphosphate
dAMP	2′-deoxyadenosine 5′-monophosphate
DAN	2,3-diaminonaphthalene
DANABITC	4′-*N,N*-dimethylnaphthyl-4-azobenzene isothiocyanate
DAP	2,3-diaminopropionic acid
DAPMP	2,6-diacetylpyridine bis(*N*-methylenepyridiniohydrazone)
DAPMT	2,6-diacetylpyridine bis(*N*-methylene-*N,N,N*-trimethylammonio-hydrazone)
DATD	*N,N′*-diallyltartardiamide
dATP	2′-deoxyadenosine 5′-triphosphate
DC	direct current
DCA	dichloroacetic acid
DCCC	droplet countercurrent chromatography
DCE	1,2-dichloroethane
dCMP	2′-deoxycytidine 5′-monophosphate
DCP	direct-current plasma
DCPAES	directly coupled plasma atomic emission spectroscopy
DCTA	dodecyltrimethylammonium
dCtd	2′-deoxycytidine
dCTP	2′-deoxycytidine 5′-triphosphate
DDC	2′,3′-dideoxycytidine
DEAE	diethylaminoethyl
DEDTC	diethyldithiocarbamate
DEDTP	diethyldithiophosphate
DEHPA	di(2-ethylhexyl)phosphoric acid
DG	diacylglycerol
DGDG	digalactosyl diacyl glycerol
DGMG	digalactosyl monoacyl glycerol
dGMP	2′-deoxyguanosine 5′-monophosphate
dGTP	2′-deoxyguanosine 5′-triphosphate
dGuo	2′-deoxyguanosine
DHEBA	*N,N′*-(1,2-dihydroxyethylene)bisacrylamide
disc	discontinuous
DIITC	diphenylindenonyl isothiocyanate
dIno	2′-deoxyinosine

DITH	diphenylindenonyl thiohydantoin
DMA	dimethylaniline
DMAC	*N,N*-dimethylacetamide
dm6Ado	2′-deoxy-N^6-methyladenosine
dm6AMP	2′-deoxy-6-methyladenosine 5′-monophosphate
dm5CMP	2′-deoxy-5-methylcytidine 5′-monophosphate
DMAPI	*p-N,N*-dimethylaminophenyl isothiocyanate
DMF	dimethylformamide
DMOX	dimethyloxazoline
DMP	dimethylphenol
DMSO	dimethylsulfoxide
DNBS	2,4-dinitrobenzenesulfonic acid
DNP	dinitrophenyl
DNS	dimethylaminonaphthalene-5-sulfonyl (dansyl)
Do	dopamine
DP	degree of polymerization
DPA	diphenylamine
DRI	differential refractive index
DRIFT	diffuse reflectance Fourier transform (infrared spectroscopy)
dThd	2′-deoxythymidine
dTMP	2′-deoxythymidine 5′-monophosphate
dTTP	2′-deoxythymidine 5′-triphosphate
dUMP	2′-deoxyuridine 5′-monophosphate
DVB	divinylbenzene

E

E_4	leukotriene E_4
EA	elemental analysis; ethylamine
ECD	electron-capture detector
ECL	equivalent chainlength
EDA	ethylene diacrylate
EDC	electrically driven chromatography; 1-ethyl-3-(3-dimethylamino-propyl)carbodiimide
EDTA	ethylenediaminetetraacetic acid
EFFF	electrical field-flow fractionation
EFFFF	elutriation focusing field-flow fractionation
EG	ethylene glycol
EGP	ethanolamine glycerophosphatides
EHPA	2-ethylhexylphosphoric acid
EI	electron impact
ELISA	enzyme-linked immunosorbent assay
em	emission
en	ethylenediamine
EP	electrostatic precipitator
EPA	Environmental Protection Agency
Eqn.	Equation
equiv	equivalent
ESA	ethanesulfonic acid
ESD	electronic scanning densitometry
ESR	electron spin resonance
Et	ethyl
ETU	ethylenethiourea

eV	electron volt
ex	excitation

F

FA	fatty acids
FAAS	flame atomic absorption spectrometry
FAB	fast atom bombardment
FC	free cholesterol
FD	field desorption
FDA	Food & Drug Administration
FFA	free fatty acids
FFF	field-flow fractionation
FFFF	flow field-flow fractionation
FFPC	forced-flow planar chromatography
fg	femtogram = 10^{-15} g
FID	flame-ionization detector
FMOC	9-fluorenylmethyl chloroformate
fmol	femtomol = 10^{-15} mol
fod	2,2-dimethyl-6,6,7,7,8,8,8-heptafluoro-3,5-octanedione
FPD	flame-photometric detector
FPLC	Fast Protein Liquid Chromatography
FSOT	fused-silica open-tubular (column)
ft	foot = 30.48 cm
FT	Fourier transform

G

g	gram; gravity
GC	gas chromatography
GDP	guanosine 5′-diphosphate
GLC	gas/liquid chromatography
Gln	glycine; glutamine
Glu	glutamic acid
Gly	glycine
GM	gangliosides
GMP	guanosine 5′-monophosphate
GPC	gel-permeation chromatography
GSC	gas/solid chromatography
GSL	glycosphingolipid
GTP	guanosine 5′-triphosphate
Gua	guanine
Guo	guanosine

H

h	hour
HAC	hydroxyapatite chromatography
HDPE	high-density polyethylene
HECD	Hall electroconductivity detector
HEDC	bis(2-hydroxyethyl)dithiocarbamate
hep	heptane
HETE	hydroxyeicosatetraenoic acid
HETP	height equivalent to a theoretical plate

hex	hexane
HFAA	hexafluoroacetylacetone
HFBA	hexafluorobutyric acid
HFIP	hexafluoro(2-propanol)
hfa	hexafluoropentane-2,4-dione
HIBA	2-hydroxyisobutyric acid
HIC	hydrophobic-interaction chromatography
HID	helium-ionization detector
HILIC	hydrophilic-interaction chromatography
His	histidine
HMDS	hexamethyldisilazane
HPIAC	high-performance immunoaffinity chromatography
HPLC	high-performance liquid chromatography
HPPC	high-pressure planar chromatography
HPPLC	high-performance precipitation liquid chromatography
HPTLC	high-performance thin-layer chromatography
HR	high-resolution
HSA	human serum albumin; hexanesulfonic acid
HSES	hydrostatic equilibrium system
HSTLC	high-speed thin-layer chromatography
HTMAB	hexadecyltrimethylammonium bromide
Hyp	hypoxanthine

I

IA	indole-3-acetamide
IAA	indole-3-acetic acid
IAN	indole-3-acetonitrile
IBA	indole-3-butyric acid
IC	ion chromatography
ICPAES	inductively coupled plasma atomic emission spectroscopy
ICPMS	inductively coupled plasma mass spectrometry
ID	inside diameter
IDP	inosine 5′-diphosphate
IEC	ion-exchange chromatography
IEF	isoelectric focusing
IEP	immunoelectrophoresis
Ile	isoleucine
IMAC	immobilized-metal affinity chromatography
IMP	inosine 5′-monophosphate; ion-moderated partitioning
in.	inch = 2.54 cm
Ino	inosine
IP	ion pairing
IPG	immobilized pH gradients
IR	infrared
ISRP	internal-surface reversed-phase
ITP	inosine 5′-triphosphate; isotachophoresis
IUPAC	International Union for Pure and Applied Chemistry

K

K	degree kelvin
kb	kilobase

kD	kilodalton = 10^3 D
KDO	3-deoxy-D-*manno*-2-octulosonic acid
kg	kilogram
KP	potassium phosphate

L

l	liter
LALLS	low-angle laser light scattering
LC	liquid chromatography
LCP	liquid-crystal polysiloxane
LDPE	low-density polyethylene
LEC	ligand-exchange chromatography
Leu	leucine
LLC	liquid/liquid chromatography
LLDPE	linear low-density polyethylene
LPC	lysophosphatidylcholine
LPE	lysophosphatidylethanolamine
LSC	liquid/solid chromatography
LSD	lysergic acid diethylamide
LSS	linear solvent strength
LT	leukotriene
LTB_4	leukotriene B_4
Lys	lysine

M

M	molar
m	meter
mA	milliampere = 10^{-3} A
μA	microampere = 10^{-6} A
m6Ade	N^6-methyladenine
m6Ado	N^6-methyladenosine
MAK	methylated albumin on kieselguhr
MBE	moving-boundary electrophoresis
MCC	microcapillary chromatography
M-chamber	microchamber
MCPC	microchamber centrifugal planar chromatography
m5Ctd	5-methylcytidine
m5Cyt	5-methylcytosine
MDA	3,4-methylenedioxyamphetamine
MDMA	3,4-methylenedioxymethamphetamine
MDNP	methyldinitrophenyl
MDPF	2-methoxy-2,4-diphenyl-3[2*H*]-furanone
ME	methyl ester
Me	methyl
MEC	micellar electrochromatography
MECA	molecular emission cavity analysis
MEK	methyl ethyl ketone
MES	morpholinoethane sulfonate
Met	methionine
meq	milliequivalent = 10^{-3} equivalent
μeq	microequivalent = 10^{-6} equivalent

MFFF	magnetic field-flow fractionation
MGDG	monogalactosyl diacyl glycerol
MGMG	monogalactosyl monoacyl glycerol
mg	milligram = 10^{-3} g
μg	microgram = 10^{-6} g
m2Gua	N^2-methylguanine
m2Guo	N^2-methylguanosine
m2,2Guo	$N^2{}_2$-dimethylguanosine
MH	Mark-Houwink
MID	multiple-ion detection
MIKES	mass-analyzed ion-kinetic-energy spectra
MIM	metastable-ion monitoring
min	minute
m1Ino	1-methylinosine
MIPAES	microwave-induced plasma atomic emission spectroscopy
MIQ	minimum identifiable quantity
ml	milliliter = 10^{-3} l
μl	microliter = 10^{-6} l
m*M*	millimolar = 10^{-3} *M*
mm	millimeter = 10^{-3} m
μm	micrometer = 10^{-6} m
MMA	monomethylarsonate
mmol	millimol = 10^{-3} mol
μmol	micromol = 10^{-6} mol
MO	methyloxime
MOG	Mills-Olney-Gaither (method)
mol.	molecular
mp	melting point
MPa	megapascal (10^6 Pa)
MPI	methylphenanthrene index
MPLC	medium-pressure liquid chromatography
MS	mass spectrometry
MS/MS	tandem mass spectrometry
msec	millisecond = 10^{-3} sec
MSWI	municipal solid-waste incinerator
MTBE	methyl *t*-butyl ether
mV	millivolt = 10^{-3} V
MW	molecular weight
MWD	molecular weight distribution
m1Xan	1-methylxanthine
m7Xao	7-methylxanthosine

N

N	normal
NAD	nicotinamide adenine dinucleotide
NADH	nicotinamide adenine dinucleotide (reduced form)
NADP	nicotinamide adenine dinucleotide 3′-phosphate
NADPH	nicotinamide adenine dinucleotide 3′-phosphate (reduced form)
NALPE	*N*-acyl lysophosphatidylethanolamine
NAPE	*N*-acyl phosphatidylethanolamine
NBDCl	7-chloro-4-nitrobenz-2-oxa-1,3-diazole

NBDF	7-fluoro-4-nitrobenz-2-oxa-1,3-diazole
N chamber	normal chamber
NCI	negative chemical ionization
NCPC	normal-chamber centrifugal planar chromatography
ng	nanogram = 10^{-9} g
NI	negative ion
nl	nanoliter = 10^{-9} l
nm	nm = 10^{-9} m
nmol	nanomol = 10^{-9} mol
NMP	*N*-methylpyrrolidone
NMR	nuclear magnetic resonance
NOEL	no-observed-effect level
NP	sodium phosphate
NPAH	nitrogen-containing polynuclear aromatic hydrocarbon
NPC	normal-phase chromatography
NPD	nitrogen/phosphorus detector
NSAID	nonsteroidal anti-inflammatory drug
NSO	nitrogen, sulfur, and oxygen compounds
NT	nortestosterone

O

OC	organochlorine
OD	outside diameter
ODS	octadecylsilane; octadecylsilyl
OFAGE	orthogonal-field alternation gel electrophoresis
ON	organonitrogen
OP	organophosphorus
OPA	*o*-phthaldialdehyde
OPLC	overpressured-layer chromatography
OPMLC	overpressured-multilayer chromatography
ORM	overlapping resolution maps
OSA	octanesulfonic acid
OTLC	open-tubular liquid chromatography

P

P	phosphonyl
PA	phosphatidic acid
Pa	pascal = 10^{-5} bar
PAA	polyacrylamide
PACE	programable, autonomously controlled gel electrophoresis
PAD	pulsed amperometric detector
PAF	platelet activating factor
PAG	polyacrylamide gel
PAGE	polyacrylamide gel electrophoresis
PAH	polycyclic aromatic hydrocarbons
PAM	Pesticide Analytical Manual
PAR	4-(2-pyridylazo)resorcinol
PAS	photoacoustic spectrometry
PBM	probability-based matching
PC	planar chromatography; phosphatidylcholine
PCA	perchloric acid

PCB	polychlorinated biphenyls
PCDD	polychlorinated dibenzo-*p*-dioxins
PCDF	polychlorinated dibenzofurans
PCP	pentachlorophenol
PCRD	postcolumn reaction detection
PDAD	photodiode-array detector
PDMS	poly(dimethyl siloxane)
PE	phosphatidylethanolamine
PEG	polyethylene glycol
PEI	poly(ethylene imine)
PEO	poly(ethylene oxide)
PET	poly(ethylene terephthalate)
PFB	pentafluorobenzoate
PFGE	pulsed-field gel electrophoresis
PFGGE	pulsed-field gradient gel electrophoresis
PFPA	pentafluoropropionic acid
PG	phosphatidylglycerol; prostaglandin
pg	picogram = 10^{-12} g
PGB_2	prostaglandin B_2
PGF	prostaglandin F
PGI_2	prostacyclin
Ph	phenyl
Phe	phenylalanine
PI	phosphatidylinositol
p*I*	isoelectric point
PICS	paired-ion chromatographic system
PID	photoionization detector
PITC	phenylisothiocyanate
pl	picoliter = 10^{-12} l
PLOT	porous-layer open-tubular (column)
PMBP	1-phenyl-3-methyl-4-benzoylpyrazolone
PMD	programed multiple development
pmol	picomol = 10^{-12} mol
POP	1,3-dipalmitoyl-2-oleoylglycerol
PPP	tripalmitoylglycerol
ppb	parts per billion = 10^{-9} parts
ppm	parts per million = 10^{-6} parts
PPO	poly(propylene oxide)
ppt	parts per trillion = 10^{-12} parts
Pr	propyl
PRC	protein reaction cocktail
Pro	proline
PRS	protein reaction system
PrSA	propanesulfonic acid
PS	polystyrene; phosphatidylserine
PSA	pentanesulfonic acid
PSD	poresize distribution
psi	pounds per square inch = 51.77 torr
PTC	phenylthiocarbamyl
PTFE	polytetrafluoroethylene (Teflon)
PTH	phenylthiohydantoin

PTV	programed-temperature vaporization
PUFA	polyunsaturated fatty acids
PVC	polyvinyl chloride
Py	pyrolysis
Q	
Q	quaternary
QAE	quaternary aminoethyl (*N,N,N*-triethylaminoethyl)acrylamide
QSRR	quantitative structure/retention relationship
R	
Ref.	Reference No.
RFGE	rotating-field gel electrophoresis
RI	refractive index
R.I.	Retention Index
RIA	radioimmunoassay
RPC	reversed-phase chromatography
R.P.C.	rotation planar chromatography
rpm	rotations per minute
RRT	relative retention time
RSD	relative standard deviation
RT	retention time
S	
S	sulfonyl
SA	sulfonic acid
SAN	styrene/acrylonitrile
satd.	saturated
SAX	strong-anion-exchange (chromatography)
SB	styrene/butadiene
SBCD	short-bed continuous development
SCB	short-chain branching
SCE	standard calomel electrode
S chamber	sandwich chamber
SCOT	support-coated open-tubular (column)
SCX	strong-cation-exchange (chromatography)
SD	standard deviation
SDS	sodium dodecyl sulfate
SE	sulfoethyl
SEC	size-exclusion chromatography; steric-exclusion chromatography
sec	second
SEMA	styrene/ethyl methacrylate copolymer
Ser	serine
SERS	surface-enhanced Raman spectrometry
SF	supercritical fluid
SFC	supercritical-fluid chromatography
SFE	supercritical-fluid extraction
SFFF	sedimentation field-flow fractionation
SFFFFF	sedimentation/flotation focusing field-flow fractionation
SHP	shielded hydrophobic phase
SIM	selected (single)-ion monitoring

SIMS	secondary-ion mass spectrometry
SMA	styrene/methyl acrylate
SMMA	styrene/methyl methacrylate copolymer
SNS	sodium naphthoquinone 4-sulfonate
SP	sulfopropyl
SPAH	sulfur-containing polynuclear aromatic hydrocarbon
SPE	solid-phase extraction
SPH	sphingomyelin
sq.	square
SRM	selected-reaction monitoring

T

t	tertiary
TACT	*N,N′,N′′*-triallylcitrictriamide
TAFE	transverse alternating field electrophoresis
TBA	tetrabutylammonium
TBAOH	tetrabutylammonium hydroxide
TBDMS	*t*-butyldimethylsilyl
TBDMSO	*t*-butyldimethylsiloxy
TBDPS	*t*-butyldiphenylsilyl
TCA	trichloroacetic acid
TCB	1,2,4-trichlorobenzene
TCD	thermal conductivity detector
TCDD	tetrachlorodibenzodioxins
TCDF	tetrachlorodibenzofuran
TDP	thymidine 5′-diphosphate
TEAA	triethylammonium acetate
TEAF	triethylammonium formate
TEAP	triethylammonium phosphate
TEMED	*N,N,N′,N′*-tetramethylethylenediamine
temp.	temperature
TFA	trifluoroacetic acid; trifluoroacetyl
tfa	1,1,1-trifluoropentane-2,4-dione
TFFF	thermal field-flow fractionation
TG	triacyl glycerol
THC	tetrahydrocannabinol
Thd	thymidine
THF	tetrahydrofuran
ThFFF	thermal field-flow fractionation
Thr	threonine
Thy	thymine
TIC	total-ion chromatogram
TID	thermionic ionization detector
TLC	thin-layer chromatography
TMP	thymidine 5′-monophosphate
TMS	trimethylsilyl
TMSO	trimethylsiloxy
TNBS	2,4,6-trinitrobenzenesulfonic acid
TREF	temperature rising elution fractionation
Tris	tris(hydroxymethyl)aminomethane
Trp	tryptophan

TX	thromboxane
TXB	thromboxane B
Tyr	tyrosine

U

U chamber	ultramicro-chamber
UCPC	ultramicro-chamber centrifugal planar chromatography
UDP	uridine 5′-diphosphate
UKDE	United Kingdom Department of the Environment
UM	ultramicro-chamber
UMP	uridine 5′-monophosphate
Ura	uracil
Urd	uridine
UTP	uridine 5′-triphosphate
UV	ultraviolet

V

V	volt
Val	valine
Vis	visible range
vol.	volume
v/v	volume-by-volume

W

WAX	weak-anion-exchange (chromatography)
WCOT	wall-coated open-tubular (column)
WCX	weak-cation-exchange (chromatography)
wt.	weight
w/w	weight-by-weight

X

Xan	xanthine
Xao	xanthosine
XMP	xanthosine 5′-monophosphate

Z

ZE	zone electrophoresis

List of Italic Symbols

A	area
A_S	peak asymmetry factor
$B°$	specific permeability coefficient
$C\%$	grams of crosslinker/$\%T$
C_i	capacity
c	concentration
c_m	concentration in the mobile phase
c_s	concentration in the stationary phase
D	diffusion coefficient
D_m	solute diffusion coefficient in the mobile phase
D_p	solute diffusion rate in the pore
D_s	solute diffusion coefficient in the stationary film
D_T	thermal diffusion coefficient
d	mean layer thickness; Stokes diameter
d_c	column internal diameter
d_f	stationary-phase film thickness
d_o	lumen of open-tubular column
d_p	particle diameter
$d_{p,n}$	number-average particle diameter
$d_{p,w}$	weight-average particle diameter
d_t	internal column diameter
E	separation impedance
E_f	electric field
E_s	solvent/solute interaction energy
F	flowrate
F_f	focusing force
f	solute fraction
g	branching parameter
H	plateheight
h	reduced plateheight
$\mathrm{h}R_F$	100 x R_F
I_R	retention index
K	equilibrium distribution constant; partition coefficient
K_a	acid dissociation constant
K_D	distribution coefficient
$K_R°$	infinite dilution gas/liquid partition coefficient
k	Boltzmann constant
k'	capacity factor; retention factor
$\bar{k}$	average k' in gradient elution
k_a	association constant
k_d	dissociation constant

L	column length
M	mass
M_n	number-average molecular mass
M_p	peakmaximum molecular mass
M_r	molecular mass
M_v	viscosity-average molecular mass
M_w	weight-average molecular mass
M_z	centrifugation-average molecular mass
m_T	electrophoretic mobility
N	number of theoretical plates (platenumber); column efficiency
N_{eff}	effective platenumber
N_{req}	platenumber required for a separation
n	peak capacity; also number
n_{R_S}	number of peaks resolved with resolution R_s
n_s	stroke frequency
ng	centrifugal force
P	pressure
P_c	critical pressure
P_E	effective pore radius
P_i	inlet pressure
P_o	outlet pressure
P_r	reduced pressure
P^+, P^-	ion pair
Q	equilibrium quotient
Q_{inj}	amount injected
Q_{max}	loadability
R	retention; relative zone velocity
$\mathcal{R}$	Rydberg constant
$<R>$	radius of gyration
R_F	migration rate relative to the solvent front
R_h	effective hydrodynamic radius
R_i	resistance to flow
R_M	logarithm of the capacity factor
R_s	peak resolution
RRT	relative retention time
RT	retention time
r	correlation coefficient; radius
r_s	diameter of solute molecule
S_i	sensitivity; response factor
T	absolute temperature
$T\%$	(grams of acrylamide + grams of Bis)/100
T_c	critical temperature
T_g	glass-transition temperature
T_r	reduced temperature
t	time
t_o	column deadtime
t_a	retention time of the first band
t_G	gradient time
t_R	retention time
t'_R	corrected retention time
t_r	delay time required for analysis

t_z	retention time of the last band
$\underline{u}$	flow velocity
$\overline{u}$	average linear velocity
u_e	actual velocity; interstitial velocity
u_{eo}	electro-osmotic velocity
u_{nom}	nominal velocity
u_{opt}	optimum velocity
V°	channel volume
V_o	void volume
V_C	volumetric flow
V_c	column bed volume
V_d	displacement volume
V_e	elution volume
V_h	hydrodynamic volume
V_i	interstitial volume
V_{inj}	injection volume
V_M	volume of the mixing chamber
V_m	volume of mobile phase; column deadvolume
V_p	porevolume
V_R	retention volume
V_r	chamber volume
V_s	volume of stationary phase
V_t	total liquid volume
$\underline{v}$	reduced velocity
$\overline{v}$	average reduced velocity
v_z	peakwidth variance
W	bandwidth at the baseline
W_h	bandwidth at half peakheight
w	distance between channel walls
Z_f	distance traveled by the solvent front

List of Greek Symbols

α	separation factor; ratio of partition coefficients
β	packing-dependent variable
γ	activity coefficient
γ_m	obstruction factor
γ_t	pore tortuosity factor
ΔG°	standard Gibbs free energy
ΔH°	enthalpy
ΔP	pressure drop
ΔS°	entropy
$\Delta\Phi$	change in mobile-phase composition
δ	Hildebrand solubility parameter
δ_f	reduced film thickness
ε°	solvent strength parameter
ε_e	porosity of packing
ε_i	molar absorptivity
ε_m	column porosity
η	viscosity
$[\eta]$	intrinsic velocity
η_t	carrier gas velocity
λ	obstruction factor; flow inequality parameter
μ	electrophoretic mobility
μ_{eo}	electro-osmotic mobility
ρ_p	density
σ	standard deviation
σ_L	standard deviation in length units
σ_t	standard deviation in time units
σ_V	standard deviation in volume units
τ	pore tortuosity factor
Υ	pressure resistance factor
Φ	solvent strength
Φ_A	volume fraction of *A*
ϕ	shape factor
χ	compressibility
ψ	phase ratio; pseudouridine
ψ_B	association factor
ω	column packing parameter; angular velocity

Foreword

Chromatography was so named by Tswett (1906) because it was first used for separating coloured compounds (chlorophyll etc.) by passing a solution through a column of a solid adsorbent (filterpaper, Al_2O_3 etc.). The coloured zones were cut out and by elution gave pure compounds. Its forerunner was the "Kapillaranalyse" of Goppelsröder. Both techniques were forgotten for many years until revived with spectacular success by R. Kuhn and E. Lederer.

It took only few years until chromatography was applied to colourless compounds. I well remember the moment when I personally started to use it in some work on adrenal cortical hormones. We had six compounds of the $C_{21}O_5$ group differing in degree of saturation and function of two oxygen atoms. They were isolated using classical methods with fractional crystallisation as last step. Three of these compounds (E, Fa, M) by oxidation with CrO_3 (including loss of the side chain) gave adrenosterone (**I**), while the other three (A, C, D) gave the saturated triketone (**II**).

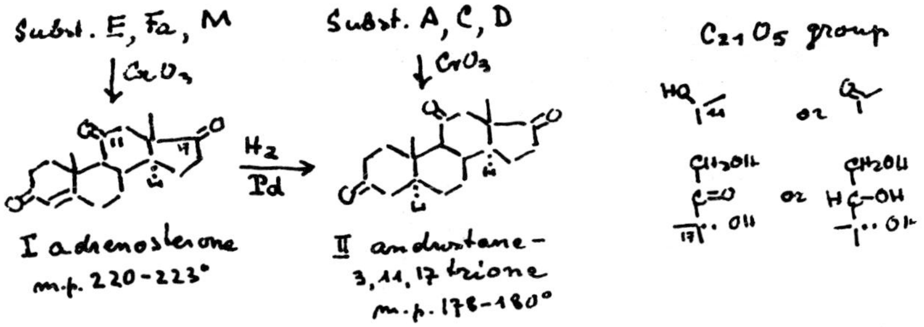

In order to show that all six compounds contain the same steroidal nucleus, with oxygen at the same position, Steiger and Reichstein (1937) hydrogenated **I** with Pd hoping to get **II**. The product was, however, a mixture which resisted all our attempts at purification by conventional methods, including fractional crystallization. We decided to make our first experiment with chromatography. Filtering our material (38 mg) in benzene through a small column of Al_2O_3 gave 8 mg of amorphous mixture. When nothing more came down we eluted the main product (28 mg) with ether. It now immediately gave pure crystals of **II**. Chromatography in this simple form or with more elaborate and less aggressive techniques (including partition chromatography with a stationary liquid phase) became indispensable to our work.

Chromatography has since these days developed enormously and is still developing today as an art of its own, both on the analytical and preparative scale. Together with electrophoretic methods, it became an indispensable tool and has changed the face and the possibilities of chemistry in an unforeseen way. The fact that isolation of individual proteins and nucleic acids became possible nearly as a routine and automatic analysers for establishing sequences are available on the market would have sounded utopic in our times.

I am confident that the authors and organisers of the Fifth Edition of *Chromatography* will be able to review the progress of this subject in a competent way and am sure they will not have to search for readers.

T. REICHSTEIN

Preface

Once again, it is my pleasure to introduce a new and completely revised edition of our standard text on chromatography and related differential migration methods. Having retired from laboratory work and being involved in editing of the Symposium Volumes of the *Journal of Chromatography* since publication of the Fourth Edition, I am no longer contributing chapters to this book. My main contribution to it is the unification of the work of 37 authors, most of them new to this project.

The remarkable success of chromatography in solving long-standing analytical and preparative chemical problems has been the result of advances in theory and instrumentation, e.g., computer-assisted optimization, automation, miniaturization, detection, identification, and special sorbents. While chromatographic methods are being perfected, related methods are beginning to reveal enticing prospects. Accordingly, I have added chapters on supercritical-fluid and countercurrent chromatography, on affinity chromatography, and on field-flow fractionation to Part A (Fundamentals and Techniques).

Since the most rapid advances are now taking place in the applications of chromatography to proteins and peptides, nucleic acids and their constituents, and pharmaceuticals, these chapters have been strengthened. Responding to the increased importance of environmental issues and of the analysis of fossil fuels and synthetic polymers, chapters on the application of chromatography to these areas have been added to Part B (Applications).

This has necessitated the omission of some chapters found in the Fourth Edition, e.g., the chapters on the history of chromatography and electrophoresis and on the chromatographic analysis of steroids and other terpenoids. The latter are now in the chapter on lipids, antibiotics have been incorporated in the chapter on pharmaceuticals, and the analysis of gases is dealt with in other chapters (e.g., inorganic and environmental analysis). Fortunately, much of the material covered in the Fourth Edition is still useable, and information omitted from the Fifth Edition can usually be located there.

As before, abbreviations and symbols used in Parts A and B and a combined subject index are given in both parts. A new feature is the list of manufacturers and dealers of chromatography and electrophoresis supplies with their addresses.

Also as before, each chapter was reviewed by two anonymous referees, usually authors of another chapter of this book. In some cases, I have also drawn on outside referees, whom I can thank only anonymously for their advice. My anonymous thanks are also expressed to the excellent editorial and production staff of Elsevier, who modestly refuse acknowledgements for work done as part of their jobs.

Orinda, California ERICH HEFTMANN

Chapter 1

Theory of chromatography

L.R. SNYDER

CONTENTS

1.1 INTRODUCTION

The theory of chromatography has evolved considerably in the past 50 years, and today it would require several volumes to address this topic thoroughly. This situation reflects the complexity of both (a) the underlying processes which contribute to and determine chromatographic separation and (b) the various physico-chemical relationships which describe different aspects of the technique. However, much of our understanding of chromatography can be encompassed in a smaller number of important concepts and relationships. As a result, my goal is a reasonable overview of the subject in this relatively short chapter – with emphasis on a simplified treatment of the more practical chromatographic concepts. Where important material could not be included, more comprehensive discussions in the literature have been noted; a number of general references of this kind can also be cited [1-11].

A further simplification of the present treatment results, if we concentrate on a special case: small-sample ("linear" or nonoverloaded) separations by elution chromatography – as opposed to preparative separations, frontal analysis [5,12], or displacement chromatography [13]. I will return to displacement chromatography and "large" samples in Section 1.11, where preparative chromatography is discussed. Planar chromatography (PC), which is covered in Chapter 3, is given only superficial treatment here; the impact of general chromatographic theory in affecting the way planar chromatography is carried out has been rather minor.

1.1.1 Elution chromatography

In elution chromatography, the column can have either of two configurations: (a) packed with particles of the sorbent or (b) wall-coated open-tubular (WCOT). WCOT columns are of practical value mainly in gas chromatography (GC) (Chapter 9) and supercritical-fluid chromatography (SFC) (Chapter 8). (For an additional discussion of WCOT columns, see these chapters also.) In this chapter it will be assumed that we are talking about packed columns, unless a WCOT column is specifically noted.

The column packing consists of a stationary phase that is coated (or bonded) onto a support. The support typically consists of small, porous particles, but other configurations (e.g., WCOT) are possible. The stationary phase can be either liquid or solid.

Some essential features of elution chromatography are illustrated in Fig. 1.1. A small volume of a 3-component sample is first applied to the column inlet (a). Under the influence of a flowing gas or liquid (the mobile phase) the sample moves through the column (b-d) and becomes successively more separated as elution proceeds. Finally (e) the sample leaves the column and is detected in the mobile phase or eluent. The final chromatogram (e) reflects the separation that was achieved on the column just prior to the elution of each band.

The separation shown in Fig. 1.1e can be characterized by the retention times (t_R) and widths (W) of the bands, zones, or peaks corresponding to the various components of the sample. The greater the difference in retention times of two adjacent bands, and the smaller their widths, the better is their separation or resolution. The remainder of this chapter will be concerned mainly with the dependence of retention and bandwidth on experimental conditions.

1.2 THE CHROMATOGRAM

1.2.1 Evaluation of the chromatogram

The separation illustrated by Fig. 1.1e will be used for purposes of discussion. The so-called column deadtime, t_o (also t_M), shown here is an important chromatographic parameter. It corresponds to the retention time for a nonretained sample component, e.g., the solvent in which the sample is dissolved. In Fig. 1.1, the arrival of the mobile-phase front ("solvent front") at the end of the column occurs at a time t_o. The column deadtime is also related to the mobile-phase flowrate, F, and the total volume, V_m, of mobile phase contained within the column

$$t_o = V_m/F \qquad (1.1)$$

A considerable number of papers deal with the precise measurement of t_o for a given chromatogram [14-16]. In gas chromatography, t_o is conveniently evaluated from the retention time for the "air peak". In liquid chromatography, any nonretained compound can be used as sample to measure t_o (see the discussion of Ref. 14).

References on p. A65

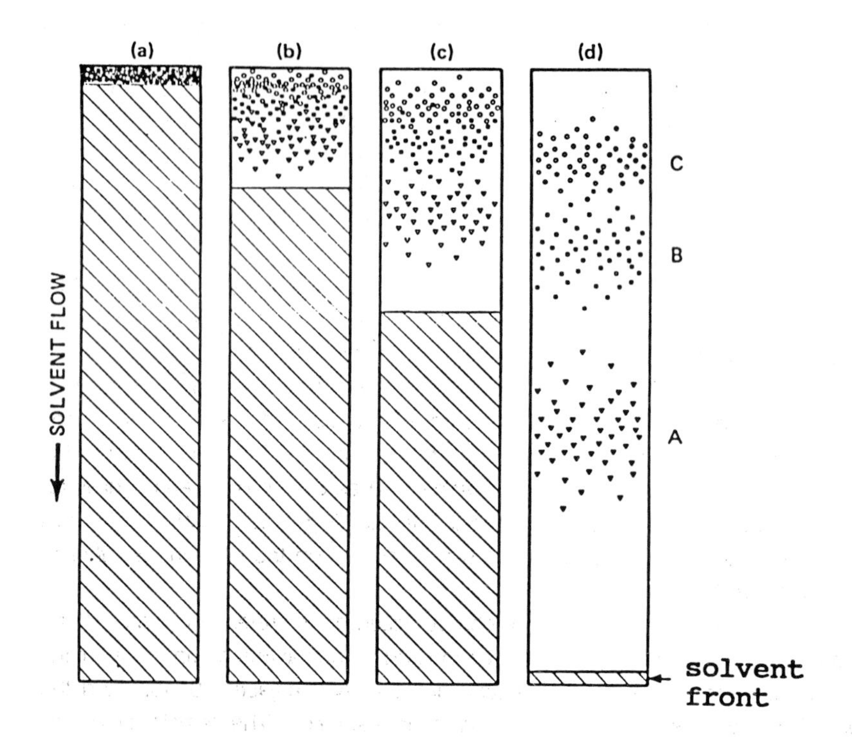

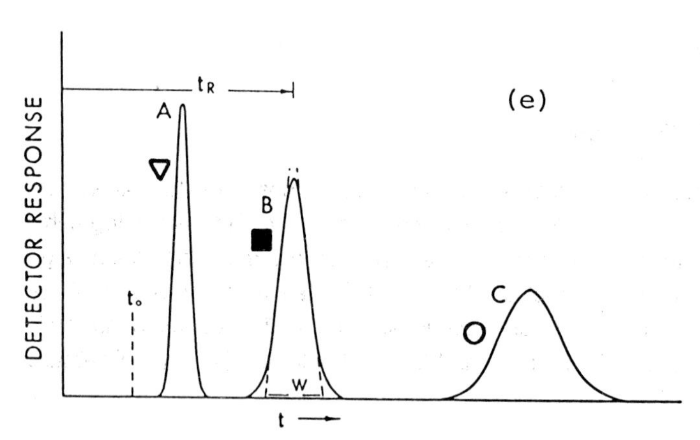

Fig. 1.1. Hypothetical representation of chromatographic separation. (a-d) Separation within the column; successive stages during elution; (e) final chromatogram corresponding to elution of sample from the column at a later time. A liquid mobile phase is assumed here, but the separation process is similar for gas chromatography.

Accurate values of t_o are mainly of interest in studies of retention thermodynamics (Section 1.5). Eqn. 1.1 assurnes that the time required for sample components to move from (a) the sample injector to the column inlet and (b) from the column outlet to the detector flowcell is small. The column deadtime can be used to determine so-called adjusted or corrected retention times t'_R

$$t'_R = t_R - t_o \tag{1.2}$$

As we will see, retention in chromatography can be understood in terms of the equilibrium distribution of the sample (solute) between the column packing (stationary phase) and the mobile phase. For this reason, a more fundamental measure of solute retention is the retention volume, V_R

$$V_R = t_R F \tag{1.3}$$

Similarly, the column deadtime, t_o, can be expressed in terms of the column deadvolume, V_m

$$V_m = t_o F \tag{1.4}$$

V_m does not include any solvent that is associated with the stationary phase, i.e. by adsorption or other sorption process.

1.2.2 Chromatographic retention

A convenient and useful measure of solute retention is given by the capacity factor (or retention factor), k'

$$k' = \frac{\text{(amount of solute in the stationary phase)}}{\text{(amount of solute in the mobile phase)}} \tag{1.5}$$

Values of k' have no simple meaning in gradient elution or temperature-programed GC and are not usually reported when these procedures are used (Section 1.9). Values of k' can be determined from the retention time, t_R

$$k' = (t_R - t_o)/t_o \tag{1.6}$$

In retentive modes of chromatography (i.e. except size-exclusion chromatography) (SEC) (Chapter 6) $k' > 0$, which means that no band is eluted before the deadtime, t_o. It is often useful to express retention time or volume in terms of the capacity factor, k'

$$t_R = t_o \ (1+k') \tag{1.7}$$

$$V_R = V_m (1+k') \tag{1.8}$$

References on p. A65

The thermodynamic equilibrium constant K for the retention process can also be defined by

$$K = \frac{\text{(solute concentration in stationary phase)}}{\text{(solute concentration in mobile phase)}} \tag{1.9}$$

This leads to a more fundamental description of retention in terms of K

$$k' = \psi K \tag{1.10}$$

and

$$V_R = V_m + KV_s \tag{1.11}$$

where ψ is the phase ratio ($\psi = V_s/V_m$) and V_s is the volume of stationary phase within the column. The determination of ψ in GC can be straightforward, since both V_m and V_s are usually measurable (at least for liquid stationary phases). In liquid chromatography (LC) there is often considerable uncertainty about the value of ψ for a given system.

The fraction of solute molecules in the mobile and stationary phases can be obtained from Eqn. 1.6.

$$\text{(Fraction in mobile phase)} = 1/(1+k') \tag{1.12}$$

$$\text{(Fraction in stationary phase)} = k'/(1+k') \tag{1.13}$$

The migration velocity, R_F, of the solute band relative to the solvent front is equal to the fraction of solute molecules in the mobile phase, so that (Eqn. 1.12)

$$R_F = 1/(1+k') \tag{1.14}$$

In PC, values of R_F are measured in terms of the distance traveled by a band relative to the solvent front

$$R_F = \frac{\text{(distance of band from start)}}{\text{(distance of solvent front start)}} = 1/(1+k') \tag{1.15}$$

Alternatively,

$$k' = (1 - R_F)/R_F \tag{1.16}$$

The logarithm of the capacity factor (R_M) is proportional to log k', which is in turn proportional to the free energy of the retention process (Section 1.5). This important quantity can be expressed as

$$R_M = \log k' = \log [(1 - R_F)/R_F] \tag{1.17}$$

The quantity R_M has important applications in the use of linear free-energy relationships (Section 1.7).

1.2.3 Flow velocity

The flow velocity, u, of the mobile phase has an important effect on the bandwidth, W. In column chromatography the flow velocity is usually maintained constant, whereas in PC u normally decreases with time. Here we will assume that flowrate and u are constant during the separation. In the case of a column packed with porous particles, two different mobile-phase velocities can be distinguished: the average velocity, u, is given by

$$u = L/t_o = LF/V_m \tag{1.18}$$

where L is the length of the column. The actual velocity of the mobile phase in the interstitial space between the particles is $u_e \approx 1.7\ u$ for columns packed with porous particles. For the case of nonporous particles of packing, or for WCOT columns in GC, $u_e = u$.

Bandspreading in chromatography as a function of the mobile-phase velocity can be better understood in terms of the so-called reduced velocity, ν. In equations dealing with the dependence of H on u, it is more accurate to replace u by u_e. However, this distinction is rarely made in practice. For packed columns

$$\nu = u\, d_p/D_m \tag{1.19}$$

Here d_p refers to particle size and D_m is the solute diffusion coefficient in the mobile phase. For open-tubular chromatography the internal column diameter, d_c, replaces the particle diameter, d_p. We will return to Eqn. 1.19 in Section 1.8 on bandbroadening.

1.2.4 Properties of Gaussian peaks

The shape of elution bands (or bands within the column) as in Fig. 1.1e is normally Gaussian. That is, the concentration, c, of solute in the effluent from the column is given as a function of time, t, by

$$c = (1/F)\ [M/\sigma_t(2\pi)^{1/2}]\ e^{-0.5\ (t_R - t)/\sigma_t^2} \tag{1.20}$$

where M is the mass of solute injected into the column, and σ_t is the standard deviation measured in time units. As seen in Fig. 1.2, the width of a Gaussian band at specified ordinate positions is proportional to the standard deviation of the band or distribution. The baseline bandwidth, W, is defined as 4σ, where σ is the standard deviation expressed in any units. A chromatographic band can also be characterized by its statistical moments

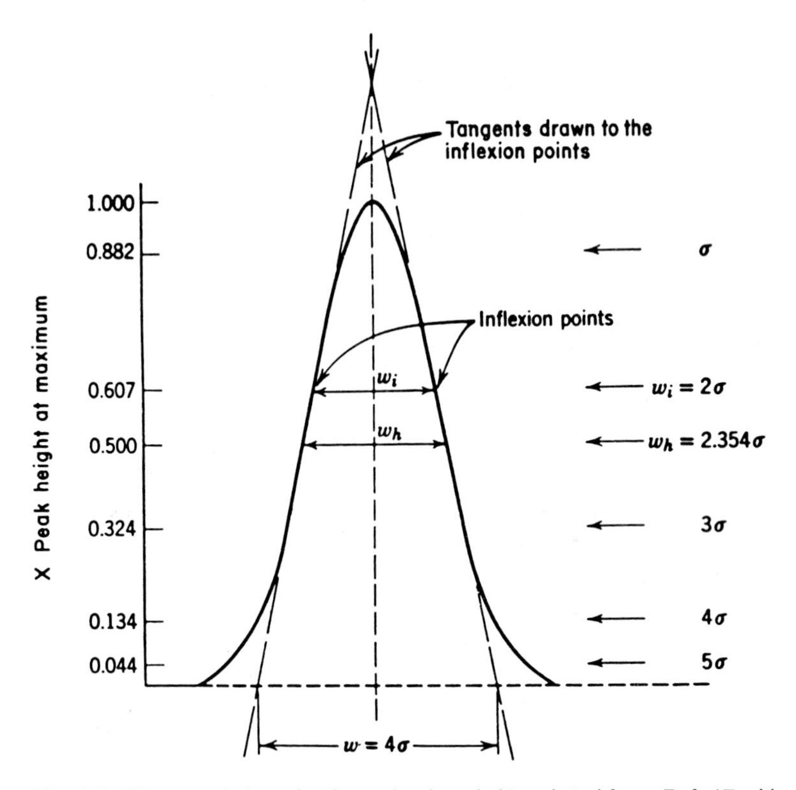

Fig. 1.2. Characteristics of a Gaussian band. (Reprinted from Ref. 17 with permission.)

[2,5,12]: the zeroth moment or band area, the first moment or retention time, and the second moment or variance, which is equal to σ^2 for a Gaussian band.

The maximum concentration of a Gaussian band (or its height) is given by

$$c_{max} = A/\sigma_t\,(2\pi)^{1/2} \tag{1.21}$$

where A is the band area (units of concentration · time).

1.2.5 Asymmetrical bands

Chromatographic separations carried out under favorable conditions generally yield bands that are close to Gaussian in shape (as in Fig. 1.2). However, a number of experimental factors can lead to a significant distortion of peakshape [14]; e.g., a poorly packed column, extracolumn bandbroadening (poorly designed equipment or bad connections), secondary retention effects, column overload, etc. Band asymmetry is usually measured in terms of the band asymmetry factor, A_s, as illustrated in Fig. 1.3a. All bands in the chromatogram should be reasonably symmetrical; i.e., $0.9 < A_s < 1.2$.

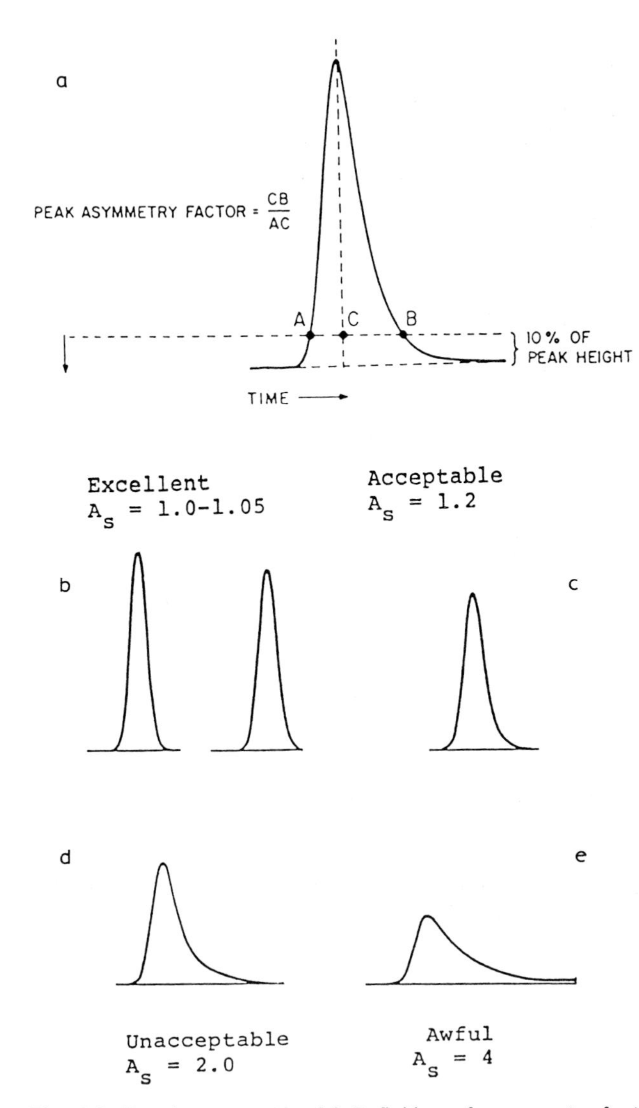

Fig. 1.3. Band asymmetry. (a) Definition of asymmetry factor, A_S; (b-e) examples of band asymmetry.

Occasionally, it will be necessary to accept somewhat less symmetrical bands (e.g., $0.8 < A_S < 1.5$), but bands with $A_S > 1.5$ are indicative of potentially severe chromatographic problems. Very asymmetric bands can cause poor separation and quantitation, and they also indicate possible problems with column-to-column reproducibility [18]. The tailing of bands, as in Figs. 1.3a, d, e, is sometimes referred to as exponential tailing. That is, the latter part of the band can be expressed by

$$C = A\,e^{-Bt} \tag{1.23}$$

rather than by Eqn. 1.20. B is a constant for a given band. Exponential tailing is predicted when band asymmetry is caused by extracolumn effects [19]. For band asymmetry due to other effects (e.g., secondary retention processes, column overload) Eqn. 1.23 is not necessarily obeyed, and bandshapes can vary substantially from the examples of Fig. 1.3.

1.3 MEASURES OF COLUMN EFFICIENCY

1.3.1 Plateheight and platenumber

Column efficiency refers to the ability of a column to provide very narrow bands in the final chromatogram (small values of W). Let us first examine bandbroadening as it occurs during migration of the solute through the column; e.g., as in Figs. 1.1a-d. At each stage during this process, a solute band has a bandwidth that can be characterized by the standard deviation of the band in length units, σ_L. It can be shown [2] that if L is the distance the band has migrated (i.e. the column length)

$$\sigma_L^2 = H\,L \tag{1.24}$$

where H is a proportionality constant defined as the column plateheight (also height equivalent to a theoretical plate or HETP). H depends on the solute, the operating conditions, and the column; from Eqn. 1.24, $H = \sigma_L^2/L$. A general goal in chromatography is to achieve narrow bands (small values of σ), which means that we desire small values of H. The form of Eqn. 1.24 and the terminology now used ("plateheight") derive from an early model of the chromatographic process [20], in which the column is mathematically divided into a number of equilibrium stages or "plates". The dimension of H is length, and its value for an efficient column is proportional to a characteristic distance within the column, i.e. the particle diameter for packed columns or the column diameter for open-tubular (capillary) columns.

The plate height, H, is a measure of bandbroadening that is normalized for column length. To compare the performance of columns of different length, it is more useful to measure the column platenumber, N. N can be defined for a band that is just about to leave the column

$$N = L^2/\sigma_L^2 \tag{1.25}$$

More commonly, N is determined experimentally for an eluted band

$$N = t_R^2/\sigma_t^2 = 16(t_R/W)^2 \tag{1.26}$$

N is also referred to as the number of theoretical plates or the platecount of the column. A large value of N means narrow bands and a better separation – other factors being equal. Platenumbers of 2000-20 000 are typical of "high-performance" liquid chromatography (HPLC); platenumbers in COT GC are usually one order of magnitude greater, while values of N for thin-layer chromatography (TLC) are in the range 1000-5000 (measured via Eqn. 1.25).

Values of the plate height, H, can be obtained from Eqn. 1.26 as

$$H = L/N \tag{1.27}$$

Referring to Fig. 1.2, bandwidth can be measured in various ways. The determination of the baseline bandwidth W (4σ) by drawing tangents to the sides of the band is not reliable, as different people will perform this operation differently. A more precise approach is to measure bandwidth at half-height (W_h in Fig. 1.2); then

$$N = 5.54\ (t_R/W_h)^2 \tag{1.28}$$

Another relationship for N that is conveniently used with many modern data processing systems is based on Eqn. 1.26 with values of σ determined as

$$\sigma = A/[h\ (2\pi)^{0.5}] \tag{1.29}$$

where A and h refer to the area and height of the band, respectively.

The column efficiency is less often described in terms of the effective platenumber, N_{eff},

$$N_{eff} = N[k'/(1+k')]^2 \tag{1.30}$$

Values of N_{eff} take into account the effect of k' on separation (see discussion of Section 1.4.2).

1.3.2 Peak capacity

Another measure of column efficiency is the peak capacity, n. This is defined as the maximum number of separated bands (resolution R_s = 1; Section 1.4.2) that can be accommodated within the chromatogram. The peak capacity for isocratic separation is given as

$$n = 1 + (N/16)^{1/2}\ln(t_z/t_a) \tag{1.31}$$

Here, t_a and t_z refer to the retention times of the first and last bands, respectively. The peak capacity is a kind of "universal" measure of separation power, which allows different chromatographic procedures to be compared. For example, peak capacity values are quite a bit larger for gradient elution than for isocratic separation, or for programed-temperature GC than for isothermal GC. Also, the maximum value of t_z/t_a is much smaller in SEC than in other LC methods, and this means that peak capacities in SEC are also smaller.

1.4 RELATIVE RETENTION AND RESOLUTION

1.4.1 Relative retention and selectivity

A typical chromatographic separation is shown in Fig. 1.4. Separation is strongly affected by the proximity of adjacent bands in the chromatogram. For example, Bands 2 and 3 are close together and marginally separated, as are Bands 4 and 5. An important measure of band proximity is the so-called separation factor, α, where for two adjacent bands, having the k' values k_1 (first band) and k_2 (second band)

$$\alpha = k_2/k_1 \tag{1.32}$$

In the example of Fig. 1.4, the α values for bandpairs 2/3 and 4/5 are 1.05-1.07, while the α values of other bandpairs in this chromatogram are 1.1-1.6. Values of α must usually be at least 1.01-1.02 for acceptable separations in GC, and 1.05-1.10 for good separation in LC.

From Eqn. 1.11 it follows that α is the ratio of the thermodynamic equilibrium constants, K, for each solute. The separation factor, α, is often identified with the selectivity of a chromatographic system, i.e. the ability of the system to provide different retention times for two specific compounds. Values of α depend on the two solutes and on (a) the

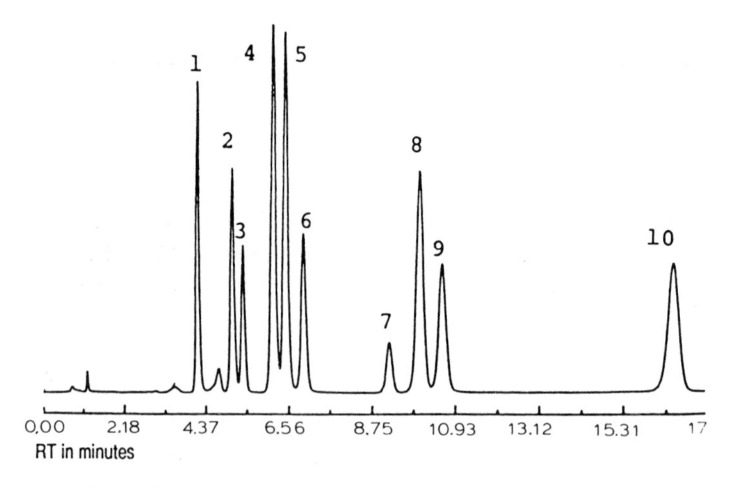

Fig. 1.4. Separation of a 10-component sample of nitro-compound by LC. Conditions: column, 25 x 0.46 cm, 5-μm C_8; mobile phase, 55% aq. methanol; flowrate, 2 ml/min. (Reprinted with permission from Ref. 21.)

chemical composition of the stationary phase, (b) the chemical composition of the mobile phase (except in GC), and (c) the temperature. In SFC, pressure can also affect α by changing the density of the mobile phase.

1.4.2 Resolution

The relative separation of two bands is often referred to as their resolution, R_S. Resolution is defined as

$$R_S = \frac{\text{(distance between the peak centers)}}{\text{(average baseline bandwidth)}} = (t_2 - t_1)/(1/2)(W_1 + W_2) \qquad (1.33)$$

Here, t_1 and t_2 refer to the retention times, t_R, of the first band (1) and second band (2), and W_1 and W_2 are their bandwidths. Fig. 1.5 illustrates the separation of two adjacent bands as a function of their R_S values and relative bandsize. It is seen that separation systematically improves for larger values of R_S, and separation is generally better for two bands of equal size (and the same R_S value). That is, larger values of R_S will normally be required for the separation of bands of quite unequal size.

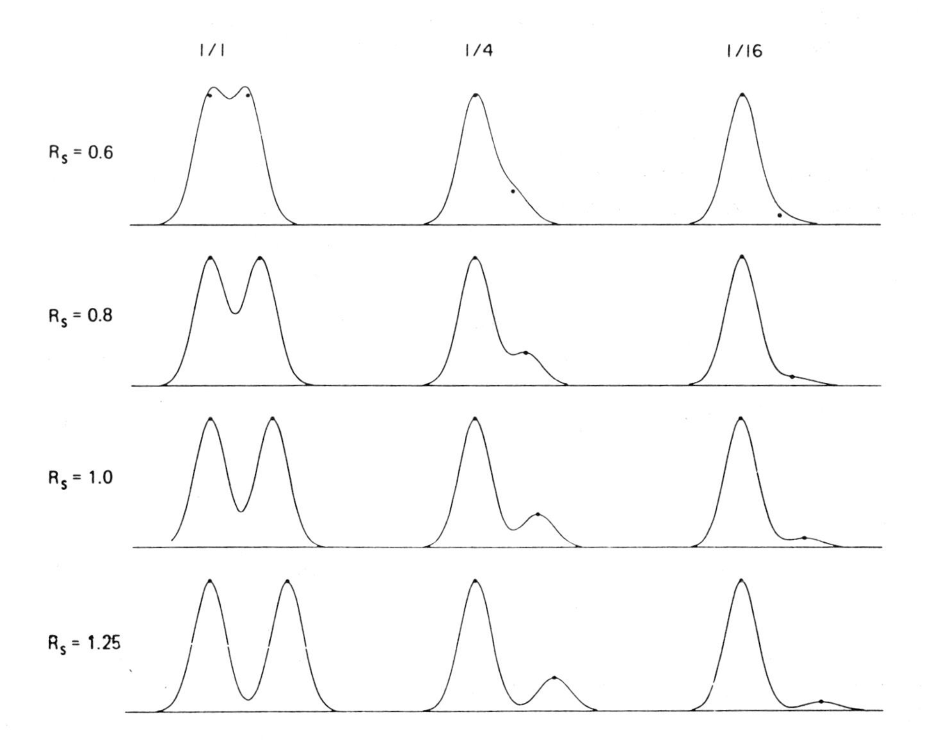

Fig. 1.5. Separation of two bands as a function of resolution (R_S) and relative band size (1/1, 1/4, 1/16). (Reprinted with permission from Ref. 26.)

References on p. A65

For quantitative analysis it is desirable (but not always practical) to achieve a resolution of at least $R_S = 1.5$, since this corresponds to baseline separation for bands of similar size, i.e. return of the detector tracing to baseline between the two bands. Baseline separation makes it easier for data systems to measure the size of each band properly, and this, in turn, means more reliable quantitation. Baseline separation is just achieved for Bands 8 and 9 in Fig. 1.4. When there are more than two bands in the chromatogram, and all bands are of interest, we are most interested in the resolution of the critical bandpair, i.e. that bandpair which has the smallest value of R_S. If we can increase R_S to an acceptable value for the critical bandpair, then the separation of other bands in the chromatogram will also be satisfactory. The critical bandpair in Fig. 1.4 consists of Bands 2 and 3, with Bands 4 and 5 a close second.

1.4.3 Optimizing resolution and separation

A general goal in chromatography is to achieve adequate resolution in the final separation. This requires that we understand resolution as a function of chromatographic conditions. Eqn. 1.32 is useful for measuring resolution, but not for telling us how to change resolution. An equation that relates resolution to separation conditions can be obtained through the combination of Eqns. 1.26, 1.32, and 1.33, with the assumption that the two bands have equal bandwidths, W. This should be true whenever their plate numbers are the same and $t_1 \approx t_2$

$$R_S = (1/4)\ (\alpha - 1)\ N^{1/2}\ [k'/(1+k')] \tag{1.34}$$

Here, k' refers to the average value of k' for the two bands. It is possible to relate changes in k', N, and α to specific changes in separation conditions, as discussed in Section 1.6.

A change in the assumptions involved in the derivation of Eqn. 1.33 leads to a slightly different expression (the Purnell equation)

$$R_S = (1/4)\ [(\alpha - 1)/\alpha]\ N^{1/2}\ (k_2/(1+k_2)] \tag{1.35}$$

Here, k_2 refers to k' for the second band. For small values of α (the case of usual interest), there is little difference between Eqns. 1.33 and 1.34. An exact expression for R_S has recently been derived [22]

$$R_S = (1/2)\ [(\alpha - 1)/(\alpha + 1)]\ N^{1/2}\ [k'/(1+k')] \tag{1.36}$$

where k' is now the average value for the two bands.

According to Eqns. 1.34-1.36, resolution depends on three separate parameters: α, N, and k'. If α is close to 1, the two compounds have very similar retention, and their separation will be difficult. If N is small, the bands will be quite wide, and again, good resolution may not be possible. Resolution is seen to be proportional to the term $[k'/(1+k')]$ of Eqns. 1.34-1.36, which corresponds to the fraction of the sample that is in

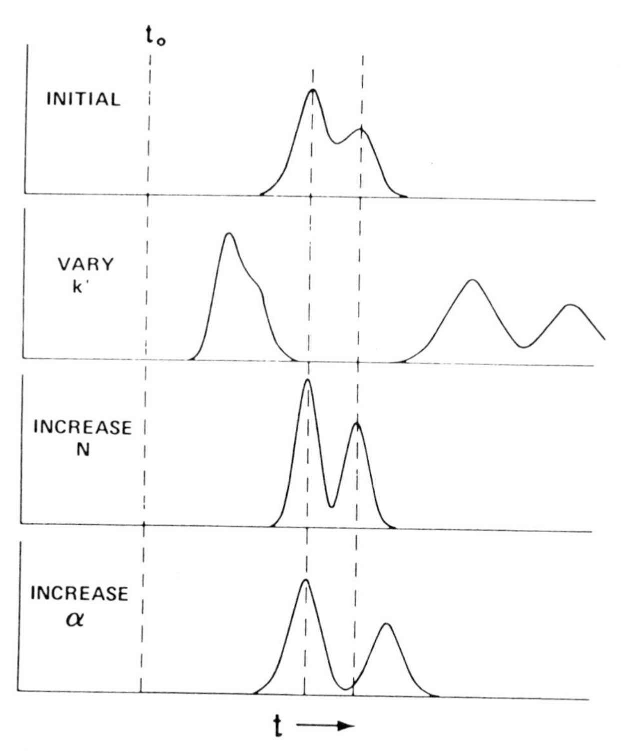

Fig. 1.6. Effects of a change in k', N, or α on the resolution of two bands. (Reprinted with permission from Ref. 26.)

the stationary phase (Eqn. 1.13). Small values of k' mean that the sample is largely in the mobile phase, and under these conditions little separation of sample can occur ($R_S \approx 0$). Fig. 1.6 illustrates the effect on resolution of changes in k', N, or α.

It is useful as a first approximation to assume that k', N, and α are independent of each other (although this is not strictly true). We can then systematically improve resolution by sequentially optimizing these three parameters. Usually, we start with an efficient column, one which provides a fairly large N value in a reasonable separation time. The next step is generally the adjustment of values of k' into an optimum range ($1 < k' < 20$), since changes in k' are normally easy to control and have a major effect on the final separation. While resolution is maximized for large values of k' (Eqns. 1.34-1.36), large values of k' also imply excessive separation times (Eqn. 1.7) and wide bands, which are hard to detect (Eqn. 1.26). This is illustrated in Fig. 1.7 for the LC separation of a 5-component sample, where the mobile-phase composition (vol. % aq. methanol) is varied to provide changes in solvent strength or k'. The chromatogram of Fig. 1.7d provides a good compromise between resolution, analysis time, and bandwidth (or detection).

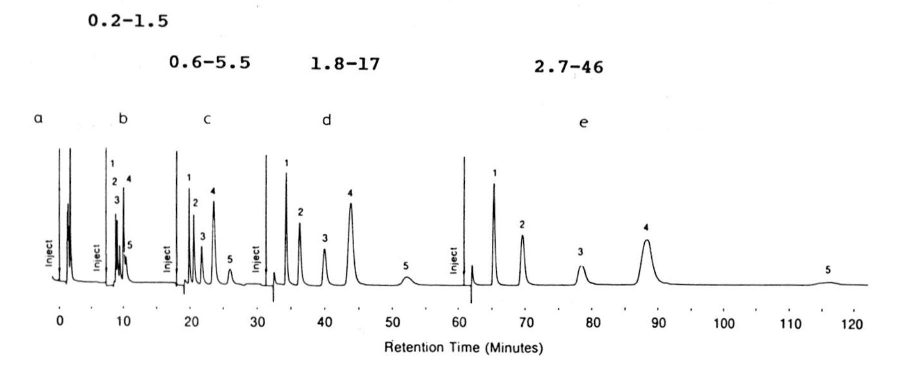

Fig. 1.7. Effect of a change in k' (by varying solvent strength) on the separation of a 5-component sample by reversed-phase LC. (a-e) 70%, 60%, 50%, 40%, and 30% aq. methanol as mobile phase. (Reprinted with permission from Ref. 26.)

When conditions have been selected for a good k' range ($1 < k' < 20$) and an efficient starting column, the resulting separation will often prove adequate. If this is not the case, then usually the most profitable approach is to vary α – or to change the bandspacing of the chromatogram. This will be discussed in more detail in Section 1.6 and in later chapters, which deal with the individual chromatographic methods.

A number of method development or optimization strategies have been developed, especially for LC [9,21,23-25]. Many of these procedures are based on the use of a computer to facilitate the final selection of experimental conditions.

1.5 RETENTION THERMODYNAMICS

The concepts and relationships that are discussed in this section mainly affect our ability to understand and predict chromatographic retention. They are important for several reasons. First, these equations allow a science of chromatography to be developed that is consistent with chemistry in general. Second, these retention relationships can, in turn, lead to a better physico-chemical picture of what is occurring inside the column; i.e., a description of the interactions (at the molecular level) of solute molecules with the mobile and stationary phases. Finally, this treatment provides a powerful basis for predicting chromatographic retention as a function of experimental conditions, and thereby systematizing the process of method development. (For a more detailed treatment, see Ref. 12.)

1.5.1 Fundamental relationships

Chromatographic retention can occur by various distinct processes. This leads to a classification of chromatographic methods on the basis of retention processes (Section

1.6). Not infrequently, however, more than one such process may be involved in a given chromatographic system. These secondary retention processes are usually undesirable (although occasionally they may prove advantageous). For purposes of discussion, we will assume that only a single process is involved in a given separation, unless otherwise stated.

Chromatographic retention is an equilibrium process that can be represented by

$$X_m \rightleftharpoons X_s \tag{1.37}$$

That is, retention can be visualized as a process (or reaction) in which a molecule of Solute X in the mobile phase (m) is transformed into a molecule of X in the stationary phase (s). We can define an equilibrium constant K for this reaction (Eqn. 1.9), which can, in turn, be related to a standard Gibbs free energy ($\Delta G°$) for the process of Eqn. 1.38 [12].

$$\log K = -\Delta G°/2.3\mathcal{R}T \tag{1.38}$$

For relatively small samples, values of K are usually independent of solute concentration (Henry's Law). Section 1.11 discusses the case where K (and k') change with solute concentration. The relation of k' to K (Eqn. 1.11) allows the transformation of Eqn. 1.38 to

$$\log k' = \log \psi - \Delta G°/2.3\mathcal{R}T \tag{1.39}$$

Here, $\mathcal{R}$ is the gas constant and T the absolute temperature. Eqn. 1.39 can also be expressed in terms of the enthalpy ($\Delta H°$) and the entropy ($\Delta S°$) of the retention process

$$\log k' = \log \psi - \Delta H°/\mathcal{R}T + \Delta S°/\mathcal{R} \tag{1.40}$$

Values of $\Delta H°$ and $\Delta S°$ are usually approximately constant (independent of temperature) over some temperature range of interest. This means that a plot of log k' vs. ($1/T$) normally yields a linear relationship (with a slope equal to $-\Delta H°/\mathcal{R}$)*. This is illustrated in Fig. 1.8. In Fig. 1.8a the log of the corrected retention volume (proportional to log k'; see Eqn. 1.8) is plotted against the reciprocal temperature for different solutes separated by GC. Fig. 1.8b shows a plot of log k' vs. reciprocal temperature for a number of solutes separated by LC. Linear dependencies are seen for each solute in these two systems. We will examine these data further in a subsequent section.

The solute concentrations referred to in Eqn. 1.9 (which are used to define values of K) should, in principle, be expressed as solute mole fractions in each phase. However, other concentration units can be used with little practical effect on the resulting predictions

---

* This is not the case for SFC, because the density (and interactive nature) of the supercritical-fluid phase can change markedly with temperature.

References on p. A65

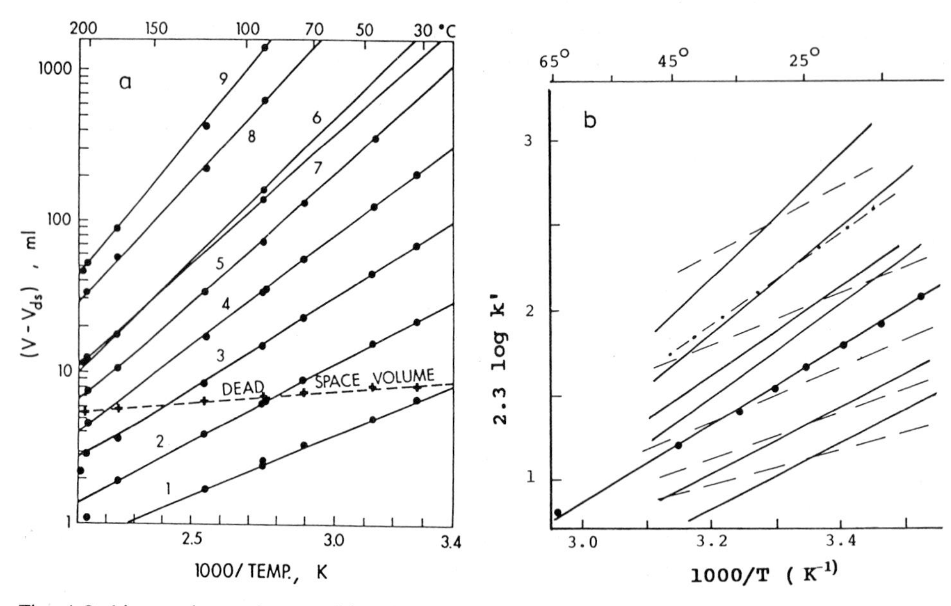

Fig. 1.8. Linear dependence of log k' on reciprocal absolute temperature. (a) GC separation on Apiezon L. (Reprinted with permission from Ref. 27.) (b) LC separation on a C_{18} column. ——, Fused-ring aromatic hydrocarbons; — • —, phenyl-substituted aromatic hydrocarbons; • • • •, 9,10-dimethyl-3,4-benzanthracene. (Redrawn from Ref. 28.)

or correlations discussed below. In some applications it is useful to express $\Delta G°$ in terms of solute activity coefficients, but we will forego any treatment of this topic. (For additional discussion of this and other aspects of the thermodynamics of chromatography, see Refs. 5 and 12).

1.5.2 Secondary chemical equilibria

When more than one distinct process, i, contributes to retention (secondary retention effects), Eqn. 1.10 becomes

$$k' = \sum^{i} \psi_i K_i \tag{1.41}$$

or

$$k' = \sum^{i} k_i \tag{1.42}$$

i.e. the k' values for each retention process are additive (for a sufficiently small sample). Likewise, if a solute is in rapid equilibrium between two (or more) forms, i and j, a single elution band will be observed, and its k' value will be given by

$$k' = f_i k_i + f_j k_j \tag{1.43}$$

Here, f_i and f_j represent the fractions of the solute in form i or j at any time, and k_i and k_j refer to the k' values for species i and j.

The commonest example of a secondary chemical equilibrium as in Eqn. 1.44 is that involving acidic (AH) or basic (B) solutes

$$AH \rightleftharpoons A^- + H^+ \tag{1.44}$$

$$B + H^+ \rightleftharpoons BH^+ \tag{1.45}$$

As the pH of the mobile or stationary phase changes, the relative concentrations of A and A^- or B and BH^+ will also change. This usually has a profound effect on the retention of compounds AH or B. Note that acid/base equilibria, as in Eqns. 1.44 and 1.45, are usually much faster than chromatographic migration. As a result, only one band will be seen in the chromatogram, representing either $(HA + A^-)$ or $(B + BH^+)$.

The fraction of an acid, HA, in the ionized form (A^-) is

$$f^- = 1/\{1 + [(H^+)/K_a]\} \tag{1.46}$$

the fraction in the nonionized form (HA) is $(1 - f^-)$. Here, $[H^+]$ is the concentration of hydrogen ion in the phase under consideration, and K_a is the acid dissociation constant of acid HA or protonated base HB^+ (equal to $[H^+][A^-]/[HA]$ or $[H^+][B]/[HB^+]$). Similarly, the fraction of a base, B, in the ionized form (BH^+) is

$$f^+ = 1/\{1 + [K_a/(H^+)]\} \tag{1.47}$$

Now, consider the chromatographic retention of HA in a LC system where the mobile-phase pH varies, but the stationary-phase pH can be considered constant (e.g., in reversed-phase LC). The retention of HA and A^- will be different in this case (A^- less retained), so let the corresponding k' values be k_{HA} and k_{A^-}. From Eqn. 1.43 we then have k' for HA (or $[HA + A^-]$ as a function of $[H^+]$)

$$k' = (1 - f^-)\, k_{HA} + f^- k_{A^-} \tag{1.48}$$

f^- being given by Eqn. 1.46. The result is illustrated in Fig. 1.9a for the reversed-phase LC separation of several acidic solutes (bile acids). These bile acids have apparent pK_a values in the mobile phase of 5.5-6.5. Therefore, for pH < 4 the acids are completely protonated and strongly retained (k_{HA} large). As the pH of the mobile phase is increased beyond pH 4, these solutes begin to ionize, so that f^- becomes significant. The ionized

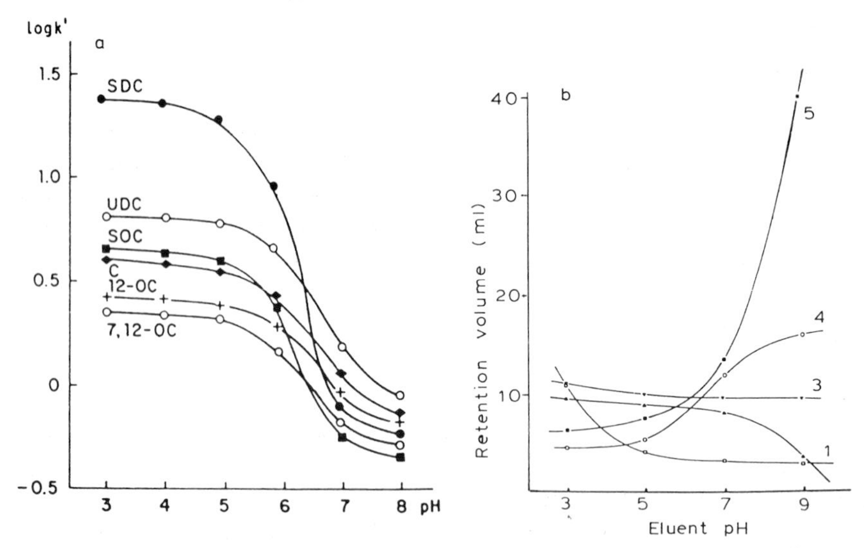

Fig. 1.9. Dependence of retention on pH in the reversed-phase LC separation of various acidic, neutral, and basic compounds. (a) Separation of several bile acids [29]. (b) Separation of (1) salicylic acid (strong acid), (2) phenobarbital (weak acid), (3) phenacetin (neutral), (4) nicotine (weak base), (5) methylamphetamine (strong base) [30].

compounds are less retained (k_{A^-} small), so that k' begins to decrease according to Eqn. 1.48. Eventually (for pH > 8), the acids are completely in the ionized form and their k' values approach a constant (small) value.

A similar pattern is seen in Fig. 1.9b for the separation of a mixture of acidic, basic, and neutral compounds – again in a reversed-phase LC system. Compound 3 in Fig. 1.9b is a neutral species, and its retention is seen to change only slightly as the pH is varied. Compound 1 is a strong acid with a p$K_a \approx 3$, and its retention pattern resembles that of Fig. 1.9a for the bile acids. Compound 4, on the other hand, is a weak base with p$K_a \approx 6$. At lower pH values this solute becomes increasingly ionized (Eqn. 1.46) and less retained, while at higher pH values the nonionized form is favored and retention is increased. The behavior of the remaining compounds in Fig. 1.9b can be rationalized from their acid/base properties (see caption of Fig. 1.9). In every example of Fig. 1.9 the maximum rate of change of k' with pH occurs at the apparent pK_a value of the solute, so that retention vs. pH for a given solute resembles the acid/base titration curve for that solute.

Other examples of secondary chemical equilibria exist and can be usefully applied to control chromatographic separation. (For a fuller account of secondary equilibria in LC see Ref. 31.)

1.6 RETENTION FOR DIFFERENT CHROMATOGRAPHIC METHODS

1.6.1 Intermolecular interactions

A number of chromatographic procedures or methods have been described, which can be classified on the basis of different types of mobile and/or stationary phases. Some simplification of this list can be achieved, if we group procedures according to the retention process and the nature of the interactions between solute molecules and the mobile or stationary phase. Table 1.1 is a highly simplified summary, which we will use to organize the discussion of the present section. Chromatographic methods exhibit marked differences in selectivity, i.e. the ability to separate different kinds of molecules. It is this feature which we will emphasize here.

TABLE 1.1

CHARACTERISTICS OF DIFFERENT CHROMATOGRAPHIC METHODS

Method	Mobile/stationary phases	Retention varies with
Gas/liquid chromatography	gas/liquid	molecular size, polarity
Gas/solid chromatography	gas/solid	molecular size, polarity
Supercritical-fluid chromatography	supercritical fluid/solid	molecular size, polarity
Reversed-phase chromatography	polar liquid/nonpolar liquid or solid	molecular size, polarity
Normal-phase chromatography	less polar liquid/more polar liquid or solid	molecular size, polarity
Ion-exchange, ion-pair chromatography	polar liquid/ionic solid	molecular charge
Size-exclusion chromatography	liquid/solid	molecular size
Hydrophobic-interaction chromatography	polar liquid/nonpolar liquid or solid	molecular size, polarity
Affinity chromatography	water/binding sites	specific structure

The procedures of Table 1.1 can also be organized in other ways:

(a) based on the nature of the mobile phase: gas, supercritical-fluid, or liquid chromatography;

(b) based on the nature of the stationary phase and the retention process: adsorption or partition chromatography;

(c) based on the physical arrangement of the chromatographic system: column, planar, or countercurrent chromatography;

References on p. A65

(d) based on the contents of the column: packed-column or capillary (open-tubular) chromatography.

Only GC and liquid adsorption and partition chromatography will be discussed in this chapter, other classifications being treated in later chapters. Here, we will focus on the retention process, as represented by the distribution constant, *K*.

In GC, it can be assumed as a first approximation that sample molecules do not interact with surrounding gas molecules. It is the interaction of the sample with the stationary phase that determines retention. In SFC or LC, sample molecules usually interact significantly with both phases. In terms of intermolecular interactions, there is no fundamental distinction between a liquid and a supercritical-fluid phase; we will use liquids to represent both cases in this chapter (but see Chapter 8). The same kinds of interaction between a sample molecule and the phase surrounding it can exist in either phase. These interactions are usefully classified into nonpolar, polar, and ionic (see Table 1.2 [5,32,33]).

TABLE 1.2

INTERMOLECULAR INTERACTIONS OF CHROMATOGRAPHIC SIGNIFICANCE

Interaction	Complementary molecular properties*
Nonpolar	
Dispersion (London forces)	polarizability/polarizability
Polar	
Dipole induction	polarizability/dipole moment**
Dipole orientation	dipole moment**/dipole moment**
Hydrogen bonding	acidity/basicity***
Ionic	
Ionic solution	ionic charge/dielectric constant
Coulombic	ionic charge/ionic charge

* Of the solute molecule and chromatographic phase with which it interacts.
** Permanent dipole moment.
*** Acidity or basicity in hydrogen bonding, not ionization in water.

Nonpolar or dispersion interactions involve the attraction of two polarizable molecules. Since every molecule is polarizable, dispersion interactions occur between every solute molecule and atoms or molecules in the surrounding liquid or solid phase. Polar interactions include dipole induction (between a polarizable molecule and one having a permanent dipole moment), dipole orientation (between two molecules, each having a permanent dipole moment), and hydrogen bonding (between a proton-donating and proton-accepting molecule; e.g., an acid and a base). The polarity of a compound is a measure of its ability to enter into polar interactions; i.e. polarity increases as the dipole moment, acidity, and/or basicity of a molecule increases. Table 1.2 summarizes these various interactions.

The polarity of sample molecules and of the chromatographic phases used to separate that sample have a pronounced effect on retention and separation. Nonpolar interactions are relatively nonselective, since these interactions occur between every pair of adjacent molecules (i.e., the solute and the surrounding mobile or stationary phase). Polar interactions can be varied selectively (in the case of polar samples) by changing the mobile or stationary phase, thus allowing a considerably control over bandspacing and values of α. It is therefore important to be able to classify or measure the polar characteristics of different chromatographic phases and various sample components.

For the case of nonionic molecules, we can distinguish: (a) the total polarity of the molecule and the contributions to total polarity from (b) the permanent dipole moment, (c) acidity, and (d) basicity of the molecule. Various theoretical models have been developed for the determination of these important characteristics of molecules of (a) the sample, (b) the mobile phase, and (c) the stationary phase. These are discussed further in Section 1.7. Table 1.3 summarizes these polarity characteristics for several common solvents.

TABLE 1.3

CONTRIBUTIONS OF DIFFERENT EFFECTS TO SOLVENT POLARITY [33]

Solvent	Solubility parameter, δ	Polarity contribution*		
		Dipole moment	Acidity	Basicity
n-Hexane	7.3	0.0	0.0	0.0
Diethyl ether	7.5	2.4	0.0	3.0
Triethylamine	7.5	0.0	0.0	4.5
Propyl chloride	8.4	2.9	0.0	0.7
Ethyl acetate	8.9	4.0	0.0	2.7
Benzene	9.2	0.0	0.0	0.6
Chloroform	9.3	3.0	6.5	0.5
Acetone	9.6	5.1	0.0	3.0
Propanol	12.0	2.6	6.3	6.3
Dimethyl sulfoxide	12.0	6.1	0.0	5.2
Acetonitrile	12.1	8.2	0.0	3.8
Phenol	12.1	2.3	9.3	2.3
Ethanol	12.7	3.4	6.9	6.9
Methanol	14.5	4.9	8.3	8.3

* Contribution to δ (cal/ml$^{1/2}$).

References on p. A65

1.6.2 Gas/liquid chromatography

The mobile phase in gas/liquid chromatography (GLC) is usually an inert carrier gas, such as helium or nitrogen, and the stationary phase is a nonvolatile liquid that covers an (ideally) inert support material. Retention occurs as a result of the partitioning of a sample component, X, between the mobile-phase gas (g) and the stationary-phase liquid (l)

$$X_g \rightleftharpoons X_l \tag{1.49}$$

There is normally no significant interaction of solute molecules with the carrier gas, and therefore retention is entirely determined by the sample molecules and their interactions with the stationary-phase liquid.

A useful first approximation is to assume that the sample and stationary-phase liquid form ideal mixtures. Then the partial pressure of the dissolved sample (and the amount of sample in the mobile phase) will be proportional to the vapor pressure of the pure sample. Under these conditions, sample components will leave the column in (approximately) the order of their boiling points. That is, sample retention time increases for higher-boiling solutes. This behavior is usually observed for stationary phases composed of nonpolar liquids, although more polar samples will be somewhat less retained than predicted by their boiling points. As the polarity of the stationary-phase liquid is increased, polar samples will be increasingly retained relative to nonpolar samples. (For a more quantitative treatment of retention in GLC, see Chapter 8 of Ref. 5.)

1.6.3 Gas/solid adsorption chromatography

In this technique, an inert gas is employed as mobile phase, and the stationary phase is a high-surface-area solid. Retention occurs by adsorption of the sample onto the surface of the stationary phase. Gas/solid chromatography (GSC) is not commonly used for chromatographic separation, except in the case of permanent gases and highly volatile hydrocarbons, because of the lower platenumbers and nonsymmetrical bands that are often observed (compared to gas/liquid chromatography, GLC). However, GSC can be used to characterize the surface of the column packing (inverse GSC) [34,35]. Retention on nonpolar adsorbents, such as charcoal or various polymeric materials usually parallels the boiling point of the solute – as in GLC with nonpolar stationary phases. Similarly, polar adsorbents result in increased retention of more polar samples relative to less polar samples of similar boiling point. (For a more detailed discussion of the theory of GSC, see Chapter 9 of Ref. 3, and Ref. 36.)

1.6.4 Liquid/liquid chromatography

In liquid/liquid chromatography (LLC) two immiscible liquids are used as mobile phase and stationary phase, respectively. For various practical reasons, this technique is only used in countercurrent chromatography (Chapter 2). Retention occurs as a result of partitioning of the solute between the two phases (Eqn. 1.37). The equilibrium distribution constant, K, is given as

$$K = \frac{\text{solute concentration in stationary phase}}{\text{solute concentration in mobile phase}} = \gamma_{xm}/\gamma_{xs} \tag{1.50}$$

Here, γ_{xm} and γ_{xs} are the activity coefficients of the solute, X, in the mobile (m) and stationary (s) phases, respectively. For less soluble solutes (a common sample type), solute solubility is inversely proportional to the activity coefficient, so that

$$K \approx \frac{\text{(solute solubility in stationary phase)}}{\text{(solute solubility in mobile phase)}} \tag{1.51}$$

Thus, if the stationary-phase liquid is more polar than the mobile phase, more polar solutes will be preferentially retained. LLC systems of this type are referred to as normal-phase LLC (NPC). If the stationary-phase liquid is less polar than the mobile phase, more polar solutes are less retained, and the system is referred to as reversed-phase LLC (RPC). (For a further discussion of retention in LLC, see Chapter 10 of Ref. 5, and Ref. 36.)

1.6.5 Reversed-phase chromatography

In this technique an aqueous/organic solvent mixture is commonly used as the mobile phase, and a high-surface-area nonpolar solid is employed as the stationary phase. The latter is usually an alkyl-bonded silica packing, e.g., with C_8 or C_{18} groups covering the silica surface. RPC is presently the most popular LC method; more than 70% of all HPLC separations are today carried out in this mode.

The basis of solute retention in RPC is still somewhat controversial; some workers favor an adsorption process, while others believe that the solute partitions into the nonpolar stationary phase. Probably both processes are important for many samples. Retention of a solute, X, by competitive adsorption onto the surface of the stationary phase can be represented by

$$X_m + z\,M_s \rightleftharpoons X_s + z\,M_m \tag{1.52}$$

where M is a mobile phase molecule. Subscripts m and s refer to molecules in the mobile or stationary phase, respectively. Eqn. 1.52 assumes that a competition between solute and mobile-phase molecules exists for a place on the stationary-phase surface. That is, an adsorbed molecule, X, will displace some number, z, of previously adsorbed molecules, M, as visualized in Fig. 1.10. A partition process for solute retention would be described by Eqn. 1.37.

The evidence which favors a partition process in RPC is of various kinds.

First, the retention of a series of substituted alkanes on a C_n-substituted RPC packing (e.g., C_8 or C_{18}) shows a distinct discontinuity when the solute alkyl group is equal in size to the C_n group of the stationary phase [37]. (See also Ref. 38 for the similar behavior of

References on p. A65

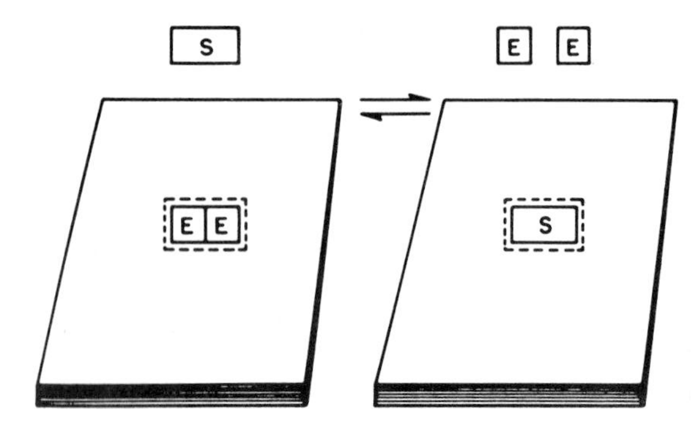

Fig. 1.10. Retention of a solute molecule, *S*, by adsorption, with displacement of some number of previously adsorbed mobile-phase molecules, *E*. (Reprinted with permission from Ref. 26.)

homologous triglycerides.) This suggests that *n*-alkyl groups on a solute molecule can penetrate (partition) into the stationary phase, as long as they are not too long [37] or are not precluded from partitioning by the structure of the solute molecule [38]. (Conversely, this observation suggests that larger molecules are incapable of a "pure" partitioning process.)

Second, it has been demonstrated [39] that the variation of solute retention in RPC as mobile-phase composition is varied closely tracks the change in the solute activity coefficient, γ_{xm}, in the mobile phase. This is in agreement with Eqn. 1.50 for a liquid/liquid partition process, assuming that γ_{xs} remains constant as the mobile phase is varied.

Third, values of *K* in RPC exhibit a maximum for intermediate concentrations of the bonded phase, which suggests a partition process [40]. (Small bonded-phase concentrations do not allow much partitioning; high concentrations are densely packed and resist solute penetration.)

Finally, it has been shown in a number of studies [41] that retention for a wide range of solutes in RPC closely parallels the retention of these same solutes in "corresponding" liquid/liquid extraction systems; e.g., water/octanol.

The evidence in favor of an adsorption or displacement process for RPC retention is equally compelling.

First, the process represented by Eqn. 1.52 should result in a solute isotherm (Section 1.11) of the Langmuir type. That is, a continued increase in the mobile-phase concentration of a solute should lead eventually to the complete (monolayer) coverage of the stationary-phase surface. Isotherms of this type are generally observed in RPC [42,43].

Second, solute retention (k') would be expected to be proportional to the quantity of bonded phase in the column, if partitioning is the dominant retention process. Instead, it is generally observed [44] that values of k' level off as the alkyl chainlength (and the amount of bonded phase) increases.

Third, it is also observed [45] that retention in RPC as mobile-phase composition is varied can be explained nicely in terms of Eqn. 1.52. Thus, for a binary-solvent mobile phase, A/B, where B is the more strongly retained solvent, Eqn. 1.52 leads to an expression for retention of the form

$$k' = \psi K (\theta_b/c_b)^n \approx (\text{constant})\, c_b^{-n} \qquad (1.53)$$

Here, θ_b is the concentration of B in the stationary phase, which for large enough concentrations of B in the mobile phase (c_b) will be constant (completion of an adsorbed monolayer). Plots of log k' vs. log c_b yield straight lines of slope $-n$, and n is observed to be proportional to the size of the solute molecule (e.g., as in Fig. 1.10, for $n = 2$). This is further evidence for an adsorption process as an explanation of RPC retention.

Finally, it should be noted that some of the evidence cited above in favor of a partition process (based on the behavior of homologous solutes) actually suggests that the retention process is a mixture of partitioning and adsorption (i.e., implying both penetration and nonpenetration of the solute into the stationary phase). In view of the foregoing observations, it would appear that a quantitative treatment of retention in RPC is somewhat premature. So far, the most widely cited attempt in this direction is the solvophobic theory of Horváth et al. [12,46].

1.6.6 Normal-phase chromatography

In this chromatographic method the stationary phase is a high-surface-area polar adsorbent, e.g., silica or a bonded silica with polar surface groups, such as cyanoalkyl. The mobile phase (a mixture of organic solvents) is less polar than the stationary phase. Consequently, more polar solutes are preferentially retained; there is often little difference in the retention of different homologs of a particular compound class. This has led to the use of NPC for so-called compound-class (group-type) separations, where, e.g., alcohols are separated as a group from monoesters and other compound classes.

The basis of NPC retention is an adsorption/displacement process, as described by Eqns. 1.52 and 1.53; this is now rather well established, on the basis of a considerable body of experimental work [3,47]. Eqn. 1.53 was first proposed for NPC by Soczewinski [48], and is commonly referred to as the Soczewinski equation. Numerous studies (e.g., Ref. 49) have shown Eqn. 1.53 to be an accurate and useful relationship.

Another feature of NPC retention is the so-called localization of adsorbed solute and mobile-phase molecules on the stationary-phase surface. Localization refers to the formation of discrete bonds (by dipole/dipole or hydrogen-bonding interactions) between polar sites on the adsorbent and polar substituents in the solute molecule. Localization, in turn, confers a high degree of specificity to the interaction of solute isomers with the adsorbent surface, leading to typically better separations of isomers by NPC than by other chromatographic methods. Other effects due to localization have been well documented, leading to a quantitative theory of solute retention as a function of NPC conditions [47]. (For additional information on NPC retention, see Refs. 3, 47, and 50.)

1.6.7 Ion-exchange chromatography

This topic is treated in more detail in Chapter 5. The column packing in IEC is usually an organic matrix (e.g., polystyrene), which is substituted with ionic groups, R^- or R^+; e.g., sulfonate ($-SO_3^-$) or trimethylammonium ($-N-(CH_3)_3^+$). The mobile phase typically consists of water plus buffer and/or salt. The retention of a solute ion, X^{n+} or X^{n-}, occurs via ion exchange with a mobile phase ion, M^+ or M^-, of similar (positive or negative) charge

$$X_m{}^{n-} + n\,R^+M^- \rightleftharpoons n\,M_m{}^- + [R^+]_n X^{n-} \tag{1.54}$$

or

$$X_m{}^{n+} + n\,R^-M^+ \rightleftharpoons n\,M_m{}^+ + [R^-]_n X^{n+} \tag{1.55}$$

Here, a subscript m designates a molecule in the mobile phase. Eqns. 1.54 and 1.55 are of the same form as Eqn. 1.52 for retention via adsorption. Therefore, retention in ion exchange is given by an equation similar to Eqn. 1.53. The only difference is that c_b now refers to the concentration of the mobile-phase ion, M^+ or M^-. If the charge of the solute ion is $\pm r$ and that of the mobile-phase ion is $\pm s$, then n in Eqn. 1.53 equals r/s. This treatment neglects the activity coefficients of the solute and mobile-phase ions, but Eqn. 1.53 is nevertheless generally reliable.

IEC is often applied to the separation of acidic or basic samples, whose charge varies with pH. In the simple case of solute molecules bearing a single acidic or basic group, the solute will be present as some mixture of charged and neutral species. The fraction of solute molecules that are ionized (f^+ or f^-; Eqn. 1.46 or 1.47) then determines retention according to Eqn. 1.48 (for an acid, HA; a similar expression holds for a base, B). In the case of ion exchange, the retention of the uncharged species (HA or B) can be ignored, so that k' is proportional to f^+ or f^-, respectively. The resulting dependence of k' on the concentration, c_b, of salt and/or buffer and fractional ionization, f^+, is then

$$k' = (\text{constant})\, f^+\, c_b{}^{-n} \tag{1.56}$$

From the foregoing treatment it is seen that solutes can be separated by IEC according to net charge, degree of ionization (or the pK_a value), and other factors. (For a further discussion of IEC retention, see Refs. 5, 12, 51, and 52.)

1.6.8 Ion-pair chromatography

In this chromatographic method the column packing is usually the same as in reversed-phase chromatography; e.g., a C_8 or C_{18} silica. The mobile phase is likewise similar to that used in RPC: an aqueous/organic solvent mixture containing a buffer plus a so-called ion-pair reagent. The ion-pair reagent will be positively charged (P^+) for the retention and separation of sample anions (X^-) and negatively charged (P^-) for the retention of sample cations (X^+). We will use the example of an anionic solute, X^-, and a cationic ion-pair reagent (P^+), but a similar discussion can be used for the opposite case (X^+ and P^-). Typical examples of ion-pair reagents are hexane sulfonate (P^-) and tetrabutylammonium (P^+).

The basis of retention in ion-pair chromatography (IPC) is still controversial, two different processes being possible: (a) adsorption of ion pairs ($X^- P^+$) or (b) formation of an in situ ion exchanger. (For a good discussion of the evidence for one or the other of these two possibilities, see Refs. 53 and 54.) The former retention process can be expressed by

$$X_m^- + P_m^+ \rightleftharpoons (X^- P^+)_m$$

and

$$(X^- P^+)_m \rightleftharpoons (X^- P^+)_s \qquad (1.57)$$

That is, an ion pair ($X^- P^+$) forms in the mobile phase and is then retained (adsorption or partition) by the stationary phase.

The latter IPC retention process involves initial retention of the ion-pair reagent

$$(P^+ M^-)_m \rightleftharpoons (P^+ M^-)_s \qquad (1.58)$$

followed by ion exchange between the solute (X^-) and the mobile-phase ion, M^-

$$X^- + (P^+ M^-)_s \rightleftharpoons M^- + (P^+ X^-)_s \qquad (1.59)$$

Although these two processes (Eqns. 1.58 and 1.59) appear somewhat different, they lead to quite similar predictions of retention as a function of experimental conditions. For this reason, we need only consider the second possibility here (Eqn. 1.59).

The formation of the in situ ion exchanger $(P^+ M^-)_s$ is similar to the retention of any solute in RPC. An increase in the concentration of P^+ in the mobile phase leads to increasing retention of P^+ in the stationary phase. Thus, in the absence of P^+ from the mobile phase, the retention of solute molecules occurs by a reversed-phase process (Section 1.6.4). As the concentration of P^+ is increased, some fraction of the stationary-phase surface is covered by P^+ and converted to an in situ ion exchanger. At large enough concentrations of P^+, the surface is completely covered by P^+, and retention occurs solely by an ion-exchange process (Section 1.6.7).

Retention in IPC can thus be continuously varied from a reversed-phase process (low concentrations of P^+) to an ion-exchange process (high concentrations of P^+). This capability provides a number of practical advantages. For example, variation of the mobile-phase composition (pH, concentration of P, etc.) allows a considerable control over the retention of individual sample ions, more so than is possible in either RPC or IEC alone. This can be used to separate particularly difficult samples, e.g., mixtures of anionic,

References on p. A65

cationic, and/or neutral molecules. (For a further discussion of the in situ ion-exchange model of IPC, see Refs. 55 (simple) or 56 and 57 (more complex). For a general review of all aspects of IPC, see Ref. 58.)

1.6.9 Size-exclusion chromatography

This chromatographic method (which is covered in more detail in Chapter 6) differs fundamentally from the other procedures in Sections 1.6.2-1.6.8. In the latter, retention is based on an attraction between solute molecules and the stationary phase, so that the capacity factor, k', is always >0, i.e., retention times $t_R > t_0$. In SEC, solute molecules experience no attraction for the stationary phase, or – which is equivalent – the enthalpy of retention is zero. Thus, SEC is said to be an entropically driven process.

Separation and retention in SEC are determined solely by the size of the solute molecule (its hydrodynamic diameter) relative to the size of the pores of the column packing. Molecules whose hydrodynamic diameter is greater than the diameter of any of the pores of the column packing are essentially excluded from these pores; these solute molecules leave the column first. Molecules whose hydrodynamic diameter is very much smaller than that of the pores of the column packing have the same concentration (at equilibrium)

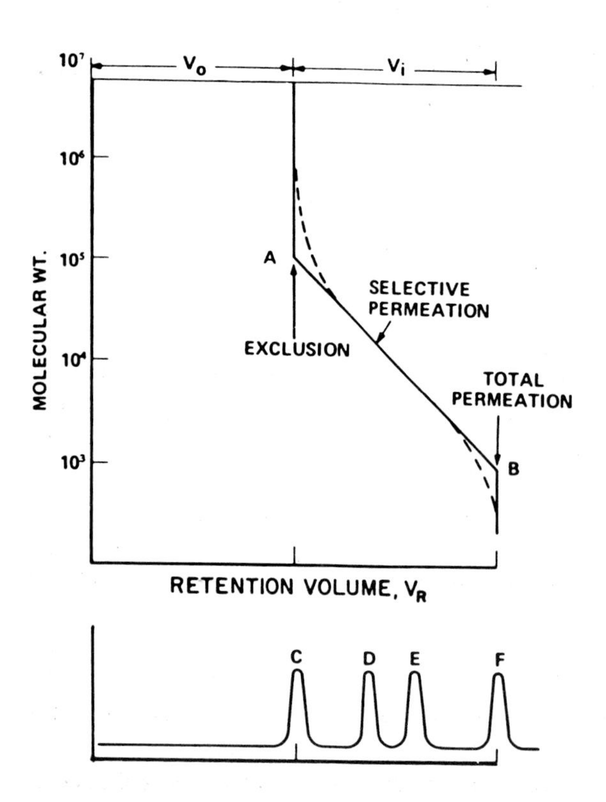

Fig. 1.11. Calibration plot for solute molecular weight vs. SEC retention. (Reprinted with permission from Ref. 26.)

inside and outside of the pores; these solute molecules have a retention time equal to t_o. Solutes of intermediate molecular size have intermediate retention times, and are separated on the basis of molecular size (i.e. their hydrodynamic diameters).

These relationships are illustrated in Fig. 1.11, where a SEC calibration plot is shown for a hypothetical column. In this case, solutes having molecular weights $>10^5$ are excluded from the pores of the column packing and are eluted in a retention volume V_o that is equal to the volume of mobile phase within the column [but excluding that contained in any pores; i.e., $V_o < V_m$. Note that V_o is not the same as (t_o F)]. Similarly, solutes with molecular weights <1000 are eluted in a volume $V_m = V_o + V_i$ (V_i is the volume of mobile phase inside the pores). In the chromatogram shown in Fig. 1.11, Compound C (MW $> 10^5$) is eluted at the exclusion limit of the column, Compound F (MW <1000) is eluted at the total permeation limit of the column, and Compounds D and E ($10^3 <$ MW $< 10^5$) are eluted in the fractionation range of the column. (For a further discussion of retention in SEC, see Refs. 12 and 59.)

1.7 EXTRA-THERMODYNAMIC RELATIONSHIPS

With some notable exceptions, exact thermodynamic relationships are somewhat limited in their ability to provide reliable (and useful) predictions of chromatographic retention as a function of the molecular structure of samples and experimental conditions. However, it is possible to formulate simple physical models of chromatographic retention in thermodynamic terms, and this can, in turn, provide an alternative (extra-thermodynamic) basis for the development of quantitative equations that can be used to predict retention.

1.7.1 Additive free-energy relationships

1.7.1.1 Martin equation

Eqns. 1.39 and 1.40 describe retention in terms of the standard free energy of retention, $\Delta G°$, where the latter is given as $\Delta H° - T\Delta S°$. The retention entropy, $S°$, can be ignored as a first approximation, leading (Eqn. 1.40) to

$$\log k' \approx \text{(constant)} - \Delta H°/\mathcal{R}T \tag{1.60}$$

The enthalpy of retention, $\Delta H°$, can be visualized as arising from the interactions of the mobile and stationary phase with individual substituents of a sample molecule. Thus, for the total molecule

$$\Delta H° = \Sigma\ \Delta\Delta H_i° \tag{1.61}$$

where $\Delta\Delta H_i°$ is the contribution to $\Delta H°$ of a substituent group i. In this simple picture, a given group, i, will always have the same value of $\Delta\Delta H_i°$, as long as the mobile and stationary phases do not change. That is, it is assumed that a given group, i, interacts with the surrounding mobile and stationary phases independently of other groups, j, in the solute molecule (or their position relative to i).

Combining Eqns. 1.60 and 1.61 with the definition of R_M (Eqn. 1.17) then yields

$$\log k' = (\text{constant}) + \Sigma\, \Delta R_M(i) \tag{1.62}$$

Here, $\Delta R_M(i)$ refers to the contribution of a group, i, to log k'. Now, if a parent (unsubstituted) solute, P, is defined with $k' = k_p$, we can write

$$\log k' = \log k_p + \Sigma\, \Delta R_M(i) \tag{1.63}$$

where groups i now refer to the various substituents on the parent compound, P. Eqn. 1.63 is known as the Martin equation [60]. If values of $\Delta R_M(i)$ are measured for various substituents i (for a given chromatographic system; i.e. the same mobile and stationary phases and the same temperature), values of k' can be predicted for any derivative of solute P by means of Eqn. 1.63. Table 1.4 summarizes ΔR_M values for a number of common substituent groups in reversed-phase LC (two different mobile-phase compositions).

A common example of Eqn. 1.63 is for a series of homologs, where group i is defined as a methylene group. In this case, for a solute containing n methylene groups (n =

TABLE 1.4

CONTRIBUTIONS OF DIFFERENT SUBSTITUENT GROUPS TO SOLUTE RETENTION IN RPC [61]

Substituent group*	$10^{\Delta R_M}$ **	
	50% aq. methanol	40% aq. acetonitrile
$-CONH_2$	0.06	0.04
$-OH$	0.10	0.08
$-CN$	0.19	0.26
$-CHO$	0.22	0.21
$-CO_2CH_3$	0.39	0.37
$-COC_2H_5$	0.46	0.47
$-Cl$	0.83	0.84
$-Br$	1.05	1.00
$-CH_2-$	2.18	1.90

* Substituent on benzene.

** Equals factor by which k' is changed due to addition of substituent to benzene as solute; e.g., phenol (addition of $-OH$) has a decrease in k' vs. benzene, by a factor of 0.10 with 50% aq. methanol as mobile phase, and by a factor of 0.08 with 40% aq. acetonitrile as mobile phase.

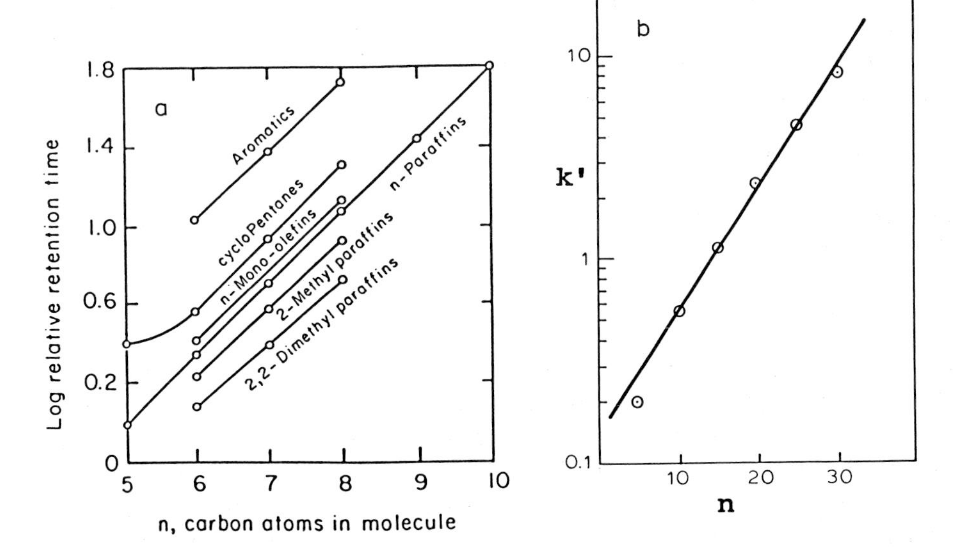

Fig. 1.12. Example of the Martin equation. (a) Relative retention time (proportional to k') vs. number of methylene groups in various solute homologous series, separated by GC (Ref. 62); (b) k' vs. number of styrene residues in a series of polystyrene oligomers, separated by normal-phase LC on a silica packing. (Reprinted from Ref. 26.)

number of alkyl carbons minus 1), we have

$$\log k' = \log k_p + n \, \Delta R_M(\text{-CH}_2\text{-}) \tag{1.64}$$

A plot of log k' vs. n for a series of homologs should therefore give a straight line. This is illustrated in Fig. 1.12a for several homologous series of compounds, separated by GC. Eqn. 1.64 applies similarly to any series of compounds that differ only in the number of some repeating unit in the sample molecule. Fig. 1.12b shows a plot of log k' vs. the number of styrene residues in the oligomers of a polystyrene sample, for a normal-phase LC separation. The Martin equation has been found to be broadly applicable to the prediction of retention in various chromatographic systems [10,26].

1.7.1.2 Retention index systems

By far the most common retention index system is that for GC by Kováts [63], who defined an index, I, for a given solute, i, and GC system as

$$I_i = 100 \frac{\ln(V_i/V_n)}{\ln(V_{n+1}/V_n)} + 100\,n \tag{1.65}$$

where V_i, V_n, and V_{n+1} are the corrected retention volumes of the solute, i, and n-alkanes of carbon number n and $n+1$, which are eluted just before and just after i. There are

References on p. A65

several advantages to expressing GC retention in terms of these I_i values. First, retention times and k' values tend to vary from column to column in GC, and this complicates solute identification via a retention measurement. The use of retention indices minimizes this problem, due to the use of the *n*-alkanes as "internal standards".

Second, the Martin equation can be used directly with I_i values. That is, the addition of a given substituent, *k*, to a parent compound, *P*, results in a predictable change (usually an increase) in I_i, facilitating the tentative identification of unknown bands in the chromatogram. Finally, differences in I_i (ΔI) for a series of selected solutes on different stationary phases are commonly used to define the separation characteristics of each stationary phase (the schemes of Rohrschneider [64] and McReynolds [65]). (For a further discussion, see Refs. 12 and 66.)

1.7.2 Linear free-energy relationships

Many predictive equations of this type are based (ultimately) on the rigorous equations [67,68] that describe different kinds of interactions between adjacent molecules, i.e., a solute molecule and molecules of the surrounding mobile or stationary phase. In general, the energies, E_S, of these various interactions are of the form

$$E_S = \text{(constant)}\ F_S F_p \qquad (1.66)$$

where F_S is some property of the sample molecule and F_p is some corresponding property of a molecule in the mobile or stationary phase. Table 1.2 summarizes some of these common intermolecular interactions and the associated quantities F_S or F_p.

The free energy of retention can be expressed as

$$-\Delta G^\circ \approx \Delta E_S = \text{(constant)}\ F_S \Delta F_p \qquad (1.67)$$

where ΔF_p is F_p for the stationary phase minus F_p for the mobile phase.

At a given temperature, Eqns. 1.39 and 1.67 combine to give

$$\log k' = A + B F_S \Delta F_p \qquad (1.68)$$

where A and B are constants for a given sample and mobile or stationary phase. Eqn. 1.68 assumes that a single property of the sample molecule (F_S) and of the chromatographic phases (ΔF_p) determines chromatographic retention. If k' values are now compared for the same samples and two different chromatographic systems, I and II (F_S remains the same for a given sample in either system), we can write

$$\log k_I = A_I + B F_S \Delta F_I \qquad (1.69)$$

and

$$\log k_{II} = A_{II} + B F_S \Delta F_{II}$$

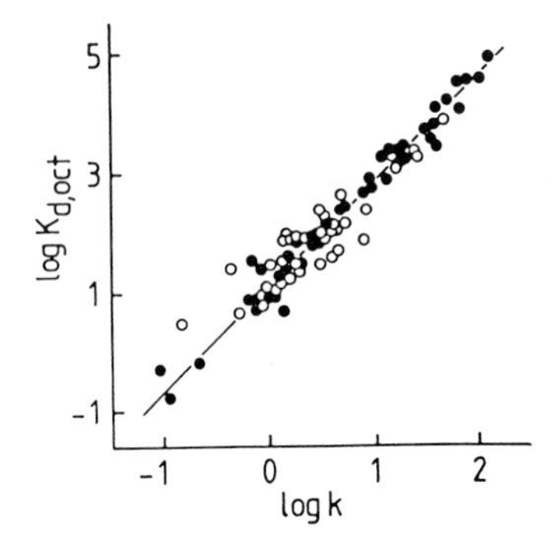

Fig. 1.13. Correlation of octanol/water P values with RPC k' values, according to Eqn. 1.71. o = Acids and alcohols; • = neutrals, bases, and amphiprotics. All solutes are nonionized under the conditions of separation. (Redrawn from Ref. 41.)

and therefore

$$\log k_{\mathrm{I}} = a + b \log k_{\mathrm{II}} \tag{1.70}$$

According to Eqn. 1.70, a plot of log k' for a series of samples in one chromatographic system (I) vs. another (II) will yield a straight line.

Eqn. 1.70 assumes that a single kind of intermolecular interaction will largely define sample retention in the two systems (I and II) being compared. This is not likely to be true when the two chromatographic systems are quite different, e.g., GLC vs. NPC. However, liquid/liquid partition with octanol and water as the two immiscible phases is often compared with RPC (aqueous/organic mixtures as mobile phase, bonded C_8 or C_{18} as stationary phase) in terms of Eqn. 1.70. For this case (see Eqn. 1.11) we can replace k_{II} in Eqn. 1.70 by the equilibrium constant, P, for the water/octanol system and obtain

$$\log k_{\mathrm{I}} = a + b \log P \tag{1.71}$$

where k_{I} refers to a k' value for RPC. Numerous studies [41] have demonstrated the approximate validity of Eqn. 1.71, which is of considerable practical importance because of the use of P values in pharmacological studies and drug design. An example is given in Fig. 1.13 for various neutral, acidic, and basic compounds as samples.

1.7.2.1 Solvent polarity scales

The relative polarity of the solvents used either as the stationary phase in GC or the mobile phase in LC is of practical importance (Section 1.6.1). As a first approximation, we can identify solvent polarity with the parameter F_p of Eqn. 1.68. (This assumes that the molecular size or the solute is roughly constant; otherwise the constant B of Eqn. 1.68 will vary with solute molecular size.) Solvent polarity can be ranked in various ways, as

TABLE 1.5

SCALES OF SOLVENT POLARITY [3,5,32,47,64,68]

Solvent	Hildebrand parameter δ [64]	Polarity Index [68]	Solvent strength parameter ε° [32]*
n-Hexane	7.3	0.1	0.00
Diethyl ether	7.4	2.8	0.38
Cyclohexane	8.2	0.2	0.04
Ethyl acetate	8.6	4.4	0.60
Chloroform	9.1	4.1	0.36
Benzene	9.2	2.7	0.32
Acetone	9.4	5.1	0.58
Dichloromethane	9.6	3.1	0.40
Tetrahydrofuran	9.9	4.0	0.51
1-Propanol	10.2	4.0	0.82
Ethanol	11.2	4.3	0.89
Acetonitrile	11.8	5.8	0.55
Methanol	12.9	5.1	0.95
Formamide	17.9	9.6	–
Water	23	10.2	–

* Solvent strength parameter for NPC with alumina as adsorbent; defined in Refs. 3 and 47 ($\varepsilon^\circ = \varepsilon''$).

illustrated in Table 1.5. The Hildebrand solubility parameter, δ, is the oldest such measure of solvent polarity [68]. It has the advantage that it can be estimated from the boiling point of the solvent (which is easily known). However, the solubility parameter fails to distinguish among the different contributions to solvent polarity (Table 1.3). As a result, it often gives a misleading indication of relative solvent polarity (e.g., the polarity of basic solvents, such as diethyl ether is underestimated). This can be seen in Table 1.5, where the ranking of solvents according to δ differs appreciably from the other two scales. The solubility parameter concept has been expanded by Hansen [69] and by Karger et al. [36,70] in an attempt to correct for this limitation.
The polarity index P' of Table 1.5 is probably the best general measure of solvent polarity, since it is derived from data on actual solute/solvent mixtures [71]. The solvent-strength parameter ε° is derived from LC retention data for alumina as column packing [3,47]; it is a good measure of solvent polarity for this particular LC system.

1.7.2.2 Multi-parameter solvent polarity scales

No single-parameter solvent polarity scale can accurately predict the effect of the solvent on chromatographic retention, because different intermolecular interactions con-

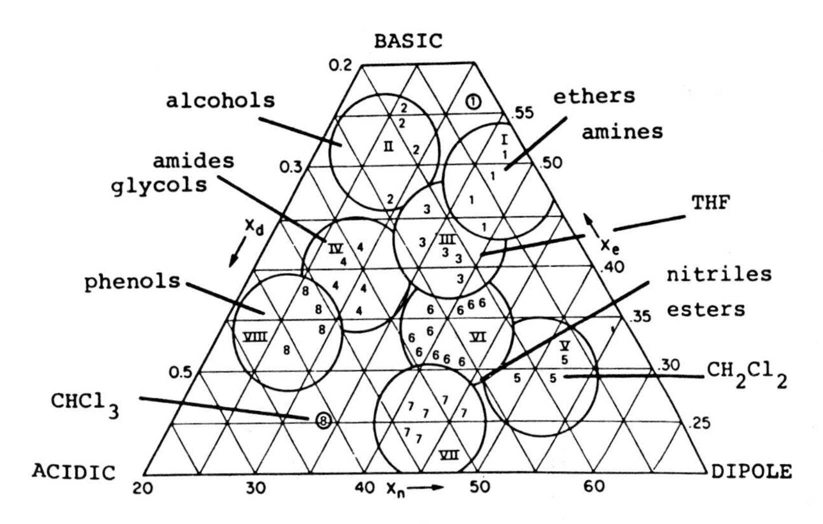

Fig. 1.14. Solvent selectivity triangle. (Redrawn from Ref. 67.)

tribute to solvent polarity and solute retention. Various attempts have been made to get around this problem, usually by assuming that the solute/solvent interaction energy, E_S, of Eqn. 1.66 can be more accurately expressed by a relationship of the form

$$E_S = \Sigma A_i(F_S F_p)_i \qquad (1.72)$$

Here, each term i is for a specific interaction, as summarized in Table 1.2.

Eqn. 1.72 has been used to expand the solubility parameter treatment for chromatographic retention, as noted above [33,67,70]. Eqn. 1.72 is also the basis of the polarity index treatment of Ref. 71. The polarity index P' can, in turn, be subdivided into contributions from the more important components of solvent polarity: dipole moment, acidity, and basicity. The fraction of P' contributed by dipole moment (x_n), acidity (x_d), and basicity (x_e) then defines solvent selectivity, on a normalized, total-polarity basis. That is, we can speak of the relative "dipolarity", acidity, and basicity of a solvent in terms of these parameters (x_n, x_d, x_e). This gives rise to the so-called solvent selectivity triangle, shown in Fig. 1.14.

The solvent selectivity triangle positions different solvents according to their relative dipole moment, acidity, and basicity. The total polarity, P', of solvent mixtures A/B can be maintained constant by varying the concentration of a less polar solvent, A, while changing the more polar solvent, B. If the different solvents B are chosen for differences in dipole moment, acidity, and basicity, this is expected to result in pronounced differences in relative retention (change in selectivity). For example, acidic solutes will preferentially interact with a more basic solvent B, leading to earlier elution if A/B is the mobile phase. (For a further discussion of the applicability of the solvent selectivity triangle in chromatography, see Ref. 26 for LC and Refs. 72 and 73 for GC.)

A similar, more quantitative approach to the prediction of chromatographic retention as a function of solvent polarity is that of Kamlet et al. [74,75]. In their so-called solvatochromic model retention is expressed as

$$\log k' = C + mV_I/100 + s\pi^* + b\beta_m + a\alpha_m \quad (1.73)$$

Here, C is a constant, m is proportional to the solubility parameter of the mobile phase, V_I is the molar volume of the solute, s and π^* measure the combined polarizability and dipole moment of the solute and mobile phase, respectively, a and b measure the basicity and acidity of the solute, and α_m and β_m measure the acidity and basicity of the mobile phase, respectively. Eqn. 1.73 appears to have considerable promise as a means of accurately predicting retention for at least some HPLC systems [76,77].

Returning to the solvent selectivity triangle of Fig. 1.14, the "dipolarity" (x_n), acidity (x_d), and basicity (x_e) parameters of this treatment can be correlated with π^*, α, and β of the solvatochromic model [78]. The latter study [78] also makes clear that the solvent-triangle parameters (x_n, x_d, x_e) are less clear-cut measures of dipole moment, acidity, and basicity than are π^*, α, and β. This means that two solvents having somewhat different values of relative dipole moment, acidity, and basicity may nevertheless be (inaccurately) treated as equivalent in the classification of the solvent triangle.

It should also be pointed out that the extension of the solvent-triangle approach to reversed-phase systems (by several authors) is further limited by (a) the nonideality of aqueous/organic solvents as mobile phases, (b) by the fact that the more polar solvent (water) normally dominates solute/solvent interactions and selectivity, and (c) by the uptake of organic solvent from the mobile phase by the stationary phase. This may account in part for reported discrepancies [79,80] between predictions from the solvent triangle and experimental data for RPC.

1.7.2.3 Hammett equation

The Martin equation (Eqn. 1.63) states that k' for a substituted parent compound, P, can be related to k' for Solute P and the group retention factors, ΔR_M, for the various substituents on P. This additive free-energy relationship assumes that the interactions of various solute groups with the mobile and stationary phases are not affected by the presence of other groups in the solute molecule. For the case of aromatic solute molecules this is often not the case. Typically, a substituent, j, on an aromatic ring affects the retention effect of a second substituent, k, on the same ring by intramolecular electronic interaction. The Hammett equation accounts for deviations, Δ, in $\log k'$ from Eqn. 1.63 as

$$\Delta = \sigma\rho \quad (1.74)$$

Here, σ is a property of the aromatic substituent, k, and ρ is a function of the parent compound, P, and the chromatographic system. (For a review of applications of the Hammett equation to chromatographic retention, see Ref. 10.)

1.7.2.4 Other extra-thermodynamic parameters

A large number of publications deal with correlations of retention with molecular structure, in which various parameters are used to describe different structural features of the solute molecule. These parameters include various quantum-chemical indices, molecular-shape descriptors, the connectivity index, and other topological indices. Often such an approach provides a reasonably reliable description of retention for compounds with limited structural variations. However, it is likely that many such reported correlations are fortuitous, since different parameters of this type often correlate with each other. (For a good review of this area, see Ref. 10.)

1.8 MOBILE-PHASE FLOW AND BANDBROADENING

The flow of the mobile phase through a sorbent bed or column has a number of important effects or consequences. In the case of a column, a certain pressure drop across the column results, and this is often of practical interest. The rate of flow also determines the time of separation. Finally, the flow of mobile phase has a profound effect on bandbroadening and resolution.

1.8.1 Flow through tubing and columns

The flowrate, F, in laminar flow through open tubing or capillary columns can be described by a simple, rigorous relationship

$$F = (P_i - P_o)\, d_c^{\,4}\, \pi / 32\, \eta\, L \qquad (1.75)$$

Here, P_i and P_o are the inlet and outlet pressures, d_c and L are the diameter and length of the tube or capillary, and η is the viscosity of the mobile phase. The velocity, u, of the mobile phase varies with position, being equal to zero at the wall and to a maximum value, u_{max}, at the center of the tube. These relationships hold only for straight tubes of a certain minimum length [81,82]. It should also be noted that F varies from the column inlet to outlet for the case of compressible mobile phases (GC, SFC).

Darcy's law governs the dependence of flowrate and pressure drop through a packed column

$$u_e = B^\circ (P_i - P_o)/\varepsilon\, \eta\, L \qquad (1.76)$$

Here, B° is the specific permeability coefficient, ε is the porosity of the column packing (equal to ≈ 0.4 for tightly packed columns), and u_e is the interstitial (actual) velocity of the mobile phase. If the column packing is nonporous, the average mobile-phase velocity, u, is equal to u_e. The coefficient B° is, in turn, given by the Kozeny-Carman equation

$$B^\circ = d_p^{\,2}\, \varepsilon_e^{\,3} / 180\, (1-\varepsilon_e)^{\,2} \approx d_p^{\,2} / 1000 \qquad (1.77)$$

Here, d_p refers to the diameter of the particles of column packing. For the case of porous column packings, the average velocity, u, is related to the interstitial velocity u_e as

$$u = u_e / x \tag{1.78}$$

where x is the fraction of the mobile phase (inside the column) that is outside of the pores of the packing.

In liquid chromatography the column pressure drop ($P_i - P_o$) is of special interest. For the usual porous column packings this pressure drop is given as

$$P_i - P_o \approx 150\, L\, \eta\, F / {d_p}^2\, {d_c}^2 \tag{1.79}$$

Here, the column length, L, and internal diameter, d_c, are in cm, the flowrate is in ml/min, the particle diameter, d_p, is in μm, the viscosity, η, is in centipoise, and the pressure is in atm. (For a further discussion of flow through packed columns as well as in thin-layer chromatography, see Ref. 12.)

1.8.2 Bandbroadening in the column

There are a number of detailed and comprehensive reviews of bandbroadening in chromatography [2,5,12,19,26,83-85] that treat the subject at any level of complexity the reader may desire. This section will cover only the more practical aspects of this subject.

1.8.2.1 Additivity of variances

Bandwidth can be defined in various ways, the most common of which is the standard deviation of the Gaussian band, σ, in either time (σ_t) or volume (σ_v) units

$$\sigma_v = \sigma_t F \tag{1.80}$$

During a chromatographic separation, solute bands exhibit broadening as a result of many different physical effects. The final (observed) bandwidth, σ, can be related to (mutually independent) contributions to bandbroadening, σ_i, from various causes, i, as

$$\sigma^2 = \Sigma\, {\sigma_i}^2 \tag{1.81}$$

where the quantity σ^2 is referred to as the variance of the band. The form of Eqn. 1.81 is such that if the variance due to a particular effect, k, is large, then $\sigma \approx \sigma_k$.

The various bandbroadening contributions, σ_i, can arise during (a) the introduction of the sample into the column (σ_{inj}) or (b) the passage of the sample through various connectors and tubing that connect the column to the sample-injection system and the detector (σ_{conn}), (c) the detector (σ_{det}), and (d) the column (σ_{col}). The final bandwidth is given (Eqn. 1.81) as

$$\sigma^2 = \underbrace{\sigma_{inj}^2 + \sigma_{conn}^2 + \sigma_{det}^2}_{\text{extra-column effects}} + \sigma_{col}^2 \tag{1.82}$$

The usual goal in chromatography is to minimize the extra-column contributions, so that $\sigma \approx \sigma_{col}$. (For a good, general discussion of these extra-column effects in LC or GC, see Refs. 19, 26, and 86-89.)

The bandbroadening that occurs within the column (σ_{col}) is of major concern, inasmuch as extra-column bandbroadening can, in principle, be reduced to an insignificant level. Different physical processes, j, also contribute to the value of σ_{col}

$$\sigma_{col}^2 = \sum^{j} \sigma_j^2$$

or (Eqn. 1.24)

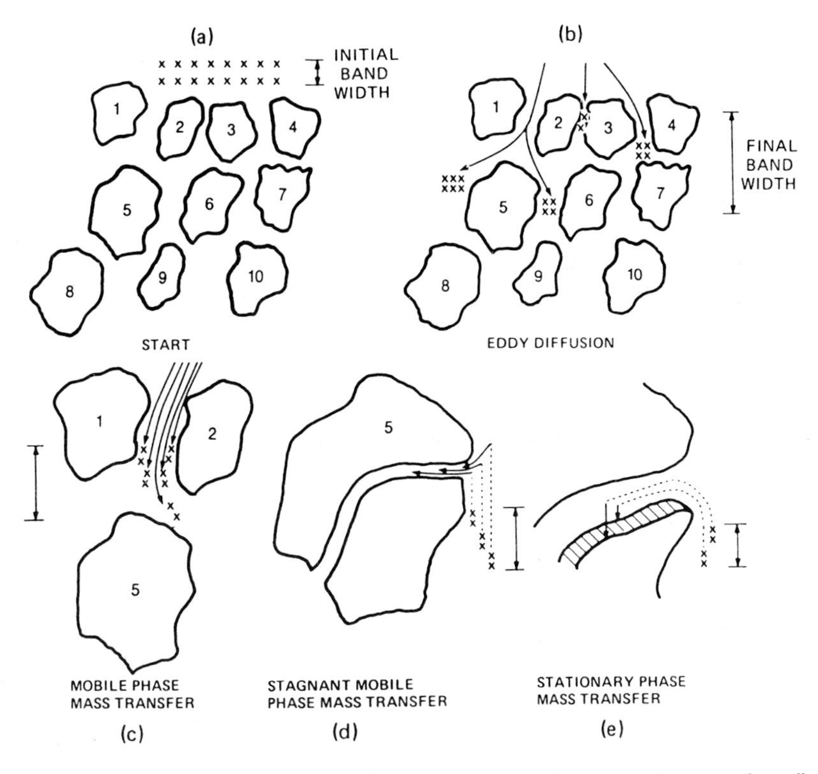

Fig. 1.15. Various processes within the column that contribute to bandbroadening. x, Molecules of the solute; numbers refer to individual particles within the column; (a) situation at the time of sample injection; (b-e) bandbroadening processes that occur after injection. (Reprinted with permission from Ref. 12.)

$$H_{col} = \sum^{j} H_j \tag{1.83}$$

We will next discuss these various contributions, j, to bandbroadening within the column. Some of these are illustrated in Fig. 1.15 for a packed column.

1.8.2.2 Longitudinal molecular diffusion

Solute molecules undergo random molecular diffusion during their passage through the column or sorbent bed. The variance contribution, σ_{md}, due to diffusion in the mobile phase (not shown in Fig. 1.15) is given by the Einstein equation

$$\sigma_{md} = (2\,\gamma_m\,D_m\,t)^{1/2} \tag{1.84}$$

where D_m is the solute diffusion coefficient in the mobile phase, γ_m is an obstruction factor $\approx$ 0.5-1.0, and t is the time spent by the band during its passage through the column. The time, t, is equal to L/u, and $H = \sigma^2/L$, so that the corresponding plateheight contribution due to diffusion in the mobile phase is

$$H_{md} = 2\,\gamma_m\,D_m/u \tag{1.85}$$

In LC the solute can also diffuse in the stationary phase, especially (as in liquid/liquid or reversed-phase chromatography) when the solute is not tightly bound to specific sites within the stationary phase [90,91]. This leads to an additional bandbroadening term (stationary phase, diffusion)

$$H_{sd} = 2\,k'\,\gamma_m\,D_s/u_e \tag{1.86}$$

The value of H_{sd} is usually small in IEC or NPC.

1.8.2.3 Bandbroadening in the mobile phase: capillary columns

In capillary columns, the flow is usually laminar – having a parabolic flow profile (i.e., faster flow in the center of the capillary; see discussion of Eqn. 1.75). This variation in flow velocity in the cross section of the capillary, combined with slow diffusion of the solute in the mobile phase, yields a plateheight contribution, H_f (flow), that can be rigorously derived [92]

$$H_f = \frac{(1+6k'+11k'^2)d_t^{\,2}\,u}{96(1+k')^2 D_m} \tag{1.87}$$

Here, d_t is the inner diameter of the capillary. Since u varies with the position within the column for compressible mobile phases, this relationship must be corrected to account for this effect in GC or SFC.

1.8.2.4 Bandbroadening in the mobile phase: packed columns

This contribution to bandbroadening is made up of two effects: multiple flowpaths and mobile-phase mass transfer (see Figs. 1.15b and c). Due to the different channels within a packed column (between the particles), solute molecules can follow different paths during their migration through the column. In some of the channels flow will be impeded by geometric constraints or the narrowness of the channel, and solute molecules will move at a lesser velocity (and faster in wider, unimpeded channels). This leads to bandbroadening by what has been called an eddy diffusion process. The corresponding plateheight contribution (flow, eddy diffusion) is

$$H_{fed} = 2\,\lambda\, d_p \qquad (1.88)$$

The parameter λ is a measure of flow inequality in the packed column.

Bandbroadening due to mobile-phase mass transfer (Fig. 1.15c) occurs for the reason discussed above for capillary columns (Eqn. 1.87): faster flow in the center of a channel, coupled with slow diffusion of solute molecules across the width of the channel. However, the channels in packed columns are highly irregular, so that Eqn. 1.87 will not apply exactly. The actual expression for this plateheight contribution (flow, diffusion) is

$$H_{fd} = \omega\, d_p^{\,2}\, u_e/D_m \qquad (1.89)$$

where ω is a column packing parameter. The similarity of Eqns. 1.87 and 1.89 should be noted. It results from the similarity of the two processes described by these equations.

The two contributions to bandbroadening represented by H_{fed} and H_{fd} are not independent of each other but actually combine (are "coupled" [93]) so as to reduce the combined bandbroadening, H_f, due to each effect

$$1/H_f = (1/H_{fed}) + (1/H_{fd}) \qquad (1.90)$$

This can be seen to yield a zero- to first-order dependence of H_f on mobile-phase velocity, u.

1.8.2.5 Bandbroadening in the stationary phase: diffusion in pores

In the case of columns packed with porous particles, the solute must diffuse in and out of the particle via the pores in the particle. This is illustrated in Fig. 1.15d. The resulting contribution to bandbroadening (pores) is given by

$$H_p = (1/30\,\gamma_t)\,(k''/[1+k'])^2\,(x/[1-x])\,(d_p^{\,2}/D_m)\,u \qquad (1.91)$$

where γ_t is a pore tortuosity factor, x represents the fraction of mobile phase outside the pores (equal to V_o/V_m), and k'' equals $(V_R - V_o)/V_o$. V_R is the retention volume and V_o is

References on p. A65

the volume of mobile phase inside the column and outside the pores (see Fig. 1.11). There has been some confusion in the past about the correct expression for H_p; an excellent discussion is given by Knox and Scott [91].

1.8.2.6 Bandbroadening in the stationary-phase liquid

For a liquid or polymeric stationary phase that is coated onto the pore walls of a porous particle, slow diffusion within this liquid film can lead to further bandbroadening, as illustrated in Fig. 1.15e. In well-designed packing materials for use in LC, this contribution to bandbroadening is usually negligible, because the diffusional distances are usually small (compared to the particle diameter), and the solute diffusion coefficient in the stationary liquid is similar in magnitude to D_m (the value in the mobile phase).

The situation can be quite different in GC, where the diffusion coefficient in the stationary-phase liquid is always much greater than in the mobile (gas) phase. For capillary GC, this contribution (stationary-phase liquid) to bandbroadening is

$$H_{spl} = (2/3)\,(2k'/[1+k']^2)\,(d_f^2/D_s)\,u \qquad (1.92)$$

Here, d_f refers to the thickness of the stationary-phase film, and D_S is the solute diffusion coefficient in this film.

1.8.2.7 Other bandbroadening processes

These processes are less important in well-designed separations. The actual binding of a solute molecule to a stationary-phase site (as in IEC or NPC) may be a slow process, especially in separations based on a bioaffinity retention process. If a small number of very retentive sites and a large number of less retentive sites are present in the stationary phase, bandbroadening may result from the saturation of the very retentive sites – even by a comparatively small sample. This is similar to the case of preparative chromatography (Section 1.11) for larger samples and an absence of strong retention sites in the column packing. Residual silanols are generally present in silica-based reversed-phase packings, and these serve as strong retention sites for basic compounds [94]. These, in turn, can lead to excess bandbroadening as a result of either or both of the above bandbroadening effects.

1.8.3 Plateheight equations

The dependence of the plateheight, H, on flow velocity, u, is important for a number of reasons. For instance, it provides a better understanding of column platenumber as a function of experimental conditions, it allows the performance of the column to be more accurately described, and it facilitates the optimization of a given separation in terms of various goals (resolution, run time, etc.). An early expression for H as a function of u is the so-called Van Deemter equation [95]

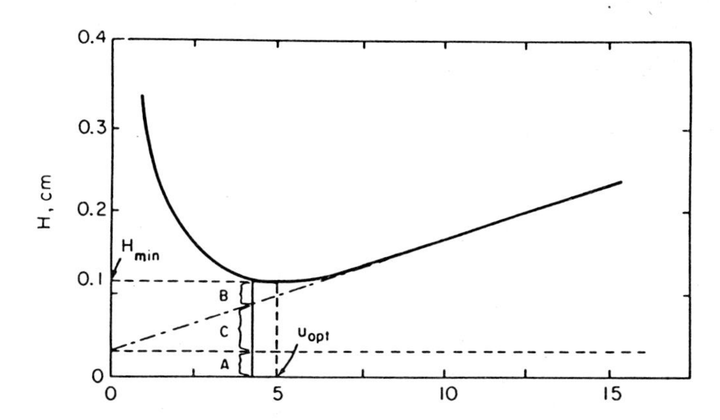

Fig. 1.16. Van Deemter plot (Eqn. 1.93) of plateheight, H, vs. mobile-phase velocity, u. (Reprinted with permission from Ref. 12.)

$$H = A + B/u + Cu \tag{1.93}$$

Here, the terms A, B, and C can be obtained from the various contributions to H given by Eqns. 1.88, 1.85, and 1.91, with Eqn. 1.93 then following from Eqn. 1.83. A plot of H vs. u according to Eqn. 1.93 is shown in Fig. 1.16.

Fig. 1.16 shows that there is a minimum value of H (maximum value of the platenumber, N) corresponding to maximum column efficiency at a so-called optimum velocity, u_{opt}. This optimum flow velocity can be derived as

$$u_{opt} = (B/C)^{1/2} \tag{1.94}$$

At lower velocities, u, the B term of Eqn. 1.93 dominates (molecular diffusion, see Eqn. 1.85), while at higher velocities the C term is most important (stationary-phase mass transfer, see Eqn. 1.91).

More general plateheight equations can be formulated, based on so-called reduced parameters: reduced plate height, h, and reduced mobile-phase velocity, ν (Eqn. 1.19)

$$h = H/d_p \tag{1.95}$$

This, in turn, leads to several more or less complicated expressions for h (and therefore H and N) [83,96,97]. Probably the most widely used equation of this type is the Knox equation [83]

$$h = A\nu^{1/3} + B/\nu + C\nu \tag{1.96}$$

The great value of Eqn. 1.96 and other, similar relationships is that column performance can be related to a large number of important variables, e.g., solute molecular weight (as reflected in the value of D_m; see below), column-packing particle size, temperature, flowrate, and column characteristics. It is, in fact, possible to use Eqn. 1.96 to predict values of H and N quite accurately for many chromatographic systems, e.g., RPC and NPC [90]. This, in turn, has led to number of general treatments for optimizing column performance and resulting chromatographic separations [2,21,26,89,98-101]. It should be noted that Eqn. 1.96 is inapplicable when a contribution to H is included which does not allow cancellation of the particle size and solute diffusion coefficient (e.g., Sections 1.8.2.6 and 1.8.2.7).

1.8.3.1 Solute diffusion coefficients

The solute diffusion coefficient in the mobile phase (either gas or liquid), D_m, plays an important role in determining the column platenumber and separation. A number of different equations have been proposed for predicting values of D_m as a function of solute molecular structure and experimental conditions [5]. A convenient relationship for estimating the value of D_m in GC is [5]

$$D_m = \frac{10^{-3}\, T^{1.75}}{P\,[(\sum v_i)_A^{\;1/3} + (\sum v_i)_B^{\;1/3}]} \left[\frac{1}{M_A} + \frac{1}{M_B}\right]^{1/2} \tag{1.97}$$

TABLE 1.6

SPECIAL ATOMIC DIFFUSION VOLUMES [101]

Atomic and structural diffusion volume increments			
C	16.5	(Cl)	19.5
H	1.98	(S)	17.0
O	5.48	Aromatic or heterocyclic rings	
(N)	5.69		– 20.2

Diffusion volumes of simple molecules			
H_2	7.07	CO_2	26.9
D_2	6.70	N_2O	35.9
He	2.88	NH_3	14.9
N_2	17.9	H_2O	12.7
O_2	16.6	(CCl_2F_2)	114.8
Air	20.1	(SF_6)	69.7
Ne	5.59		
Ar	16.1	(Cl_2)	37.7
Kr	22.8	(Br_2)	67.2
(Xe)	37.9	(SO_2)	41.1
CO	18.9		

Here, T is the absolute temperature, P is the pressure (atm), and M_A and M_B are the molecular weights of the mobile-phase gas, A, and solute, B. The v_i values are empirical atomic diffusion volumes (Table 1.6), which are summed for the molecules of the solute and mobile phase (e.g., H_2, He) to give the appropriate molecular volumes.

The most common method of estimating values of D_m in LC is by means of the Wilke-Chang equation [102]

$$D_m = 7.4 \times 10^{-8} (\psi_B M_B)^{1/2} T/\eta \ V_A{}^{0.6} \tag{1.98}$$

Here, V_A is the solute molar volume (ml/g-mol), M_B is the molecular weight of the solvent, B, η is the mobile-phase viscosity (cP) and ψ_B is an "association factor" for the mobile phase. The latter is 1.0 for nonpolar solvents, 1.5 for ethanol, 1.9 for methanol, and 2.6 for water.

Eqn. 1.98 is restricted to small solute molecules, those with molecular weights less than a few thousand D. [For a review of corresponding equations for large molecules (synthetic polymers, natural polymers, such as proteins, etc.) see Refs. 59, 89 and 98.] The values of D_m for large molecules in LC are much smaller than for typical organic compounds with molecular weights of 100-1000. This, in turn, means that the reduced velocity, ν, will be larger, other factors equal, and N will be smaller. This is illustrated by the general plot of reduced plate height, h, vs. reduced mobile-phase velocity, ν, in Fig. 1.17.

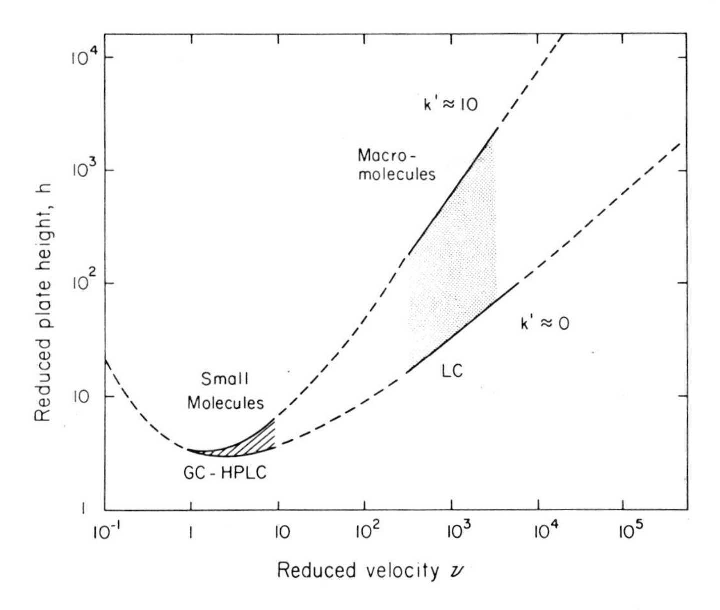

Fig. 1.17. Plot of reduced plate height, h, vs. reduced mobile-phase velocity, ν (Eqn. 1.96). (Reprinted with permission from Ref. 12.)

Alternatively, equivalent platenumbers require much smaller flowrates than are typical for small-molecule solutes.

1.9 VARIATIONS ON ISOCRATIC ELUTION

So far, we have assumed that experimental conditions (e.g., flowrate, temperature, mobile-phase composition) remain constant during chromatographic separation (so-called isocratic elution). However, there are many practical situations where a stepwise or continuous change in some experimental condition during the separation will be highly beneficial. This gives rise to several anisocratic procedures that we will discuss in this section: temperature programing, flow programing, gradient elution, and column switching. These different procedures may involve one or more stepwise changes in operating conditions and/or a continuous change in some experimental parameter.

The most common reason for using an anisocratic procedure in chromatography is for samples where the retention range is too large. That is, for any given set of isocratic separation conditions, the k' value of the first band is <1 and/or the k' value of the last band is >20. This is illustrated in Figs. 1.18a-c for a hypothetical LC separation, carried out with mobile phases of different composition. The mobile-phase composition varies from 10%B (weak) in Fig. 1.18a to 30%B (strong) in Fig. 1.18c. However, no single composition results in an acceptable separation. Either the front-end resolution is poor, because the k' values of these bands are too low (Figs. 1.18b and c), or else later bands take too long to leave the column and are therefore too wide for easy detection (Figs. 1.18a and b; note that Band 8 is not eluted in Fig. 1.18a in 150 min; $k' > 150$).

If we change the mobile-phase composition (%B) during the separation of the sample in Fig. 1.18, all of these problems disappear, because each band moves through the column at an average %B that results in $k' \approx$ 2-5. This is shown in Fig. 1.18d, where the mobile phase is changed from 0 to 80%B over a period of 20 min. Now, each band is well separated from adjacent bands, the width of each band is relatively narrow (and therefore the band is easily detected), and the separation is completed in a short time.

The problem of wide-range samples can occur in either GC or LC, and the solution is usually similar to the example of Fig. 1.18. That is, conditions must be varied during separation so as to cause a continuous decrease in sample retention (k'). In GC, the column temperature is usually increased during the separation (temperature programing), while in LC the concentration of the stronger solvent (%B) is normally increased (gradient elution). In each case, we can calculate the migration of a solute through the column as a function of time, t, after sample injection, if we know the isocratic retention of the solute (k') as a function of experimental conditions. That is, anisocratic separation can be visualized as a series of small isocratic steps, and this allows us to predict accurately the retention time in gradient elution. For this reason the anisocratic separation of a given sample also bears a close resemblance to isocratic separation [103,104], as we will see next.

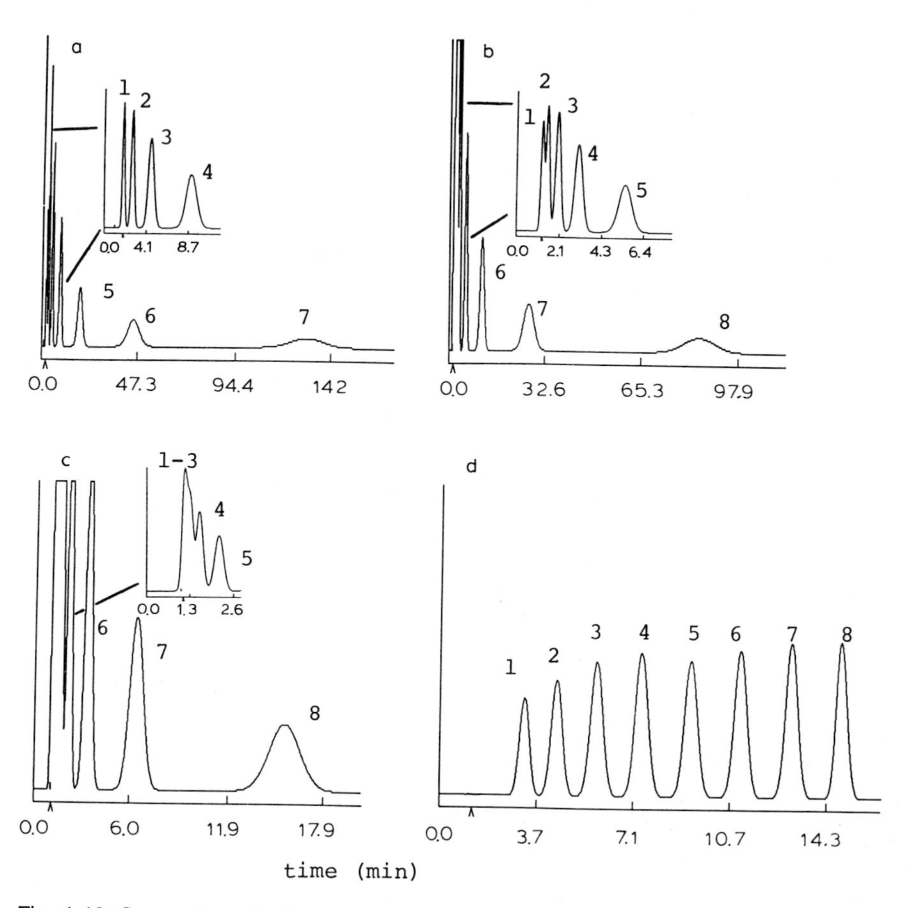

Fig. 1.18. Separation of a hypothetical 8-component sample having a wide retention range by reversed-phase LC. (a-c) Isocratic elution with mobile phases of composition 10%B, 20%B, and 30%B, respectively. (d) Gradient elution from 0 to 80%B in 20 min.

1.9.1 Gradient elution

A better appreciation of how separation in gradient elution takes place can be obtained from Fig. 1.19, where the migration of different sample components from the column inlet to the outlet is visualized. In Fig. 1.19a, the migration of the first (X) and last (Z) bands in a chromatogram is shown (solid lines). Initially (just after sample injection), the various solutes in the sample enter the column in a weak mobile phase (k' large for all bands), and all solutes are thereby retained at the column inlet. As separation proceeds and the mobile phase becomes progressively stronger (k' decreases), the k' value of the least-retained compound, X, is eventually lowered (dashed curve in Fig. 1.19a) so that at time t_1 Solute X begins to move through the column as indicated. During the migration of X, its k' value continues to decrease so that X leaves the column (fractional migration $x/L = 1$)

References on p. A65

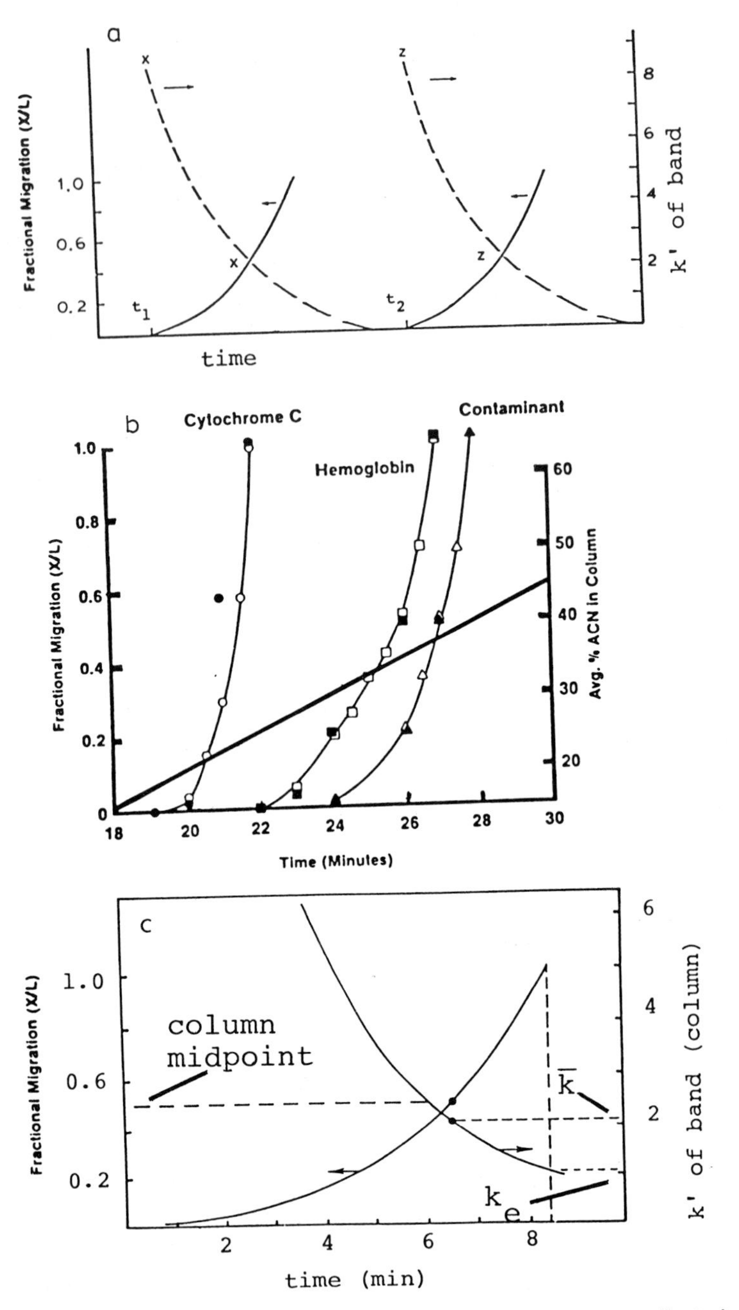

Fig. 1.19. Migration of solutes through the column during gradient elution. (a) Hypothetical representation for the first (X) and last (Y) bands in a sample; (———) fractional migration (x/L) of each band as a function of time; (— — —) value of k' for each band as a function of time; (b) experimental plots obtained for colored proteins during reversed-phase LC [105]; (c) detail for Band X in Fig. 1.19a, showing values of average k' ($\bar{k}$) and k' at elution (k_e).

with a much smaller value of k' than at the beginning of the separation.

The hypothetical example of Fig. 1.19a is confirmed by the experimental plots of Fig. 1.19b for the gradient elution of two proteins. Curved migration plots are seen, the result of the increasing speed of migration caused by the increase in %B (and decrease in k') during elution. Fig. 1.19c shows the migration of Solute X in more detail. Of special interest is the value of k' for the solute when it has migrated halfway through the column ($x/L = 0.5$). This value of k' (defined as $\bar{k}$) in gradient elution provides the same resolution as a "corresponding" isocratic separation, where all separation conditions (and sample) are the same, except %B is constant. When the band leaves the column, its value of k' (k_e) is equal to about half of $\bar{k}$. This means that the bands in gradient elution are typically narrower than in isocratic elution, and detection sensitivity is thereby enhanced.

Returning to Fig. 1.19a, it is seen that the migration of the last band, Z, through the column proceeds in the same way as for the first band, X. The average value of k' during migration ($\bar{k}$) is similar, and the solute leaves the column with a similar value of k' (k_e) as for Band X. Thus, in gradient elution the migration and separation of every band in the sample can be optimized with respect to k' by choosing appropriate gradient conditions.

The theory of gradient elution separations is now well understood [98,104,106] to the point where quantitative predictions can be made of retention, bandwidth, and resolution. When gradient conditions are selected so that log k' for each band varies linearly with time during the gradient (so-called linear solvent-strength or LSS gradients), the resulting equations take a particularly simple form [98,104]. A number of computer programs have also been described that permit so-called computer simulation [107], the development of a final gradient elution procedure on the basis of a few preliminary experiments, followed by further "experiments" with a computer.

An important general relationship for gradient elution can be derived

$$\bar{k} = 1/1.15\, b \tag{1.99}$$

where for reversed-phase HPLC

$$b = t_G\, F/V_m\, \Delta\Phi\, S \tag{1.100}$$

Here, t_G is the gradient time, $\Delta\Phi$ is the change in mobile phase composition, Φ, during the gradient (Φ = %B/100), and S measures the change in k' for some change in Φ (S equals $-$ d(log k')/dΦ). Similar expressions apply to other HPLC methods (IEC, NPC) [98,106].

Eqns. 1.99 and 1.100 allow the application of many equations for isocratic retention to gradient elution as well [98,104]; e.g., Eqn. 1.33 with k substituted for k'. According to Eqn. 1.33 (see Fig. 1.7), resolution should increase as $\bar{k}/(1+\bar{k})$, and this is illustrated in Fig. 1.20 for the change in separation as the gradient time, t_G, is increased. For a rather steep gradient (t_G = 5 min, hypothetical sample in Fig. 1.18), sample resolution is marginal. When the gradient time is increased to 20 min (Fig. 1.20b) and 80 min (Fig. 1.20c), resolution is much improved as a result of an increasingly shallow gradient and smaller values of b (larger k). Resolution is plotted vs. gradient time in Fig. 1.20d, showing a

References on p. A65

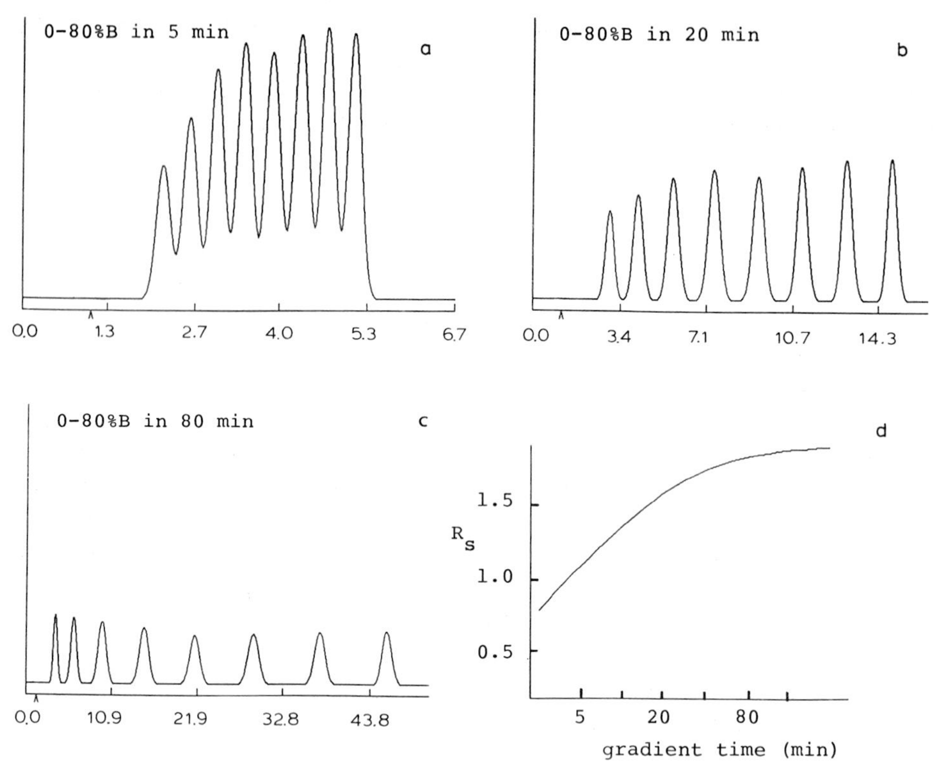

Fig. 1.20. Separation of sample of Fig. 1.18 by gradient elution as a function of gradient steepness (or time). (a-c) Chromatograms for varying gradient times; (d) plot of sample resolution as a function of gradient time.

characteristic rapid rise in resolution as gradient time is increased from an initially small value, followed by a leveling off in resolution due to the approach of $\bar{k}/(1+\bar{k})$ to 1.0.

As long as the gradient volume (equal to $t_G\,F$) is maintained constant for the separation of a given sample (and other conditions remain the same), sample retention (expressed as retention volume) does not change. This has important practical implications, inasmuch as retention (expressed as $\bar{k}$) is often first adjusted for satisfactory separation, following which the column platenumber can be increased for better resolution by changing column length, L, or flowrate, F. If the gradient volume is allowed to vary as a result of adjustments in L or F, the resulting change in values of $\bar{k}$ can result in a completely unpredictable separation. It is best to first optimize retention, and then to maintain values of $\bar{k}$ constant while optimizing column conditions (L, F, and particle size).

Baseline bandwidth, W, in gradient elution is given as [98]

$$W = 4\,G\,t_o\,(1+k_e)\,N^{-1/2} \tag{1.101}$$

which can be compared with Eqns. 1.5 and 1.26 for isocratic separation. The parameter G in Eqn. 1.101 is a gradient compression factor ($0.6 < G < 1.0$), which reflects the compression of eluted bands by the mobile-phase gradient. Since $k_e \approx 0.5\,\overline{k}$, the bandwidth decreases with an increase in b and decrease in gradient time (Eqn. 1.99); peakheight therefore increases for steeper gradients (cf. Fig. 1.20).

1.9.2 Temperature programing

An increase in temperature during the separation has been used in both GC and LC. For various practical reasons this technique is commonly applied in GC but hardly ever in LC. (A comprehensive treatment of temperature-programed GC separation is given in Ref. 93.) The basic theory of programed-temperature GC is remarkably similar to the corresponding theory of gradient elution, discussed in the preceding section. Almost all of the discussion in Section 1.9.1 therefore applies equally to temperature programing in GC, if we replace mobile phase composition (in gradient elution) by temperature (for temperature programing). One minor difference exists, however, that affects quantitative predictions of retention for these two different procedures. In temperature programing, the temperature of the entire column changes at the same time; i.e., if the column temperature is raised from 100 to 300°C in 20 min, the temperature in the column is everywhere the same at any given time. In gradient elution, there is a gradient of mobile-phase composition along the column, so that typically the mobile phase entering the column is stronger than that leaving the column. (see Refs. 98, 103, and 104 for a further discussion of this and other aspects of gradient elution and temperature programing.)

1.9.3 Flow programing

In flow programing, the flowrate, F, is varied as a function of time during the separation; usually, F increases with time after sample injection. Flow programing is a much more limited technique for solving the problem of samples with a wide retention range, in that it has no effect on sample k' values (except in SFC). Therefore, it can provide only minor improvement in sample resolution (cf. Fig. 1.20d for gradient elution, similar for temperature programing). Today it is used hardly at all in LC, but to some extent in GC and SFC.

1.9.4 Column switching and related procedures

Column switching refers to the collection of a sample fraction from an initial chromatographic separation and the on-line injection of this fraction into a second column. This is illustrated by the example in Fig 1.21. Fig. 1.21a shows a schematic drawing of a column-switching apparatus; a sample is first separated on Column 1, and the effluent from Column 1 goes to the detector (or to waste). At some time during the separation, the 3-way valve is turned to allow the mobile phase to flow from Column 1 to Column 2. At some later time, the valve is turned again so that Column 2 is bypassed. Finally, the valve is switched again to allow flow of the mobile phase through Columns 1 and 2 to complete

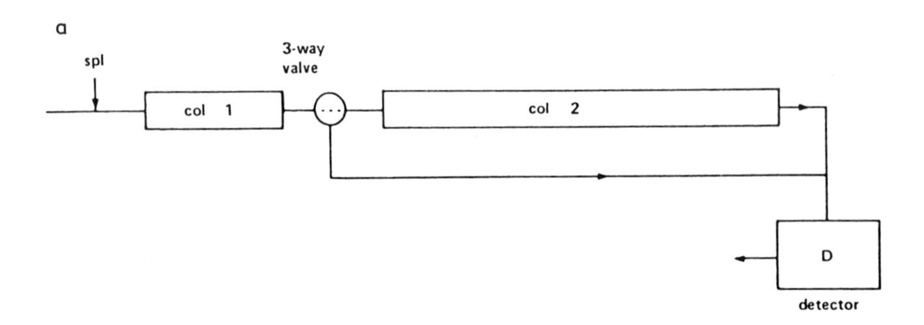

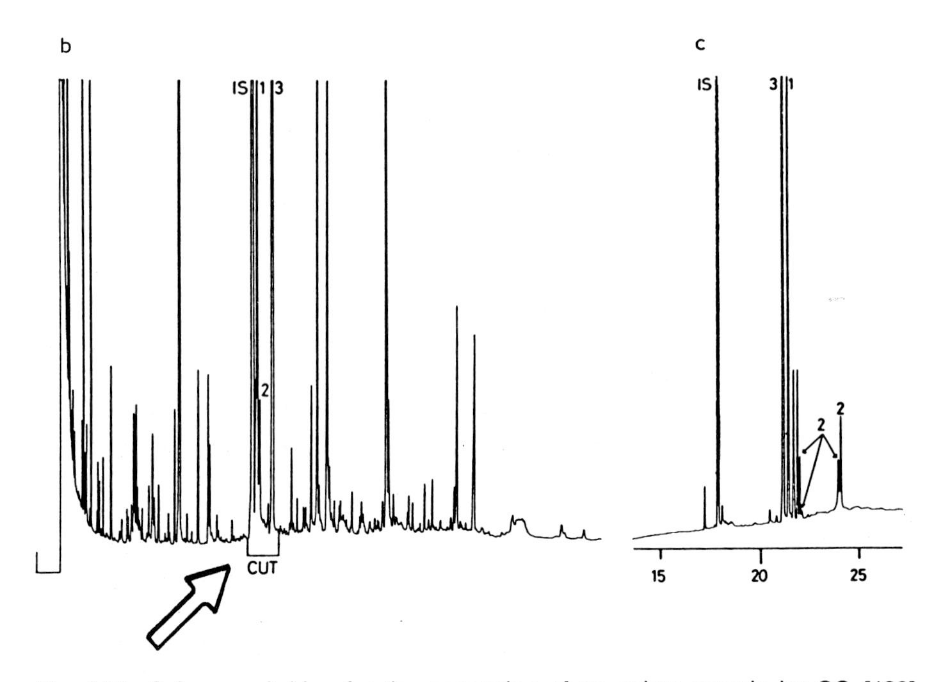

Fig. 1.21. Column switching for the separation of an opium sample by GC [108]. (a) General column arrangement for column switching; "col" = column, D = detector; (b) separation of the sample on Column 1 with transfer of part of the sample to Column 2 (arrow); (c) separation of the transferred fraction on Column 2.

the separation of the fraction originally diverted to Column 2.

Figs. 1.21b and c illustrate the GC separation of an opium sample. The chromatogram in Fig. 1.21b shows the separation on Column 1 (OV-1), where Bands 1-3 and the internal standard (IS) are of interest. From this column, a group of 4 compounds is eluted as a single band (No. 2), which it is desired to separate and quantitate. Diversion of this fraction to Column 2 (OV-17) yields a second chromatogram (Fig. 1.21c), where the

TABLE 1.7

APPLICATIONS OF COLUMN SWITCHING

Procedure	Application	Ref.
Selectivity switching	Separation of selected compounds from complex mixtures (LC, GC)	105, 111, 112
Pre-column venting	Elimination of solvent band and/or early-eluted material (LC, GC)	113, 114
"Boxcar" chromatography	Repetitive HPLC analysis at high sampling frequency (LC, GC)	115, 116
Sample pretreatment	Removal of components that are detrimental to Column 2; on-column concentration of trace analytes (LC)	116-120
Recycle chromatography	Increasing the column platenumber in preparative LC	121, 122
Venting of later-eluted material	Alternative to gradient elution for samples with a wide k' range	123, 124

various components of Band 2 (and the other compounds of interest) are well resolved, because of a difference in separation selectivity. Had the sample been injected directly into Column 2, the separation in Fig. 1.21c would still have occurred, but other bands of the sample would have overlapped these peaks of interest. The use of column switching with two columns of different selectivity allows an effective increase in the peak capacity of the chromatogram.

Column switching is a very versatile, powerful technique that has found application in both LC and GC. In most cases, the physical system is more complex than shown in Fig. 1.21a. Often two separate chromatographic systems are coupled together, and the action of the valve is usually controlled automatically (by a timer or computer). There are many possible applications of this procedure, some of which are summarized in Table 1.7. (For an additional discussion of column switching, see Refs. 26 and 109-124.)

1.10 LARGE-MOLECULE SEPARATIONS

Liquid chromatography has played an important role in the separation and analysis of large-molecule samples, including both synthetic and natural polymers (cf. Chapters 14, 17, and 22). In some respects, the chromatography of these compounds appears to differ from that of "small" molecules, i.e., compounds with molecular weights < 1000. The special characteristics of large-molecule chromatography therefore merit some discussion.

1.10.1 Unique features of large molecules

1.10.1.1 Molecular size

The molecular weight range of the compounds under discussion is roughly 10^3 to 10^6. The diffusion coefficient, D_m, decreases with increasing solute molecular weight, so that two orders of magnitude can separate the diffusion rates of small and large molecules in the mobile phase outside of the particles. This is illustrated in Fig. 1.22 by the solute molecules myoglobin (a protein) and a small, substituted benzene derivative. The arrows adjacent to each molecule represent the relative distance each molecule can diffuse in a given time. As seen in Fig. 1.17, this difference in solute diffusion rates means that the separation of large-molecule samples generally is carried out at much higher reduced velocities than that of small-molecule solutes, with a corresponding increase in h and decrease in the platenumber, N; i.e., poorer separation. This can be partially counteracted by using lower flowrates for the separation of large molecules. (For a further discussion of the diffusion coefficients of large molecules, see Refs. 59, 125, and 126.)

The diffusion of large solute molecules in the pores of the column packing can be even slower than in the mobile phase outside the pores, if the size of the molecule is comparable to the size of the pore. This is illustrated in Fig. 1.22 for the molecule myoglobin (in its native, folded conformation) adjacent to a 60-Å pore. It is seen that a myoglobin molecule will barely be able to enter the pore, because its cross-sectional diameter is about the size of the pore diameter.

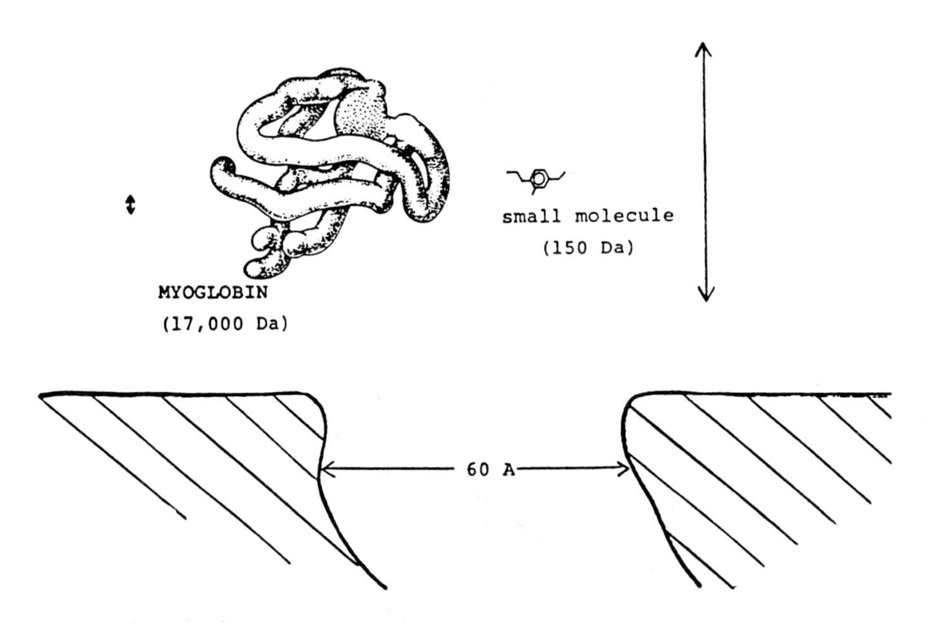

Fig. 1.22. Hypothetical representation of a large and small solute in the mobile phase outside a pore of the column packing. Arrows represent relative distances migrated by each molecule in some given time (by diffusion).

When the diameter of a solute molecule is a significant fraction of the diameter of a pore, its diffusion coefficient in the pore is lowered significantly, because of "frictional drag" at the pore wall [127]. A simple model of diffusion in pores predicts that the diffusion coefficient in the pore, D_p, will be related to D_m (diffusion in the mobile phase) as

$$D_p/D_m = 1 - 2.10\, r_s + 2.09\, r_s^3 - 0.95\, r_s^5 \qquad (1.102)$$

where r_s refers to the diameter of the solute molecule divided by the diameter of the pore. Eqn. 1.102 has been verified for the diffusion of protein solutes in different HPLC systems [126,128]. One consequence of the example in Fig. 1.22 and of Eqn. 1.102 is that porous column packings for large-molecule solutes require larger pores (30-100 nm, compared to the 6- to 10-nm pores commonly used for small-molecule solutes).

1.10.1.2 Molecular conformation

Another characteristic of natural polymers is that they often adopt a highly ordered conformation. This is illustrated for the folded (globular) protein myoglobin in Fig. 1.22. It is now well documented that the comformation of a protein can change during LC separation [129,130]. This is represented schematically in Fig. 1.23a, where the native (folded) protein enters the column, and the denatured (random-coil) protein leaves the column, followed (in some cases) by spontaneous refolding of the protein. If the goal of separation

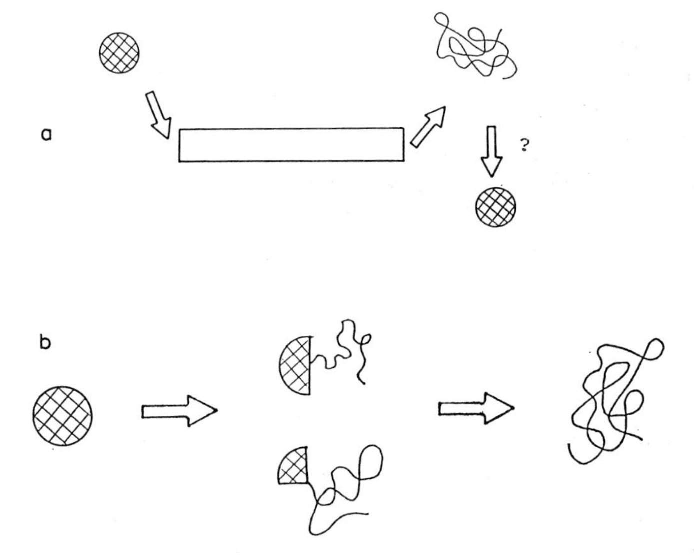

Fig. 1.23. Representation of the unfolding of a protein molecule during chromatographic separation. (a) Folded protein enters the column, leaves in the unfolded (random-coil) state, and (maybe) refolds at a later time; (b) unfolding of a protein to yield multiple conformations.

is recovery of the native form of the protein, then chromatographic conditions should be selected to favor the desired conformation.

Because proteins can have parts of the molecule folded (secondary structure consisting of α-helices and/or β-sheets), as well as the whole molecule, a variety of protein conformations can arise during LC separation, as represented in the reaction process of Fig. 1.23b. The time required for interconversion of these various more-or-less folded species is often comparable to the time of chromatographic separation [130], and this has several practical consequences. Usually, the resulting protein band is quite wide and often misshapen, resulting in a poor separation of the protein from other components of the sample [98,131]. Again, it is desirable to select separation conditions that will avoid this problem.

1.10.2 Theories of retention

Numerous workers have noted certain peculiarities in the separation of large molecules by either isocratic or gradient elution. This has led to the proposal that separations of such samples differ in some fundamental way from similar (same LC method) separations of small molecules. Among the examples of "peculiar" separation are the following:

(a) When the mobile-phase composition is varied for isocratic elution, it is found that the sample never migrates normally, but is either retained completely or eluted at t_0.

(b) When the column length is varied in gradient elution, the resulting chromatogram is observed to change very little.

(c) In gradient elution, a large-molecule solute is observed to be eluted at the same mobile-phase composition regardless of conditions.

One theory of large-molecule retention which is often cited to explain the above observations is the "on-off" model. According to this hypothesis, the solute initially sticks at the inlet of the column (at the start of the gradient), then at some "critical" mobile-phase composition during gradient elution the solute is totally desorbed and migrates through the column without any further interaction with the stationary phase.

A closer examination of the above experimental findings shows that they can be explained in terms of conventional chromatographic theory, as presented in this chapter [98,132]. That is, there is no need to invoke special retention processes, such as the "on-off" theory. This has been further confirmed by the direct observation of the migration of proteins during RPC [105]. The primary reason for the *apparently* unusual behavior of large molecules in many chromatographic systems is the rapid change in solute retention with small changes in mobile-phase composition. This can be explained in terms of a displacement process, as described by Eqn. 1.52, where the number, n, of displaced solvent molecules, B, is quite large for a large-solute molecule. In the case of large-molecule separations by RPC [133,134] it is often found that a 1-2% change in the mobile-phase composition results in a change in retention (k') by a factor of 2 to 10 (or more) (cf. Eqn. 1.53).

Our understanding of the LC separation of large molecules is now reasonably well developed, and it is possible in many cases to make quantitative predictions of retention and bandwidth as a function of experimental conditions. (For a further discussion of the fundamental features of large-molecule chromatography, see Ref. 98. Some examples of the computer-assisted design of optimized large-molecule separations are provided in Ref. 133.)

1.11 PREPARATIVE CHROMATOGRAPHY

The preceding discussion of chromatographic theory has assumed that the sample to be separated is small, so that sample retention and bandwidth do not change with sample size. However, when the objective of chromatographic separation is the recovery of purified material, it is often advantageous to apply a much larger sample to the column. It is then (generally) observed that with increasing sample size the retention times of all compounds decrease, their bandwidths increase, and resolution becomes worse. A semi-quantitative understanding of how separation changes with sample size and other experimental conditions has recently emerged and will be summarized in this section.

1.11.1 Behavior of single bands: touching-band separations

An understanding of preparative chromatography begins with the changes that occur in band retention and width when a large sample of a single compound is applied to a column. It is commonly found that the injection of several samples of increasing size leads to a set of overlapping ("nested") bands of roughly right-triangular shape; this is illustrated in Fig. 1.24 for the separation of diethyl phthalate by NPC. For a large enough sample, the band of interest (the product we are trying to purify) will widen to the point where it reaches the adjacent, less retained band. This is illustrated in Fig. 1.25b for the case of "touching-band" separation; the band of interest (product) is indicated by an asterisk. Touching-band separation allows a maximum sample size with 100% recovery of the desired product in almost 100% purity.

As a first approximation, the sample size for touching-band separation can be approxi-

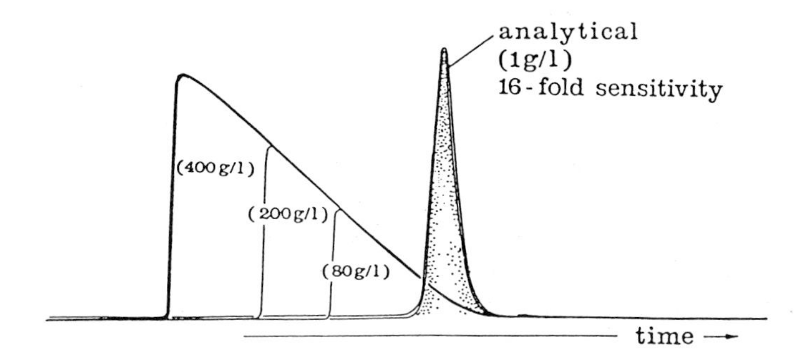

Fig. 1.24. Change in the elution band with sample size to give "nested" right-triangle peaks. Normal-phase separation of diethyl phthalate. (Redrawn from Ref. 134.)

References on p. A65

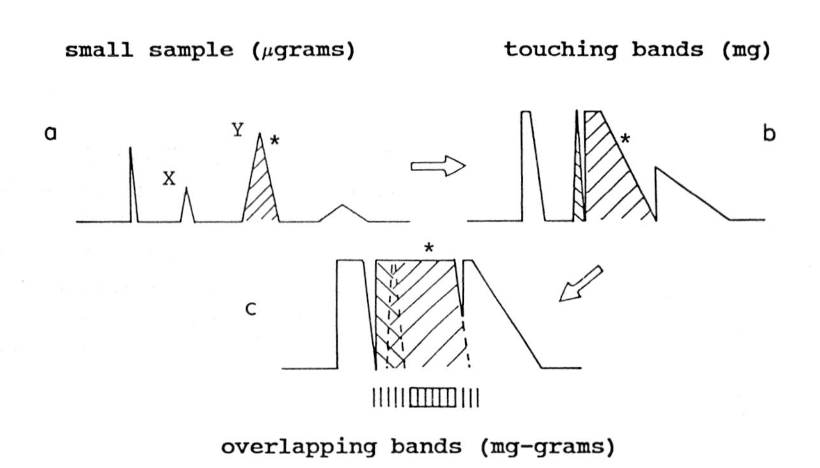

Fig. 1.25. Preparative separation carried out for three different sample sizes. (a) Small-sample separation; (b) touching-band separation; (c) overlapping-band separation. * = Product band.

mated from the bandwidth vs. sample size behavior of a single-solute sample. Thus, consider Bands X and Y in Fig. 1.25a, where Y is the desired product. The distance between the two bands is the difference in their (small-sample) retention times: $t_y - t_x$. When the width of band Y (W_y) becomes equal to $t_y - t_x$, the two bands just touch. This ignores the small-sample bandwidths of each band, which is a reasonable approximation, and it assumes that band broadening takes place in one direction (Fig. 1.24). The width of a large-sample band can be derived, if we assume that solute retention is governed (approximately) by a Langmuir isotherm [135]. The latter assumes that there is a competition between solute and mobile-phase molecules for adsorption sites on the stationary phase (see discussion of Eqn. 1.52), and that equal numbers of solute or mobile-phase molecules are found in the stationary phase when the latter is saturated (i.e., $z = 1$ in Eqn. 1.52). Further assumptions of this model are that stationary-phase saturation corresponds to monolayer coverage of the column-packing surface, and that there are no interactions between molecules of solute or mobile phase in the stationary phase.

Bandwidth as a function of sample size is given [135,136] as

$$W^2 = W_o{}^2 + W_{th}{}^2 \qquad (1.103)$$

where W is the baseline width of a mass-overloaded band, W_o is the corresponding width of a small-sample band, and W_{th} is the contribution to band broadening from a large sample. Values of W_o and W_{th} can be expressed further as

$$W_o = (16/N_o)^{1/2}\, t_o\, (1+k_o) \qquad (1.104)$$

and

$$W_{th} = (6)^{1/2}\, t_o\, k_o\, (w/w_s)^{1/2} \tag{1.105}$$

Here, k_o and N_o are the small-sample values of k' and N, respectively, w is the weight of the injected solute, and w_s is the column saturation capacity (the maximum amount of solute that can be retained by the stationary phase in a given column).

From the preceding relationships for the broadening of a single band as a function of sample size, experimental conditions (values of N_o and k_o) and the separation factor, α, for Bands X and Y, it is possible to derive some important relationships. Thus, there will be an optimum plate number, N_o, for every touching-band separation (that provides a maximum amount of purified product per unit time), given by [135, 137]

$$N_{opt} \approx 50\ [1 + k_y)/k_y]^2\ [\alpha/(\alpha - 1)]^2 \tag{1.106}$$

and there will be an optimum sample size, given by

$$w_y \approx (w_s/9)\ [(\alpha - 1)/\alpha]^2 \tag{1.107}$$

Here, k_y refers to the small-sample k' value of Band Y, and w_y is the weight of Y injected. The optimum value of N_o (Eqn. 1.106) can also be determined by the requirement that the small-sample bands X and Y have a resolution $R_s \approx 1.7$.

The above relationships can provide the basis for a systematic approach to method development for preparative LC separation. The separation factor, α, for Bands X and Y has a major effect on sample size (Eqn. 1.107) and the yield of purified product obtained, so conditions are first adjusted to maximize α. The mobile-phase strength is next adjusted to give $k' \approx 1$ for Band X. Finally, column length, flowrate, and the particle size of the column packing are varied to give a small-sample resolution for Bands X and Y of $R_s \approx$ 1.7.

1.11.2 Behavior of multi-component samples: overlapping-band separations

The preceding discussion of touching-band separations provides an approximate description of separations that do not involve significant overlap of the product band with adjacent impurities (as in Fig. 1.25b). However, when larger amounts of sample must be purified, it is found that overlapping-band separations (Fig. 1.25c) can provide a 3- to 10-fold yield of purified product. For these larger samples it is also found that the resulting separations can no longer be approximated very well by the behavior of individual solute bands. That is, for large enough samples (Fig. 1.25c), the elution of one band can be strongly affected by the presence of adjacent bands. This is illustrated in Fig. 1.26 for separations where the two bands overlap significantly, and the shape of the individual solute bands is seen to have been changed significantly (compared to Fig. 1.24).

References on p. A65

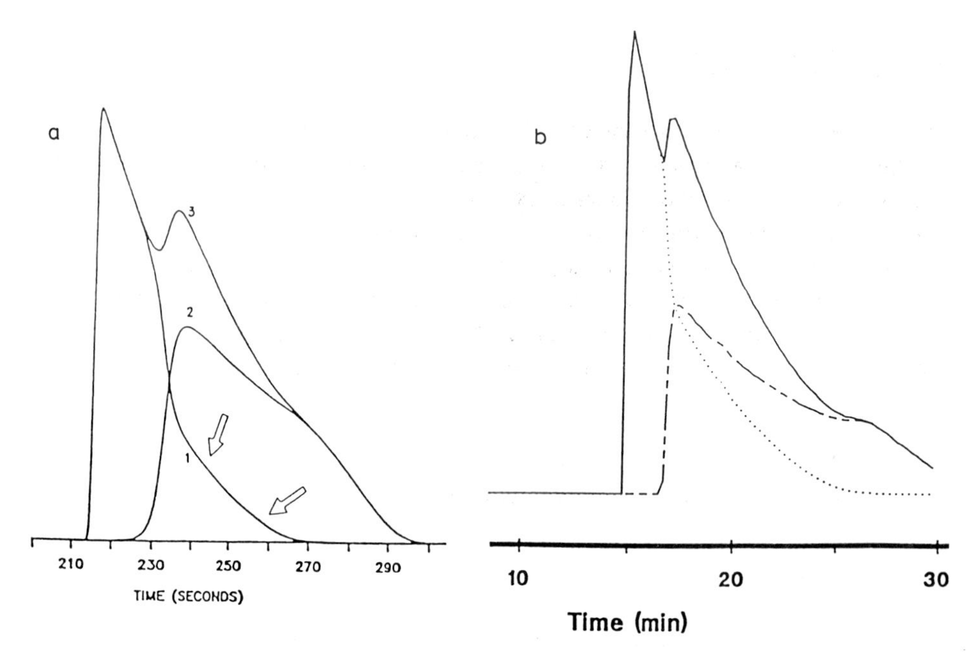

Fig. 1.26. Separation of two compounds with overlapping bands. (a) Computer simulation for Bands 1 and 2 [139]; (b) experimental separation of two compounds with similar conditions as used in (a).

The interaction of two adjacent bands that overlap significantly (as in Fig. 1.26) is now well documented and rather well understood. This is mainly the result of computer simulations (based on the Langmuir isotherm) that have been used to mimic such separations [136-141]. An example of this is seen in a computer simulation of the separation in Fig. 1.26a and the related experimental separatioan in Fig. 1.26b. The similarity of the two results demonstrates the adequacy of the computer model, as have other similar comparisons. On the basis of a large number of "experiments" carried out by computer simulation, a number of generalizations concerning overlapping-band separation (as in Fig. 1.25b) can now be made. Some of these are summarized in the plots of Fig. 1.27.

As in the case of touching-band separation, there exists an optimum platenumber, N_0, for every overlapping-band separation (Fig. 1.27b). This optimum platenumber depends on the α value for the two bands (X and Y) and the desired recovery of pure product (50%, 95%, and 99.8% in Fig. 1.27b). For smaller values of α and a higher percent recovery of pure product, larger values of N_0 will be required. Alternatively, the resolution of a small-sample separation can also be used to assess the best value of N_0 for an overlapping-band separation: R_s should equal about 1.2 for 95% recovery of pure product, and 0.9 for 50% recovery.

The (maximum) weight of product that can be injected is mapped vs. α and percent recovery of pure product in Fig. 1.27a. Here, w_x or w_y refers to the weight of Solutes X or

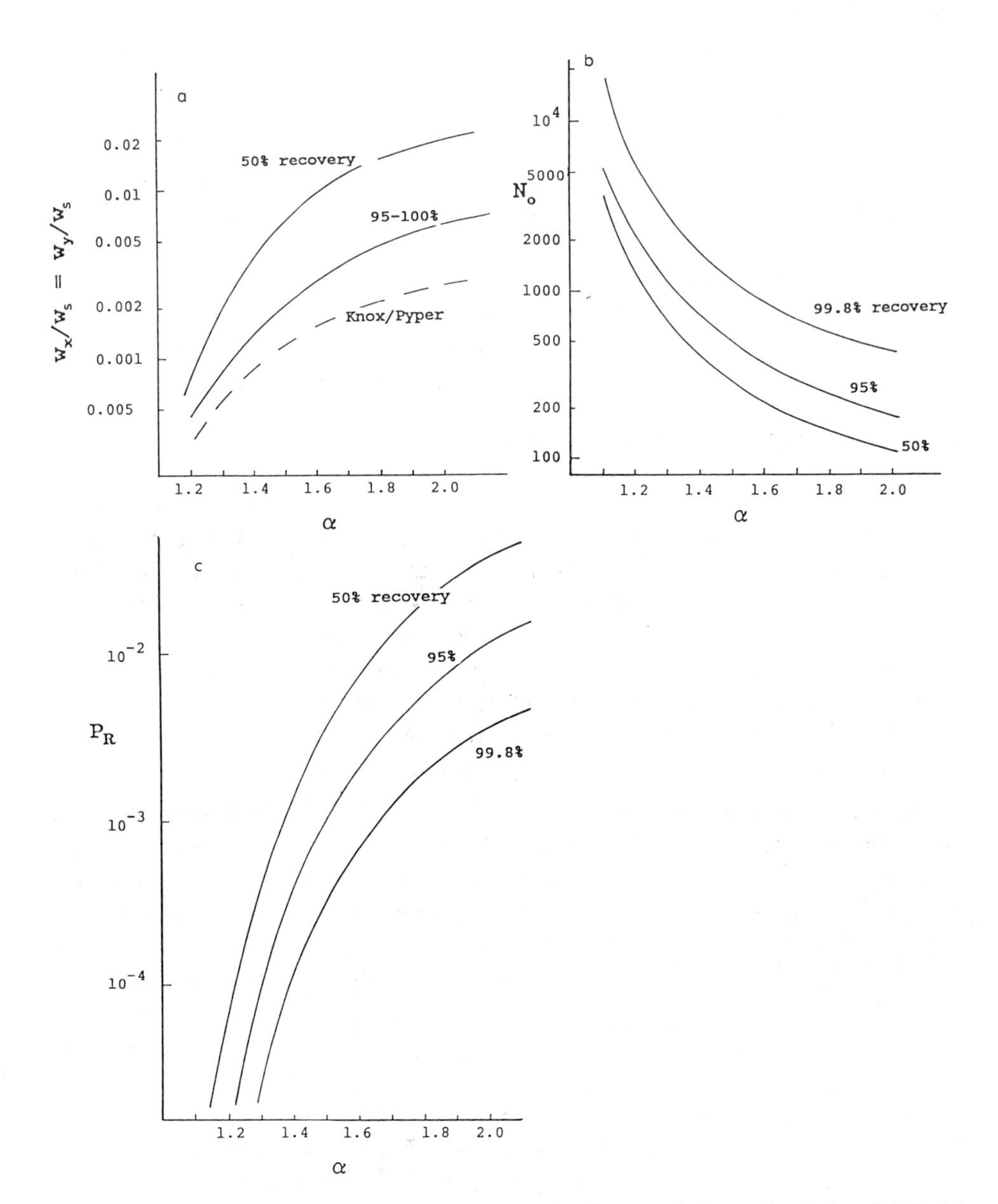

Fig. 1.27. Predicted conditions for maximum production rate in isocratic elution [137]. (a) Sample size (fractional saturation of either Solute X or Y) vs. separation factor, α, and desired recovery of 99% pure product; (b) optimum value of the column platenumber vs. α and percent recovery of pure product; (c) relative production rate, P_R (g/h of purified product), as a function of α and percent recovery. All data assume small-sample k' for Compound X equal 1.0, and that the sample is composed of equal concentrations of X and Y.

Y in the sample (equal weights assumed) and w_S is the column saturation capacity. Larger values of α and lower percent recoveries allow larger samples to be separated. The dashed curve in Fig. 1.27a ("Knox/Pyper") is predicted on the basis of "touching-band" separation (Section 1.11.1). It is erroneously low (vs. "95-100%" curve) due to solute interference, as in Fig. 1.26. The relative production rate, P_R (g purified product/h), is similarly plotted vs. α and percent recovery in Fig. 1.27c. It is seen that the g/h of pure product produced rises sharply with an increase in α. It is also seen that the acceptance of 95% recovery of pure product provides ca. 3 times as much purified product per hour as 99.8% recovery ("touching-band" conditions). Likewise, a 50% recovery of pure product allows a further 3-fold increase in production rate, compared to the case of 95% recovery. For 50% recovery of pure product, the remaining (impure) product from each run would be recycled to allow recovery of most of the starting product.

1.11.3 Other preparative separation modes

The preceding discussion applies to isocratic elution, the main application of preparative LC today. An increasing number of preparative separations are being carried out in a gradient elution mode, especially in the case of large molecules of biochemical interest (e.g., peptides, proteins, oligonucleotides, etc.). The similarities between isocratic and gradient elution pointed out in Section 1.9 are also applicable to preparative separations by isocratic or gradient elution. Thus, the relationships of Fig. 1.27 also apply to some extent to gradient elution. (For a further discussion, see Ref. 138.)

Another preparative separation mode of increasing interest is displacement chromatography, illustrated in Fig. 1.28. In this procedure, the sample is first introduced into the column in a weak solvent (high k' values for all components). A very strong solvent

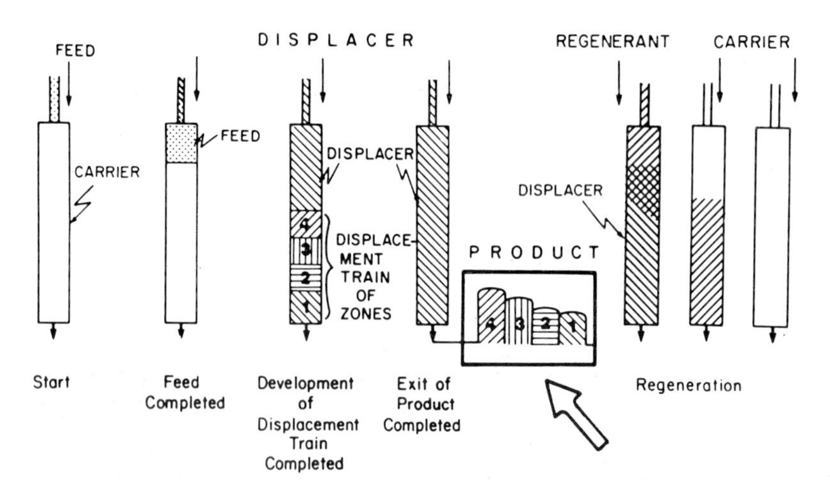

Fig. 1.28. Schematic illustration of displacement chromatography. (Reprinted with permission from Ref. 142.)

("displacer") is next introduced into the column, and during the continued flow of the displacer, the sample is pushed ahead of the displacer front. In the process, the various sample components separate into zones (bands) that are touching. The final chromatogram (No. 4 = displacer; Nos. 1-3 = sample components) is indicated in Fig. 1.28 by the arrow. Following the separation, a regenerant solvent is used to restore the column to its initial condition, allowing the entire process to be repeated.

Displacement chromatography has a number of potential advantages: larger samples and higher production rates, more concentrated fractions with less mobile phase to remove from purified products, etc. On the debit side, method development for this procedure is more complicated and difficult than in the case of elution (Figs. 1.25b and c). The relative importance of displacement vs. elution chromatography for preparative separation is yet to be decided. (For a further account of displacement chromatography, see Ref. 142.)

REFERENCES

1 H. Purnell, *Gas Chromatography*, Wiley, New York, 1962.
2 J.C. Giddings, *Dynamics of Chromatography, Part I, Principles and Theory,* Dekker, New York, 1965.
3 L.R. Snyder, *Principles of Adsorption Chromatography,* Dekker, New York, 1968.
4 F. Helfferich and G. Klein, *Multicomponent Chromatography,* Dekker, New York, 1970.
5 B.L. Karger, L.R. Snyder and Cs. Horváth, *An Introduction to Separation Science,* Wiley, New York, 1973.
6 J.R. Conder and C.L. Young, *Physicochemical Measurement by Gas Chromatography,* Wiley, New York, 1979.
7 A.S. Said, *Theory and Mathematics of Chromatography,* Huethig, Heidelberg, 1981.
8 S.T. Balke, *Quantitative Column Liquid Chromatography,* Elsevier, Amsterdam, 1984.
9 P.J. Schoenmakers, *Optimization of Chromatographic Selectivity,* Elsevier, Amsterdam, 1986.
10 R. Kaliszan, *Quantitative Structure-Chromatographic Retention Relationships,* Wiley-Interscience, New York, 1987.
11 E. Katz (Editor), *Quantitative Analysis Using Chromatographic Techniques,* Wiley, New York, 1987.
12 Cs. Horváth, in E. Heftmann (Editor), *Chromatography,* Part A, Elsevier, Amsterdam, 1983, Ch. 3.
13 J. Frenz and Cs. Horvath, in Cs. Horvath (Editor), *High-performance Liquid Chromatography. Advances and Perspectives,* Vol. 5, Academic Press, New York, 1988, p. 211.
14 J.H. Knox and R. Kaliszan, *J. Chromatogr.,* 349 (1985) 211.
15 B. Wronski, L.M. Szczepaniak and Z. Witkiewicz, *J. Chromatogr.,* 364 (1986) 53.
16 E. sz. Kovats, in F. Bruner (Editor), *The Science of Chromatography,* Elsevier, Amsterdam, 1985, p. 205.
17 C.F. Poole and S.A. Schuette, *Contemporary Practice of Chromatography,* Elsevier, Amsterdam, 1984, p. 9.
18 J.W. Dolan and L.R. Snyder, *Troubleshooting LC Systems,* Humana Press, Clifton, NJ, 1989, Ch. 14.
19 J.C. Sternberg, *Adv. Chromatogr.,* 2 (1966) 205.
20 A.J.P. Martin and R.L.M. Synge, *Biochem. J.,* 35 (1941) 1358.
21 L.R. Snyder, J.L. Glajch and J.J. Kirkland, *Practical HPLC Method Development,* Wiley-Interscience, New York, 1988, p. 98.

22 K. Suematsu and T. Okamoto, *J. Chromatogr. Sci.*, 27 (1989) 13.
23 J.C. Berridge, *Techniques for the Automated Optimization of HPLC Separations*, Wiley-Interscience, New York, 1985.
24 Sz. Nyiredy (Editor), *Optimization of Mobile Phase, J. Liq. Chromatogr.*, 12 (1989).
25 J.L. Glajch and L.R. Snyder (Editors), *Computer-assisted HPLC Method Development, J. Chromatogr.*, 485 (1989).
26 L.R. Snyder and J.J. Kirkland, *Introduction to Modern Liquid Chromatography*, Wiley-Interscience, New York, 2nd Edn., 1979.
27 W.E. Harris and H.W. Habgood, *Programmed Temperature Gas Chromatography*, Wiley, New York, 1967, p. 54.
28 J. Chmielowiec and H. Sawatsky, *J. Chromatogr. Sci.*, 245 (1979) 245.
29 D.S. Lu, J. Vialle, H. Tralongo and R. Longeray, *J. Chromatogr.*, 268 (1983) 1.
30 P.J. Twitchett, P.L. Williams and A.C. Moffat, *J. Chromatogr.*, 149 (1978) 683.
31 B.L. Karger, JN. LePagae and N. Tanaka, in Cs. Horvath (Editor), *High-performance Liquid Chromatography. Advances and Perspectives*, Vol. 1, Academic Press, New York, 1980, p. 113.
32 A.F.M. Barton, *Chem. Rev.*, 75 (1975) 731.
33 B.L. Karger, L.R. Snyder and C. Eon, *J. Chromatogr.*, 125 (1976) 71.
34 G. DiPaola-Baranyi, *Polym. Mater. Sci. Eng.*, 58 (1988) 735.
35 D.R. Lloyd, T.C. Ward and H.P. Schreiber (Editors), *Inverse Gas Chromatography: Characterization of Polymers and Other Materials* (ACS Symposium Series 391), American Chemical Society, Washington, DC, 1989.
36 B.L. Karger, L.R. Snyder and C. Eon, *Anal. Chem.*, 50 (1978) 2126.
37 A. Tchapla, H. Colin and G. Guiochon, *Anal. Chem.*, 56 (1984) 621.
38 M. Martin, G. Thevenon and A. Tchapla, *J. Chromatogr.*, 452 (1988) 157.
39 W.J. Cheong and P.W. Carr, *J. Chromatogr.*, 499 (1990) 373.
40 K.B. Sentell and J.G. Dorsey, *Anal. Chem.*, 61 (1989) 930.
41 T.L. Hafkenscheid and E. Tomlinson, *Adv. Chromatogr.*, 25 (1986) 1.
42 J. Jacobson, J. Frenz and Cs. Horváth, *J. Chromatogr.*, 316 (1984) 53.
43 J.E. Eble, R.L. Grob, P.E. Antle and L.R. Snyder, *J. Chromatogr.*, 384 (1987) 45.
44 N. Tanaka and E.R. Thornton, *J. Am. Chem. Soc.*, 99 (1977) 7300.
45 X. Geng and F.E. Regnier, *J. Chromatogr.*, 332 (1985) 147.
46 W.R. Melander and Cs. Horvath, in Cs. Horvath (Editor), *High-performance Liquid Chromatography. Advances and Perspectives*, Vol. 2, Academic Press, New York, 1980, p. 113.
47 L.R. Snyder, in Cs. Horvath (Editor), *High-performance Liquid Chromatography. Advances and Perspectives*, Vol. 3, Academic Press, New York, 1980, p. 157.
48 E. Soczewiński, *Anal. Chem.*, 41 (1969) 179.
49 E. Soczewiński and J. Kuczmierczyk, *J. Chromatogr.*, 150 (1978) 53.
50 H. Engelhardt and H. Elgass, in Cs. Horvath (Editor), *High-performance Liquid Chromatography. Advances and Perspectives*, Vol. 2, Academic Press, New York, 1980, p. 57.
51 L.R. Snyder and J.J. Kirkland, *Introduction to Modern Liquid Chromatography*, Wiley-Interscience, New York, 2nd Edn., 1979, Ch. 11.
52 W. Rieman and H.F. Walton, *Ion Exchange in Analytical Chemistry*, Pergamon Press, New York, 1970.
53 J.H. Knox and R.A. Hartwick, *J. Chromatogr.*, 204 (1981) 3.
54 W.R. Melander and Cs. Horváth, in M.T.W. Hearn (Editor), *Ion-pair Chromatography*, Dekker, New York, 1985, Ch. 2.
55 A.P. Goldberg, E. Nowakowaska, P.E. Antle and L.R. Snyder, *J. Chromatogr.*, 316 (1984) 241.
56 A. Bartha and G. Vigh, *J. Chromatogr.*, 395, (1987) 503.
57 J. Stahlberg and A. Bartha, *J. Chromatogr.*, 456 (1988) 253.
58 M.T.W. Hearn (Editor), *Ion-pair Chromatography*, Dekker, New York, 1985.
59 W.W. Yau, J.J. Kirkland and D.D. Bly, *Modern Size-exclusion Liquid Chromatography*, Wiley-Interscience, New York, 1979.
60 A.J.P. Martin, *Biochem. Soc. Symp.*, 3 (1949) 4.

61 R.M. Smith and C.M. Burr, *J. Chromatogr.,* 481 (1989) 71.
62 D.H. Desty and B.H.F. Whyman, *Anal. Chem.,* 29 (1957) 320.
63 E. sz. Kováts, *Helv, Chim. Acta,* 41 (1958) 1915.
64 L. Rohrschneider, *J. Chromatogr.,* 22 (1966) 6.
65 W.O. McReynolds, *J. Chromatogr. Sci.,* 8 (1970) 685.
66 J.K. Haken, in J.C. Giddings, E. Grushka, J. Cazes and P.R. Brown (Editors), *Advances in Chromatography,* Vol. 14, Dekker, New York, 1977, Ch. 8.
67 R.A. Keller, B.L. Karger and L.R. Snyder, in R. Stock and S.G. Perry (Editors), *Gas Chromatography 1970. Proc. 8th Intern. Symp. Gas Chromatogr.,* Institute of Petroleum, London, 1971, p. 125.
68 J.H. Hildebrand and R.L. Scott, *The Solubility of Nonelectrolytes,* Dover Publishers, New York, 3rd Edn., 1964.
69 C. Hansen, *Ind. Eng. Chem. Prod. Res. Dev.,* 8 (1969) 2.
70 B.L. Karger, L.R. Snyder and C. Eon, *J. Chromatogr.,* 125 (1976) 71.
71 L.R. Snyder, *J. Chromatogr. Sci.,* 16 (1978) 223.
72 M.S. Klee, M.A. Kaiser and K.B. Laughlin, *J. Chromatogr.,* 279 (1983) 681.
73 B.R. Kersten, S.K. Poole and C.F. Poole, *J. Chromatogr.,* 468 (1989) 235.
74 M.J. Kamlet, J.M. Abboud and R.W. Taft, Jr., *Prog. Phys. Org. Chem.,* 13 (1981) 485.
75 M.J. Kamlet and R.W. Taft, *Acta Chem. Scand.,* B39 (1985) 611.
76 M.J. Kamlet, M.H. Abraham, P.W. Arr, R.M. Dopherty and R.W. Taft, *J. Chem. Soc. Perkin Trans. II,* (1988) 2087.
77 J.H. Park, P.W. Carr, M.H. Abraham, R.W. Taft, R.M. Doherty and M.J. Kamlet, *Chromatographia,* 25 (1988) 373.
78 S.C. Rutan, P.W. Carr, W.J. Cheong, J.H. Park and L.R. Snyder, *J. Chromatogr.,* 463 (1989) 21.
79 S.D. West, *J. Chromatogr. Sci.,* 25 (1987) 122.
80 S.D. West, *J. Chromatogr. Sci.,* 27 (1989) 2.
81 K. Hofmann and I. Halasz, *J. Chromatogr.,* 173 (1979) 211.
82 J.G. Atwood and M.J.E. Golay, *J. Chromatogr.,* 218 (1981) 97.
83 J.H. Knox, *J. Chromatogr. Sci.,* 15 (1977) 352.
84 S.G. Weber and P.W. Carr, in P.R. Brown and R.A. Hartwick (Editors), *High Performance Liquid Chromatography,* Wiley-Interscience, New York, 1989, p. 1.
85 R.P.W. Scott, *J. Chromatogr.,* 468 (1989) 99.
86 M. Martin, C. Eon and G. Guiochon, *J. Chromatogr.,* 108 (1975) 229.
87 R.P.W. Scott, *Contemporary Liquid Chromatography,* Wiley-Interscience, New York, 1976.
88 P. Kucera (Editor), *Micro-column High-performance Liquid Chromatography,* Elsevier, Amsterdam, 1984.
89 G. Guiochon, in Cs. Horvath (Editor), *High-performance Liquid Chromatography. Advances and Perspectives,* Vol. 2, Academic Press, New York, 1980, p.1.
90 R.W. Stout, J.J. DeStefano and L.R. Snyder, *J. Chromatogr.,* 282 (1983) 263.
91 J.H. Knox and H.P. Scott, *J. Chromatogr.,* 282 (1983) 297.
92 M. Golay, in D.H. Desty (Editor), *Gas Chromatography 1958,* Butterworth, London, 1958, p. 36.
93 J.C. Giddings, *Anal. Chem.,* 35 (1963) 1338.
94 M.A. Stadalius, J.S. Berus and L.R. Snyder, *LC.GC Mag.,* 6 (1988) 494.
95 J.J. van Deemter, F.J. Zuiderweg and A. Klinkenberg, *Chem. Eng. Sci.,* 5 (1956) 271.
96 Cs. Horvath and H.-J. Lin, *J. Chromatogr.,* 149 (1978) 43.
97 J.F.K. Huber, *Ber. Bunsenges. Phys. Chem.,* 77 (1973) 179.
98 E.D. Katz, K.L. Ogan and R.P.W. Scott, *J. Chromatogr.,* 270 (1983) 51.
99 L.R. Snyder and M.A. Stadalius, in Cs. Horvath (Editor), *High-performance Liquid Chromatography. Advances and Perspectives,* Vol. 4, Academic Press, New York, 1986, p. 195.
100 G. Guiochon, in Cs. Horvath (Editor), *High-performance Liquid Chromatography. Advances and Perspectives,* Vol. 2, Academic Press, New York, 1980, p. 1.
101 E.N. Fuller, P.D. Schettler and J.C. Giddings, *Ind. Eng. Chem.,* 58 (1966) 19.
102 C.R. Wilke and P. Chang, *Am. Inst. Chem. Eng. J.,* 1 (1955) 264.

103 W.E. Harris and H.W. Habgood, *Programmed Temperature Gas Chromatography*, Wiley, New York, 1966.
104 L.R. Snyder, in Cs. Horvath (Editor), *High-performance Liquid Chromatography. Advances and Perspectives*, Vol. 1, Academic Press, New York, 1980, p. 207.
105 J.M. Di Bussolo and J.R. Gant, *J. Chromatogr.*, 327 (1985) 67.
106 P. Jandera and J. Churacek, *Gradient Elution in Column Liquid Chromatography*, Elsevier, Amsterdam, 1985.
107 J.W. Dolan, D.C. Lommen and L.R. Snyder, *J. Chromatogr.*, 485 (1989) 91.
108 H. Neumann and H.-P. Meyer, *J. Chromatogr.*, 391 (1987) 442.
109 K.A. Ramsteiner, *J. Chromatogr.*, 456 (1988) 3.
110 D. Westerlund (Editor), *Int. Symp. Coupled Column Separations, Uppsala, Oct. 26-27, 1988, J. Chromatogr.*, 473, No. 2 (1989).
111 J.F.K. Huber, I. Fogy and C. Fioresi, *Chromatographia*, 13 (1980) 408.
112 Y. Tapuhi, N. Miller and B.L. Karger, *J. Chromatogr.*, 205 (1981) 325.
113 K.-G. Wahlund and U. Lund, *J. Chromatogr.*, 122 (1976) 269.
114 W. Blass, K. Riegner and H. Hulpke, *J. Chromatogr.*, 172 (1979) 67.
115 L.R. Snyder, J.W. Dolan and Sj. van der Wal, *J. Chromatogr.*, 203 (1981) 3.
116 A. Nazareth, L. Jaramillo, B.L. Karger, R.W. Giese and L.R. Snyder, *J. Chromatogr.*, 309 (1984) 357.
117 B.L. Karger, R.W. Giese and L.R. Snyder, *Trends Anal. Chem.*, 2 (1983) 106.
118 C.J. Little, D.J. Tompkins, O. Stahel, R.W. Frei and C.E. Werkhoven-Goewie, *J. Chromatogr.*, 264 (1983) 183.
119 W. Roth, *J. Chromatogr.*, 278 (1983) 347.
120 D.L. Conley and E.J. Benjamin, *J. Chromatogr.*, 257 (1983) 337.
121 G.J. Fallick, *Am. Lab.*, 5 (1973) 19.
122 R.A. Henry, S.H. Byrne and D.R. Hudson, *J. Chromatogr. Sci.*, 12 (1974) 197.
123 D.W. Patrick and W.R. Kracht, *J. Chromatogr.*, 318 (1985) 269.
124 F. Erni, H.P. Keller, C. Morin and M. Schmitt, *J. Chromatogr.*, 204 (1981) 65.
125 G. Guiochon and M. Martin, *J. Chromatogr.*, 326 (1985) 3.
126 B.F.D. Ghrist, M.A. Stadalius and L.R. Snyder, *J. Chromatogr.*, 387 (1987) 1.
127 C.N. Satterfield, C.K. Colton and W.H. Pitcher, *AIChE J.*, 19 (1973) 628.
128 M.A. Stadalius, B.F.D. Ghrist and L.R. Snyder, *J. Chromatogr.*, 387 (1987) 20.
129 S.A. Cohen, K. Benedek, S. Dong and B.L. Karger, *Anal. Chem.*, 56 (1984) 217.
130 S. Lin and B.L. Karager, *J. Chromatogr.*, 499 (1990) 89.
131 K.D. Nugent, W.G. Burton, T.K. Slattery, B.F. Johnson and L.R. Snyder, *J. Chromatogr.*, 443 (1988) 381.
132 M.A. Stadalius, M.A. Quarry, T.H. Mourey and L.R. Snyder, *J. Chromatogr.*, 358 (1986) 17.
133 B.F.D. Ghrist, B.S. Cooperman and L.R. Snyder, *J. Chromatogr.*, 459 (1988) 1, 25, 43.
134 F. Eisenbeiss, S. Ehlevding, A. Wehrli and J.F.K. Huber, *Chromatographia*, 20 (1985) 657.
135 J.H. Knox and H.M. Pyper, *J. Chromatogr.*, 363 (1986) 1.
136 G.B. Cox, L.R. Snyder and J.W. Dolan, *J. Chromatogr.*, 484 (1989) 409.
137 L.R. Snyder, J.W. Dolan and G.B. Cox, *J. Chromatogr.*, 483 (1989) 63.
138 L.R. Snyder, J.W. Dolan and G.B. Cox, *J. Chromatogr.*, in press.
139 S. Ghodbane and G. Guiochon, *J. Chromatogr.*, 444 (1988) 275.
140 G.B. Cox and L.R. Snyder, *J. Chromatogr.*, 483 (1989) 95.
141 S. Golshan-Shirazi and G. Guiochon, *Anal. Chem.*, 61 (1989) 1368.
142 J. Frenz and Cs. Horvath, in Cs. Horvath (Editor), *High-performance Liquid Chromatography. Advances and Perspectives*, Vol. 5, Academic Press, New York, 1988, p. 212.

Chapter 2

Countercurrent chromatography

Y. ITO

CONTENTS

2.1 INTRODUCTION

Countercurrent chromatography (CCC) has a distinct feature among all chromatographic systems in that the method utilizes no solid support matrix [1-3]. As indicated by its name, CCC may be considered as a hybrid of two classical partition methods, countercurrent distribution (CCD) and liquid partition chromatography, and inherits all the merits from both parent methods. As in CCD, CCC is a genuine liquid partition method that can eliminate all the complications arising from the use of solid supports, such as adsorptive sample loss and deactivation, tailing of the solute peaks, and contamination. Unlike CCD,

References on p. A105

CCC uses a continuous partition process and, therefore, the separation is efficiently performed by adapting elution systems developed for liquid chromatography.

The partition process in CCC takes place in an open column space where the mobile phase continuously elutes through the stationary phase, retained in the column. The retention of the stationary phase in the open column can be accomplished by a combination of the column geometry and the applied force field, either gravitational or centrifugal in nature. As a result, the existing CCC systems display a variety of mechanical designs which are quite different from those of other chromatographic systems.

2.2 PRINCIPLE OF COUNTERCURRENT CHROMATOGRAPHY

2.2.1 Two basic CCC systems

All existing CCC schemes have been derived from two basic systems: One is called the hydrostatic equilibrium system and the other, the hydrodynamic equilibrium system. The mechanisms of these basic CCC systems are illustrated in Fig. 2.1 [3].

The hydrostatic system (Fig. 2.1, left) uses a stationary coiled tube. The coil is first entirely filled with the stationary phase, and then the mobile phase is introduced through one end of the coil. Then, by the effect of gravity, the mobile phase percolates through the stationary-phase segments on one side of the coil. The process is repeated until the mobile phase reaches the other end of the coil. Thereafter, the introduced mobile phase only displaces mobile phase, and a large amount of the stationary phase is retained in each turn of the coil.

The hydrodynamic system (Fig. 2.1, right) uses a similar coil which is slowly rotated about its axis. This simple motion creates amazingly complex effects on the hydrodynamic motion of the two immiscible solvent phases present in the coil. Rotation of the coil produces an Archimedean screw force which drives all objects of different density competitively toward one end of the coil. This end is called the head and the other end is the tail.

When a drop of the heavier phase is introduced into a rotating coil filled with the lighter phase, it migrates toward the head of the coil. Similarly, a drop of the lighter phase introduced in the coil filled with the heavier phase also migrates toward the head of the coil. In fact, any suspended object, whether heavier or lighter than the medium, migrates toward the head of the coil. If the coil is initially filled with roughly equal amounts of the two immiscible solvents, rotation of the coil results in competitive migration of the two phases toward the head of the coil, where either phase occupying less space in a given portion of the coil becomes the suspended phase to migrate toward the head. Eventually, the two solvent phases establish a hydrodynamic equilibrium starting at the head of the coil, while any excess amount of either phase remains at the tail end of the coil. The volume ratio of the two phases at this equilibrium state will mainly be determined by the centrifugal force field acting on the coil. In a slowly rotating coil under unit gravity, as in the present case, the volume ratio of the two phases is roughly 1, as reported elsewhere [4]. In this situation, the mobile phase (either lighter or heavier phase), introduced through the head

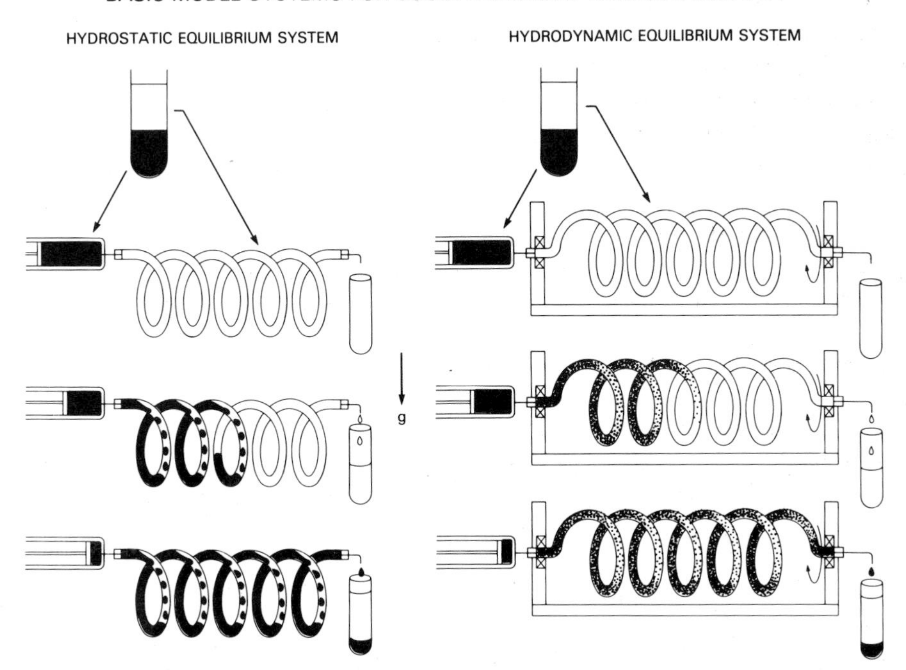

Fig. 2.1. Mechanism of two basic CCC systems, drawn for a mobile phase (black) of higher density than the stationary phase (white); g = gravity. Left: In the hydrostatic equilibrium system (HSES), a stationary coil is filled with the stationary phase, and the mobile phase is introduced through one end of the coil. By the effect of gravity, the mobile phase percolates through the stationary-phase segment on the side of the coil. This process is repeated until the mobile phase reaches the other end of the coil. Thereafter, the mobile phase only displaces the mobile phase, and a large amount of the stationary phase is retained in each turn of the coil. Right: In the hydrodynamic equilibrium system (HDES), the coil is slowly rotated around its own axis to induce an Archimedean screw effect, which drives all objects competitively toward the head (left end, in this case) of the coil. Consequently, the mobile phase introduced through the head of the rotating coil is mixed with the stationary phase to establish a hydrodynamic equilibrium, where each phase occupies roughly equal space in each helical turn. Once the mobile phase reaches the tail of the coil, it replaces the mobile phase only, leaving a large volume of the stationary phase in the coil. In both basic CCC systems, the mobile phase of lower density can also produce similar effects.

of the coil, would become an excess phase and migrates toward the tail, thus, the hydrodynamic equilibrium being automatically maintained on the head side of the coil.

In Fig. 2.1 (right), the coil is first entirely filled with the stationary phase, and then the mobile phase is introduced into the rotating coil through the head (left end). As soon as the two solvent phases meet in the coil, they are vigorously mixed to establish a hydrodynamic equilibrium at the head side of the coil, where each phase occupies nearly equal volumes in each helical turn. This process continues until the mobile phase reaches the tail of the coil (right end). Thereafter, the introduced mobile phase displaces the mobile phase, leaving a large volume of the stationary phase in the coil.

Consequently, in both basic CCC systems, solutes introduced locally at the inlet of the coil are subjected to an efficient partition process between the percolating mobile phase and the retained stationary phase in the coil and separated according to their partition coefficients, as in liquid chromatography, but in the absence of the solid support.

Each system has its specific merit for performing countercurrent chromatography. The hydrostatic equilibrium system enables stable retention of the stationary phase in the column, while the hydrodynamic equilibrium system provides efficient mixing of the two solvent phases to yield a high partition efficiency. These two basic CCC systems have produced a variety of efficient CCC schemes, as described in the material that follows.

2.2.2 Principle of hydrostatic CCC schemes

Because of its simplicity, the basic hydrostatic equilibrium system was quickly developed to produce various useful CCC schemes in the early Seventies. As illustrated in Fig. 2.2, the modification of the basic system was carried out in two different directions.

For large-scale preparative separations (Fig. 2.2, top, and Fig. 2.7), one side of the coil in the basic system, which is entirely occupied by the mobile phase (hence forming an inefficient deadspace), was replaced with a fine transfer tube, while the other half of the coil (providing the efficient column space for partitioning) was modified to a large-diameter straight tube. In droplet CCC [2,3,5], the mobile phase (heavier than the stationary phase) introduced at the top of the vertical tube, filled with the stationary phase, forms multiple droplets at regular intervals. Because each droplet occupies an area in the cross section of the tube, an efficient partition process takes place between the droplet and a thin layer of the stationary phase as each droplet travels downward through the column of the stationary phase. The system requires formation of droplets of suitable dimensions, and this limits the choice of the solvent systems in this method. In order to extend the capability of droplet CCC, a locular column was developed by placing centrally perforated discs at regular intervals to form multiple partition units, called locules, in each column unit. In rotation locular CCC [2,3], the locular column is tilted at a proper angle and slowly rotated around its axis. The mobile phase introduced at the top of the column, initially filled with the stationary phase, displaces about half the volume of the stationary phase in each locule through the effect of gravity, while the two phases are efficiently mixed by the rotation of the column. In practice, a set of multiple column units is connected in series and arranged around a tilted rotary cylinder to perform CCC separation. Under this column arrangement, the partition efficiency increases with rotational speed up to about 200 rpm, when the radially acting centrifugal force field created by the rotation begins to interfere with the mixing of the two phases in the locule to lower the partition efficiency. This limitation was removed by the gyration locular CCC [2,3], where the vertical column units gyrate around a vertical central axis. The rotating centrifugal force field produced by the gyration of the column vigorously mixes the two phases up to the point where they are emulsified.

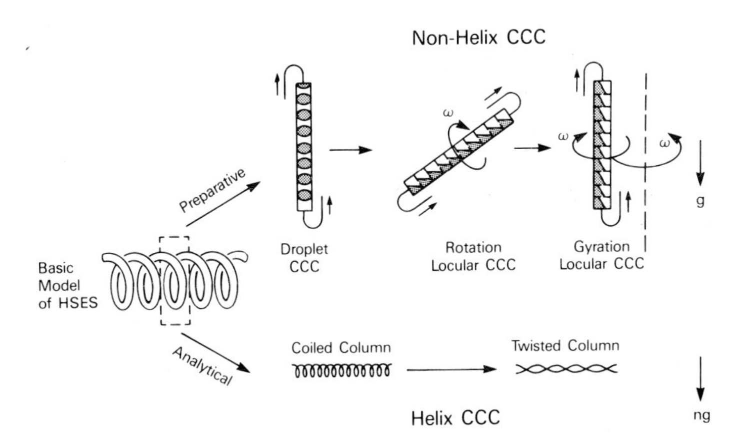

Fig. 2.2. Development of various hydrostatic CCC systems. The basic hydrostatic CCC system (left) was modified in two directions: For preparative-scale separations (top), one side of the coil, which is entirely occupied by the mobile phase, was replaced with a fine transfer tube, while the other half of the coil was modified to a large-diameter straight tube. In droplet CCC, the mobile, heavier phase (black) introduced at the top of the vertical tube, filled with the stationary, lighter phase (white), forms multiple droplets in the open-tubular space at regular intervals. By the effect of gravity, these droplets move toward the bottom of the column at a uniform rate without coalescence. Because each droplet occupies an area in the cross section of the tube, efficient partition takes place between the droplet and a thin layer of the stationary phase. The system requires formation of droplets of suitable size. This limitation is removed by the locular column, made by placing centrally perforated discs in the column at regular intervals, thus forming multiple compartments, called locules. In rotation locular CCC [2,3], the locular column is tilted at a proper angle and slowly rotated around its axis. The mobile phase (black) introduced at the top of the column, initially filled with the stationary phase (white), displaces about half the volume of the stationary phase in each locule through the effect of gravity, while the two phases are efficiently mixed by the rotation of the column. In gyration locular CCC [2,3], the vertical locular column is gyrated around a vertical central axis. The rotating centrifugal force field, generated by the gyration of the column, vigorously mixes the two phases up to the point where they are emulsified. The system eliminates the need for the rotary seal for continuous elution (note that the gyration is identical to the type I synchronous planetary motion in Fig. 2.3). In all these systems, the lighter phase can also be the mobile phase if introduced from the bottom end of the column, previously filled with the heavier, stationary phase. For microscale analytical separations, the basic system was modified by reducing the dimensions of the coil, which is then subjected to a strong centrifugal force field (bottom). The hydrostatic CCC process takes place in the narrow opening of the coil, as in the HSES in Fig. 2.1. The hydrostatic pressure produced by the centrifugal force is substantially reduced by the use of a twisted column orientation. g = gravity; ng = centrifugal force; ω = angular velocity.

For microscale analytical separations, the basic system was modified simply by reducing the dimensions of the coil, as shown in Fig. 2.2 (bottom). In order to prevent plug flow of the two solvent phases in a narrow-bore tube, a strong centrifugal force field is generated by rotating the coil, mounted tangentially at the periphery of the centrifuge bowl. Accordingly, the method is called helix CCC [1-3] or toroidal (the term comes from a torus formed by the circular arrangement of the coil) coil CCC. In this method, the mobile phase introduced into the coil, initially filled with the stationary phase, displaces about half the volume of the stationary phase in each helical turn; an efficient partition process between the stationary-phase segment and the flowing mobile phase takes place within a narrow space in each turn of the coil. A strong centrifugal force field applied to the column tends to create a high hydrostatic pressure in the column, and this becomes the limiting factor in this method. The column pressure can be substantially reduced by the use of a twisted column, illustrated in the diagram. The recently introduced centrifugal droplet CCC (centrifugal partition chromatography) [6] is considered to be a combination of helix CCC and droplet or locular CCC, described above.

2.2.3 Principle of hydrodynamic CCC schemes

Development of useful CCC schemes from the basic hydrodynamic equilibrium system necessitated a suitable rotating mechanism, providing a leak-free flow-through system. In the Seventies, various flow-through centrifuge systems were developed for performing CCC. In many of these systems, the use of the conventional rotary seal was eliminated, and solvents can be freely passed through the rotating column without the risk of leakage and contamination, which often occurs at the site of the rotary seal. Varieties of these rotary-seal-free flow-through centrifuge schemes are classified into three categories: synchronous, nonplanetary, and nonsynchronous (Fig. 2.3).

In Fig. 2.3, each diagram shows the orientation and motion of a cylindrical coil holder with a bundle of flexible flow tubes, the end of which is tightly supported on the centrifuge axis. In the synchronous schemes, the holder undergoes a synchronous planetary motion, i.e., one revolution around the centrifuge axis for each rotation about its own axis. In type I synchronous planetary motion (left, top), the vertical holder revolves around the centrifuge axis and simultaneously rotates about its own axis at the same angular speed but in the opposite direction. This counter-rotation of the holder steadily unwinds the twists of the tube bundle caused by revolution, thus eliminating the need for a rotary seal [7]. Without disturbing this twist-free principle, the orientation of the holder can be modified in the following two different ways.

In the I-L-J series (synchronous varieties, left column), the holder is tilted toward the central axis of the centrifuge to form type I-L. The continued movement of the holder produces type L, type J-L, and finally type J, which has an inverted orientation of the holder. In the I-X-J series (synchronous varieties, right column), the holder is tilted sideways to form types I-X, X, J-X, and J in a similar fashion. If viewed from the top of the centrifuge, the axis of the holder is oriented radially in type L and tangentially in type X. Type J is considered to be a transitional form to the nonplanetary system. When the holder of type I is moved to the central axis of the centrifuge, the rotation of the holder cancels out the revolution, causing the holder to become stationary (nonplanetary, top). However, when the holder of type J is moved to the centrifuge axis, without being inverted, the rotation and revolution add up to rotate the holder at twice the angular velocity (2ω), as indicated in the diagram [8]. This nonplanetary scheme provides the basis for the nonsynchronous series. In the nonplanetary scheme, the holder is again shifted to the periphery to undergo synchronous planetary motion, as in the synchronous

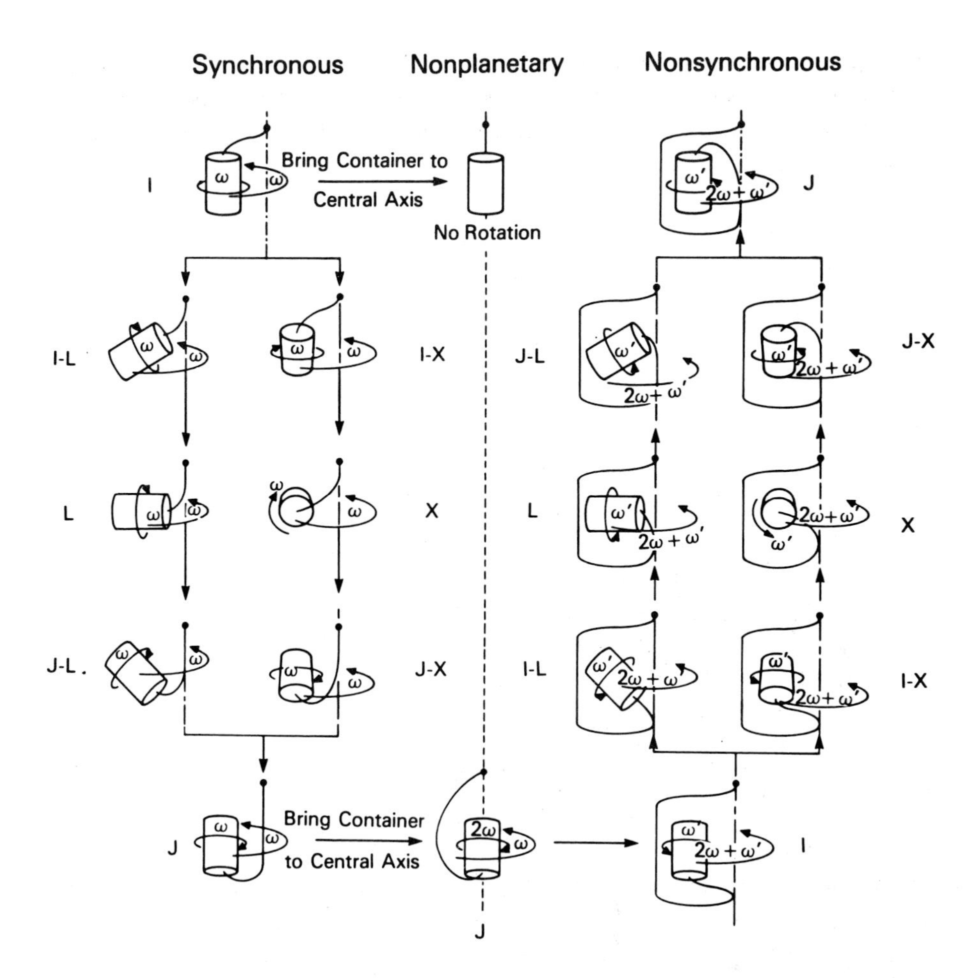

Fig. 2.3. Rotary-seal-free flow-through centrifuge system for performing CCC. Each diagram shows orientation and motion of a cylindrical coil holder with a bundle of flexible flow tubes with one end tightly supported on the centrifuge axis. These schemes are divided into three main categories according to the motion of the holder. In the synchronous group, the holder rotates about its own axis once per one revolution about the centrifuge axis, whereas in the nonsynchronous group the rotation and revolution of the holder is freely adjustable. The nonplanetary centrifuge provides simple rotation of the holder. The synchronous and nonsynchronous groups are further divided each into 8 types - I, L, X, J, and their hybrids - according to the orientation of the holder. Types I and J have vertical orientation of the holder, while the holder of type J is inverted. Types L and X have horizontal orientation of the holder where the holder of type L shows radial orientation and that of type X, tangential orientation to the centrifuge axis. In all these schemes, the holder can be rotated without twisting the bundle of flow tubes.

series, to produce various schemes. In the nonsynchronous schemes, the rotation and revolution of the holder are independently regulated to provide versatility of the system at the expense of greater complexity in the mechanical design of the apparatus.

Among these schemes, synchronous types J and X were found to be most useful for performing CCC because of their ability to provide bilateral hydrodynamic distribution of the two solvent phases in a rotating coil, which can be applied to high-speed CCC, the most advanced form of CCC.

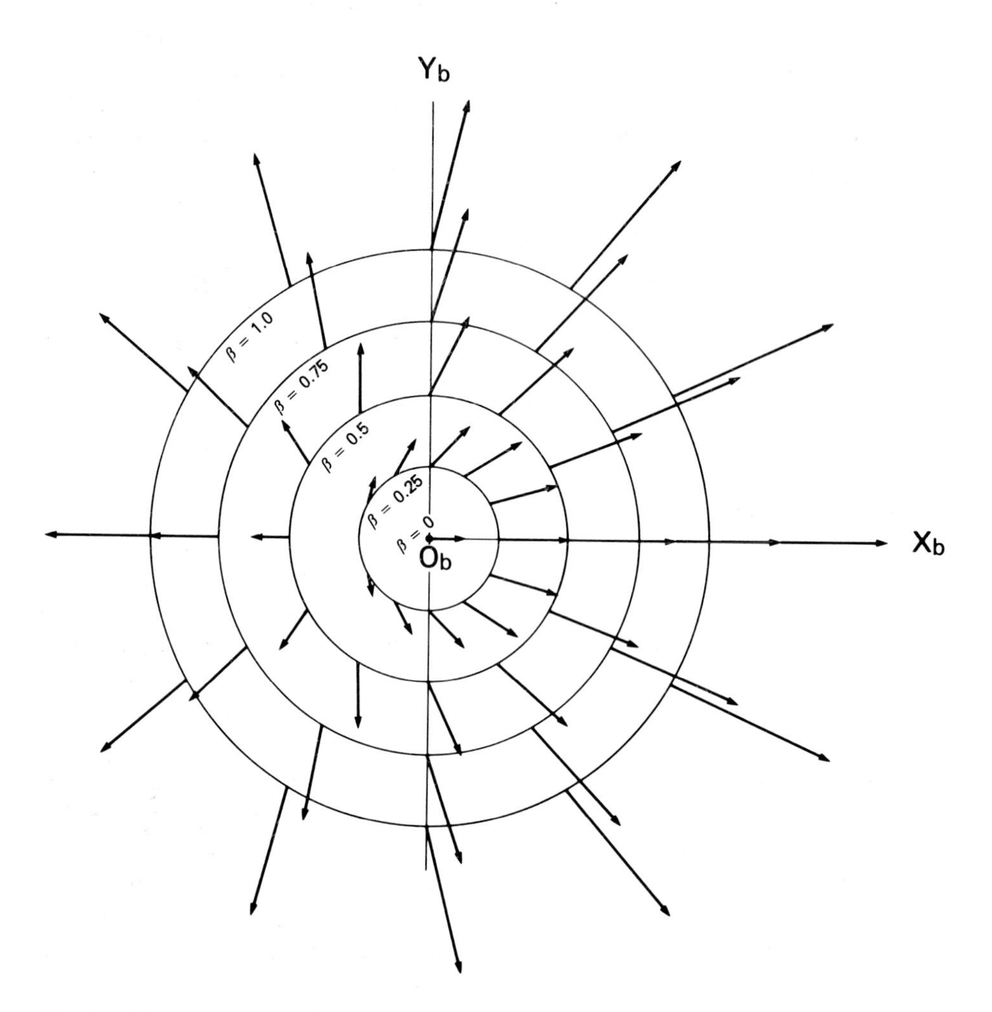

Fig. 2.4. Force distribution diagram of type J synchronous planetary motion. O_b = center of rotation (column holder axis); X_b-Y_b = body coordinate system for analysis; $\beta = r/R$, where r = distance from the holder axis to each circle and R = distance from the holder axis to the centrifuge axis (located at the cross section between the outermost circle of $\beta = 1$ and the X_b-axis on the left side). The mathematical analysis of the type J synchronous planetary motion reveals a complex force distribution pattern. Because the holder rotates around point O_b in the force field affixed to the X_b-Y_b coordinate system, every point in the rotating holder is exposed to the successive change of the force vectors during each revolution cycle. Consequently, this force field can produce different hydrodynamic effects on the two solvent phases in the coiled column, depending on the location and orientation of the coil inside the holder.

Analyses of the centrifugal force field produced by these planetary motions were performed to understand the hydrodynamic motion and distribution of the two solvent phases in the rotating coil [9,10]. In these analyses, the centrifugal force field computed from each type of the synchronous planetary motion was expressed as a set of vectors in a three-dimensional body coordinate system where the Z_o-axis coincides with the axis of the holder. Fig. 2.4 shows the force distribution diagram generated by the type J synchronous planetary motion, where the axis of the coil holder with vertical orientation to the X_b-Y_b plane is located at the center of the coordinate system, O_b, and the concentric circles indicate various locations inside the holder. Beta values on each circle indicate the ratio r/R, where r is the radius of the circle and R the distance from the holder axis (O_b) to the centrifuge axis (at the intersection between the outermost circle and the X_b-axis on the left).

Here, it is important to note that the centrifugal force field in the coordinate system is stationary while the holder rotates around point O_b at a uniform angular velocity, ω. In a relative sense, every point present in the holder is subjected to a rotating centrifugal force field: At locations close to the holder axis, the centrifugal force vectors are short and always pointing toward the periphery of the centrifuge, whereas at locations remote from the holder axis the centrifugal force vectors increase in strength and always point away from point O_b.

This force distribution pattern can produce different hydrodynamic distributions of the two immiscible solvent phases in the coil, depending on the orientation of the coil inside the holder. When the coil is placed near the periphery of the holder (eccentric location), either parallel to the holder axis or around the holder in the form of a toroidal coil, the centrifugal force field separates the two solvent phases in such a way that the heavier phase is distributed at the outer half and the lighter phase at the inner half of each helical turn to form alternating segments of the two phases along the length of the coil; at the same time, the centrifugal force field, which is fluctuating, since the coil is rotating with respect to the holder, produces efficient mixing of the two phases in the coil. This scheme is considered to be a modified hydrostatic CCC system. On the other hand, when the coil is directly wound around the periphery of the holder hub (coaxial orientation), the outwardly directed force field separates the two solvent phases to form two continuous layers along the length of the coil in such a way that the heavier phase occupies the outer portion and the lighter phase the inner portion of the coil. In this particular situation, the rotating force field strangely acts to drive one of the phases (head phase) toward the head of the coil and, as a result, the other phase (tail phase) accumulates at the tail side of the coil. Either lighter or heavier phase can become the head phase, depending on various factors, such as the beta value, physical properties of the solvent system, and revolution speed [11-13]. Eventually, the two solvent phases confined in the rotating coil become completely separated along the length of the coil, the head phase entirely occupying the head side, and the tail phase the tail side of the coil. It should be mentioned here that this bilateral phase distribution is observed only in a closed coil; in an open coil both lighter and heavier phases come out together from the head of the coil. The above hydrodynamic phenomenon can be efficiently utilized for performing CCC, as illustrated in Fig.

References on p. A105

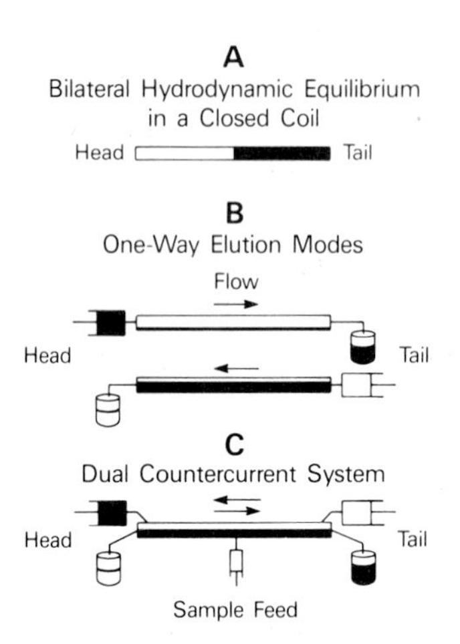

Fig. 2.5. Mechanisms of high-speed CCC and dual CCC. A: Bilateral phase distribution of the head phase (white) and the tail phase (black) in an end-closed rotating coil. B: Two different unilateral percolation modes for performing high-speed CCC. The upper coil is first entirely filled with the white phase, and the black phase passes from the head toward the tail. The lower coil is first filled with the black phase, and the white phase passes from the tail toward the head. In either case, the system can retain a large volume of the stationary phase in the coil against a high flowrate of the mobile phase. C: The system allows dual (true) countercurrent movement of the two phases through the coil, if the two phases are simultaneously introduced through the respective terminal of the coil. This dual CCC operation necessitates an additional flowtube at each end of the coil and a sample feedline at the middle portion of the coil.

2.5, where four coiled tubes are schematically drawn uncoiled to show the overall distribution of the two solvent phases.

In Fig. 2.5A, two immiscible solvent phases confined in the rotating coil form a bilateral distribution, the head phase (white) entirely occupying the head side, and the tail phase (black) the tail side of the coil. Fig. 2.5B illustrates two different unilateral elution modes. In one case, the coil is first entirely filled with the white phase, and the black phase is eluted from the head toward the tail. In the other case, the coil is first filled with the black phase and the white phase is passed from the tail toward the head. In either case, the system can retain a large volume of the stationary phase in the coil against a high flowrate of the mobile phase. The system can also perform dual CCC, where two solvent phases undergo true countercurrent movement through a coiled tube, as illustrated in Fig. 2.5C. In this case, the two solvent phases are simultaneously introduced into the rotating coil from the respective terminals, i.e., the head phase from the tail and the tail phase from the head of the coil. The system requires an additional flowtube at each terminal of the coil to collect the effluents, and a sample feedline is installed near the middle of the coil. This dual countercurrent system has been successfully applied to foam CCC [14-16] and liquid/liquid dual CCC [17].

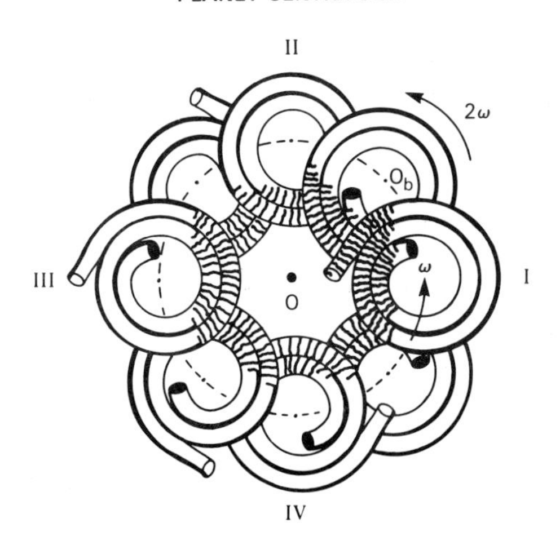

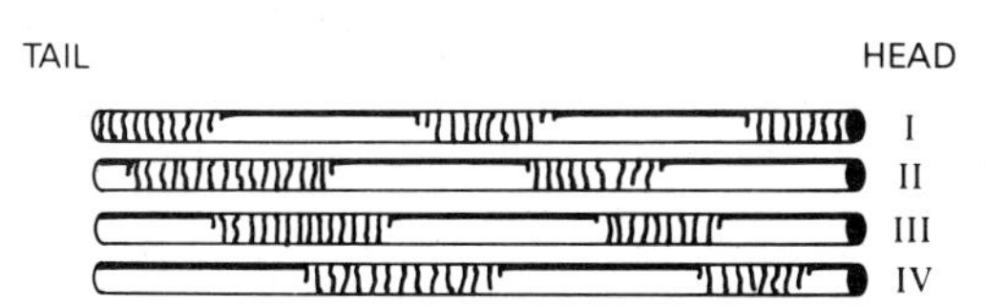

Fig. 2.6. Motion of mixing zones in a concentric coil planet centrifuge. Top: Successive positions of the concentric spiral coil undergoing the type J synchronous planetary motion around the central axis of the centrifuge, O. Under the centrifugal force field in Fig. 2.4, the area of the spiral column shows distinct mixing zones near the center of the centrifuge, where the centrifugal force becomes minimal. These mixing zones are affixed around the center of the centrifuge, while the spiral column rotates about its own axis (O_b) at an angular velocity, 2ω. Bottom: Uncoiled columns, numbered I to IV, correspond to the column position indicated in the top diagram and show movement of the mixing zones through the spiral column. Each mixing zone is traveling toward the head of the spiral column at a rate equal to the revolution speed of the column. Consequently, the two solvent phases at any portion of the column are subjected to repetitive mixing and settling at the rate of over 13 times per sec (at 800 rpm).

The unique hydrodynamic motion of the two immiscible solvent phases observed in the rotating coil [18] is schematically illustrated in Fig. 2.6, where the top diagram shows successive positions of a spiral column during one complete revolution around the central axis of the centrifuge (O). Under a unilateral elution of the mobile phase, the area of the spiral column is distinctly divided into mixing and settling zones, where the white area indicates the lighter (stationary) phase and the black area, the heavier (mobile) phase. The mixing zone is always located near the center of the centrifuge (which corresponds to the reduced force field of the force distribution diagram shown in Fig. 2.4), while the spiral column rotates about its own axis (O_b) twice during each revolutional cycle. In the bottom diagram, uncoiled columns, numbered I to IV, correspond to the column position indicated in the top diagram and show movement of the mixing zones through the spiral column. Each mixing zone is traveling toward the head of the spiral column at a rate equal to the revolution speed of the column. This indicates the extremely important fact that the two solvent phases at any portion of the column are subjected to a typical partition process of repetitive mixing and settling at an enormously high rate of over 13 times per second (at 800 rpm of revolution), while the mobile phase is steadily passing through the stationary phase. Consequently, this finding explains the high partition efficiency of the present system under a high flowrate of the mobile phase.

Because of the rapid and efficient chromatographic separations, the centrifugal CCC utilizing the bilateral phase distribution is named "high-speed CCC" [19]. Other types of synchronous planetary motion (Fig. 2.3), such as types J-L, X, and X-L can also form similar bilateral phase distributions and have been successfully applied for performing high-speed CCC [9,10,20-27].

2.3 APPARATUS

Among the existing CCC instruments, the design and capability of ten selected models are briefly described below. General applications of these instruments are summarized in Table 2.1 and their performance is compared in terms of sample size, separation time, and theoretical plate number in the separation of dinitrophenyl (DNP) amino acids (Table 2.2).

2.3.1 Helix countercurrent chromatograph (toroidal coil centrifuge)

The helix countercurrent chromatograph is the earliest model of the CCC instrument, reported in 1970 [1,2]. The original instrument was fabricated by modifying an International centrifuge (Model PR-2) (International Equipment Co.). A toroidal coiled column was prepared from PTFE (polytetrafluoroethylene) tubing with an ID of 0.3 mm and a length of 100 m and mounted in the periphery of a modified Helix Tractor centrifuge head (International Equipment Co.). The mobile phase was fed from a rotating syringe, mounted at the center of the rotor, while the effluent was collected through a rotary seal. Despite this primitive design, a partition efficiency of 8000 theoretical plates was achieved in DNP amino acid separations [2]. Later, the system was redesigned to eliminate the rotating syringe and the rotary seal by adopting the nonplanetary system J (Fig. 2.3, center, bottom) [28]. This improved apparatus was successfully applied to separations of *E. coli* strains with polymer phase systems [29]. The apparatus has a high potential for performing analytical CCC.

2.3.2 Droplet CCC instrument

The droplet CCC instrument, a simple preparative apparatus, can produce a partition efficiency of up to 900 theoretical plates. The commercial model, distributed by Tokyo Rikakikai Co., consists of 300 partition units, each 60 cm long and 2-4 mm in ID. Since

1980, the apparatus has been extensively used for separations of various natural products [30]. Two-phase solvent systems, composed of chloroform, methanol, and water in various ratios, were the main ones used for separations because they form mobile-phase droplets of suitable size. The recent improvement in the design of the apparatus, the flow system divided into five sections by valves, has radically shortened the time required for both filling and washing the column from 7.5 h to 1.5 h [31].

2.3.3 Rotation locular CCC instrument

The rotation locular CCC instrument, also distributed by Tokyo Rikakikai Co., consists of 16 column units, connected in series, each unit measuring 50 cm x 11 mm ID and containing about 40 locules [32]. The column units can be rotated up to 120 rpm in column holders, which are tilted at the desired angle of 30-40° from the horizontal plane. The method permits the use of most conventional two-phase solvent systems for performing preparative-scale separations.

2.3.4 Centrifugal droplet CCC instrument (centrifugal partition chromatograph)

The centrifugal droplet CCC apparatus has been developed and manufactured by Sanki Engineering Ltd. since the early Eighties [6]. It holds a chain of cartridges circularly arranged in a centrifuge bowl. Each cartridge contains a number of partition units, connected in series with narrow transfer channels, as shown in Fig. 2.7 [33]. The apparatus is equipped with a pair of rotary seals said to withstand pressures up to 60 kg/cm^2.

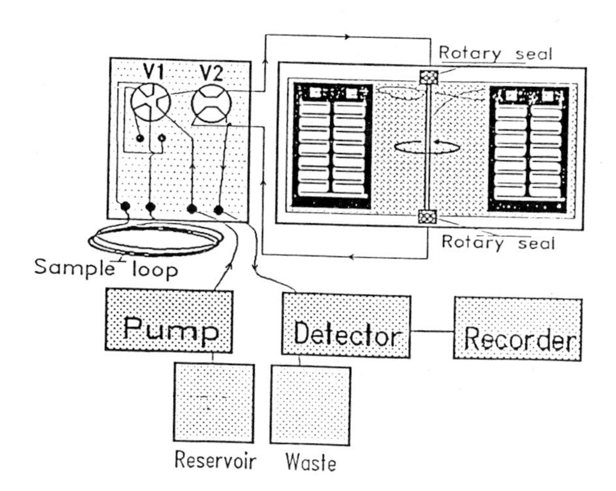

Fig. 2.7. Design of a centrifugal droplet CCC instrument (centrifugal partition chromatograph). V1 = injection valve; V2 = switching valve. Each cartridge contains 400 channels, of which 14 channels are shown in the diagram. Each channel and connecting duct is 12.4 mm long and 1.1 mm deep. The width of the channel is 2.4 mm, while that of the duct is 0.9 mm.

TABLE 2.1

VARIOUS CCC INSTRUMENTS AND THEIR APPLICATIONS

Apparatus (planetary motion)	Sample size		Polymer phase systems	Other applications	Special features
	Analytical	Preparative			
Toroidal coil centrifuge (nonplanetary J)	+	−	+	cell elutriation	extremely high partition efficiency with a small-ID column
Droplet CCC* (stationary)	−	+	−		simple design
Rotation locular CCC (nonplanetary, rotary seal)	−	+	−		reliable retention of the stationary phase
Centrifugal droplet CCC (nonplanetary, rotary seal)	+	+	+		reliable retention of the stationary phase; equipped with well-developed software

Rotating coil assembly (nonplanetary J)	–	+	–	continuous extraction	simple design; large-scale preparative separations
Horizontal flow-through CPC** (synchronous J)	+	+	+		high partition efficiency; reliable retention of the stationary phase
Multilayer coil CPC (synchronous J)	+	+	–	foam CCC, dual CCC, continuous extraction	reliable retention of hydrophobic solvents; high efficiency and rapid separation
Cross-axis CPC (synchronous X)	–	+	–	foam CCC	large preparative separation; high partition efficiency
Nonsynchronous flow-through CPC (nonsynchronous I)	+	+	+	cell elutriation	freely adjustable coil rotation

*CCC: Countercurrent chromatograph.

**CPC: Coil planet centrifuge.

TABLE 2.2

PERFORMANCE OF VARIOUS CCC INSTRUMENTS IN SEPARATION OF DNP AMINO ACIDS WITH CHLOROFORM/ACETIC ACID/0.1 *N* HCl (2:2:1)

Apparatus	Sample mixture				Mobile phase	Flowrate	Elution time	TP***	Ref.
	Sample No.*	Wt.	Vol.	$K(C_m/C_s)$**					
Toroidal coil centrifuge (helix CCC)	1-9	385 μg	5 μl	> 100-0.18	upper	125 μl/h	40 h	2500-5200	1,2
Droplet CCC	1-7	50 mg	3 ml	> 100-0.45	upper	16 ml/h	70 h	900	2,5
Rotation locular CCC[a]	2-10	2.5 mg	40 μl	> 50-0.18	upper	5 ml/h	68 h	3000	2
Centrifugal droplet CCC[b]	1,4,7,11	36 mg	-	> 100-0.45	upper	240 ml/h	1.6 h	400-700	6
Rotating coil assembly	3,6	1 g	30 ml	1.9-0.56	upper	120 ml/h	10 h	250	37
	3,6	1 g	30 ml	1.8-0.53	lower	120 ml/h	13 h	170	37
Horizontal CPC									
Preparative column	1-4, 6-8	30 mg	10 ml	> 100-0.26	upper	60 ml/h	12 h	720-2500	43
	2-4, 6,9	24 mg	10 ml	5.6-0.26	lower	60 ml/h	10 h	550-2500	43
Analytical column	1-4, 6-8	1.5 mg	50 μl	> 100-0.26	upper	6 ml/h	10 h	1900-4000	43
	2-4, 6,9	1.2 mg	50 μl	5.6-0.26	lower	6 ml/h	7 h	3700-4000	43
Multilayer CPC									
Semianalytical column	2-4,6	10 mg	1 ml	3.8-0.56	upper	180 ml/h	2.3 h	2900-5300	53
	3,4,6,8	10 mg	1 ml	3.6-0.53	lower	180 ml/h	2.9 h	2600-4000	53

Semipreparative column	2-4,6	50 mg	5 ml	3.8-0.56	upper	240 ml/h	2.9 h	1700-2700	51
	3,4,6,8	50 mg	5 ml	3.6-0.53	lower	240 ml/h	3.3 h	1400-2000	51
Preparative column	3,4,6,8	4.2 g	80 ml	3.6-0.53	lower	420 ml/h	8 h	1000-1800	64
Cross-axis CPC									
Preparative column	1-4, 6	100 mg	2 ml	> 100-0.56	upper	120 ml/h	6 h	600-1400	21
	3,4,6,7,9	100 mg	2 ml	5.6-0.53	lower	120 ml/h	7 h	700-1100	21
Large preparative column	1-4,6	10 g	100 ml	> 100-0.56	upper	120 ml/h	35 h	500-1000	25
	3-5,7,9	10 g	100 ml	5.6-0.53	lower	120 ml/h	47 h	400-800	25

*1: *N*-δ-2,4-DNP-L-ornithine; 2: *N*-2,4-DNP-L-aspartic acid; 3: *N*-2,4-DNP-DL-glutamic acid; 4: *N,N*-di(2,4-DNP)-L-cystine; 5: *N*-2,4-DNP-β-alanine; 6: *N*-2,4-DNP-L-alanine; 7: *N*-2,4-DNP-L-proline; 8: *N*-2,4-DNP-L-valine; 9: *N*-2,4-DNP-L-leucine; 10: *N*-2,4-L-arginine; 11: N-2,4-DNP-L-threonine.

**Partition coefficient expressed as solute concentration in the mobile phase divided by that in the stationary phase.

***Partition efficiency in terms of theoretical plates.

a: These data are based on the original unit built at the Laboratory of Technical Development, NIH. The commercial model has a much lower efficiency but a larger sample load capacity.

b: Also called "Centrifugal Partition Chromatograph".

The unit includes well-developed, expensive software. The system permits the use of various two-phase solvent systems, including aqueous/aqueous polymer phase systems.

2.3.5 Rotating coil assembly

The rotating coil assembly, a simple modification of the basic hydrodynamic equilibrium system, lets a large coil assembly rotate slowly in the gravitational field [34-36]. It has been observed that the use of multilayer coils coaxially mounted on the rotary shaft produces excellent peak resolution [37]. More recent studies have shown that the system works well with a large-bore coil of 2 cm ID and 20 cm helical diameter at relatively slow rotation of 80 rpm for a variety of conventional two-phase solvent systems [4]. The overall results also suggest that dimensions of the multilayer coil may be further increased without significantly disturbing the required hydrodynamic conditions. Because of its simplicity and minimum mechanical constraints, the present system could be applied to large-scale industrial separations, ranging from 100-g to 1-kg quantities of samples. The nonplanetary seal-free mechanism (Fig. 2.3) was used in this system. The apparatus will be particularly useful for large-scale preparative separations in industrial plants.

2.3.6 Horizontal flow-through coil planet centrifuge

In the original apparatus, ten coiled columns were connected in series and arranged around the horizontal rotary shaft [38,39]. Each column consisted of 100 helical turns of 2.6-mm-ID PTFE tubing, wound onto a 50-cm-long, 1.25-cm-OD core with a capacity of 25 ml. The column assembly undergoes type J synchronous planetary motion (Fig. 2.3) around the central axis of the centrifuge at a maximum speed of 400 rpm. The second model holds a pair of column holders on the rotary frame, one for preparative separations (type J synchronous planetary motion) and the other one for analytical separations (type I synchronous planetary motion) [40-43]. The apparatus has been extensively used for separation and purification of synthetic peptides by Knight [44]. The improved version of the apparatus, recently manufactured by Varex Corp., is equipped with a pair of short multilayer coil assemblies, which are connected in series and mounted one on each side of the rotary frame to balance the centrifuge system. This design not only doubles the column capacity but also improves partition efficiency at a higher revolutional speed of 800 rpm.

An analytical variety of the apparatus, called the toroidal coil planet centrifuge, was constructed by modifying the coil geometry [45,46]. A long, coiled tube (50 m x 0.55 mm ID) was completely wound on a flexible core (13 m x 1.5 or 5.0 mm OD), which, in turn, was coiled onto the holder hub (15 cm OD), making multiple turns to form a coiled coil. This apparatus has been used for analytical-scale separations of a variety of samples, including DNP amino acids [45,46], oligopeptides [46], and various plant hormones [47] with organic/aqueous two-phase solvent systems, and for partitioning cell organelles with aqueous/aqueous polymer phase systems [48].

2.3.7 Multilayer coil planet centrifuge for high-speed CCC

The multilayer coil planet centrifuge for high-speed CCC holds a multilayer coil coaxially around the coil holder, which undergoes type J synchronous planetary motion (Fig. 2.3) around the central axis of the centrifuge [19,49]. The original model for semipreparative separations has a 10-cm revolutional radius and holds a multilayer coil of 130-m-long, 1.6-mm-ID PTFE tubing with a total capacity of about 300 ml, wound around the holder hub, 10 cm in diameter (Fig. 2.8). Various commercial models are now available from P.C. Inc. and also from Pharma-Tech Research Corporation. Recently, Shimadzu Corp. began manufacturing the instruments. The most recent design of the high-speed CCC centrifuge eliminates the need for the counterweight by mounting two or more column holders symmetrically around the rotary frame. This new design provides a perfect centrifuge balance and increases both the partition efficiency and the sample load

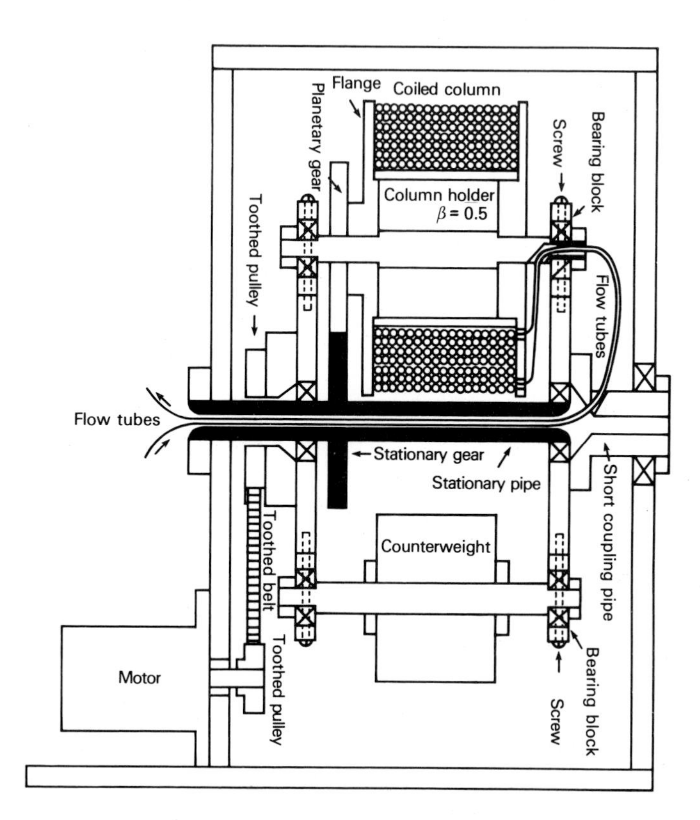

Fig. 2.8. Cross-sectional view of the multilayer coil planet centrifuge for high-speed CCC. The planetary gear mounted on the column holder shaft is coupled to the identical stationary sun gear (shaded), mounted around the central stationary pipe (shaded). This gear arrangement produces the desired planetary motion of the column holder, i.e., rotation about its own axis and revolution around the centrifuge axis at the same angular velocity in the same direction.

capacity by connecting all columns in series [50-53]. The system allows the multiple columns on the rotary frame to be interconnected without the twisting of flowtubes. The apparatus can produce partition efficiencies of several thousand theoretical plates in a few hours of elution [52,53]. Fig. 2.9A shows a preparative chromatogram of flavonoids from a crude ethanol extract (100 mg) of sea buckthorn (*Hippophae rhamnoides*), obtained by a high-speed CCC centrifuge, equipped with three multilayer coils (300 m x 1.1 mm ID, 270-ml capacity) connected in series (see figure caption for details) [53].

Because of its efficient, speedy separations high-speed CCC is also suitable for analytical-scale separations. For this application, the dimensions of the column and the centrifuge radius are proportionally reduced while the revolution speed is increased to enhance countercurrent movement of the two solvent phases through the narrow lumen

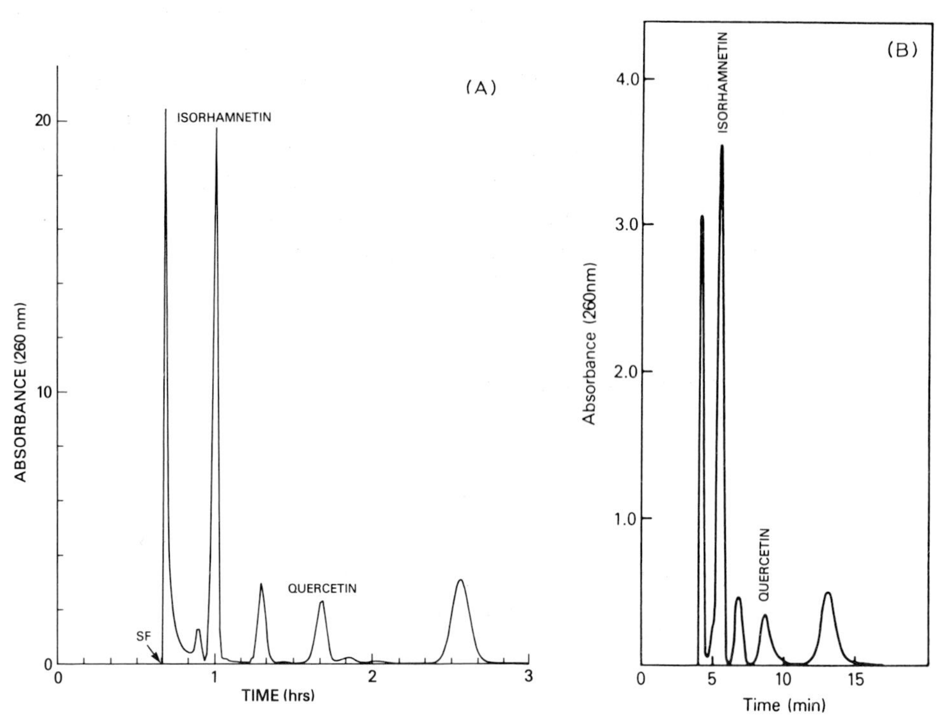

Fig. 2.9. Chromatograms of flavonoids from a crude ethanol extract of *H. rhamnoides*, obtained with a semipreparative (A) and analytical (B) high-speed CCC centrifuge. SF = solvent front. Experimental conditions were as follows: (A) Apparatus: multilayer coil planet centrifuge with 7.5-cm revolution radius, 3 multilayer coils, 300 m x 1.07 mm ID, 270-ml capacity; sample: crude ethanol extract of *H. rhamnoides*, 100 mg in 4.8 ml solvent mixture; solvent system: chloroform/methanol/water (4:3:2), lower, nonaqueous phase mobile; flowrate: 3 ml/min; speed: 1200 rpm; retention of stationary phase: 60%. (B) Apparatus: analytical multilayer coil planet centrifuge with 6.35-cm revolution radius, single multilayer coil, 70 m x 0.85 mm ID, 40-ml capacity; sample: crude ethanol extract of *H. rhamnoides*, 3 mg in 0.5 ml solvent; solvent system: chloroform/methanol/water (4:3:2), lower, nonaqueous phase mobile; flowrate: 5 ml/min; speed: 1800 rpm; retention of stationary phase: 68%.

of the column. Fig. 2.9B illustrates a typical analytical-scale (3 mg) separation of flavonoids from the crude extract of *H. rhamnoides* obtained with a high-speed CCC centrifuge with a 6.35-cm revolution radius (see figure caption for experimental details) [54]. The analytical high-speed CCC centrifuge with a 2.5-cm revolution radius, recently designed in Ito's laboratory, produced rapid and efficient separations of microgram quantities of samples in less than 10 min [55]. A combination of CCC with MS has been effected by interfacing with a thermospray device [56].

2.3.8 Foam CCC centrifuge

Foam separation is based on the unique parameter of foam affinity and/or foam-producing capacity of samples in an aqueous solution, and it is therefore ideal for application to biological samples [57]. Nevertheless, the utility of the method has been extremely limited, mainly due to a lack of efficient instruments. Recently, Ito has developed a foam CCC centrifuge, which provides a strong centrifugal force field to induce a rapid countercurrent movement between the gas and liquid phases through a long, narrow, coiled tube [14]. The instrument is a modified high-speed CCC centrifuge, equipped with a coiled column specially designed for performing dual or true CCC (see Fig. 2.5C). The original apparatus (synchronous type J) has a large revolutional radius of 20 cm, and the column consists of 10-m-long, 2.6-mm-ID PTFE tubes with 50-ml capacity. The coil is equipped with 5 flow channels, as illustrated in Fig. 2.10. A surfactant solution is fed from one end of the coil and collected at the other end, while nitrogen at 80 psi flows from a gas feedline in

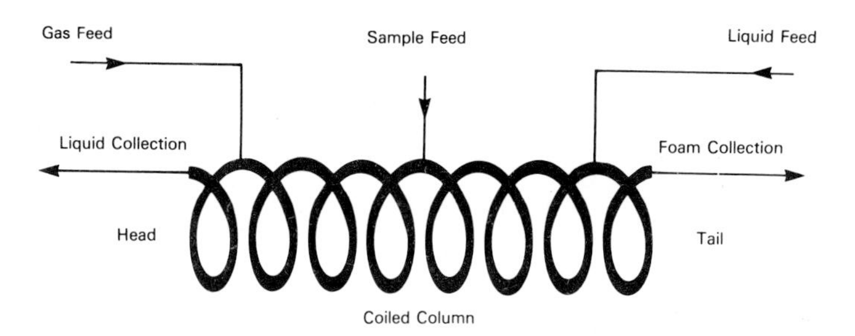

Fig. 2.10. Column design for foam CCC. The coiled column is equipped with 5 flow lines, as shown. Surfactant solution is introduced through the tail, and nitrogen is simultaneously fed through the head of the coil, while the coil is rotated at 500 rpm. Then, foam is generated by the countercurrent process between the surfactant solution and nitrogen through the narrow opening of the coil. A sample mixture, injected through the sample feedline at the middle portion of the coil, is separated according to the foam affinity: molecules adsorbed on the foam are carried with the foaming stream and quickly collected through the foam collection line, while the remainder is carried with the liquid stream in the opposite direction and collected through the liquid collection line. Samples having foam-producing capacity can be separated without the use of surfactants.

References on p. A105

the opposite direction through the coil and is discharged at the foam collection line. The sample solution, introduced near the middle of the coil, either batchwise or continuously, is fractionated, and the components are collected with the foam. The capability of this method had been demonstrated in the separation of various dyes by ionic surfactants [14,15] and in the separation of proteins with a phosphate buffer solution [14]. More recently, bacitracin components have been separated according to their hydrophobicity with nitrogen and distilled water [58,59]. In both protein and bacitracin separations, the solutes provided the surface activity so that no additional surfactants were needed.

2.3.9 Cross-axis synchronous flow-through coil planet centrifuge (X-axis coil planet centrifuge)

The cross-axis synchronous flow-through coil planet centrifuge was developed for performing gram-quantity preparative separations. It is based on the synchronous planetary motion, type X, as illustrated in Fig. 2.3. The first prototype kept a column holder and a counterweight holder symmetrically at a distance of 10 cm from the main axis of the centrifuge [10,21]. In the most recent model, a pair of large column holders are kept, one on each side of the rotary frame, in the lateral positions, shown in Fig. 2.11 [26]. A series of studies has shown that the lateral coil position can yield higher retention of the stationary phase in the column than the central coil position, as in the first prototype. The X-axis

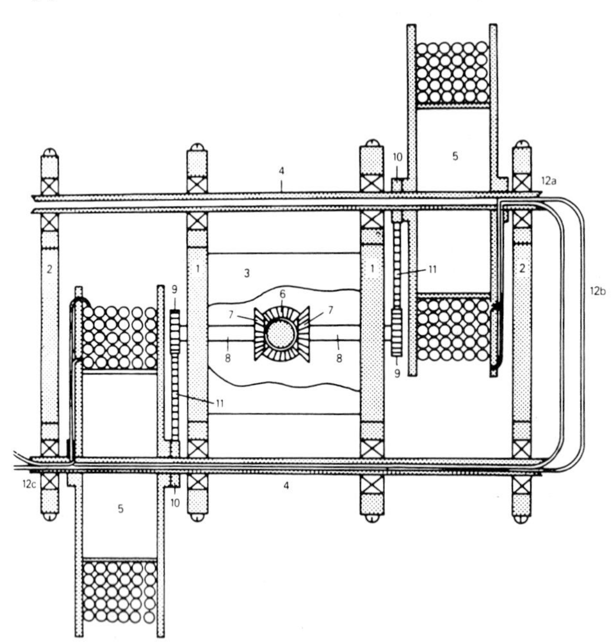

Fig. 2.11. Design of the improved cross-axis synchronous flow-through coil planet centrifuge (horizontal cross section). Two multilayer coils are serially connected without twisting flow tubes. 1, 2 = Side plates; 3 = horizontal plate; 4 = column holder shafts; 5 = column holders; 6 = stationary miter gear; 7 = planetary miter gears; 8 = countershafts; 9, 10 = toothed pulleys; 11 = toothed belts; 12a-c: flowtubes.

coil planet centrifuge has a unique capability of retaining a large volume of the stationary phase against heavy sample loads. The preparative capability of the method has been successfully demonstrated in multigram (2-10 g) separations of various biological samples, such as DNP amino acids [23-25,27], peptides [23,25], auxins [24,27], steroids [24], bacitracin [22], and flavonoids [24]. Fig. 2.12 illustrates a preparative chromatogram of steroid intermediates, obtained with the X-axis coil planet centrifuge [24]. A 2.4-g quantity of a crude reaction mixture was resolved into multiple peaks in 5 h. Five steroids, corresponding to Peaks 1 to 5, were determined by NMR analysis, as indicated in the diagram. The desired products were found at Peak fraction 5 where over 300 mg of crystallized material was recovered in high purity.

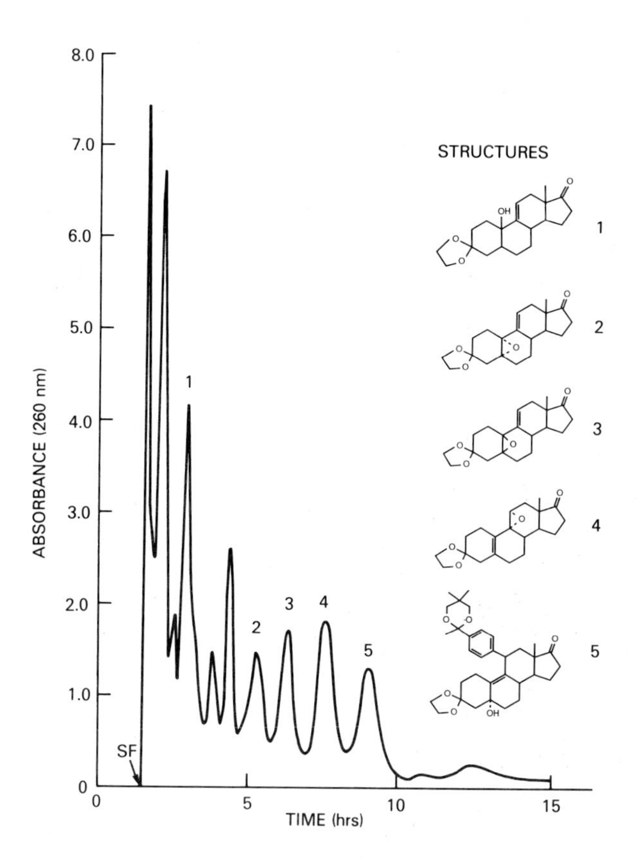

Fig. 2.12. Chromatogram of synthetic steroids, obtained by cross-axis coil planet centrifuge. Fractions corresponding to Peaks 1 to 5 were analyzed by NMR to determine the chemical nature. The desired products were found in Peak fraction 5, where over 300 mg of crystalline materials were obtained in high purity. The experimental conditions were as follows: Apparatus: Cross-axis coil planet centrifuge with a 20-cm revolution radius, a pair of multilayer coils, 300 m x 2.6 mm ID, 1600-ml capacity; sample: crude steroid intermediates 2.4 g in 20 ml solvent; solvent system: *n*-hexane/ethyl acetate/methanol/water (6:5:4:2), lower, aqueous phase mobile; flowrate: 240 ml/h; speed: 450 rpm; retention: 71.3%.

References on p. A105

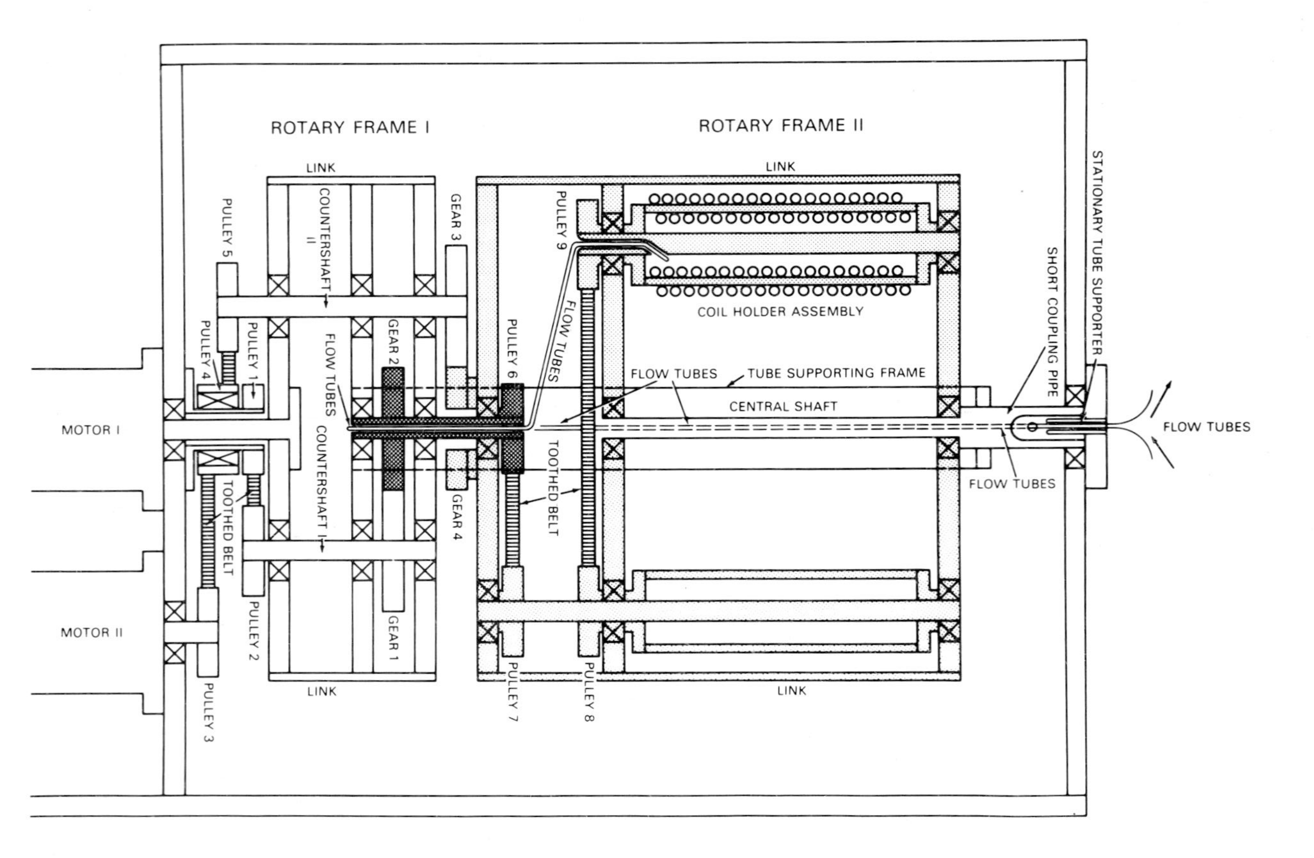

Fig. 2.13. Cross-sectional view of the nonsynchronous flow-through coil planet centrifuge. Motor I produces a high-speed revolution, while Motor II provides slow rotation of the coil assembly around its own axis. The design is based on the nonsynchronous type I planetary motion, illustrated in Fig. 2.3, right bottom.

2.3.10 Nonsynchronous flow-through coil planet centrifuge

The nonsynchronous flow-through coil planet CCC centrifuge system is the most versatile among all CCC systems, but it requires a very elaborate mechanical design. Several models have been constructed with [60] or without [61-63] the use of rotary seals for performing CCC and cell elutriation (separation of cells based on their size and density under a continuous flow of liquid against the applied force field). Among these models, the most successful prototype is based on the nonsynchronous type I planetary motion illustrated in Fig. 2.3. The design of that instrument is shown in Fig. 2.13. The centrifuge is equipped with two motors, one for the revolution of the rotary frame and the other for the rotation of the coil holder around its own axis, which can be regulated independently up to 50 rpm. The column holder, mounted at a distance of ca. 14 cm from the central axis of the centrifuge, holds a set of serially connected coiled columns (1 mm ID, 15-ml capacity) symmetrically around and parallel to the holder axis. The capability of the apparatus was demonstrated in the elutriation of mammalian erythrocytes and rat hepatic cells with physiological saline solutions and also in the partition of macromolecules and cells with polymer-phase systems, represented by the isolation of *E. coli* plasmid DNA and the resolution of *Salmonella typhimurium* mutant bacteria, coated with short and long lipo-polysaccharide chains [62,63].

2.4 SELECTION OF TWO-PHASE SOLVENT SYSTEMS

The first and most important step in designing a CCC separation is to select a two-phase solvent system suitable for the separation of the desired compounds. The basic requirements for the solvent selection are as follows:

1. It must not alter the sample;
2. It must provide adequate sample solubility;
3. It should give suitable partition coefficient values for the components to be resolved;
4. It must have a stationary phase that is adequately retained.

Among these requirements, the first three are shared with conventional liquid chromatography while the third and fourth requirements are of particular importance to CCC technology.

2.4.1 Partition coefficient

The partition coefficient (K) is the ratio of the solute concentration between two immiscible solvent phases which have been mutually equilibrated. Here, K may be defined in two different ways: as $K(C_U/C_L)$, the solute concentration in the upper phase divided by that in the lower phase, and as $K(C_m/C_s)$, the solute concentration in the mobile phase divided by that in the stationary phase. In CCC, as in the countercurrent distribution method, the most efficient separation usually occurs when $K(C_m/C_s) = 1$. However, in some CCC schemes, such as the toroidal coil centrifuge and the horizontal flow-through coil planet centrifuge (in which the retention of the stationary phase is considerably less

References on p. A105

than 50% of the total column capacity), the best results are obtained with the lower $K(C_m/C_s)$ values of 0.3-0.5.

The partition coefficients can be determined by various means. When a pure standard sample is available, $K(C_U/C_L)$ can be determined by simple spectrophotometric measurement of the solute absorbance in the upper and lower phases. However, the sample often contains various impurities or components which interfere with the determination of individual compounds. In this case, the partition coefficient of each component can be conveniently determined by HPLC or TLC analysis of aliquots of the upper and the lower phases, equilibrated with the sample [65].

In CCC, the partition coefficients are conveniently used to predict the retention volume (or time) of solutes according to the formula

$$R = R_{SF} + (V_C - R_{SF})/K(C_m/C_s) \tag{2.1}$$

where R is the retention volume of the peak maximum (distance between the starting point and the peak maximum); R_{SF}, the retention volume of the solvent front (distance from the starting point to the solvent front); V_C, the total column capacity or the void volume (distance from the starting point to the point where the total column capacity of the mobile phase is eluted); and $K(C_m/C_s)$, the partition coefficient, expressed as the solute concentration in the mobile phase divided by that in the stationary phase. It is interesting to note that the solute with $K = 1$ is always eluted at V_C, regardless of the volume of the stationary phase retained in the column (or location of the solvent front).

By transforming Eqn. 2.1, the partition coefficient of each peak can be computed from the chromatogram

$$K(C_m/C_s) = (V_C - R_{SF})/(R - R_{SF}) \tag{2.2}$$

The partition coefficient $K(C_m/C_s)$ is a useful parameter for predicting the resolution between two adjacent peaks, provided that the column efficiency (N), total column capacity (V_C), and retention volume of the solvent front (R_{SF}) are known, i.e.,

$$R_S = 0.5\,N\,(\alpha - 1)/[(\alpha + 1) + 2K_1R_{SF}/(V_C - R_{SF})] \tag{2.3}$$

where R_S is the peak resolution ($R_S = 1$ corresponding to 4σ and $R_S = 1.5$ to baseline separation); N, the column efficiency expressed in terms of theoretical plates (see Eqn. 2.4); α, the ratio of partition coefficients between the two adjacent peaks or K_1/K_2, $K_1 > K_2$. The column efficiency, N, is computed from the chromatographic formula

$$N = (4R/W)^2 \tag{2.4}$$

where R is the retention volume of the peak maximum, and W is the peak width expressed in the same units as R. The derivation of Eqn. 2.3 and a more detailed description of peak resolution are given elsewhere [66].

2.4.2 Retention of stationary phase

One of the essential requirements in CCC is that a sufficient amount of the stationary phase, either upper or lower phase, be retained in the open column, as the mobile phase continuously percolates through the column. In CCC, the amount of the stationary phase retained in the separation column greatly influences the resolution of solute peaks. In general, the greater the volume of the stationary phase retained in the column, the higher the peak resolution achieved [66].

The degree of stationary-phase retention varies with the CCC schemes and also depends on various factors, such as the physical properties of the two-phase solvent system, flowrate of the mobile phase, and applied centrifugal force field. In droplet CCC, where the separation is performed in a stationary column, a large density difference between the two solvent phases becomes the predominant factor for the retention of the stationary phase, and the majority of the successfully performed separations were with chloroform solvent systems [30]. In other CCC schemes, various types of two-phase solvent systems can be used under optimized experimental conditions.

Among the various centrifugal CCC systems, the toroidal coil centrifuge, the centrifugal droplet CCC (centrifugal partition chromatography), and the horizontal flow-through coil planet centrifuge, equipped with eccentric multilayer coils, provide universal applications of two-phase solvent systems, including the aqueous/aqueous polymer phase systems with an extremely low interfacial tension. For these CCC systems, satisfactory stationary-phase retention of polymer phase systems can be attained simply by applying a slow flow of the mobile phase in a strong centrifugal force field. In the nonsynchronous flow-through coil planet centrifuge system, the aqueous/aqueous polymer phase systems can be retained in the column under optimized operational conditions of a low flowrate, slow coil rotation, and high revolutional speed. In other hydrodynamic CCC systems, such as the high-speed CCC centrifuge and the X-axis coil planet centrifuge, the use of two-phase solvent systems has so far been limited to the conventional organic/aqueous solvent systems, which provide a reasonably short settling time of less than 30 sec, as measured in a 5-ml-capacity graduated cylinder [13].

2.4.3 Search for two-phase solvent systems

Success in CCC separation largely depends upon choosing the two-phase solvent system that provides the proper partition coefficient values for the desired compounds and satisfactory retention of the stationary phase. When the nature of the sample to be separated is known, one may find a suitable solvent system by searching the literature for solvent systems that have been successfully applied to similar compounds. A variety of solvent systems that have been used for the separation of a broad spectrum of compounds are listed in Table 2.3 (for earlier application data, see Refs. 98-100).

If the nature of the sample is unknown or previous data for similar compounds are not available, the search for a suitable solvent systems must rely on a tedious trial-and-error method. However, one can save a considerable amount of time and effort by carrying out

TABLE 2.3

RECENT APPLICATIONS OF CCC

Sample[a]	Wt.	Solvent system[b] (volume ratio)	Elution time	Apparatus[c]	Ref.
Amino acids, peptides, proteins					
DNP amino acids	2-30 mg	$CHCl_3$/AcOH/0.1 *M* HCl (2:2:1)	3-6 h	ACPC	20
DNP amino acids	0.1-2 g	$CHCl_3$/AcOH/0.1 *M* HCl (2:2:1)	6 h	XCPC	21
DNP amino acids	2-5 g	$CHCl_3$/AcOH/0.1 *M* HCl (2:2:1)	30 h	XCPC	23
DNP amino acids	4 g	$CHCl_3$/AcOH/0.1 *M* HCl (2:2:1)	22-30 h	XCPC	24
DNP amino acids	10 g	$CHCl_3$/AcOH/0.1 *M* HCl (2:2:1)	35-50 h	XCPC	25
DNP amino acids	10-250 mg	$CHCl_3$/AcOH/0.1 *M* HCl (2:2:1)	3.3 h	MLCPC	51
DNP amino acids	4 g	$CHCl_3$/AcOH/0.1 *M* HCl (2:2:1)	24 h	XCPC	27
DNP amino acids	10 mg	$CHCl_3$/AcOH/0.1 *M* HCl (2:2:1)	3 h	MLCPC	53
DNP amino acids	4.2 g	$CHCl_3$/AcOH/0.1 *M* HCl (2:2:1)	7.5 h	MLCPC	64
Dipeptides	0.1-1 g	*n*-BuOH/DCA/0.1 *M* HCO_2NH_4 (1:0.01:1)→(1:1)	8 h	XCPC	21
Trp-Leu, Val-Tyr	50 mg	*n*-BuOH/AcOH/H_2O (4:1:5)	3-8 h	MLCPC,HCPC	67
Trp-Tyr, Val-Tyr	1.3 g	*n*-BuOH/AcOH/H_2O (4:1:5)	20 h	XCPC	23
Trp-Tyr, Leu-Tyr, Val-Tyr	2.5 g	*n*-BuOH/AcOH/H_2O (4:1:5)	30 h	XCPC	25
Dipeptides	30 mg	*n*-BuOH/AcOH/H_2O (4:1:5)	5 h	HCPC	68
Pentadecapeptides	1 g	*n*-BuOH/AcOH/H_2O (4:1:5)	1 h	HCPC	69
Bombesin-antagonist analogs	95-400 mg	$CHCl_3$/AcOH/H_2O (2:2:1)	2 h	HCPC	69
CCK analog	100 mg	$CHCl_3$/AcOH/H_2O (2:2:1)	1 h	MLCPC	69
CCK fragments	225-500 mg	*n*-BuOH/AcOH/H_2O (4:1:5)	3-9 h	HCPC	68
Bovine insulin	100 mg	*n*-BuOH/0.5 and 2% DCA (1:1)	3-7 h	MLCPC,HCPC	67
Bovine insulin	100 μg	1 m*M* SDS/N_2	10 min	Foam CCC	15
Bovine insulin	100 mg	*sec*-BuOH/1% DCA (1:1)	5 h	MSMSC	70
Cytochrome *c*, lysozyme	200 mg	5%(w/w)PEG8000, 5%(w/w)PEG1000, 10%K_2HPO_4	5.4 h	MSMSC	70
Cytochrome *c*, lysozyme	200 mg	12%(w/w)PEG1000, 12%(w/w)K_2HPO_4 in H_2O	5 h	HCPC	71
Endogenous modulator of Ca^{2+}	150-200 μg	*n*-BuOH/AcOH/H_2O (4:1:5)	6 h	HCPC	72

Antibiotics					
Antitrypanosomal factor	59-175 mg	*n*-BuOH/MeOH/satd. NH_4HCO_3 sol. (7:3:10)	3-4 h	MLCPC	73
Bacitracin	5 g	$CHCl_3$/95%EtOH/H_2O (5:4:3)	28 h	XCPC	27
Bacitracin	100 mg	$CHCl_3$/95%EtOH/H_2O (5:4:3)	2 h	MLCPC	52
Bacitracin	100 mg	$CHCl_3$/95%EtOH/H_2O (5:4:3)	3 h	MLCPC	53
Bacitracin	5 mg	H_2O/N_2	11 min	Foam CCC	57
Tetracyclines	50 mg	EtOAc/*n*-BuOH/0.25 *M* NH_4OAc (1:1:2)	1.2-3.2 h	MLCPC	52,53
Sporaviridin	15 mg	*n*-BuOH/Et_2O/H_2O (5:2:5)	3.5 h	MLCPC	74
2-Norerythromycins	0.5 g	*n*-hep/C_6H_6/Me_2CO/2-PrOH/0.01 *M* citrate buffer (pH 6.3) (5:10:2:3:5)	–	MLCPC	75
2-Norerythromycins	280-392 mg	CCl_4/MeOH/0.01 *M* KP (pH 7.0) (1:1:1)	–	MLCPC	75
2-Norerythromycins	342 mg	CCl_4/MeOH/0.01 *M* KP (pH 7.3) (1:1:1)	–	MLCPC	75
2-Norerythromycins	256 mg	*n*-hex/EtOAc/0.01 *M* KP (pH 7.5) (1:1:1)	–	MLCPC	75
2-Norerythromycins	172 mg	$CHCl_3$/MeOH/0.01 *M* citrate buffer (pH 6.0) (1:1:1)	–	MLCPC	75
Nidamycins	200 mg	CCl_4/MeOH/0.01 *M* KP (pH 7.0) (2:3:2)	–	MLCPC	75
Tiacumycins	200 mg	CCl_4/$CHCl_3$/MeOH/H_2O (7:3:7:3)	–	MLCPC	75
Coloradocin	400 mg	$CHCl_3$/MeOH/H_2O (1:1:1)	–	MLCPC	75
Medicinal herbs					
Flavonoids from *Hippophae rhamnoides*	3 mg	$CHCl_3$/MeOH/H_2O (4:3:2)	15-90 min	MLCPC	54
Flavonoids from *Hippophae rhamnoides*	100 mg	$CHCl_3$/MeOH/H_2O (4:3:2)	20 h	XCPC	24
Flavonoids from *Hippophae rhamnoides*	120 μg	$CHCl_3$/MeOH/H_2O (4:3:2)	8 min	MLCPC	55
Flavonoids from *Hippophae rhamnoides*	100 mg	$CHCl_3$/MeOH/H_2O (4:3:2)	1.8 h	MLCPC	52
Flavonoids from *Hippophae rhamnoides*	100 mg	$CHCl_3$/MeOH/H_2O (4:3:2)	2.8 h	MLCPC	53
Flavonoids from *Hippophae rhamnoides*	20 mg	$CHCl_3$/MeOH/H_2O (4:3:2)	4.5 h	HCPC	76
Vincamine, vincine	40 μg	*n*-hex/EtOH/H_2O (6:5:5)	45 min	MLCPC	56
3′-Hydroxygenkwanin, apigenin, luteolin	30 mg	$CHCl_3$/MeOH/H_2O (4:3:2)	2 h	MLCPC	77
Alkaloids from *Anisodus tangulicus* (Maxin) Pasch	400 mg	$CHCl_3$/0.07 *M* NaP (pH 6.4) (1:1)	3 h	MLCPC	77

(Continued on p. A98)

TABLE 2.3 (continued)

Sample[a]	Wt.	Solvent system[b] (volume ratio)	Elution time	Apparatus[c]	Ref.
Tetrandrine, fangchinoline, cyclanoline from *Stephania tetrandra* S. Moore	3 mg	*n*-hex/EtOAc/MeOH/H_2O (1:1:1:1)	1.8 h	MLCPC	78
	3 mg	*n*-hex/EtOAc/MeOH/H_2O (3:7:5:5)	1.5 h	MLCPC	78
Hydroxyanthraquinones from *Rheum palmatum* L.	1 mg	*n*-hex/EtOAc/MeOH/H_2O (9:1:5:5)	1.3 h	MLCPC	79
Lignans from *Schisandra rubriflora* Rhed et Wils	125 mg	*n*-hex/EtOAc/MeOH/H_2O (10:5:5:1)	1 h	Dual CCC	80
	–	*n*-hex/EtOAc/MeOH/H_2O (6:5:5)	2 h	MLCPC	81
	125 μg	*n*-hex/EtOAc/MeOH/H_2O (6:5:5)	1.2 h	MLCPC	82
Triterpenoic acids from *Boswellia carterii*	500 mg	*n*-hex/EtOH/H_2O (6:5:2)	3 h	MLCPC	52
	100 mg	*n*-hex/EtOH/H_2O (6:5:2)	2.8 h	MLCPC	53
Alkaloids from *Sophora flavescens* Ait	10 mg	$CHCl_3$/0.07 *M* NaP (pH 6.4) (1:1)	3.3 h	MLCPC	83
Alkaloids from *Datura mete* L.	10 mg	$CHCl_3$/0.07 *M* NaP (pH 6.5) (1:1)	2.5 h	MLCPC	83
Agrochemicals					
IA, IAA, IBA, IAN	2.5 mg	*n*-hex/EtOAc/MeOH/H_2O (3:7:5:5)	1.5 h	MLCPC	84
IA, IAA, IBA	3 g	*n*-hex/EtOAc/MeOH/H_2O (3:7:5:5)	12 h	XCPC	24
IA, IAA, IBA	3 mg	*n*-hex/EtOAc/MeOH/H_2O (3:7:5:5)	2.5 h	MLCPC	85
IA, IAA, IBA	3 g	*n*-hex/EtOAc/MeOH/H_2O (3:7:5:5)	24 h	XCPC	27
IA, IAA, IBA, IAN	100 mg	*n*-hex/EtOAc/MeOH/H_2O (1:1:1:1)	50 min	MLCPC	52
IA, IAA, IBA, IAN	100 mg	*n*-hex/EtOAc/MeOH/H_2O (1:1:1:1)	2.2 h	MLCPC	53
IA, IAA, IBA	50-200 μg	*n*-hex/EtOAc/MeOH/H_2O (1:1:1:1)	8-18 min	MLCPC	86
IA, IAA, IBA	50-200 μg	*n*-hex/EtOAc/MeOH/H_2O (1:1:1:1), (4:5:4:5) and (3:5:3:5)	7-36 min	TCC	86
IAA or ABA	100 μg	1 m*M* CPCl/N_2	10 min	Foam CCC	15
s-Triazine herbicides	800 μg	*n*-hex/EtOAc/MeOH/H_2O (8:2:5:5)	1 h	MLCPC	87
Azadirachtin from *Azadiracta indica*	0.1-0.5 g	*n*-hex/EtOAc/MeOH/H_2O (3:5:3:5)	2-3 h	MLCPC	88
	5 g	*n*-hex-EtOAc/MeOH/H_2O (3:5:3:5)	15 h	XCPC	89

20-Hydroxyecdysone, ajugasterone C from *Vitex madiensis*	2.65 g	$CHCl_3$/MeOH/H_2O (13:7:4)	3 h	DCCC	90
Phytoecdysteroids from *Vitex thyrsflora*	2.05 g	$CHCl_3$/MeOH/H_2O (13:7:4)	2.5 h	DCCC	90
Ecdysone, 20-hydroxyecdysone from *Bombyx mori*	0.065 g	$CHCl_3$/MeOH/H_2O (13:7:4)	5 h	DCCC	90
Pigments					
Apocarotinoids from *Cochlospermum tinctorium*	500 mg	CCl_4/MeOH/H_2O (5:4:1)	2 h	MLCPC	91
Azo dyes	0.2-0.5 g	*n*-BuOH/pyridine/H_2O (5.0:3.5:1.5)	8 h	MLCPC	92
Rhodamine B, Evans Blue	250 μg	1 m*M* SDS, 5 m*M* NaCl/N_2	5 min	Foam CCC	16
Rhodamine B, Methylene Blue	250 μg	0.5 m*M* SDS, 0.5 *M* Na_2HPO_4/N_2	75 min	Foam CCC	16
Acidic and basic dyes	100 μg	1 m*M* SDS or 1 m*M* CPCl/N_2	3-10 min	Foam CCC	15
Lipids					
Steroid intermediates	250 mg	*n*-hex/EtOH/H_2O (6:5:4)	1.5 h	Dual CCC	17
Steroids	120-225 mg	*n*-hex/EtOAc/MeOH/H_2O (6:5:5:5)	2.5-4 h	Dual CCC	17
Steroid reaction mixture	2.4 g	*n*-hex/EtOAc/MeOH/H_2O (6:5:4:2)	15 h	XCPC	24
Unsaturated fatty acids	300 g	*n*-hex/MeCN (1:1)	70 min	CDCCC	93
Other organic compounds					
2-(2'-Hydroxyethoxy)terephthalic acid	2.36 g	$CHCl_3$/MeOH/H_2O (37:37:26)	2 h	MLCPC	94
Naphthalene, benzophenone, *o*-nitrophenol	–	*n*-hex/MeOH/H_2O (15:10:8)	56 min	MLCPC	95
Naphthalene, benzophenone, *o*-nitrophenol, acetophenone	–	*n*-hex/MeOH/H_2O (3:3:2)	1-2 h	MLCPC	95
Acetophenone, benzoic acid, *p*-nitrophenone, phenol	–	$CHCl_3$/MeOH/HCl (pH 2) (3:1:3)	1-2 h	MLCPC	95
Inorganic elements					
Rare-earth elements	3-5 mg	0.1 *M* DEHPA in hep/0.03-0.15 *M* HCl (1:1)	3-6 h	CDCCC	96
Rare-earth elements	150 μg	0.02 *M* EHPA in kerosene/0.1 *M* (H,Na)DCA in 20% EG-H_2O (pH 1.35-2.5)	5.5 h	CDCCC	97
$LaCl_3$, $PrCl_3$, $NdCl_3$	75 μg	0.02 *M* DEHPA in *n*-hep/0.02 *M* HCl (1:1)	1.5 h	MLCPC	53

(Continued on p. A100)

TABLE 2.3 (continued)

Abbreviations:

[a]ABA, abscisic acid; IA, indole-3-acetamide; IAA, indole-3-acetic acid; IAN, indole-3-acetonitrile; IBA, indole-3-butyric acid; CCK, cholecystokinin; DNP, dinitrophenyl.

[b]AcOH, acetic acid; BuOH, butanol; CPCl, cetylpyridinium chloride; DCA, dichloroacetic acid; DEHPA, di(2-ethylhexyl)phosphoric acid; EHPA, 2-ethylhexylphosphoric acid; EG, ethylene glycol; EtOAc, ethyl acetate; EtOH, ethanol; Et_2O, diethyl ether; hep, heptane; hex, hexane; Me_2CO, acetone; MeOH, methanol; KP, potassium phosphate; NaP, sodium phosphate; PEG, polyethylene glycol; PrOH, propanol; SDS, sodium dodecyl sulfate.

[c]CPC, coil planet centrifuge; ACPC, angle rotor CPC; CDCCC, centrifugal droplet countercurrent chromatograph; DCCC, droplet countercurrent chromatograph; HCPC, horizontal flow-through CPC; MLCPC, multilayer CPC; MSMSC, multistage mixer-settler centrifuge; TCC, toroidal coil centrifuge; XCPC, cross-axis synchronous flow-through CPC.

a systematic search, as illustrated in Table 2.4, where various nonionic two-phase solvent systems are arranged according to the hydrophobicity of the nonaqueous phase. All listed two-phase solvent systems provide nearly equal volumes of the upper and the lower phases and reasonably short settling times, so that they can be applied to all centrifugal CCC schemes, including high-speed CCC.

The search for a solvent system may begin with the chloroform solvent system shown on the left in Table 2.4. When the partition coefficient value, $K(C_U/C_L)$, in chloroform/methanol/water (2:1:1) falls somewhere between 0.2 and 5, the desirable K value may be obtained by further modifying the volume ratio of each component, substituting acetic acid for methanol, and/or partially replacing chloroform by tetrachloromethane or dichloromethane. If the sample is more unevenly partitioned into one of the phases, the chloroform solvent systems may be inadequate, and the search must be directed to other solvent systems which provide broader ranges of hydrophobicity and polarity, as illustrated on the right in Table 2.4.

When the sample is mostly distributed into the lower (nonaqueous) phase of the chloroform solvent system, a slightly more hydrophobic solvent system of *n*-hexane ethyl acetate/methanol/water (1:1:1:1) should be tested, as indicated by an upward arrow in the table. The further search should be directed upward, if the sample is still largely distributed into the upper (nonaqueous) phase, or downward if the sample is more concentrated in the lower (aqueous) phase. If the continued search leads to the top solvent system of *n*-hexane/methanol/water (2:1:1) and indicates the need for a more hydrophobic solvent system, the solvent composition can be further modified by reducing the volume of water and/or replacing methanol by ethanol. Some useful solvent systems for extremely nonpolar compounds are *n*-hexane/ethanol/water (6:5:2) and *n*-hexane/methanol (2:1). Various nonaqueous/nonaqueous solvent systems have been successfully used for separations of nonpolar compounds and/or compounds that are unstable in aqueous solutions [100].

On the other hand, if the sample is mostly distributed into the upper aqueous phase of the chloroform solvent system, the search should be directed in the opposite direction, toward the polar solvent systems, as indicated by the downward arrow. If the most polar solvent system of *n*-butanol/water listed at the bottom of the table still partitions the sample largely to the lower aqueous phase, the *n*-butanol solvent system may be modified by adding a small amount of an acid and/or neutral salt. The most commonly used among these modified solvent systems are *n*-butanol/acetic acid/water (4:1:5), *n*-butanol/trifluoroacetic acid/water (1:0.001-0.01:1), and *n*-butanol/0.2 *M* ammonium acetate (1:1), all of which have been extensively used for the separation of polar peptides [44]. Even more polar organic/aqueous solvent systems include ethanol/30% ammonium sulfate (1:2), ethyl or butyl cellosolve/phosphate aqueous solution [101], etc.

Partitioning of macromolecules and cell particles can be performed with a variety of aqueous/aqueous polymer phase systems [102]. Among the various polymer phase systems available, the following two types are the most versatile for performing CCC: The polyethyleneglycol (PEG)/potassium phosphate systems provide convenient means of adjusting the partition coefficient of macromolecules by changing the molecular weight of

References on p. A105

TABLE 2.4

SEARCH FOR A SUITABLE TWO-PHASE SOLVENT SYSTEM

	n-Hexane	/	EtOAc	/	MeOH	/	*n*-BuOH	/	H_2O	
	10		0		5		0		5	← HYDROPHOBIC
	9		1		5		0		5	
	8		2		5		0		5	
	7		3		5		0		5	
	6		4		5		0		5	
	5		5		5		0		5	
	4		5		4		0		5	
$CHCl_3$/MeOH/H_2O (2:1:1)	3		5		3		0		5	
	2		5		2		0		5	
	1		5		1		0		5	
	0		5		0		0		5	
	0		4		0		1		5	
	0		3		0		2		5	
	0		2		0		3		5	
	0		1		0		4		5	
	0		0		0		5		5	POLAR →

PEG and/or pH of the phosphate buffer. The PEG 6000 (3.5-4%, w/w)/Dextran 500 (5%, w/w) systems provide a physiological environment, suitable for separation of mammalian cells by optimizing osmolarity and pH with electrolytes.

On many occasions, the partition behavior of ionizable compounds is sensitively affected by the pH of the aqueous phase. For separating such compounds, the desirable partition coefficient values with high selectivity can be obtained by manipulating the pH of the solvent system. The solvent systems listed in Table 2.4 may be modified by adding an acid (HCl, acetic acid, etc.), base (NaOH, NH_4OH, etc.) or buffer solution (phosphate, borate, citrate, etc.) depending on the range of the pH desired (see Table 2.3 for examples).

2.5 SEPARATION PROCEDURE

2.5.1 Preparation of two-phase solvent systems and sample solution

CCC utilizes two immiscible solvent phases, which are mutually equilibrated at room temperature. The solvent mixture is delivered into a separatory funnel and thoroughly equilibrated by repeated shaking and degassing, and the two phases are separated shortly before use.

The sample solution is prepared by dissolving the sample in the upper and/or lower phase of the two-phase solvent system used for the separation. If a relatively small amount of the sample is to be separated, it may be dissolved in the stationary phase to preserve the resolution of the early-eluted peaks [103]. However, when a large sample contains both polar and nonpolar components, it may require a large volume of solvent to dissolve all components in one phase. In this case, it is advantageous to dissolve the sample in both phases to reduce the volume of the sample solution. The use of the two solvent phases for sample solution also prevents the formation of a single phase in the sample compartment at the beginning of the separation column, which would cause a detrimental loss of stationary phase from the column. When the two-phase mixture is used as a sample diluent, the formation of a single phase is readily observed in the sample bottle, and this can be corrected by further diluting the sample solution with the two solvent phases until two phases are formed.

2.5.2 Elution

The CCC separation is initiated by filling the separation column entirely with the stationary phase without trapping air in the column. This is followed by the injection of the sample through the sample port. (If desired, the column may be equilibrated with the mobile phase before sample charge, as in HPLC.) Then, the mobile phase is passed through the column, while the apparatus (excluding droplet CCC) is rotated at the desired speed. The percolation of the mobile phase should be performed in a suitable mode according to the hydrodynamic trend (see below) of each CCC scheme to establish retention of the stationary phase in the column. In the nonhelical hydrostatic CCC

References on p. A105

schemes, such as droplet CCC, centrifugal droplet CCC (centrifugal partition chromatography), and locular CCC, the upper phase is passed through each column unit in the ascending mode and the lower phase in the descending mode with respect to gravity or to the applied centrifugal force field. In the hydrodynamic CCC schemes, such as high-speed CCC and X-axis coil planet centrifuge, usually the upper phase should be passed through the coil in the tail-to-head elution mode and the lower phase in the head-to-tail elution mode, whereas the type I synchronous and nonsynchronous coil planet centrifuge requires head-to-tail elution, regardless of the choice of the mobile phase. The X-axis coil planet centrifuge with a lateral coil orientation (type X-L) requires the application of a proper mode of planetary motion in addition to the head-tail elution mode, as described elsewhere [22].

In the hydrodynamic CCC schemes, countercurrent flow of the two phases through the rotating coil creates a pressure gradient along the length of the coil in such a way that the pressure on the head side becomes higher than that on the tail side. Consequently, the elution of the mobile phase from the tail toward the head may produce a negative pressure at the inlet of the column, resulting in an increased flowrate of the mobile phase due to an extra amount of solvent that is sucked from the reservoir through the metering pump, equipped with one-way check valves. This adverse phenomenon is enhanced by several factors, such as a large density difference between the two solvent phases, application of high revolutional speed, and slow flowrate. However, this problem can be easily controlled by restricting the flow through the column with a piece of narrow-bore tubing, typically 50-100 cm x 0.5 mm ID, inserted at the outlet of the detector [98].

The CCC separation also permits the use of stepwise and gradient elutions, as in other liquid chromatographic systems. However, the method requires a particular choice of solvent system and mobile phase to maintain a steady-state hydrodynamic phase equilibrium in the running column: The key element producing the gradient (such as acid or neutral salt) should be partitioned almost entirely to the mobile phase and at the same time should not significantly alter the volume ratio of the two solvent phases in the column. Solvent systems, such as *n*-butanol/aq. solution for a gradient of dichloroacetic acid or trifluoroacetic acid concentration and *n*-butanol/phosphate buffer for a pH gradient will satisfy the above requirements, if the lower (aqueous) phase is used as the mobile phase [98].

Elution can be performed at an elevated temperature to increase the partition efficiency. This method is particularly effective for separations of polar compounds, such as peptides with viscous butanol solvent systems [13,104]. For this purpose, the apparatus should be equipped with a temperature-regulating system, and the two-phase solvent system should be pre-equilibrated at the temperature to be applied during the separation.

2.5.3 Detection

The effluent from the CCC column can be continuously monitored with various detectors. As in liquid chromatography, a UV monitor is most commonly used as a CCC detector. In order not to trap the stationary phase droplets, it is advisable to use a flowcell with a straight vertical flowpath, where the mobile phase passes downward, if it is the upper phase, and passes upward, if it is the lower phase. In the former case, the flowcell tends to trap gas bubbles that are generated in the mobile phase under reduced pressure. Formation of gas bubbles in the flowpath can be prevented in most cases by thoroughly degassing the solvent, as in HPLC, and/or applying a fine tube at the outlet of the monitor to create a backpressure. Restricting the flow with a fine tube also prevents the elution of an excess amount of the mobile phase from the reservoir in the tail-to-head elution mode in high-speed CCC, as described earlier. In thermolabile solvent systems, such as the various chloroform/methanol/water systems, a subtle change of temperature produces a turbidity of the mobile phase, which causes intense noise with an upward shift of the baseline in the UV tracing. This problem can be avoided by heating the effluent to 30°C at the inlet of the monitor [105].

In addition to the UV and visual-wavelength detectors, the CCC effluent can be directly monitored by other combinations, such as CCC/FTIR [106,107] and CCC/MS [56]. CCC/FTIR provides advantages over LC/FTIR in that it gives higher concentrations of solute and less background absorbance of IR in the nonaqueous mobile phase. In CCC/MS, the thermospray capillary tube at the interface tends to create a high backpressure and, therefore, the effluent from the CCC column may have to be split into two streams, one pumped into the mass spectrometer and the other one passed through the UV monitor and then fractionated or discarded.

REFERENCES

1 Y. Ito and R.L. Bowman, *Science,* 167 (1970) 281.
2 Y. Ito and R.L. Bowman, *J. Chromatogr. Sci.,* 8 (1970) 315.
3 Y. Ito, *J. Biochem. Biophys. Methods,* 5 (1981) 105.
4 Y. Ito, *J. Liq. Chromatogr.,* 11 (1988) 1.
5 T. Tanimura, J.J. Pisano, Y. Ito and R.L. Bowman, *Science,* 169 (1970) 54.
6 W. Murayama, Y. Kobayashi, Y. Kosuge, H. Yano, Y. Nunogaki and K. Nunogaki, *J. Chromatogr.,* 239 (1982) 643.
7 Y. Ito and R.L. Bowman, *Science,* 173 (1971) 420.
8 Y. Ito, J. Suaudeau and R.L. Bowman, *Science,* 189 (1975) 999.
9 Y. Ito, *J. Chromatogr.,* 358 (1986) 313.
10 Y. Ito, *Sep. Sci. Technol.,* 22 (1987) 1971.
11 Y. Ito, *J. Chromatogr.,* 301 (1984) 377.
12 Y. Ito, *J. Chromatogr.,* 301 (1984) 387.
13 Y. Ito and W.D. Conway, *J. Chromatogr.,* 301 (1984) 405.
14 Y. Ito, *J. Liq. Chromatogr.,* 8 (1985) 2131.
15 M. Bhatnagar and Y. Ito, *J. Liq. Chromatogr.,* 11 (1988) 21.
16 Y. Ito, *J. Chromatogr.,* 403 (1987) 77.
17 Y.-W. Lee, C.E. Cook and Y. Ito, *J. Liq. Chromatogr.,* 11 (1988) 37.
18 W.D. Conway and Y. Ito, *Pittsburgh Conference and Exposition on Analytical Chemistry, 1984,* Abstract 472.
19 Y. Ito, J. Sandlin and W.G. Bowers, *J. Chromatogr.,* 244 (1982) 247.
20 Y. Ito, *J. Chromatogr.,* 358 (1986) 325.
21 Y. Ito, *Sep. Sci. Technol.,* 22 (1987) 1989.
22 Y. Ito and T.-Y. Zhang, *J. Chromatogr.,* 449 (1988) 135.
23 Y. Ito and T.-Y. Zhang, *J. Chromatogr.,* 449 (1988) 153.

24 T.-Y. Zhang, Y.-W. Lee, Q.-C. Fang, C.E. Cook, R. Xiao and Y. Ito, *J. Chromatogr.,* 454 (1988) 185.
25 Y. Ito and T.-Y. Zhang, *J. Chromatogr.,* 455 (1988) 151.
26 Y. Ito, H. Oka and J.L. Slemp, *J. Chromatogr.,* 463 (1989) 305.
27 M. Bhatnagar, H. Oka and Y. Ito, *J. Chromatogr.,* 463 (1989) 317.
28 Y. Ito and R.L. Bowman, *Anal. Biochem.,* 85 (1978) 614.
29 I.A. Sutherland and Y. Ito, *J. High Resolut. Chromatogr. Chromatogr. Commun.,* 10019 (1978) 171.
30 K. Hostettmann, Droplet countercurrent chromatography, in J.C. Giddings, E. Grushka, J. Cazes and P.R. Brown (Editors), *Advances in Chromatography, Vol. 21,* Dekker, New York, 1983, p. 165.
31 F.J. Hanke and I. Kubo, *J. Chromatogr.,* 329 (1985) 395.
32 J.K. Snyder, K. Nakanishi, K. Hostettmann and M. Hostettmann, *J. Liq. Chromatogr.,* 7 (1984) 243.
33 A. Berthod and D.W. Armstrong, *J. Liq. Chromatogr.,* 11 (1988) 547.
34 Y. Ito and R.L. Bowman, *Anal. Biochem.,* 78 (1977) 506.
35 Y. Ito and R.L. Bowman, *J. Chromatogr.,* 136 (1977) 189.
36 Y. Ito, *J. Chromatogr.,* 196 (1980) 295.
37 Y. Ito and R. Bhatnagar, *J. Liq. Chromatogr.,* 7 (1984) 257.
38 Y. Ito and R.L. Bowman, *Anal. Biochem.,* 82 (1977) 63.
39 Y. Ito and R.L. Bowman, *J. Chromatogr.,* 147 (1978) 221.
40 Y. Ito, *Anal. Biochem.,* 100 (1979) 271.
41 Y. Ito, *J. Chromatogr.,* 188 (1980) 33.
42 Y. Ito, *J. Chromatogr.,* 188 (1980) 43.
43 Y. Ito and G.J. Putterman, *J. Chromatogr.,* 193 (1980) 37.
44 M. Knight, Countercurrent chromatography for peptides, in N.B. Mandava and Y. Ito (Editors), *Countercurrent Chromatography,* Dekker, New York, 1988, p. 583.
45 Y. Ito, *Anal. Biochem.,* 102 (1980) 150.
46 Y. Ito, *J. Chromatogr.,* 192 (1980) 75.
47 N.B. Mandava and Y. Ito, *J. Chromatogr.,* 247 (1982) 315.
48 I.A. Sutherland, D. Heywood-Waddington and T.J. Peters, *J. Liq. Chromatogr.,* 8 (1985) 2315.
49 Y. Ito, *J. Chromatogr.,* 214 (1981) 122.
50 Y. Ito and F.E. Chou, *J. Chromatogr.,* 454 (1988) 382.
51 Y. Ito, H. Oka and J.L. Slemp, *J. Chromatogr.,* 475 (1989) 219.
52 Y. Ito, H. Oka and Y.-W. Lee, *J. Chromatogr.,* 498 (1990) 169.
53 Y. Ito, H. Oka, E. Kitazume, M. Bhatnagar and Y.-W. Lee, *J. Liq. Chromatogr.,* 13 (1990) 2329.
54 T.-Y. Zhang, R. Xiao, Z.-Y. Xiao, L.K. Pannell and Y. Ito, *J. Chromatogr.,* 455 (1988) 199.
55 H. Oka, F. Oka and Y. Ito, *J. Chromatogr.,* 479 (1989) 53.
56 Y.-W. Lee, R.D. Voyksner, Q.-C. Fang, C.E. Cook and Y. Ito, *J. Liq. Chromatogr.,* 11 (1988) 153.
57 P. Somasundaran, *Sep. Purif. Methods,* 1 (1972) 117.
58 H. Oka, K. Harada, M. Suzuki, H. Nakazawa and Y. Ito, *Anal. Chem.,* 61 (1989) 1998.
59 H. Oka, K. Harada, M. Suzuki, H. Nakazawa and Y. Ito, *J. Chromatogr.,* 482 (1989) 197.
60 Y. Ito, P. Carmeci and I.A. Sutherland, *Anal. Biochem.,* 94 (1979) 249.
61 Y. Ito, P. Carmeci, R. Bhatnagar, S. Leighton and R. Seldon, *Sep. Sci. Technol.,* 15 (1980) 1589.
62 Y. Ito, G.T. Bramblett, R. Bhatnagar, M. Huberman, L. Leive, L.M. Cullinane and W. Groves, *Sep. Sci. Technol.,* 18 (1983) 33.
63 L. Leive, L.M. Cullinane, Y. Ito and G.T. Bramblett, *J. Liq. Chromatogr.,* 7 (1984) 403.
64 Y. Ito, E. Kitazume and J.L. Slemp, *J. Chromatogr.,* in press.
65 W.D. Conway and Y. Ito, *J. Liq. Chromatogr.,* 7 (1984) 291.
66 W.D. Conway and Y. Ito, *J. Liq. Chromatogr.,* 8 (1985) 2195.
67 J.L. Sandlin and Y. Ito, *J. Liq. Chromatogr.,* 11 (1988) 55.

68 M. Knight and Y. Ito, *J. Chromatogr.,* 484 (1989) 319.
69 M. Knight, J.D. Pineda and T.R. Burke, Jr., *J. Liq. Chromatogr.,* 11 (1988) 119.
70 Y. Ito and T.-Y. Zhang, *J. Chromatogr.,* 437 (1988) 121.
71 Y. Ito and H. Oka, *J. Chromatogr.,* 457 (1988) 393.
72 I. Hanbauer, A.G. Wright, Jr. and Y. Ito, *J. Liq. Chromatogr.,* 13 (1990) 2363.
73 T.I. Marcado, Y. Ito, M.P. Strickler and V.J. Ferrans, *J. Liq. Chromatogr.,* 11 (1988) 203.
74 K. Harada, I. Kimura, A. Yoshikawa, M. Suzuki, H. Nakazawa, S. Hattori, K. Komori and Y. Ito, *J. Liq. Chromatogr.,* 13 (1990) 2373.
75 R.H. Chen, J.E. Hocklowski, J.B. McAlpine and R.R. Rasmussen, *J. Liq. Chromatogr.,* 11 (1988) 191.
76 T.-Y. Zhang, Y. Hua, R. Xiao and S. Kong, *J. Liq. Chromatogr.,* 11 (1988) 233.
77 T.-Y. Zhang, D.-G. Cai and Y. Ito, *J. Chromatogr.,* 435 (1988) 159.
78 T.-Y. Zhang, L.K. Pannell, D.-G. Cai and Y. Ito, *J. Liq. Chromatogr.,* 11 (1988) 1661.
79 T.-Y. Zhang, L.K. Pannell, Q.-L. Pu, D.-G. Cai and Y. Ito, *J. Chromatogr.,* 422 (1988) 445.
80 Y.-W. Lee, Q.-C. Fang, C.E. Cook and Y. Ito, *J. Nat. Prod.,* 52 (1989) 706.
81 Y.-W. Lee, R.D. Voyksner, Q.-C. Fang, T.W. Pack, C.E. Cook and Y. Ito, *Anal. Chem.,* 62 (1990) 244.
82 Y.-W. Lee, Q.-C. Fang, C.E. Cook and Y. Ito, *J. Chromatogr.,* 477 (1989) 434.
83 D.-G. Cai, M.-J. Gu, J.-D. Zhang, G.-P. Zhu, T.-Y. Zhang, N. Li and Y. Ito, *J. Liq. Chromatogr.,* 13 (1990) 2399.
84 Y. Ito and Y.-W. Lee, *J. Chromatogr.,* 931 (1987) 290.
85 Y. Ito and F.E. Chou, *J. Chromatogr.,* 454 (1988) 382.
86 H. Oka, Y. Ikai, N. Kawamura, M. Yamada, K. Harada, M. Suzuki, F.E. Chou, Y.-W. Lee and Y. Ito, *J. Liq. Chromatogr.,* 13 (1990) 2301.
87 Y.-W. Lee, Y. Ito, Q.-C. Fang and C.E. Cook, *J. Liq. Chromatogr.,* 11 (1988) 75.
88 H.E. Hummel, H. Oka and Y. Ito, *Prep-90, 7th International Symposium on Preparative Chromatography, Ghent, Belgium, April 8-11, 1990,* Abstract.
89 H. Oka, H.E. Hummel and Y. Ito, unpublished data.
90 I. Kubo, F.J. Hanke and G.T. Marshall, *J. Liq. Chromatogr.,* 11 (1988) 173.
91 B. Diallo and M. Vanhaelen, *J. Liq. Chromatogr.,* 11 (1988) 227.
92 H.S. Freeman, Z. Hao, S.A. McIntosh and K.P. Mills, *J. Liq. Chromatogr.,* 11 (1988) 251.
93 W. Murayama, Y. Kosuge, N. Nakaya, Y. Nunogaki, K. Nunogaki, J. Cazes and H. Nunogaki, *J. Liq. Chromatogr.,* 11 (1988) 283.
94 Y. Miura, C.A. Panetta and R.M. Metzger, *J. Liq. Chromatogr.,* 11 (1988) 245.
95 R.J. Romañach and J.A. de Haseth, *J. Liq. Chromatogr.,* 11 (1988) 91.
96 T. Araki, T. Okazawa, Y. Kubo, H. Ando and H. Asai, *J. Liq. Chromatogr.,* 11 (1988) 267.
97 K. Akiba, S. Sawai, S. Nakamura and W. Murayama, *J. Liq. Chromatogr.,* 11 (1988) 2517.
98 Y. Ito, *CRC Crit. Rev. Anal. Chem.,* 17 (1986) 65.
99 Appendix, in N.B. Mandava and Y. Ito (Editors), *Countercurrent Chromatography,* Dekker, New York, 1988, pp. 781-795.
100 W.D. Conway and Y. Ito, *LC Magazine,* 2 (1984) 368.
101 R.R. Porter, *Biochem. J.,* 53 (1953) 320.
102 P.Å. Albertsson, *Partition of Cell Particles and Macromolecules,* Wiley-Interscience, New York, 3rd Edn., 1986.
103 J.L. Sandlin and Y. Ito, *J. Liq. Chromatogr.,* 7 (1984) 323.
104 M. Knight, Y. Ito, P. Peters and C. diBello, *J. Liq. Chromatogr.,* 8 (1985) 2281.
105 H. Oka and Y. Ito, *J. Chromatogr.,* 475 (1989) 229.
106 R.J. Romañach, J.A. de Haseth and Y. Ito, *J. Liq. Chromatogr.,* 8 (1985) 2209.
107 R.J. Romañach and J.A. de Haseth, *J. Liq. Chromatogr.,* 11 (1988) 133.

Chapter 3

Planar chromatography

SZABOLCS NYIREDY

CONTENTS

3.1 INTRODUCTION

Planar chromatography (PC) is a collective term including all analytical, micropreparative, and preparative separation methods where the mobile phase moves through the stationary phase (paper or porous sorbent) in a planar arrangement. The movement of compounds to be separated by PC is the result of two opposing forces, the driving force of the solvent system and the retarding action of the stationary phase.

The planar geometry (flat bed) has several advantages, such as simplicity, flexibility, parallel analysis of a large number of samples, various development modes, as well as the applicability of selective and specific chemical and biological detection methods. In his recent book "Fundamentals of Thin-Layer Chromatography (Planar Chromatography)" Geiss [1] stated that the cost of quantitative analysis by PC is one-third of the cost in column liquid chromatography (CLC). A disadvantage of PC is that a certain amount of skill and experience are required to derive full benefit from its possibilities, due to the relatively large number of parameters influencing the result.

In the present chapter on planar chromatography only brief mention is made of paper chromatography, since this technique has been eclipsed by the faster, more sensitive method of thin-layer chromatography (TLC), and because both TLC and paper chromatography are very closely related in their technical aspects [2]. Because it is difficult to find a sharp borderline between classical TLC and high-performance TLC (HPTLC), both are discussed side by side in this chapter. HPTLC plates have a smaller average particle size (6 μm) and a narrower particle-size range. The term mobile phase is more appropriate in forced-flow planar chromatography (FFPC), where it migrates in a closed system; hence the term solvent system [1] should be used for paper chromatography and classical TLC/HPTLC.

Three topics are discussed in greater detail in this chapter: The various chamber types and development modes, which are only practicable in PC, and the newer FFPC methods. The topic of PC has been surveyed in detail for paper chromatography [3,4], classical TLC [5-11], and HPTLC [12-14]. The principles and possibilities of the FFPC techniques

[15,16] and the state of the art of modern PC [17-21] have also been reviewed in recent years. The fundamentals of modern PC are discussed in detail in the textbook by Geiss [1].

3.2 CLASSIFICATION OF PLANAR CHROMATOGRAPHIC TECHNIQUES

Planar chromatography can be classified from different points of view, e.g., (a) the flow of the mobile phase, (b) the nature of the stationary phase, (c) the mechanism of the separation, (d) the polarity relation between the mobile and stationary phase, (e) the aim of the separation, and (f) whether the principal steps are performed as separate operations or not.

(a) The solvent can migrate through the stationary phase by capillary action or under the influence of forced flow. Forced flow can be achieved either by application of external pressure {overpressured-layer chromatography (OPLC)} [22-25], an electric field (high-speed TLC) [15,26], or by centrifugal force {CPC or rotation PC (R.P.C.)} [27-30].

Fig. 3.1 schematically demonstrates the superior efficiency of FFPC techniques by comparing their analytical properties with those of classical TLC and HPTLC [23]. FFPC techniques permit the advantage of optimal mobile-phase velocity to be exploited over practically the whole separation distance without loss of resolution. Needless to say, this effect is independent of the type of forced flow and the layer thickness used.

(b) PC can also be classified as paper and thin-layer chromatography. In the first category, the solvent migrates through paper, and in the second through a layer of porous stationary phase, which is generally bound to an inert support in a planar configuration.

(c) Depending on its nature, the stationary phase promotes the separation of compounds by adsorption, partition, ion-exchange, or size-exclusion separation processes. Some types of stationary phases do not fit into these four basic classes, and separations actually involve a combination of these four basic mechanisms.

(d) In normal-phase (NP) PC the sorbent is more polar than the solvent system, whereas in reversed-phase (RP) PC the stationary phase is less polar than the mobile phase.

(e) An aim of a separation may be analysis for the qualitative assay or quantitative determination of the separated compounds. Another goal of PC may be a preparative separation for the isolation or purification of substances.

(f) PC may also be classified as an off-line or an on-line separation technique. Classical TLC for analytical purposes is a typical fully off-line process, where the principal steps of sample application, development, evaporation of the solvent system, and densitometric evaluation are performed as separate operation steps. In preparative applications the separated substances are not subjected to in situ quantitation, but instead, the zones are scratched off the support and the separated compounds are eluted from the sorbent, using a solvent of high solvent strength.

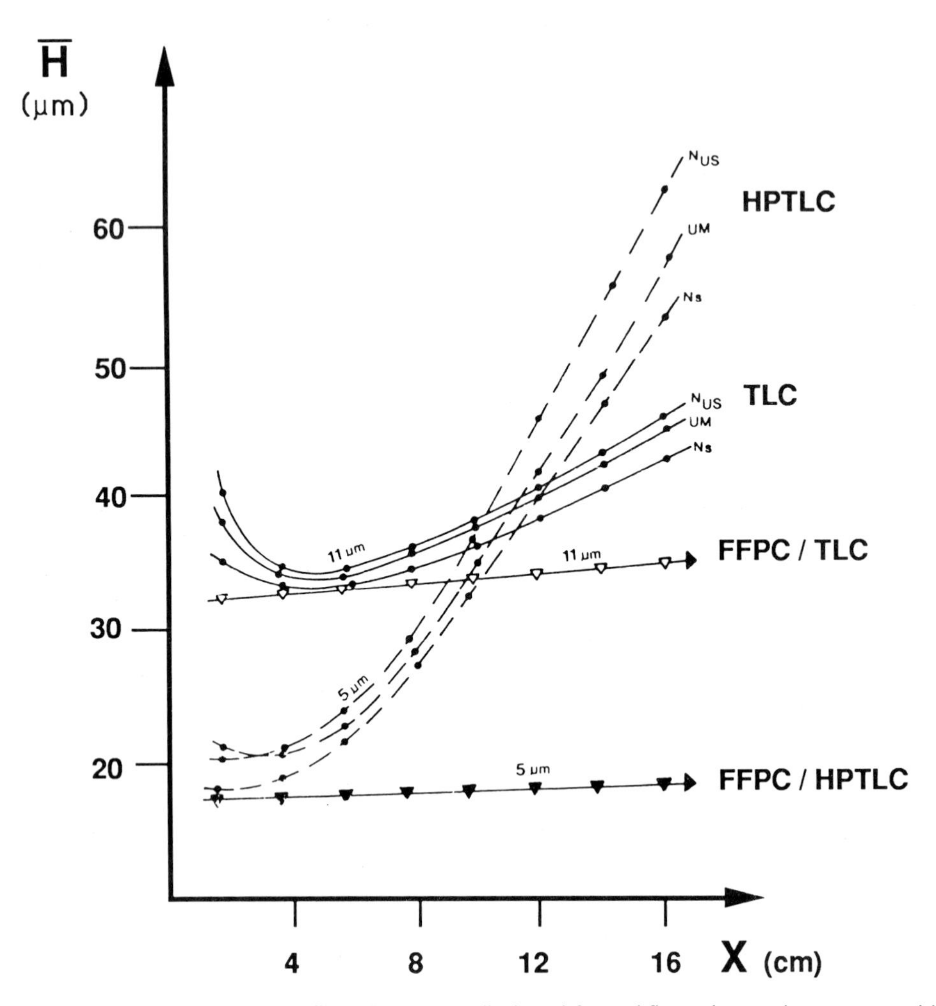

Fig. 3.1. Comparison of capillary-flow-controlled and forced-flow planar chromatographic methods for different plates and chamber types. The developing distance from the start to the solvent front (X) is plotted against plate height ($\overline{H}$). N_{US} = normal, unsaturated chamber, UM = ultramicro-chamber, N_S = normal, saturated chamber.

In the fully on-line mode the principal steps are not performed as separate operations; rather, the separated compounds are drained from the layer [31]. If a FFPC system is equipped with an injector and a detector, the various off-line and on-line operations can also be combined. Possible combinations are off-line sample application and on-line separation/detection, or on-line sample application/separation and off-line detection. OPLC and R.P.C. may be used as fully on-line preparative techniques [14] by connection to a flow-through detector, and/or collection of separated compounds with a fraction collector.

3.3 PRINCIPLES OF PLANAR CHROMATOGRAPHIC METHODS

3.3.1 Capillary-flow-controlled planar chromatography

3.3.1.1 Paper chromatography

In paper chromatography differences in migration occur because the components of the mixture are selectively distributed between the stationary phase, or attached to cellulose, and the solvent system, which may be organic or aqueous or a mixture of the two [2,6]. For many applications, the properties of ordinary cellulose paper can be modified to increase resolution.

3.3.1.2 Thin-layer chromatography

The driving force for solvent migration in TLC is the decrease in free energy of the liquid as it enters the porous structure of the layer, and the transport mechanism is the result of the capillary effects [2,5,32]. Classical TLC or HPTLC is a fully off-line process, because the principal steps are performed as separate operations; quantitative determination is usually performed in situ on the plate.

3.3.2 Forced-flow planar chromatography

3.3.2.1 Overpressured-layer chromatography

Apart from capillary action, the driving force for solvent migration in OPLC is external pressure [22-25,31,33]. Depending on the desired mobile-phase velocity, low (2-5 bar), medium (10-30 bar), and high (50-100 bar) operating pressures can be used [31]. With this type of forced flow, the advantage of the optimal mobile-phase velocity can be exploited over practically the whole separation distance without loss of resolution. In OPLC the vapor phase is completely eliminated, the chromatoplate being covered with an elastic membrane under external pressure (Fig. 3.2a). Thus, the separation can be carried out under controllable conditions. The separation may be started with a dry layer, as in classical TLC, but the closed system also permits carrying out a fully on-line separation, where chromatography can be started with a mobile-phase-equilibrated stationary-phase system, as in high-performance column liquid chromatography.

A new, radial version of OPLC, viz. high-pressure planar chromatography (HPPC) has been developed by Kaiser and Rieder [34,35]. In this method the sample application and mobile-phase inlet are identical, i.e. the center of the chromatoplate. Only one sample is generally chromatographed on a 10 x 10-cm HPTLC plate at ca. 30 bar external pressure. The main fields of use of HPPC are the mobile-phase transfer to HPLC and direct coupling with HPLC, as well as fast, routine, single-sample analysis with high accuracy, including calibration [36].

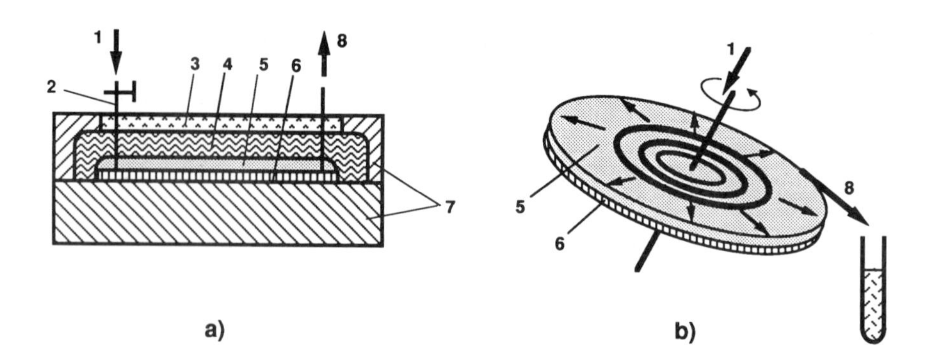

Fig. 3.2. Principle of on-line preparative (a) OPLC and (b) R.P.C. 1 = Mobile-phase inlet, 2 = valve for mobile-phase inlet, 3 = safety glass, 4 = water cushion, 5 = stationary phase, 6 = glass support, 7 = chromatographic chamber, 8 = eluate outlet.

3.3.2.2 Rotation (centrifugal) planar chromatography

In R.P.C. the driving force for solvent migration is the centrifugal force, in addition to capillary action. The samples are applied to the rotating stationary phase near the center. The centrifugal force drives the mobile phase through the sorbent from the center to the periphery of the plate. For analytical purposes, up to 72 samples can be applied, and quantitation can be carried out in situ on the plate. For micropreparative and preparative purposes only one sample is applied as a circle. The separations can be carried out either in the off-line or on-line mode. In the latter, the separated compounds are eluted from the stationary phase by the centrifugal force and collected in a fraction collector, as shown in Fig. 3.2b [29,30].

3.4 PRINCIPAL FACTORS IN PLANAR CHROMATOGRAPHY

The chromatographic processes in analytical or preparative PC are basically the same, regardless of whether the driving force is capillary action alone or augmented by additional forces. Among the factors which may influence PC separations, the chamber types and development modes deserve special treatment.

3.4.1 Stationary phase

3.4.1.1 Paper chromatography

Cellulose paper consists of a partially oriented collection of cellulose fibers. These fibers are composed of approximately parallel carbohydrate chains, strongly crosslinked by hydrogen bonding in some regions, to give a partly crystalline and partly amorphous structure. In the amorphous regions, water or other hydrophilic solvents are adsorbed by the cellulose. This leads to the formation of pools of liquid connected by crystallite bridges. Water in these regions is of two types, one being chemically bonded to the cellulose fibers and the other being more loosely bound and available for partitioning the solutes [6]. Chromatographic paper may be either pure, chemically modified, impregnated, or loaded cellulose paper or glass fiber paper [3]. The papers generally used have ion-exchange properties due to the presence of a small number of carboxyl groups. Paper chromatography in systems with stationary water involves a combination of adsorption, partition, and ion-exchange mechanisms [2].

3.4.1.2 Thin-layer chromatography

Most users prefer to buy commercially available precoated layers, rather than prepare their own. Besides saving time, precoated layers have the advantage of much higher reproducibility than can be obtained with self-prepared plates. Generally, two types of precoated plates are available, viz. unmodified sorbents and modified silicas in layer thicknesses of 0.1-2 mm. For analytical separations the layer thickness should be between 0.1 and 0.25 mm, whereas for preparative purposes 0.5-mm, 1-mm, and 2-mm layer thicknesses are available. A comparison of different precoated plates is given by Brinkman and De Vries [37].

The unmodified stationary phases include silicas, aluminas, Kieselguhr, silicates, controlled-porosity glass, cellulose, starch, gypsum, polyamides, and chitin [38]. For TLC and HPTLC the most frequently used stationary phase is silica. It is prepared by spontaneous polymerization and dehydration of aqueous silicic acid, which is generated by adding acid to a solution of sodium silicate. The product of this process is an amorphous, porous solid, the specific surface area of which can vary over a wide range (200 to more than 1000 m^2/g), as can the average pore diameter (10-1500 Å) [1,39].

Modified silicas may be nonpolar or polar sorbents. The former class includes silicas bearing alkane chains or phenyl groups, while the polar modified silicas contain cyano, diol, amino, and thiol groups or substance-specific complexing ligands [40,41]. Chiral separation possibilities in PC via ligand exchange, charge transfer, and diastereomeric compound formation, as well as with cyclodextrins were summarized by Han and Armstrong [42]. The structures of some chemically modified silicas are shown in Fig. 3.3.

It is generally accepted that resolution is higher on a thinner layer (0.1 mm), but this effect is also simulated by the detection mode [43]. Commercially available analytical thin-layer plates have dimensions of 10 x 10, 10 x 20, or 20 x 20 cm. The silica materials commonly used for precoated plates have an average particle size of ca. 11 μm, ranging from 3 to 18 μm; for self-prepared analytical layers the average particle size is 15 μm and the range of particle sizes is much greater. The average particle size of precoated HPTLC plates is now 5-6 μm with a very narrow range of particle sizes. Precoated plates are available either with glass, alufoil or plastic sheets as supports. The quality of the stationary phase is similar for each type of support, but there is some difference in the resolution due to the binder used. At the moment, many chemically modified silicas in HPTLC quality are commercially available only on glass plates.

References on p. A146

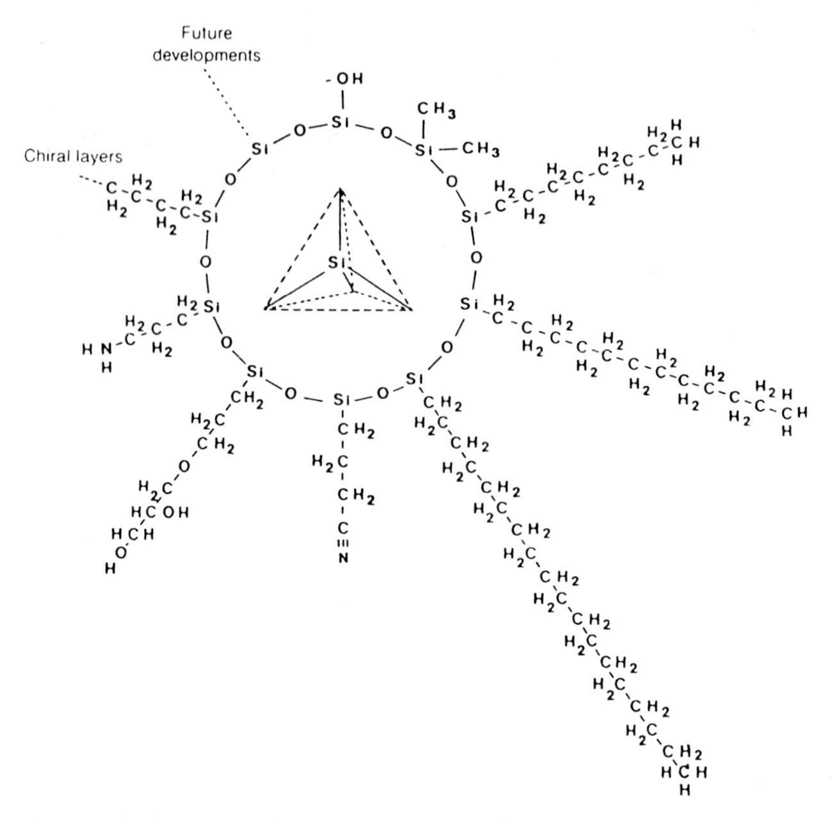

Fig. 3.3. Structures of some commercially available surface-modified silicas. (Reproduced from Ref. 99, with permission.)

Various precoated analytical layers with a preadsorbent zone are also commercially available for linear development. This zone serves to hold the sample until development begins. Compounds soluble in the solvent system pass through the preadsorbent zone and are concentrated in a narrow band before entering the chromatographic layer proper, and this improves their resolution. Just about all the stationary phases used in NP- and RP-HPLC are now also available for PC [19,44].

3.4.2 Solvent system

Solvent system optimization can be performed by trial and error, solvent selection being bases both on the analyst's experience and intuition and on modifications of published data. However, as the sample composition becomes more complex, systematic solvent optimization becomes more important. The methods used for optimizing isocratic mobile phases in HPLC (Chapter 1) are generally also applicable, with some modifications, to PC. For example, window diagrams have been successfully applied [45-47] to the optimization of PC solvent systems. Similarly, overlapping resolution maps were used as criterion by Issaq et al. [48] and by Nurok et al. [49], who also extended this approach to continuous development. The fruitful application of the sequential simplex method was also reported [50,51].

Geiss [1] has suggested a structural approach, which assumes that selectivity and solvent strength are independent variables. For this optimization process he used the Vario KS chamber (Section 3.4.3.2) with three strong solvents (methyl *t*-butyl ether, acetonitrile, and methanol). All three solvents were diluted with a suitable amount of a weaker, fourth solvent (F-113 or 1,2-dichloroethane) to obtain a series of solutions spanning the solvent strength (ε°) range from 0.0 to 0.70 in increments of 0.05 ε°. In the next step, the appropriate solvent strength must be determined. Once this has been identified, fine-tuning is accomplished by blending solvent mixtures of this strength but of different selectivity. Many elegant separations have been achieved in this manner, but this method reduces the number of solvents available for optimization.

On the basis of Snyder's solvent characterization [52], a new solvent optimization method, called the "PRISMA" system [53-57] has recently been developed. It consists of three parts: In the first part, the basic parameters, such as the stationary phase and the individual solvents are selected by TLC. In the second part, the optimal combination of these selected solvents is selected by means of the "PRISMA" model. The third part of the system includes selection of the appropriate FFPC technique (OPLC or R.P.C.) and HPTLC plates, selection of the development mode, and, finally, application of the optimized mobile phase in the various analytical and preparative chromatographic techniques.

After selection of the stationary phase, the recommended initial optimization step in the first part of the "PRISMA" system is TLC. Ten neat solvents, miscible with hexane and representing Snyder's eight selectivity groups (see Table 3.1) are tested, using TLC plates. For the separation of nonpolar compounds the solvent strength can be decreased with hexane; for the separation of polar compounds the solvent strength can be increased by adding water or another polar solvent in low concentration, to bring the R_F values of the compounds to be separated into a range of 0.2-0.8. Generally, the solvents giving the best resolutions are then selected for further optimization.

After selection of the solvents, construction of the "PRISMA" model is begun. Generally, between two and five solvents may be selected; modifiers may also be added. The tripartite "PRISMA" model is a three-dimensional geometrical design which correlates the solvent strength with the selectivity of the solvent system. As a rule, in NP-PC, the upper frustrum is used for solvent optimization for polar compounds, the center portion of the prism is used for solvent optimization for nonpolar compounds, and the lower part symbolizes the modifiers. For details the reader is referred to Refs. 56 and 57. Strategies for optimizing the solvent systems for PC, including two-dimensional TLC separations, have been summarized by Geiss [1] and Nurok [58,59].

References on p. A146

TABLE 3.1

COMMONLY USED SOLVENTS FOR MOBILE-PHASE OPTIMIZATION IN PC

Group	Solvent strength	Solvent
—	ca. 0	*n*-Hexane
I	2.1	*n*-Dibutyl ether
	2.4	Diisopropyl ether
	2.7	Methyl *t*-butyl ether*
	2.8	Diethyl ether*
II	3.9	*n*-Butanol
	3.9	2-Propanol*
	4.0	1-Propanol
	4.3	Ethanol*
	5.1	Methanol
III	4.0	Tetrahydrofuran*
	5.3	Pyridine
	5.5	Methoxyethanol
	6.4	Dimethylformamide
IV	6.0	Acetic acid*
	9.6	Formamide
V	3.1	Dichloromethane*
	3.5	1,1-Dichloroethane
VI	4.4	Ethyl acetate*
	4.7	Methyl ethyl ketone
	4.8	Dioxane*
	5.1	Acetone
	5.8	Acetonitrile
VII	2.4	Toluene*
	2.7	Benzene
VIII	4.1	Chloroform*
	10.2	Water

*Proposed solvents for the first experiments for mobile-phase optimization when using the "PRISMA" optimization system.

3.4.3 Chamber type and vapor phase

Selection of the chamber type and vapor space is a variable offered only by PC, because the separation process occurs in a three-phase system of stationary, mobile, and vapor phases, all of which interact with one another until equilibrium is reached [60]. Basically, one can distinguish between the normal (N) chamber and the sandwich (S) chamber. In the common N chamber there is a distance of more than 3 mm between the layer and the wall (or between the layer and the lid of the chamber in horizontal development) of the chromatographic tank. If this distance is smaller, the chamber is said to have the S configuration. Both types of chromatographic chamber can be used for unsaturated or saturated systems. Although the chambers used for FFPC separation can be also assigned to the above two categories, their special features warrant a separate discussion of FFPC chambers.

The best results in classical TLC/HPTLC are achieved by using saturated S chambers. The same solvent systems can be used without modification in microchamber R.P.C. separations, since the vapor phase in that apparatus is fully saturated. For use of the optimized solvent system in OPLC or in ultramicrochamber R.P.C. it must be optimized in unsaturated S chambers, since there is no vapor phase. Unsaturated or S chambers are also required for optimization [30]. The Vario KS chamber, which allows selection of the optimal vapor phase conditions for all separation problems, is best suited for working with the "PRISMA" system.

3.4.3.1 N chambers

Although a wide variety of chromatographic chambers are available, rectangular glass N chambers with internal dimensions of 23 x 23 x 8 cm or 13 x 13 x 5 cm are most frequently used in PC for development of two 20 x 20-cm or 10 x 10-cm plates, respectively. Starting the separation with unsaturated chromatographic tanks generally gives higher R_F values for NP systems because of the evaporation of the solvents from the surface of the layer. Two disadvantages of using unsaturated tanks deserve mention. The first is that the reproducibility of the R_F values may be poor, and the second, that a concave solvent front can occur, leading to higher R_F values for solutes near the edges.

If the layer is placed in the chamber immediately after introducing the solvent systems into the chromatographic tank, separation starts in an unsaturated system which will become progressively more saturated in the course of the separation. A chamber is saturated when all components of the solvent are in equilibrium with the entire vapor (gas) space before and during the separation [1]. Therefore, the N chamber is lined on all four sides with filter paper, thoroughly soaked with the solvent systems. A tank prepared in this way should stand for 60-90 min to allow the internal atmosphere to become saturated with the vapor of the solvent system. Each plate must lean against a side wall, so that they do not touch each other. The saturated tanks have the advantage that the front of the solvent system (α-front) is much more regular, and also the separation efficiency is higher when a development distance of 18 cm is used.

References on p. A146

A versatile version of the N tank is the twin-trough chamber [61], which has a raised glass ridge along the center that effectively separates the chromatographic tank into two separate compartments. To obtain a saturated vapor space, the solvent system is placed in one of the compartments, while the plate to be developed is put in the other one. When the layer has become equilibrated with the vapor phase, the tank is carefully tipped so as to transfer the solvent system from one compartment to the other and to start the separation [1].

The short-bed continuous-development (SBCD) chamber is a flat N chamber [62,63], designed for short separation distances and high solvent velocities. Low-strength mobile phases will then resolve slow-moving solutes in a reasonable time. The bottom of the SBCD chamber has five ridges, serving as stop positions for the plate, which leans against the rim, protruding from the chamber. The solvents migrating up the chromatoplate evaporate from the protruding portion. Because the migration distance, which depends on the angle of the plate, is short and migration is slow, diffusion is decreased, and extremely high resolution can be obtained. If saturated conditions are attained by lining the walls and the lid with filter paper, solvent demixing is precluded.

3.4.3.2 S chambers

The S chambers are very narrow, unsaturated tanks, with a distance between the chromatoplates and the glass cover plate of less than 3 mm. Saturation can be established with a facing chromatoplate that has been soaked with the solvent system. Part of the stationary phase of the plate to be developed is scraped off, so that the solvent can initially only reach the level of the facing plate [1]. After sorptive and capillary saturation of this plate, the level of the solvent system is increased to start the separation.

The Brenner and Niederwieser (BN) chamber [64,65] is a special case of the S chamber for linear continuous development, where the solvent system and the analytes advance practically continuously on the plate at a constant velocity. The top part of the chromatoplate does not face another plate but rests on a heated metal block. Once the α-front has reached the heated block, the solvent system starts to evaporate. The BN chamber is suitable for the separation of compounds which otherwise would have very low R_F values.

The U chamber [66] belongs to the S chamber type, where the vapor space is decreased. The chromatoplate is placed face down in a very small flat chamber above a layer of solvent. All basic development modes (Section 3.4.5) allow both equilibration before development and a choice of flowrates for separation.

The most versatile S chamber, which can also be used in the N chamber configuration, is the Vario-KS chamber, devised by Geiss et al. [1,67]. It is suitable for evaluation of the effects of different solvents, solvent vapors, and relative humidities. The chromatoplate is placed face down over a glass conditioning tray, containing 5, 10, or 25 compartments to hold the required conditioning solvents. The design of the device ensures that saturating and developing solvents are kept separate. A major advantage of this chamber is that up to ten activity and/or saturating conditions can be compared on the same chromatoplate with the same solvent for solvent optimization purposes. Another advantage of the Vario-KS chamber is that the same chromatoplate can be simultaneously developed with five different solvents. Additionally, the chamber can also be used with a heating accessory for continuous development. The optimum conditions developed in this chamber may easily be transferred to a regular twin-trough tank for routine work or to FFPC separations for more complex separation problems.

3.4.3.3 Chambers for forced flow

The chambers used for OPLC separations are unsaturated S chambers, theoretically and practically devoid of any vapor space. This must be considered in the optimization of the solvent system, especially in connection with the disturbing zone [68] and multifront effect [69], which are specific features of the absence of a vapor phase.

The main difference between the chamber types used in R.P.C. lies in the size of the vapor space, which is an essential criterion in rotation planar chromatography [30]. Therefore we use an additional symbol to indicate the vapor space {normal-chamber, micro-chamber, ultramicro-chamber, and column R.P.C. (N-R.P.C., M-R.P.C., U-R.P.C., and C-R.P.C., respectively)}.

In N chamber R.P.C. the layer rotates in a stationary N chamber, where the vapor space is extremely large. Due to extensive evaporation, this chamber is practically unsaturated. The M and U chambers in R.P.C. belong to the S-chamber type, the difference between these two chambers being that the former is saturated, while the latter is unsaturated. Since in M-R.P.C. the chromatoplate rotates together with the small chromatographic chamber, where the distance between the layer and the lid of the chamber is smaller than 2 mm, the vapor space is rapidly saturated. In the case of the ultramicro-chamber, the lid of the rotating chamber is placed directly on the chromatoplate [29] so that practically no vapor space exists.

3.4.4 Flow velocity

In capillary-flow-controlled PC the solvent velocity is the parameter which, in principle, cannot be influenced by the chromatographer. In TLC this depends entirely on the characteristics of the stationary phase and the solvent chosen. The solvent velocity constant can be determined from the time required for the solvent to travel between fixed points, marked on the TLC plate, as the slope of a plot of Z_f^2 against time, i.e. based on the linear equation

$$Z_f^2 = \kappa t \qquad (3.1)$$

where Z_f is the distance traveled by the solvent front from the solvent level in the chromatographic tank (in cm), κ the solvent system velocity constant (in cm^2/sec), and t the time (in sec) [1,32]. When FFPC is used in the linear development mode, with a constant-flow pump, a linear relationship exists between the migration distance, Z_f, the mobile phase velocity, u (in cm/sec) and the time, t, required [70].

$$Z_f = u\,t \quad (3.2)$$

One possibility of exerting an influence on the flow velocity is to avoid solvents of higher viscosity in solvent optimization. Also, saturated chromatographic systems have the advantage that development is much faster than in unsaturated chambers. This means that the solvent velocity increases in going from the unsaturated N chamber via the unsaturated S(U) chamber to the saturated S chamber. The highest mobile-phase velocity can be achieved by using FFPC techniques. In OPLC the upper limit of velocity depends on the applied external pressure, besides the viscosity. In R.P.C., the higher the rotational speed, the faster is the migration of the mobile phase. The flowrate is limited by the amount of solvent that may be kept in the layer without skimming over the surface. The amount of migrating solvent can also be increased by scratching a round hole into the center of the layer, varying its perimeter according to the optimal mobile-phase velocity [71].

The local mobile-phase velocity can be influenced by the selection of the development mode (Section 3.4.5). In continuous development (BN and SBCD chambers), migration of the solvent system starts off relatively fast, but always falls off along the separation distance. Once the α-front has reached the heated block, the solvent system can be evaporated, and the migration rate of the solvent system becomes constant.

3.4.5 Development mode

PC differs from CLC not only in having a vapor space, but also in permitting a selection of the optimal development mode.

3.4.5.1 Linear development

The ascending mode is most frequently used for capillary-flow-controlled PC conditions. The use of descending development has also been reported, especially in paper chromatography [3,4], but it is rarely applied because it is more complicated in practice while having no significant advantages with regard to resolution. Because ascending development has no theoretical advantage over horizontal development, the latter, being more adaptable, has become increasingly common in the recent years.

Continuous development can be achieved by allowing the end of the plate to remain uncovered in horizontal development (BN chamber) or to protrude from a slot in the cover of the chamber in the ascending mode (SBCD chamber). In either case, the solvent system flows continuously and evaporates from the uncovered area, which can be warmed to accelerate the process.

In OPLC, the most frequent development modes are the linear one- and two-directional (Fig. 3.4a,b) modes. However, the linear-type OPLC technique requires a special chroma-

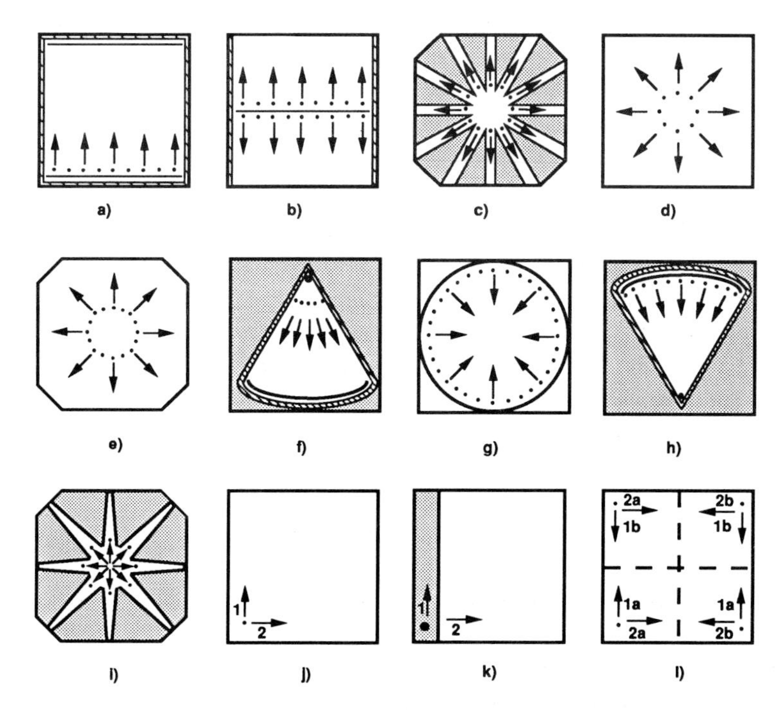

Fig. 3.4. Development modes. (a) One-directional linear in OPLC; (b) two-directional linear in OPLC, (c) linear in R.P.C., (d) radial in TLC, OPLC, and HPPC, (e) radial in R.P.C., (f) radial in OPLC, (g) antiradial in TLC and HPPC, (h) antiradial in OPLC, (i) antiradial in R.P.C., (j) two-dimensional TLC, (k) two-dimensional TLC on a bilayer plate, (l) two-dimensional TLC of 4 samples.

toplate, sealed along the edge by impregnation to prevent the solvent from flowing off the layer.

In R.P.C. with a M or an U chamber the mobile-phase movement can also be linearized (linear development mode) by scraping the layer to form lanes (Fig. 3.4c) [30].

3.4.5.2 Radial development

The advantage of radial development (Fig. 3.4d), where the solvent system migrates radially from the center of the plate to the periphery, is well known for the separation of compounds in the lower R_F range [1,66,72]. Using the same solvent system, the resolution, particularly in the lower R_F range, is about 4-5 times greater in the radial than in the linear development mode. It can be stated that the separating power of the radial development mode can be better exploited, if the samples are spotted near the center [35]. As the distance between the mobile-phase inlet and sample application increases, the resolution begins to approach that of linear development. For off-line radial OPLC and R.P.C. no preparation of the plate is necessary (Fig. 3.4d,e); for on-line radial OPLC a sector must be isolated by scratching off the surrounding sorbent and impregnation of the sector (Fig. 3.4f).

3.4.5.3 *Antiradial development*

Antiradial development is a widely accepted approach in analytical TLC, if the resolution must be increased in the higher R_F range [1,66,72]. In this development mode, the solvent system enters the layer at a circular line and flows towards the center (Fig. 3.4g). Since the solvent flow velocity decreases with the square of the distance, but the area wetted also decreases with the square of the distance traveled, the rate of solvent system migration is practically constant. Therefore, this developing mode is the fastest with respect to separation distance. However, the flowrate cannot be controlled. Antiradial on-line OPLC development is carried out similar to the radial mode, except the mobile phase enters the plate from the opposite side (Fig. 3.4h). For antiradial U-R.P.C. separations the plate must be prepared specially; lanes must be scraped into the layer, such that the surface of the stationary phase decreases along the radius (Fig. 3.4i).

3.4.5.4 *Multiple development*

Multiple development includes all linear development procedures, where development is repeated after a development is completed and the stationary phase has been carefully evaporated.

Unidimensional multiple development can be used to increase the resolution. The locations of the compounds to be separated, and hence the ΔR_F values, are influenced by the number of developments. After the first development (${}^1R_F$) the R_F values of a multiply developed solute can be predicted by the following equation [73]

$$ {}^n(R_F) = 1 - (1 - {}^1R_F)^n \tag{3.3} $$

where n is the number of developments. In this way, the R_F values and thus also the ΔR_F values can be calculated for all compounds of interest. Multiple development may also be carried out with different mobile phases in the same direction. It is also possible to develop the plates with the same or different mobile phases at different distances. Needless to say, unidimensional multiple development can also be carried out with FFPC techniques. In this case, the spot capacity increases linearly as a function of the square root of the development distance, instead of going through a maximum, as it does for capillary-controlled-flow conditions.

A square layer is always used for two-dimensional development (Fig. 3.4j) [74,75]. The sample is spotted at the corner of the layer and developed linearly. After evaporation of the first solvent system, the second development is carried out at a right angle to the first one. Generally, the method is used for the separation of complex mixtures. For this, different types of solvent systems are used in the two directions. If the same solvent system is used for both directions, the sample components will be distributed along a line diagonally from the origin to the farthest corner. In this case, the resolution will only be increased by a factor of $\sqrt{2}$, corresponding to the increased migration distance for the sample, but this method is useful for finding out whether the analyte has undergone some chemical change. Compounds that are unchanged by the separation process will lie on the diagonal of the plate.

Potential methods for utilizing two different retention mechanisms in orthogonal directions were recently summarized by Poole et al. [21]. For practical reasons, only the use of bilayer plates should be mentioned. Two stationary phases with different selectivities are used (Fig. 3.4k). The sorbent layer for the first development is a narrow strip that borders on the much larger area used for the second development. Similarly, four samples can be developed on a single plate, developed from all four sides (Fig. 3.4l). The first two developments (*1a* and *b*) are carried out with the same solvent system, and the following two with another solvent system (*2a* and *b*).

Programed multiple development (PMD) was introduced by Perry et al. [76-78] in 1973. In this technique, the chromatoplate is automatically cycled through a preset number of unidimensional developments with the optimized solvent system. Based on this idea, Burger [79-81] recently derived the automated multiple development (AMD) mode. In AMD, the chromatoplate is developed repeatedly in the same direction over an increasing migration distance. The developing solvent for each successive run differs from the one used before so that a stepwise gradient can be obtained. In contrast to the situation in CLC, the gradient used starts with the most polar solvent, for which the shortest developing distance is employed, and is varied towards decreasing polarity. The longest migration distance is used with the most nonpolar solvent. Another unique feature of the multiple development technique, which is also used in AMD, is the zone reconcentration mechanism [78,82], which ensures that if a sufficiently large number of developments are used, the zone will be compressed to a thin band.

3.4.6 Separation distance and time

In PC the separation distance improves with the square root of the separation distance. However, the optimum depends on the quality of the plate, the vapor space, the development mode, and the properties of the compounds to be separated.

In capillary-flow-controlled PC the separation distance and time depend on the average particle size and size distribution of the stationary phase, the vapor space, and the development mode. The first-mentioned parameter cannot be influenced by the user of precoated plates. The maximal length of commercially available precoated plates is 20 cm. Thus, the maximal separation distance is 18 cm in the linear development mode. To demonstrate the main difference between PC separation methods, the various chamber types and the degree of saturation, the distances traveled by the solvent system at various moments during linear and radial development were measured. Plots of the migration of the solvent front as function of time for a distance between 3 and 10 cm are compared in Fig. 3.5. It can be clearly seen that the function is almost linear for OPLC, U-R.P.C. and M-R.P.C. in the linear development mode, whereas for radial FFPC methods

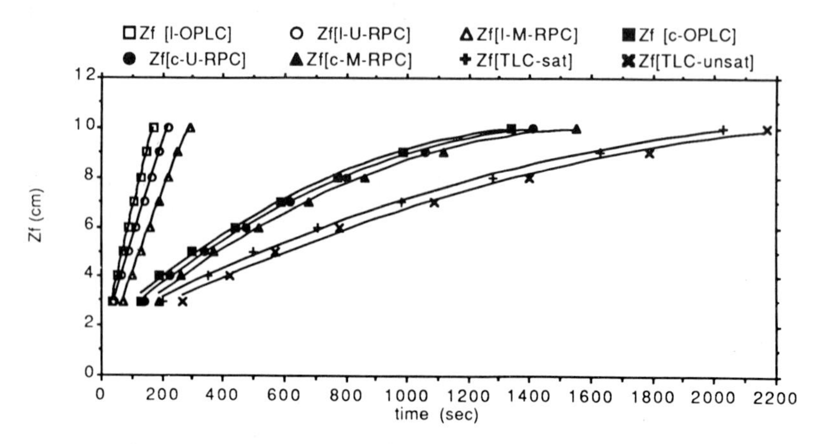

Fig. 3.5. Migration of the α-front as a function of time for vaious PC techniques. (Reproduced from Ref. 71, with permission.)

the lines are curved, because the volume of the stationary phase increases along the radius. In linear development the migration of the solvent system is faster on TLC plates in a saturated N chamber than in an unsaturated tank [71].

Commercially available U chambers can accommodate 10 x 10-cm chromatoplates for analyses by the radial and antiradial development modes in capillary-flow-controlled PC. Therefore, the maximal separation distances are 5 cm and 4 cm, respectively. These two development modes are rarely used for preparative separations, because accurate and efficient solvent system inlets are difficult to produce [83].

In linear OPLC the maximum separation distances are 18 cm and 36 cm for the 20 x 20-cm and 20 x 40-cm chromatoplates, respectively. Working with off-line radial OPLC, the maximal separation distance is 10 cm; in this case, only one sample can be analyzed. If the distance between the mobile-phase inlet and the sample application is 2 cm, then a separation distance of 8 cm can be achieved, which allows more samples to be applied. Conventional antiradial separation cannot be carried out, due to the large perimeter (ca. 60 cm for a 20 x 20-cm plate) of the mobile-phase inlet; an even distribution of the mobile phase by means of one or two inlet valves is impossible because of the decreasing mobile-phase inlet pressure. Recently, it was reported [84] that, after a suitable preparation of the plate by scraping a segment out of the layer and sealing the segment with a polymer suspension, radial as well as antiradial fully off-line and on-line separations can be carried out over a separation distance of 18 cm or 38 cm.

In HPPC 10 x 10-cm plates can be used. The maximum separation distance is 5 cm for a single sample; application of more samples reduces the separation distance. In analytical R.P.C. the separation distance and number of samples depend on the development mode. In the radial mode, which is the most commonly used, up to 36 samples may be applied as points on a circle of 2 cm radius; in this case, the separation distance is 8 cm. In the separation of enantiomers, baseline separation of 72 samples on a single chromatoplate with 5 cm separation distance was reported [85]. Since special preparation of the plate is necessary (by scribing lines into the layer), for linear and antiradial R.P.C. the sample number is correspondingly reduced. A separation distance of up to 11 cm [29] is possible, when 10 x 20-cm chromatoplates are used.

Generally, it can be stated that for the separation of nonpolar compounds, by FFPC techniques and with silica as the stationary phase, an extremely short separation time (1-2.5 min, over a separation distance of 18 cm) can be used without great loss in resolution [69]. In contrast, longer separation times are extremely important when silica is used for the separation of polar compounds [86].

3.4.7 Sample amount and application

The process of sample application is one of the most important steps for a successful PC separation [87]. The aim of sample application is quantitative transfer of the sample to the layer in such a way that the sample penetrates the layer to form a compact zone of minimum size without causing damage to the layer. The sample should be dissolved in a nonpolar, volatile solvent in such a concentration that the components of the sample are adsorbed throughout the whole thickness of the plate and not only on the surface of the layer. Local overloading may distort the applied bands, since the rate of dissolution of the components in the solvent system will then become a limiting factor [10].

The amount of applied sample depends on the determination method (e.g., UV, fluorescence, radiometry, postchromatographic derivatization). Generally, μg and ng quantities of sample can be determined, but even less than 100 pmol compound per chromatogram zone has been reported [88].

If the samples are applied as spots, the diameter is usually 5-10 mm in paper chromatography and 1-4 mm in TLC. In HPTLC, optimum results are obtained when 2 nl is applied over an area 0.1 mm in diameter [1], but in practice considerably larger volumes are spotted (100-200 nl) in a diameter of 2-6 mm [19]. Fenimore has shown [82,89] that the size of developed zones correlates more closely with the quality of the layer than with the size of the starting zone, provided that the latter is maintained within reasonable limits.

Among the various applicators designed for the deposition of small sample volumes on chromatoplates, the microcapillary (especially Pt-Ir capillaries) is one of the simplest and most useful. An alternative to these capillaries is the microsyringe, which offers greater flexibility in the choice of sample volumes.

The solid-phase sample transfer involves in situ solvent evaporation of the sample to form an oily residue on a PTFA tape, which can be forced into contact with the surface of the stationary phase by applying pressure to the top side of the tape.

Although all of the methods mentioned so far deal with the deposition of the sample as a spot, application of a continuous streak is also possible with modern instrumentation. With automated instruments for sample streaking the plate moves beneath a syringe containing the sample solution. An atomizer, operating by a controlled stream of nitrogen, sprays the sample from a syringe, forming narrow, homogeneous bands on the stationary

References on p. A146

phase [87]. Sample preparation has been reported in many handbooks; the reader is referred to Refs. 90 and 91.

3.4.8 Temperature

Temperature is not an effective parameter for modifying selectivity and maximizing resolution under normal circumstances [1]. Stahl [5] developed the first thermostated chamber, suitable for temperatures up to about 50°C [82]. Using a hotplate [92], the chromatoplate was heated to ca. 50°C, which focused the zones and lowered the detection limit, because the volatile solvents evaporated during the separation process. TLC at elevated temperature (150°C) was successfully used for the separation of slightly soluble substances [93].

Generally, it can be stated that in saturated chromatographic chambers, which are the most commonly used, the temperature does not exert a great influence on separations. A change of ± 5°C results in a change of the ΔR_F of less than 0.03 [1]. Nevertheless, in the interest of reproducibility in repetitive separations it is important to note the working temperature.

Normally, if two compounds are unresolved at a given temperature, they will remain unseparated at other temperatures, regardless of whether N or S chambers are used for the separations. Remarkably, temperature is now being found to play an important role also in the selectivity and efficiency of OPLC separations [94]. Preliminary experiments show that development by temperature-programed OPLC with a suitable control unit will bring further improvement to PC separations.

3.5 INSTRUMENTATION

3.5.1 Overpressured-layer chromatography

OPLC separations can be performed on two commercially available instruments, the Chrompres 10 and Chrompres 25 (Laboratory Instruments Works Co.). The first type offers different separation distances (18 and 36 cm) in the on-line mode and, in addition, off-line radial development. The Chrompres 25 accommodates a higher cushion pressure (25 bar) than Chrompres 10 (Fig. 3.6). This allows the use of more viscous mobile phases and/or higher mobile-phase velocities in the linear development mode.

Both chambers consist [95] of a bottom support block and an upper block with a poly(methyl methacrylate) support plate, fixed in an external frame. The prepared preparative layer is placed on the surface of the bottom block, and then the chamber is closed. The plate is covered with a plastic cushion, into which water is pumped through an inlet tube to maintain the desired external pressure. The mobile phase is delivered by a second pump, working in a range of 1-12 ml/min via a mobile-phase inlet valve. Further along, a mobile-phase outlet is located on the upper block. After one separation is finished, the cover plate of the instrument can be opened by means of a hydraulic system; the equipment is then ready for the next separation.

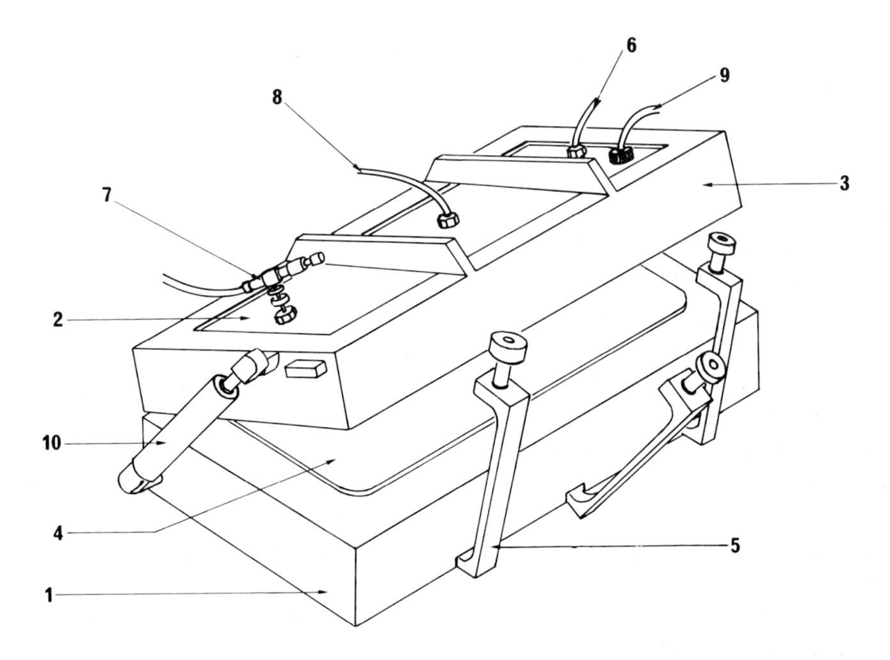

Fig. 3.6. Schematic drawing of Chrompres 10 OPLC instrument. 1 = Bottom support block, 2 = polymethacrylate support plate, 3 = external frame, 4 = surface of the bottom plate for the chromatographic plate, 5 = clamps, 6 = inlet tube for water, 7 = inlet valve for mobile phase, 8 = outlet for eluate, 9 = outlet tube for water, 10 = hydraulic system. (Reproduced from Ref. 95, with permission.)

Linear separations require specially prepared plates, with edges that are chamfered off and impregnated with a suitable polymer suspension, in order to prevent solvent leakage at overpressure. To ensure that mobile-phase migration forms a linear front, a channel is scratched out of the layer. A second channel, cut at a distance of 18 cm (for a 20 x 20-cm plate) or 36 cm (for a 20 x 40-cm plate) from the inlet channel permits collection of the eluent.

After sample application, the plate is placed in the instrument in such a way that, on closing the chamber, the mobile phase inlet is located exactly in the inlet channel, and the mobile phase outlet fits tightly into the outlet channel. The water cushion is then subjected to pressure. Before starting the separation with the optimized mobile phase, the mobile phase inlet valve is closed and the eluent pump is started to establish an appropriate solvent pressure. After this, the separation is started by opening the inlet valve, which ensures fast distribution of the mobile phase in the inlet channel to obtain linear migration of the mobile phase. No preparation of the plate is needed for off-line radial separations.

The HPPLC 3000 instrument (Institut für Chromatographie) offers analytical separation possibilities in off-line radial and antiradial development modes with a maximum separation distance of 5 cm. The pressure, applied mechanically, is 3000 kg on a 10 x 10-cm HPTLC plate.

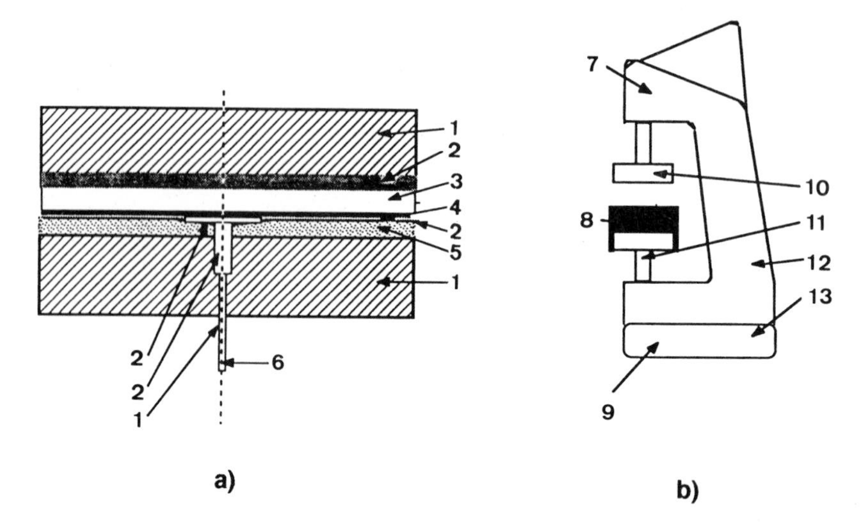

Fig. 3.7. Schematic drawing of the HPPLC 3000 instrument. 1 = Steel, 2 = PTFE, 3 = glass support, 4 = stationary phase, 5 = special rubber, 6 = sample and mobile-phase inlet, 7 = 3-ton press, 8 = chromatographic sandwich chamber, 9 = pneumatics, 10 = top part of the press, 11 = lower part of the press, 12 = support for applicator, digital pump, and electronics, 13 = connection for compressed air. (Reproduced from Ref. 35, with permission.)

The chromatographic S chamber (Fig. 3.7a) consists of a bottom support block, covered with a sheet of an elastic membrane and a sheet of PTFE. The mobile-phase inlet is located in the center. The chromatoplate is placed face down on the body of the instrument (Fig. 3.7b), and the chamber is subsequently compressed pneumatically. The mobile phase is fed under pressure with complete digital control. This allows thorough cleaning of the planar separation field and multiple use of the same glass plate.

Generally, HPPC is used in the single-sample mode [35], where one sample can be chromatographed by radial development in 1-2 min. The sample is injected with a HPLC injection valve through the mobile-phase inlet. The separation of the sample into individual compounds yields an accurate quantitation when an optical scanner is used in a radial mode of operation. The separation of 24 samples in the radial or 48 samples in the antiradial mode is also possible.

HPPC can also be used for the mobile-phase optimization of HPLC, because the k' data can be transferred between the two methods with an accuracy of 2% [36].

3.5.2 Rotation planar chromatography

Two instruments are commercially available for R.P.C.; the Chromatotron Model 7924 (Harrison Co.), and the ROTACHROM Model P rotation planar chromatograph (Petazon).

The Chromatotron (Fig. 3.8a) consists of an annular N chamber, inclined at an angle and fixed on a pedestal [95,96]. A flat glass rotor, covered with stationary phase, is mounted on an axle with the aid or a fixing screw. The motor-driven glass disk of 24 cm diameter rotates at a constant speed of 750 rpm. The mobile-phase inlet eccentrically pierces a quartz lid, which covers the chromatographic chamber. The chamber is provided with a circular channel for the collection of eluate. The solvent outlet is placed at the lowermost point of the collection channel. An inlet tube is mounted on the side of the chamber for flushing with nitrogen or other inert gases. The preparation of the preparative layers is given in Ref. 95.

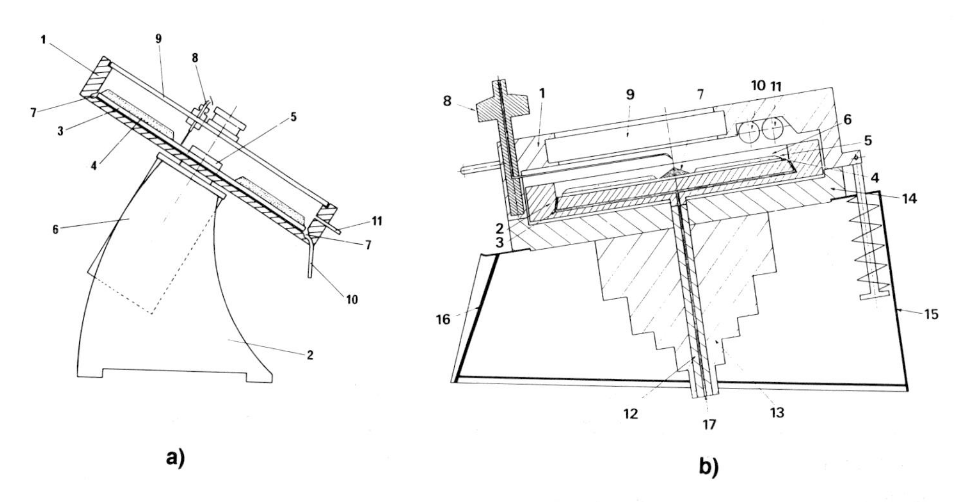

Fig. 3.8. Schematic drawing of R.P.C. instruments. (a) Chromatotron Model 7924. 1 = Annular chromatographic chamber, 2 = pedestal, 3 = flat glass rotor, 4 = stationary phase, 5 = fixing screw, 6 = motor, 7 = circular channel for eluate collection, 8 = mobile-phase inlet, 9 = quartz glass lid, 10 = eluate outlet, 11 = inlet tube for inert gas. (b) ROTACHROM Model P. 1 = Upper part of the stationary chromatographic chamber, 2 = collector, 3 = tubes in the collector, 4 = flat glass rotor, 5 = stationary phase, 6 = vapor space, 7 = fixing screw, 8 = solvent delivery system, 9 = safety glass, 10 = UV lamp (254 nm), 11 = UV lamp (366 nm), 12 = motor shaft with tube, 13 = motor, 14 = lower part of the stationary chromatographic chamber, 15 = casing of the instrument, 16 = front panel for adjusting and controlling units with keyboard, 17 = eluate outlet. (Reproduced from Ref. 96, with permission.)

A schematic drawing of the ROTACHROM in the preparative U-R.P.C. mode is shown in Fig. 3.8b. The upper part of the instrument is a stationary N chamber, which is covered with a thick safety glass; the two UV lamps (254 nm and 366 nm) are located underneath the lid. The two solvent delivery devices are horizontally and vertically adjustable; vertical adjustment is necessary because the layers have different heights; the horizontal adjustment is especially necessary for sequential R.P.C. (S-R.P.C.). The delivery needles lead from the solvent delivery systems to the center of the chromatographic chamber.

For preparative R.P.C. separations the prepared stationary phase on the glass rotor is fixed at the center of the collector. In N-R.P.C. and S-R.P.C. the layer rotates together with the collector in the instrument; no glass cover is used. Therefore, the chromatographic N chamber is practically the lower and the upper part of the stationary chamber [30].

The rotating collector is used for collection of the eluates in the various preparative modes. It may also be adapted as a chromatographic S chamber for analytical and preparative micro- and ultramicro-R.P.C. separations [29]. The inside of the collector has a special ellipsoid form [95]. Due to the centrifugal force, the eluent collects in the holes at the ends of the larger shaft. From there if flows continuously under nitrogen overpressure through the two tubes inside the collector against the centrifugal force to the center and motor shaft and thence to the eluent outlet.

The front panel of the instrument casing carries the adjusting and controlling units {keys for the UV lamps, timing and temperature unit, switch for the nitrogen overpressure and a manometer (0-2.5 bar)}: The microprocessor-controlled rotation speed may be varied between 20 and 2000 rpm, in steps of 10 and 100 rpm.

3.5.3 Automated multiple development

A schematic flow diagram of the AMD system (CAMAG) is shown in Fig. 3.9. Its central component is a closed N chamber with connections for feeding and withdrawing developing solvents, and for pumping a gas phase in and out. The solvent system is concocted from up to six reservoir bottles containing the neat solvents, which are passed via a motor valve to the gradient mixer. The gas phase is made up externally by passing nitrogen through the wash bottle into the reservoir, from where it is pumped into the chromatographic chamber at the appropriate time.

After the plate is placed in the chromatographic chamber and the solvent system is fed into the gradient mixer, the separation is started. When the preprogramed time determining the running distance has elapsed, the solvent system is withdrawn from the chamber, first to the waste-collecting bottle, and then, after the solvent system has been completely removed, vacuum is applied by a pump, drying the chromatoplates. The duration of the drying cycle can be programed. Before the next developing cycle is started, the chromatoplate is reconditioned by feeding a gas phase from the reservoir into the chamber. During the drying phase the solvent system is prepared for the next chromatographic step.

More details about the highly developed microprocessor-controlled AMD system are given in Refs. 21, 80, and 97.

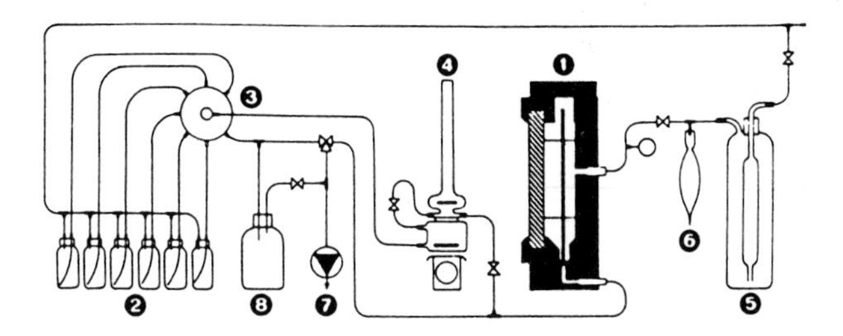

Fig. 3.9. Schematic drawing of AMD instrument. 1 = Chromatographic chamber, 2 = reservoir bottles, 3 = motor valve, 4 = gradient mixer, 5 = wash bottle, 6 = mobile-phase reservoir, 7 = vacuum pump, 8 = waste collection bottle. (Reproduced from Ref. 21, with permission.)

3.6 QUALITATIVE AND QUANTITATIVE ANALYSIS

3.6.1 Identification of separated compounds

Apart from chromatographic mobility data, the list of analytical methods that can be brought to bear upon the analysis of a chromatoplate includes visual inspection, UV/visible spectrophotometry, fluorescence spectrophotometry, optical and electron microscopy techniques, Auger and reflectance IR spectroscopy, radioimaging methods, near-IR analyses, and mass spectrometry in various forms, including secondary-ion mass spectrometry, fast-atom bombardment, and laser desorption ionization [98]. It should be remembered that the chromatographic retention data are not enough for correct identification; at least one spectroscopic method is necessary to make a valid statement.

3.6.1.1 Chromatographic data

In PC, compounds are identified primarily on the basis of their mobility in a suitable solvent system. The mobility of separated zones is described by the R_F value of each compound, where

$$R_F = \frac{\text{distance of spot migration, } Z_X}{\text{distance between start and front, } Z_F - Z_0} \qquad (3.4)$$

The R_F value bears no linear relationship to any basic PC parameter or structural element of the analyte. Such a linearity can be achieved with the R_M value, which is a logarithmic function of the R_F value.

$$R_M = \log \frac{1}{R_F} - 1 \qquad (3.5)$$

References on p. A146

For continuous and multiple development, where the solvent front is not measurable, the R_X value can be used, which is defined by the equation [11]

$$R_{X_a} = \frac{\text{distance traveled by Solute } a}{\text{distance traveled by Standard } x} \tag{3.6}$$

Instead of R_F values, it is more convenient to use the hR_F (= 100 x R_F) values. The problems surrounding the determination of correct R_F values for linear, radial, and anti-radial development, and calculation of other retention data are discussed in detail in Ref. 1.

Identification by comparison with R_F, hR_F, R_M, or R_X values in the literature is fraught with uncertainty, because they depend on many factors, such as quality of stationary phase, humidity, layer thickness, development distance, and temperature. Comparison of the retention data with those of reference substances gives better results, but still does not provide unequivocal proof of identity. If the retention data of the compound to be identified are identical with those of the reference substance in three different solvent systems but the same stationary phase or with the same solvent system but three different types of stationary phase, the two compounds can be regarded as identical with a good probability. The probability can be increased by in situ UV and/or VIS spectra, and after color reaction, with off-line VIS spectra.

3.6.1.2 Visualization of compounds

For the visualization of compounds, one can use physical, chemical, or biological detection methods. Physical detection methods are based on substance-specific properties. The most commonly employed methods are the absorption or emission of electromagnetic radiation, which is measured by detectors (Section 3.6.1.3). The β-radiation of radioactively labeled compounds can also be detected directly on the plate. These nondestructive detection methods allow subsequent isolation and can also be followed by microchemical and/or biological detection methods [99]. Since physical detection methods are frequently not sufficient to establish identity they must be complemented by specific chemical reactions (derivatization). These reactions may be carried out either before or after chromatography.

Prechromatographic derivatization can be performed either during sample preparation or on the chromatoplate at the origin. It is generally used to introduce a chromophore, leading to the formation of strongly absorbing or fluorescent derivatives, to increase the selectivity of the separation, enhance the sensitivity of detection, and improve the linearity. If includes oxidation, reduction, hydrolysis, halogenation, nitration, diazotization, esterification, etherification, hydrazone formation, and dansylation [99].

The primary aim of postchromatographic derivatization is the detection of the chromatographically separated compounds for better visual evaluation of the chromatogram. This step generally also improves the selectivity and the detection sensitivity. Postchromatographic reactions can be carried out by spraying reagents onto the chromatoplate, by dipping the layer in reagent solutions, or exposing the plate to vapors. The reagent can also be in the solvent system of in the adsorbent. In most instances, subsequent heating is necessary. Postchromatographic derivatizations have been extensively reviewed [100,101].

For biological/physiological detection the separated compounds can be transferred to the biological system. Alternatively, bioautographic analysis, reprint methods, and enzymatic test may also be applied.

Reagents and detection methods have been summarized in a practical handbook by Jork et al. [99].

3.6.1.3 Optical spectroscopy

Regardless of whether they are colored or colorless, compounds absorbing in the UV range can be detected by direct scanning at the wavelength of maximum absorption of the compound in the sample. Some types of compound are naturally fluorescent, while others can be converted to fluorescent derivatives through pre- and/or postchromatographic reactions; this is a highly specific and sensitive method. Fluorescence quenching is limited to compounds that readily absorb in the wavelength range of maximum excitation of a phosphor, incorporated into the stationary phase [8].

Instruments for scanning densitometry after the above treatments can be operated in the reflectance, transmission, or combined reflectance/transmission mode [102]. Usually, the sample beam is fixed and the plate is scanned by mounting it on a movable stage, controlled by a stepper motor. The geometry of the light beam of scanners can have the form of a slit or a spot. The most common method is slit scanning, in which the sample beam illuminates a rectangular area on the chromatoplate surface, while the plate is transported in the direction of development. With the other type, a small light spot moves in two dimensions. In this case, three different kinds of movement are possible: meandering, zig-zag, and flying-spot scanning, the last-mentioned with a sinusoidal type of movement.

Different principal optical geometries are used in scanning densitometry: the single-beam methods, which can be used in the reflectance, transmittance, or simultaneous mode, and the double-beam methods. In the latter case the two beams can be either separated in time at the same point on the chromatoplate or separated in space and recorded simultaneously by two detectors. Fig. 3.10 shows the three most commonly employed optical arrangements.

Absorption spectra are rarely sufficient for identification, except when directly compared with the spectrum of a standard substance, measured on the same chromatoplate. If possible, preference should be given to the fluorescence mode, which has a much higher spectroscopic selectivity, because two different wavelengths, an excitation and an emission wavelength, are used for each measurement.

References on p. A146

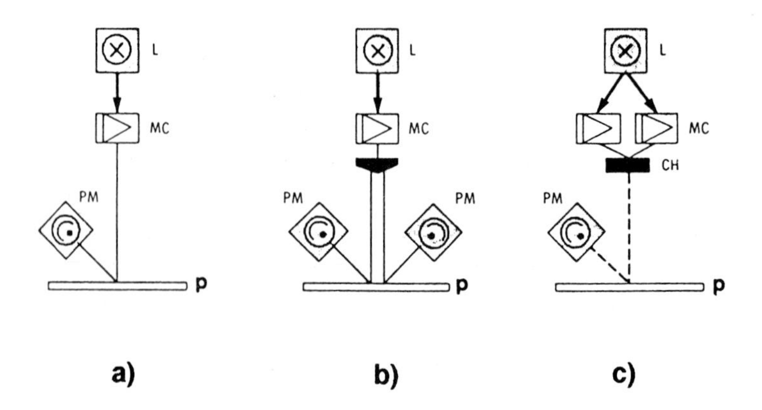

Fig. 3.10. Schematic diagrams showing the optical arrangements of different types of scanning densitometers. (a) Single-beam, (b) double-beam in space, (c) double-beam in time. L = Source, MC = monochromator, P = TLC/HPTLC plate, CH = chopper, PM = photomultiplier. (Reproduced from Ref. 82, with permission.)

3.6.1.4 Infrared and Raman spectroscopy

Depending on the nature of the compound, the IR absorption of a TLC spot can be measured either after transfer to an IR-transparent substrate or in situ [103]. The off-line method involves elution of ca. 5 μg substance, evaporation of the solvent, and pressing into a micropellet suitable for recording spectra.

IR spectra can also be recorded in situ by diffuse reflectance Fourier transform (DRIFT)-IR spectroscopy [104]. To obtain suitable IR spectra, all solvents must be carefully removed prior to measurement, and a background correction must be applied for absorption by the chromatoplate. Generally, more than 1 μg of compound is necessary for detection of individual functional groups and at least 10 μg for partial spectral recording.

Analysis of TLC spots by photoacoustic spectrometry (TLC/PAS) has been preferred over DRIFT analysis of strongly IR-absorbing samples [105]. The spot containing 1-50 μg of the sample must be physically removed from the chromatoplate. After some preparation, it is placed in the photoacoustic cell for measurement.

Reasonable spectra can also be obtained by surface-enhanced Raman spectroscopy (SERS) [106]. Recently, a new method was reported for preparing SERS-active surfaces, in which colloidal silver spheres are deposited on HPTLC plates. The sensitivity of these activated HPTLC plates is so high than in situ vibrational investigation of spots are possible at the femtogram level [107].

3.6.1.5 Mass spectrometry

Different methods have been described for obtaining mass spectra of TLC spots. The zone of the compound to be identified can be scraped from the stationary phase, and after elution and solvent evaporation, the residue is inserted into the mass spectrometer. Alternatively, the sample, together with the stationary phase, can be inserted directly. In fast atom bombardment (FAB) and secondary-ion mass spectrometry (SIMS), a high-energy ion or atom beam is used to sputter molecules from the condensed phase into the gas phase for mass spectrometric analysis. However, most mass spectrometric measurements are destructive in nature. FAB and SIMS are surface-sensitive methods in which the material actually consumed in the analysis is sputtered only from the top layers of the sample spot. The sample required for FAB and SIMS is between 1 ng and 1 μg [108-110].

3.6.2 Quantitative determination

Methods for the quantitative evaluation of thin-layer chromatograms may be divided into two categories. In the first, solutes are eluted from the stationary phase before being examined further; in the second, solutes are assayed directly on the layer.

3.6.2.1 Measurements after elution

Quantitation may be performed after scraping off the separated zone and recovery of the substance by elution from the sorbent. Thereafter, the methods used for quantitative analysis are essentially the current methods of microanalysis (e.g., GC, HPLC, liquid scintillation counting). A special possibility is on-line OPLC, where only one sample can be analyzed at a time, but all compounds are eluted from the chromatoplate and can be measured with the help of a flow-through detector, as in HPLC [111].

3.6.2.2 In situ measurements

Modern optical densitometric scanners (Section 3.6.1.3) are linked to computers and are equipped for automated peak location, multiple-wavelength scanning, and spectral comparison of fractions, and are capable of measurement in any operating mode (reflectance, absorption, transmission, fluorescence) [98].

In electronic scanning densitometry (ESD) the stationary chromatoplate is scanned electronically. This type of densitometer for image analysis requires a computer with video digitizer, light source, and appropriate optics, such as lenses, filters, and monochromators, and a vidicon tube or charge-coupled video camera. Electronic scanning always operates in the point-scanning mode, as opposed to devices with electro-mechanical scanning, which mostly use the slit-scanning mode. For two-dimensional separations, modern pattern recognition techniques can be employed. A powerful microcomputer of the IBM-AT class is usually adequate [112].

References on p. A146

For the quantitation of separated radioactive substances, autoradiography, liquid scintillation counting, and direct scanning with radiation detectors can be used. Recently, a new detector for radiochromatography was reported [113], which measures position and intensity of ionizing radiation on a two-dimensional TLC plate.

Photothermal deflection densitometers measure the refractive index gradient formed in the gas phase over a solid sample, heated by a laser [98]. In the absorption mode, the detection limit can be similar to that of optical scanning densitometers.

The applications of flame-ionization detectors are summarized in Section 3.8.1. Many books, book chapters [43,101,102,114-116] and review articles [e.g., 98,117] on quantitative PC are available to those interested in this topic.

3.7 PREPARATIVE PLANAR CHROMATOGRAPHY

The aim of preparative PC is the isolation of compounds in amounts of 10-1000 mg for structure elucidation (MS, ^{1}H-NMR, ^{13}C-NMR, IR, UV, etc.), for various analytical purposes (further chromatography), or for the determination of biological activity. The preparative chromatographic processes operative in capillary-flow-controlled PC, OPLC, and R.P.C. basically resemble those in their analytical counterparts [96]. However, some special characteristics need to be considered, such as the average particle size and thickness of the stationary phase, application of a large amount of sample, detection of the separated compounds, as well as their removal.

In preparative capillary-flow-controlled PC and in micropreparative R.P.C., only off-line methods can be used. This means that once the analytical or preparative plate has been developed and the solvent system is evaporated, the separated bands must be located and the desired compounds removed from the plate. For OPLC and R.P.C. on-line elution may also be used, where all compounds migrate over the whole separation distance. This results in a better separation in the lower R_F range and, additionally, allows the separated compounds to be directly eluted from the chromatoplate.

For self-made preparative plates the stationary phases most often applied are silica, alumina, cellulose, and gypsum. To produce layers with a thickness between 0.5 and 2 mm, so-called P-type sorbents may be used, which do not contain $CaSO_4$. Sorbents designated "P + $CaSO_4$" are suitable for preparing layers up to 10 mm thick. Slurrying the sorbent, spreading with commercial "thick-layer" spreading equipment, drying, and activating the preparative layers should be performed in compliance with the manufacturers' instructions, otherwise the layer may be damaged by pitting, cracking, or flaking. The advantage of preparing one's own plates is that any desired thickness or composition of plates becomes feasible.

Most users prefer to buy commercially available precoated preparative layers to producing their own. Besides saving time, precoated layers have the advantage of much better reproducibility than self-made plates. Generally, four types of precoated preparative plates are commercially available: silica, alumina, RP-2, and RP-18, in layer thicknesses of 0.5-2 mm. It is generally accepted that high resolution requires relatively thin layers (0.5-

1 mm). On a high-capacity (1.5-2 mm) layer the resolution is much more limited. The load capacity of a preparative layer without loss of separating power increases with the square root of the thickness. The load capacity of a 0.5-mm layer is approximately half that of a plate with a layer thickness of 2 mm.

In addition to the 20 x 20-cm or 20 x 40-cm plates, 20 x 100-cm plates are also commercially available for the separation of larger amounts of sample [10]. The silica materials commonly used for preparative separations have excessively coarse particles (average of ca. 25 μm), and their distribution is also excessive (5-40 μm). Various precoated preparative layers with a preadsorbent zone are also commercially available.

The process of sample application is an important step in preparative PC. The preferred method of placing a sample on a preparative layer is to apply it as a narrow streak across the plate. It is convenient to use precoated preparative layers with a concentrating zone. A solid-phase sample application method by Botz et al. [118] permits uniform sample application over the whole cross section of the preparative layer with the advantage of in situ sample concentration and cleanup and an extremely sharp edge. The proposed device can be applied in both capillary-action and forced-flow PC.

3.7.1 Off-line separation

For off-line separations of 5- to 15-mg samples analytical TLC and/or HPTLC plates can be used, but where the total amount of substances to be separated lies between 50 and 1000 mg preparative plates must be used.

Many methods are available for the location/detection of the separated components [5,8,10,11]. Precoated plates containing indicators fluorescent at 254 nm or 365 nm provide a general nondestructive mode of detection. If the compounds themselves are not visible or fluorescent, detection can be performed by applying specific reagents. After development, a vertical channel is scraped in the layer, ca. 0.5 cm from the beginning of the applied streak. After covering the rest of the layer with a glass plate, the uncovered part of the layer, sprayed with a suitable reagent, serves as a guide for locating the zones on the remainder.

The mechanical removal of these zones is followed by elution of the compounds from the stationary phase with a suitable solvent, separation from the residual adsorbent, and concentration of the solvent. Several commercially available devices and individually developed methods exist for eluting the compounds from the stationary phase [8-11].

3.7.2 On-line separation

Mincsovics et al. [119] summarized all possible combinations of off- and on-line OPLC separation techniques, which are also valid for R.P.C. The fastest FFPC separation can be obtained in the fully on-line mode (on-line sample application on an equilibrated stationary phase and on-line detection). This operating mode is also the simplest and most economic preparative FFPC method since, after cleaning and re-equilibration, the same plate may be used several times without loss of resolution.

References on p. A146

Since OPLC and R.P.C. may be used not only for on-line preparative separations, but also for analytical and micropreparative purposes, both analytical methods allow a direct scaleup to preparative OPLC and R.P.C., respectively.

From the TLC separation in unsaturated or saturated chromatographic tanks, the solvent system can be transferred via analytical OPLC, M- and U-R.P.C. to preparative OPLC, M- and U-R.P.C., respectively [30]. For scaleup, the sample may be applied on an analytical TLC plate and the amount of sample increased stepwise in subsequent separations. The resulting plates are scanned (off-line) to see whether the resolution is satisfactory. Thus, the maximum amount of sample for the on-line preparative separation is determined, considering the particle size and the volume of the stationary phase [29]. The flowrate of the mobile phase must be adapted to preparative separation, so that the migration of the α-front is as fast as in the analytical separation.

A special variant of on-line separation is the use of mixed sorbents, which can be accomplished with the column R.P.C. technique [67]. The stationary phase is poured into a closed radial chamber (column), which has a special geometric design. The volume of stationary phase stays constant along the separation distance, hence the name "column" R.P.C. Therefore, the flow is accelerated linearly by centrifugal force. The simplest method is to fill the major part of the planar "column" with silica and then for the last 1 cm separation distance with kieselguhr, as a preadsorbent zone [118]. It is also possible to fill the same planar "column" with more than two successive stationary phases in the order of increasing or decreasing polarity.

3.7.3 Selection of the appropriate method

Whether the use of forced-flow techniques is necessary or not depends on the kind of sample to be separated. Instrumental methods will increase preparation time and costs but also significantly improve efficiency. As a rule of thumb, if the sample contains more than 5 substances, up to 10 mg of sample can be separated by a micropreparative method and up to 500 mg by a preparative method. If the sample contains fewer than 5 substances, the amounts may be increased up to 50 mg and 1000 mg, respectively. If no more than 6 compounds are to be separated, distributed over the whole R_F range, and present in more or less the same amounts, and if the total amount of sample exceeds 150 mg, preparative capillary-flow-controlled PC can be used successfully. This is the simplest and therefore the most widely used method [90].

The potential of on-line linear OPLC on 20 x 40-cm plates with a separation distance of 36 cm as a preparative method is considerable. Because the overpressure in the existing instrument is limited and the particle size of the precoated plates is too large, not all advantages of this method can be realized yet. Generally, on-line OPLC can be used for the separation of 6 to 8 compounds in amounts of up to 300 mg [96].

The oldest forced-flow planar chromatographic method uses centrifugal force for on-line purification and isolation [15]. Generally, 15-μm particle size is used in the stationary phase with all the advantages of free selection of the size of the vapor space and development mode. Up to 10 compounds in amounts up to 500 mg can be isolated by the appropriate R.P.C. method [30,90].

3.8 SPECIAL PLANAR CHROMATOGRAPHIC TECHNIQUES

3.8.1 Combination with flame-ionization detection

TLC can also be carried out on permanent, rod-shaped layers, having mechanical and chemical properties that permit detection of the separated compounds with a flame-ionization detector (FID) [120]. It should be noted that this is a planar but not a "flat-bed" technique, i.e. the term flat-bed chromatography does not include all PC methods.

Recently, a new TLC/FID system has become available, the Iatroscan MK-5 (Iatron Laboratories). The layer, composed of a suitable sintered mixture of glass powder and adsorbent, is coated on a quartz rod with a diameter of 0.9 mm and a length of 15 cm. The glass powder (1-10 μm) is mixed with the stationary phase (5-10 μm) in a ratio between 2 and 10 to 1. At present, three types of rods are commercially available: Chromarod S (10-μm silica; layer thickness, 100 μm), Chromarod S II (5-μm silica; layer thickness, 50 μm), and Chromarod A (10-μm aluminum oxide; layer thickness, 35 μm).

Before the rods are used, they are cleaned and activated in the flame of the detector. After they have been placed in a holder, the samples are applied (1-50 μg of sample in 0.1-1 μl) and the chromatogram is developed in a saturated N chamber. After development, the solvent system is evaporated and the rods are placed in a sliding frame of the instrument, which passes through the FID at constant speed. The individual zones are ionized in a hydrogen flame, and the ionization current produced is amplified and fed into the integrator and recorder. The method, instruments, and applications have been surveyed by Ranny [120].

3.8.2 Sequential planar chromatography

The sequential development of analytical and preparative plates has the advantage that the solvent system supply is fully variable in time and location. Thus the resolution can be improved and the separation time reduced. The principle of the sequential technique is based on the fact that the solvent system velocity is much higher at the beginning of the separation than later on. After a first separation, the layer is dried, and either the same or a different suitable solvent system is applied. The supply of solvent system may be stopped at any time in order to transfer it directly to the area of the compound zones to be separated. Therefore, the high initial velocity of the solvent system is always used, and this substantially shortens the analysis time and increases the resolution. Sequential TLC can be carried out with the S-chamber-type Mobil-R_F chamber, developed by Buncak [121,122].

References on p. A146

3.8.3 Mobile-phase gradient

In PC a true mobile-phase gradient can only be used with the FFPC techniques. Whereas all forms of gradient are possible when these methods are used, until now only the use of step gradients has been reported. The positive effects of step gradients were demonstrated not only for analytical [123] and preparative OPLC separations [124], but also for analytical and preparative R.P.C. separations of various plant extracts [71].

3.8.4 Layer-thickness gradient

Use of preparative taper plates [125] greatly reduces spot elongation and overlapping, due to the gradient effect of layer thickness. The improved performance of the taper plate is similar to the improved resolution in the lower R_F range observed with radial TLC. In the taper plate, the cross sectional area traversed by the solvent increases as development progresses. Therefore, the cross sectional flow per unit stationary-phase area is always highest at the beginning of the layer, decreasing toward the mobile-phase front. As a result, the tail end of a zone moves faster than the front end, thus keeping each component focused in a narrow band. Bandbroadening is significantly reduced, especially for compounds with higher R_F values. Compounds with lower R_F values are subject to greater mobile-phase velocity relative to higher-R_F compounds than on conventional plates. This is because the amount of solvent at the front increases with migration distance. Because of this, the distance between bands at lower R_F values is increased, providing better separations.

3.8.5 Combination of radial and antiradial developments

Combination of the radial and antiradial development modes is possible by sequential development in R.P.C. In this technique, the mobile phase can be introduced at any desired place on the plate and at any time. In S-R.P.C. the solvent application system – a sequential solvent delivery device – operates by centrifugal force and with the aid of capillary action against a reduced centrifugal force (antiradial mode). Generally, the radial mode is used for the separation of zones and the antiradial mode for pushing zones back towards the center with a stronger solvent (e.g., ethanol). After drying of the plate under nitrogen at a high rotational speed, the next development with another suitable mobile phase may be started. This combination of two operating modes makes the separation pathways in S-R.P.C. theoretically unlimited [95].

3.9 COMPARISON OF VARIOUS PLANAR CHROMATOGRAPHIC TECHNIQUES

PC methods may be compared with respect to simplicity, reproducibility, efficiency, rapidity, and speed. TLC and HPTLC plates and techniques are compared in Refs. 1, 19 and 82. In this context, it should be noted that at smaller average particle size of the

TABLE 3.2

COMPARISON OF THE ANALYTICAL FFPC METHODS

Basis of comparison	Methods			
	OPLC	HPPC	U-R.P.C.	M-R.P.C.
Flowrate	depending on overpressure		depending on centrifugal force	
Vapor space	absent		practically absent	defined
Development mode	linear, radial, two-dimensional	radial (antiradial)	radial (linear, antiradial)	
Separation distance	18 cm (36 cm)	5 cm	8 cm (11 cm)	
Number of samples	up to 72	1 (24)	up to 72	
Temperature	programable	constant	controllable	
Disadvantages	disturbing zone multifront effect	multifront effect (disturbing zone)	overflow effect	
Special possibilities	on-line detection	multiple scanning	observation during separation (sequential technique)	

stationary phase and, therefore, at shorter separation distance, the detection limit is as much as 5-10 times higher in the absorption and in the fluorescence mode.

Chamber saturation is one of the most important factors in achieving reproducibility in capillary-flow-controlled PC. Therefore, the type of chromatographic chamber and the degree of saturation of the vapor phase should be stated when reporting results [1]. The possibilities of transferring optimized solvent systems between various PC methods are discussed in Ref. 126. It can be said that the most reproducible results are achieved with the vario-KS chamber, where the vapor phase can be freely selected.

Capillary-flow-controlled PC is much simpler to use than the forced-flow techniques. As a rule, FFPC techniques are of value only with the small-particle-size HPTLC plates. The linear development mode (over a 18-cm separation distance) can be recommended when pairs of peaks show incipient separation on a TLC plate with capillary-controlled flow but not sufficient resolution for quantitative evaluation. If the separation problem is in the upper R_F range, antiradial development should be tried; if it is in the lower R_F region, radial development is preferred. In both cases, the resolution can be increased further by FFPC techniques. If many compounds are to be separated (up to 72), U-R.P.C. in the radial development mode is preferred. If the separation problem is in the very-low R_F region

References on p. A146

(<0.2), HPPC in the single-sample analysis mode is the appropriate separation technique. A comparison of the analytical FFPC methods is given in Table 3.2.

In capillary-flow-controlled PC under optimized operating conditions, the spot capacity lies between 15 and 25. For two-dimensional development this value can be increased to 400, but it is almost impossible to exceed 500, except under very favorable conditions [19]. Using FFPC, the spot capacity can be between 60 and 100. If OPLC development is used in the first direction and the elution of the compounds in the second direction, a spot capacity of a few thousand could be achieved theoretically [127]. However, this has not yet been accomplished in practice, due to technical difficulties.

At the moment, the fastest separations can be achieved with OPLC and HPPC. A separation time of 2.5 min for 18 samples of nonpolar isomers has been reported [69].

A comparison of the different R.P.C. methods has been published [30]. Classical preparative layer chromatography was also compared with the various preparative FFPC techniques [96].

3.10 TRENDS IN PLANAR CHROMATOGRAPHY

3.10.1 Development of instrumentation

There is a strong trend toward instrumentation of the individual steps in PC. A significant improvement is the CAMAG sample applicator, which enables the selection of the *x*, *y*, and *z* coordinates for automatic sample application in linear, radial, and antiradial PC. Solvent front detection with a double-beam optical sensor [128] in an automated developing chromatographic tank with multiple plate holder is now commercially available. Also, a new instrument for pre- and postchromatographic derivatization is in development. Instrumentation has been recently surveyed by Szepesi [129].

A further improvement of the multidimensional detection systems [21] in the field of identification by spectroscopic methods (e.g., FAB-MS, DRIFT, SERS, SIMS) can be expected soon, based on results achieved so far. The next generation of densitometers will enable greater accuracy and more rapid quantitative determinations.

3.10.2 Development of forced-flow planar chromatographic methods

It is expected that future research will concentrate on the positive effects of forced flow, e.g., the applied pressure in OPLC and the centrifugal force in R.P.C. As a consequence, smaller particle size and a narrower distribution range will be needed for the stationary phases in order to achieve maximum resolution.

The advantage of combining on-line and off-line separations [119] as well as the two-dimensional development can also be exploited in OPLC. This development mode for planar "columns" was described by Guiochon et al. [127,130]. After development with one mobile phase, zones are eluted in perpendicular direction with a second mobile phase into a diode-array detector. Under optimized conditions a separation number (SN) > 1000 could be reached [1,19]. This would be a new record in the separating power of

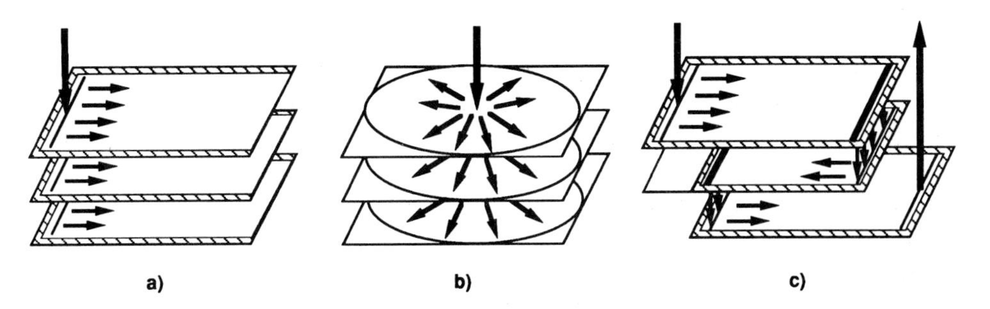

Fig. 3.11. Schematic drawing of overpressured-multilayer chromatography (OPMLC). (a) Off-line linear OPMLC, (b) off-line radial OPMLC, (c) long-distance OPMLC.

PC, but the experimental difficulties in implementing such a separation process are excessive.

A very realistic possibility of increasing the efficiency and rapidity of PC separations of complex samples is the use of multilayer OPLC (OPMLC). In the proposed version of Tyihák et al. [25] the same or different types of stationary phases can be used for the simultaneous development of several chromatoplates (Fig. 3.11a,b). This version is not only excellent for rapid off-line analytical OPMLC, but also suitable for HPPC and R.P.C. The efficiency of the method was recently demonstrated by Botz et al. [131], who developed five HPTLC plates simultaneously. By radial OPMLC 360 samples of plant extracts could be separated in 150 sec.

A novel category of OPMLC is the long-distance OPMLC, where the efficiency of the separation is increased significantly [132]. The end of the first chromatoplate has a slit-like perforation to permit the mobile phase to migrate to a second layer (Fig. 3.11c). Clearly, on this basis a very long separation distance can be achieved by adding one plate to another. Also, different stationary phases can be used so that each part of a complex mixture can reach a suitable stationary phase. The method is applicable to off-line and on-line determination of analytical samples and perhaps a combination of off-line and on-line analysis of complex samples.

3.10.3 Multimodal separations by planar chromatography

In multimodal separations [21] where PC is one of the modes, two separation techniques having complementary retention mechanism are used. It is generally accepted that in multimodal separations the second step is the PC. The first separation technique can be a thermo-extraction method, where the compounds are deposited on the plate by an inert gas or supercritical fluid, as proposed by Stahl [133]. The first separation method can also be GC or CLC. Although GC/PC was used successfully earlier, it is no longer popular.

References on p. A146

Boshoff et al. [134] were the first to describe an interface for depositing the effluent from conventional CLC on a TLC plate. Hofstraat et al. [135,136] reported the coupling of narrow-bore CLC (HPLC) and PC (HPTLC). Due to the low mobile-phase velocity (10-100 μl/min), the complete column effluent can be deposited on the chromatoplate, using a modified spray-jet band applicator from CAMAG. Using the advantage of the coupled reversed-phase microbore HPLC and normal-phase AMD techniques, Jänchen and Issaq [137] reported the separation of more than 50 compounds on a single chromatoplate.

Though only a few papers have been published so far concerning this type of multi-modal separations, successful applications of this technique can be expected in the near future. Another possibility that can be envisioned is that the effluent from a SFC separation may be coupled with the various types of forced-flow planar chromatography, like OPLC (HPPC) and R.P.C. Furthermore, it may be expected that the combination of on-line FFPC and AMD techniques will result in extremely high resolutions.

REFERENCES

1 F. Geiss, *Fundamentals of Thin Layer Chromatography (Planar Chromatography),* Hüthig, Heidelberg, 1987.
2 K. Macek, in E. Heftmann (Editor), *Chromatography. Fundamentals and Applications of Chromatographic and Electrophoretic Methods,* Elsevier, Amsterdam, 1983, pp. 162-194.
3 I.M. Hais and K. Macek, *Handbuch der Papierchromatographie, Band I, II, and III,* VEB Gustav Fischer Verlag, Jena, 1958, 1960, 1963.
4 J. Gasparic and J. Churacek, *Laboratory Handbook of Paper and Thin-Layer Chromatography,* Wiley, New York, 1978.
5 E. Stahl (Editor), *Dünnschicht-Chromatographie, Ein Laboratoriumshandbuch,* Springer, Berlin, 1967.
6 G. Zweig and J. Sherma (Editors-in-Chief), *Handbook of Chromatography,* CRC Press, Boca Raton, FL, 1972.
7 J. Kirchner and E.S. Perry, *Thin-Layer Chromatography,* Wiley, New York, 2nd Edn., 1978.
8 J.C. Touchstone and M.F. Dobbins, *Practice of Thin Layer Chromatography,* Wiley, New York, 1983.
9 C.F. Poole and S.A. Schuette, *Contemporary Practice of Chromatography,* Elsevier, Amsterdam, 1989.
10 J.C. Touchstone and J. Sherma (Editors), *Techniques and Applications of Thin Layer Chromatography,* Wiley, New York, 1985.
11 B. Fried and J. Sherma, *Thin-Layer Chromatography, Techniques and Applications,* Second Edition, Revised and Expanded, Dekker, New York, 1986.
12 R.E. Kaiser (Editor), *Einführung in die Hochleistung-Dünnschicht-Chromatographie,* Institut für Chromatographie, Bad Dürkheim, 1976.
13 R.E. Kaiser and A. Zlatkis, *High Performance Thin Layer Chromatography,* Elsevier, Amsterdam, 1977.
14 W. Bertsch, S. Hara, R.E. Kaiser and A. Zlatkis (Editors), *Instrumental HPTLC,* Hüthig, Heidelberg, 1980.
15 E. Tyihák and E. Mincsovics, *J. Planar Chromatogr.,* 1 (1988) 6.
16 Z. Witkewicz and J. Bladek, *J. Chromatogr.,* 373 (1986) 111.
17 R.E. Kaiser (Editor), *Planar Chromatography, Vol. 1,* Hüthig, Heidelberg, 1986.
18 D.E. Jänchen and H.J. Issaq, *J. Liquid Chromatogr.,* 11 (1988) 1941.
19 C.F. Poole and S.K. Poole, *Anal. Chem.,* 61 (1989) 1257A.
20 A.M. Siouffi, E. Mincsovics and E. Tyihák, *J. Chromatogr.,* 492 (1989) 471.

21 C.F. Poole, S.K. Poole, W.P.N. Fernando, T.A. Dean, H.D. Ahmed and J.A. Berndt, *J. Planar Chromatogr.,* 2 (1989) 336.
22 E. Tyihák, E. Mincsovics and H. Kalász, *J. Chromatogr.,* 174 (1979) 75.
23 E. Mincsovics, E. Tyihák and H. Kalász, *J. Chromatogr.,* 191 (1980) 293.
24 E. Tyihák, E. Mincsovics, H. Kalász and J. Nagy, *J. Chromatogr.,* 211 (1981) 45.
25 E. Tyihák, E. Mincsovics and T.J. Székely, *J. Chromatogr.,* 471 (1989) 250.
26 V. Pretorius, B.J. Hopkins and J.D. Schicke, *J. Chromatogr.,* 99 (1974) 23.
27 Z. Deyl, J. Rosmus and M. Pavlicek, *Chromatogr. Rev.,* 6 (1964) 19.
28 E. Heftmann, J.M. Krochta, D.F. Farkas and S. Schwimmer, *J. Chromatogr.,* 66 (1972) 365.
29 Sz. Nyiredy, S.Y. Mészáros, K. Dallenbach-Tölke, K. Nyiredy-Mikita and O. Sticher, *J. Planar Chromatogr.,* 1 (1988) 54.
30 Sz. Nyiredy, L. Botz and O. Sticher, *J. Planar Chromatogr.,* 2 (1989) 53.
31 E. Mincsovics and E. Tyihák, *J. Planar Chromatogr.,* 1 (1988) 309.
32 C.F. Poole, *J. Planar Chromatogr.,* 2 (1989) 95.
33 E. Tyihák, E. Mincsovics and A. Siouffi, *J. Planar Chromatogr.,* 3 (1990) 121.
34 R.E. Kaiser and R.I. Rieder, in R.E. Kaiser (Editor), *Planar Chromatography, Vol. 1,* Hüthig, Heidelberg, 1986, pp. 165-191.
35 R.E. Kaiser, *Einführung in die HPPLC, Hochdruck- Planar-Flüssig-Chromatographie (High Pressure Planar Liquid Chromatography),* Hüthig, Heidelberg, 1987.
36 R.E. Kaiser, *J. Assoc. Off. Anal. Chem.,* (1988) 123.
37 U.A.Th. Brinkman and G. de Vries, in J.C. Touchstone and J. Sherma (Editors), *Techniques and Applications of Thin Layer Chromatography,* Wiley, New York, 1985, pp. 87-107.
38 J.K. Rozylo, I. Malinowska and A.V. Musheghyan, *J. Planar Chromatogr.,* 2 (1989) 374.
39 K.K. Unger, *Porous Silica,* Elsevier, Amsterdam, 1979.
40 K. Günther, *J. Chromatogr.,* 450 (1988) 11.
41 M. Mack and H.-E. Hauck, *J. Planar Chromatogr.,* 2 (1989) 190.
42 S.H. Han and D.W. Armstrong, in N. Grinberg (Editor), *Modern Thin-Layer Chromatography,* Dekker, New York, 1990, pp. 398-427.
43 J.C. Touchstone and J. Sherma, *Densitometry in Thin Layer Chromatography, Practice and Applications,* Wiley, New York, 1979.
44 R.M. Scott, in J.C. Touchstone and J. Sherma (Editors), *Techniques and Applications of Thin Layer Chromatography,* Wiley, New York, 1985, pp. 25-35.
45 D. Nurok and M.J. Richard, *Anal. Chem.,* 53 (1981) 563.
46 D. Nurok, R.M. Becker, M.J. Richard, P.D. Cunningham, W.B. Gorman and C.L. Bush, *J. High Resolut. Chromatogr. Chromatogr. Commun.,* 5 (1982) 373.
47 E. Wang and H. Wang, *J. Planar Chromatogr.,* 3 (1990) 15.
48 H.J. Issaq, J.R. Klose, K.L. McNitt, J.E. Haky and G.M. Muschik, *J. Liq. Chromatogr.,* 4 (1981) 2091.
49 D. Nurok, R.M. Becker and K.A. Sassic, *Anal. Chem.,* 54 (1982) 1955.
50 S. Turina, in R.E. Kaiser (Editor), *Planar Chromatography, Vol. 1,* Hüthig, Heidelberg, 1986, p. 15.
51 B.M.J. De Spiegeleer, P.H.M. De Moerloose and G.A.S. Slegers, *Anal. Chem.,* 59 (1987) 59.
52 L.R. Snyder, *J. Chromatogr. Sci.,* 16 (1978) 223.
53 Sz. Nyiredy, C.A.J. Erdelmeier, B. Meier and O. Sticher, *Planta Med.,* (1985) 241.
54 K. Dallenbach-Toelke, Sz. Nyiredy, B. Meier and O. Sticher, *J. Chromatogr.,* 365 (1986) 63.
55 Sz. Nyiredy, *Application of the "PRISMA" Model for the Selection of Eluent-Systems in Overpressure Layer Chromatography (OPLC),* Labor MIM, Budapest, 1987.
56 Sz. Nyiredy, K. Dallenbach-Tölke and O. Sticher, *J. Planar Chromatogr.,* 1 (1988) 336.
57 Sz. Nyiredy, K. Dallenbach-Toelke and O. Sticher, *J. Liquid Chromatogr.,* 12 (1989) 95.
58 D. Nurok, *LC-GC Mag.,* 6 (1988) 310.

59 D. Nurok, *Chem. Rev.*, 89 (1989) 363.
60 F. Geiss, *J. Planar Chromatogr.*, 1 (1988) 102.
61 P. Petrin, *J. Chromatogr.*, 123 (1972) 65.
62 J.A. Perry, *J. Chromatogr.*, 165 (1979) 117.
63 R.E. Tecklenburg, R.M. Becker, E.K. Johnson and D. Nurok, *Anal. Chem.*, 55 (1983) 2196.
64 A. Niederwieser, *Chromatographia*, 2 (1969) 519.
65 A. Niederwieser and C.C. Honegger, in J.C. Giddings and R.A. Keller (Editors), *Advances in Chromatography, Vol. 2*, Dekker, New York, 1966, pp. 123-131.
66 D.E. Jänchen, in W. Bertsch, S. Hara, R.E. Kaiser and A. Zlatkis (Editors), *Instrumental HPTLC*, Hüthig, Heidelberg, 1980, pp. 133-164.
67 F. Geiss, H. Schlitt and A. Klose, *Fresenius Z. Anal. Chem.*, 213 (1965) 331.
68 Sz. Nyiredy, S.Y. Mészáros, K. Nyiredy-Mikita, K. Dallenbach-Toelke and O. Sticher, *J. High Resolut. Chromatogr. Chromatogr. Commun.*, 9 (1986) 605.
69 Sz. Nyiredy, C.A.J. Erdelmeier and O. Sticher, in E. Tyihák (Editor), *Proc. Intern. Symp. on TLC with Special Emphasis on OPLC*, Labor MIM, Budapest, 1986, pp. 222-231.
70 E. Tyihák and E. Mincsovics, *Hung. Sci. Instrum.*, 57 (1984) 1.
71 Sz. Nyiredy, K. Dallenbach-Toelke and O. Sticher, in F.A.A. Dallas, H. Read, R.J. Ruane and I. Wilson (Editors), *Recent Advances in Thin Layer Chromatography*, Plenum Press, London, 1988, pp. 45-54.
72 R.E. Kaiser, *J. Planar Chromatogr.*, 1 (1988) 265.
73 J.A. Perry, T.H. Jupille and L.H. Glunz, *Anal. Chem.*, 47 (1975) 65A.
74 G. Guiochon, M.F. Gonnord, A. Siouffi and M. Zakaria, *J. Chromatogr.*, 250 (1982) 1.
75 M. Zakaria, M.F. Gonnord and G. Guiochon, *J. Chromatogr.*, 271 (1983) 127.
76 J.A. Perry, K.W. Haag and L.H. Glunz, *J. Chromatogr. Sci.*, 11 (1973) 447.
77 J.A. Perry, *J. Chromatogr.*, 113 (1975) 267.
78 T.H. Jupille and J.A. Perry, *J. Chromatogr. Sci.*, 13 (1975) 163.
79 K. Burger, *Fresenius Z. Anal. Chem.*, 318 (1984) 228.
80 K. Burger, *GIT Suppl. Chromatogr.*, 4 (1984) 29.
81 K.D. Burger and H. Tengler, in R.E. Kaiser (Editor), *Planar Chromatography, Vol. 1*, Hüthig, Heidelberg, 1986, pp. 193-205.
82 D.C. Fenimore and C.M. Davis, *Anal. Chem.*, 53 (1981) 252A.
83 A. Studer and H. Traitler, *J. High Resolut. Chromatogr. Chromatogr. Commun.*, 9 (1986) 218.
84 Sz. Nyiredy, L. Botz and O. Sticher, *J. Chromatogr.*, in press.
85 Sz. Nyiredy, K. Dallenbach-Tölke and Sz. Nyiredy, *J. Chromatogr.*, 450 (1988) 241.
86 K. Dallenbach-Tölke, Sz. Nyiredy, S.Y. Mészáros and O. Sticher, *J. High Resolut. Chromatogr. Chromatogr. Commun.*, 10 (1987) 362.
87 R.E. Kaiser, *J. Planar Chromatogr.*, 1 (1988) 182.
88 A. Junker-Buchheit and H. Jork, *J. Planar Chromatogr.*, 2 (1989) 65.
89 D.C. Fenimore, in W. Bertsch, S. Hara, R.E. Kaiser and A. Zlatkis (Editors), *Instrumental HPTLC*, Hüthig, Heidelberg, 1980, p. 81.
90 Sz. Nyiredy, *Anal. Chim. Acta*, 236 (1990) 83.
91 J.C. Touchstone, in N. Grinberg (Editor), *Modern Thin-Layer Chromatography*, Dekker, New York, 1990, pp. 465-468.
92 S. Turina and K. Jamnicki, *Anal. Chem.*, 44 (1972) 1892.
93 G. Székely and P. Baumgartner, *J. Chromatogr.*, 186 (1979) 575.
94 E. Tyihák, Sz. Nyiredy, G. Verzár-Petri, S.Y. Mészáros, I. Farkas-Tompa, A. Nagy, L. Szepesy, L. Vida, E. Mincsovics, G. Kemény and Z. Baranyi, *Hung. Pat.* 189.737; *German Pat.* 3.512.547 (1986).
95 Sz. Nyiredy, C.A.J. Erdelmeier, O. Sticher, in R.E. Kaiser (Editor), *Planar Chromatography, Vol. 1*, Hüthig, Heidelberg, 1986, pp. 119-164.
96 Sz. Nyiredy, in J. Sherma and B. Fried (Editors), *Handbook of Thin Layer Chromatography*, Dekker, New York, 1990, pp. 283-315.
97 D.E. Jänchen, *Int. Lab.*, March (1987) 66.
98 C.F. Poole and S.K. Poole, *J. Chromatogr.*, 492 (1989) 539.

99 H. Jork, W. Funk, W. Fischer and H. Wimmer, *Thin-Layer Chromatography, Reagents and Detection Methods, Physical and Chemical Detection Methods, Reagents I,* Vol. 1a, VCH, Weinheim, 1990.
100 W. Funk, *Fresenius Z. Anal. Chem.,* 318 (1984) 206.
101 L.R. Treiber (Editor), *Quantitative Thin Layer Chromatography and its Industrial Applications,* Dekker, New York, 1987.
102 J.C. Touchstone and J. Sherma, *Densitometry in Thin Layer Chromatography, Practice and Applications,* Wiley, New York, 1979.
103 P.R. Brown and B.T. Beauchemin, *J. Liq. Chromatogr.,* 11 (1988) 1001.
104 G.E. Zuber, R.J. Warren, P.P. Begash and E.L. O'Donnell, *Anal. Chem.,* 56 (1984) 2935.
105 R.L. White, *Anal. Chem.,* 57 (1985) 1819.
106 E. Koglin, *J. Mol. Structure,* 173 (1988) 369.
107 E. Koglin, *J. Planar Chromatogr.,* 3 (1989) 194.
108 J.W. Fiola, G.C. Didonato and K.L. Busch, *Rev. Sci. Instrum.,* 57 (1986) 2294.
109 R.A. Flurer and K.L. Busch, *Anal. Instrum.,* 17 (1988) 255.
110 K.L. Busch, *J. Planar Chromatogr.,* 2 (1989) 355.
111 G.C. Zogg, Sz. Nyiredy and O. Sticher, *J. Planar Chromatogr.,* 1 (1988) 351.
112 V.A. Pollak and J. Schulue-Clewing, *J. Planar Chromatogr.,* 3 (1990) 104.
113 H. Filtuth, *J. Planar Chromatogr.,* 2 (1989) 198.
114 C.F. Poole and S. Khatib, in E. Katz (Editor), *Quantitative Analysis Using Chromatographic Techniques,* Wiley, New York, 1987, pp. 193-270.
115 H. Jork and H. Wimmer, *Quantitative Auswertung von Dünnschicht-Chromatogrammen,* GIT, Darmstadt, 1986.
116 S. Ebel, in F. Geiss (Editor), *Fundamentals of Thin Layer Chromatography (Planar Chromatography),* Hüthig, Heidelberg, 1987, pp. 420-436.
117 C.F. Poole, S.K. Poole, T.A. Dean and N.M. Chirco, *J. Planar Chromatogr.,* 2 (1989) 180.
118 L. Botz, Sz. Nyiredy and O. Sticher, *J. Planar Chromatogr.,* 3 (1990) 10.
119 E. Mincsovics, E. Tyihák and A.M. Siouffi, *J. Planar Chromatogr.,* 1 (1988) 141.
120 M. Ranny, *Thin-Layer Chromatography with Flame Ionization Detection,* D. Reidel Publishing Company, Dordrecht, 1987.
121 P. Buncak, *GIT Suppl. Chromatogr.,* 3 (1982) 3.
122 P. Buncak, *Fresenius Z. Anal. Chem.,* 318 (1984) 291.
123 J. Vajda, L. Leisztner, J. Pick and N. Anh-Tuan, *Chromatographia,* 21 (1986) 152.
124 Sz. Nyiredy, C.A.J. Erdelmeier, K. Dallenbach-Tölke, K. Nyiredy-Mikita and O. Sticher, *J. Nat. Prod.,* 49 (1986) 885.
125 UNIPLATE Taper Plate, *Analtech Technical Report No. 8202,* Newark, DE, 1985.
126 Sz. Nyiredy, K. Dallenbach-Toelke and O. Sticher, in H. Traitler, H. Studer and R.E. Kaiser (Editors), *Proceedings of the Fourth International Symposium on Instrumental High Performance Thin-Layer Chromatography,* Institut für Chromatographie, Bad Dürkheim, 1987, pp. 289-300.
127 G. Guiochon, M.F. Gonnord, M. Zakaria, L.A. Beaver and A.M. Siouffi, *Chromatographia,* 17 (1983) 121.
128 T. Omori, *J. Planar Chromatogr.,* 1 (1988) 66.
129 G. Szepesi, in N. Grinberg (Editor), *Modern Thin-Layer Chromatography,* Dekker, New York, 1990, pp. 285-311.
130 M.F. Gonnord and G. Guiochon, in E. Tyihák (Editor), *Proc. Intern. Symp. on TLC with Special Emphasis on OPLC,* Labor MIM, Budapest, 1986, pp. 241-250.
131 L. Botz, Sz. Nyiredy and O. Sticher, *37th Annual Congress on Medicinal Plant Research, Braunschweig, FRG, 5.-9. September 1989,* Abstr. No. 1-20.
132 L. Botz, Sz. Nyiredy and O. Sticher, *J. Planar Chromatogr.,* 3 (1990) 352.
133 E. Stahl, *J. Chromatogr.,* 142 (1977) 15.
134 P.R. Boshoff, B.J. Hopkins and V. Pretorius, *J. Chromatogr.,* 126 (1976) 35.
135 J.W. Hofstraat, M. Engelsma, R.J. van de Nesse, C. Gooijer, N.H. Velthorst and U.A.Th. Brinkman, *Anal. Chim. Acta,* 186 (1986) 247.

136 J.W. Hofstraat, S. Griffioen, R.J. van de Nesse, U.A.Th. Brinkman, C. Gooijer and N.H. Velthorst, *J. Planar Chromatogr.*, 186 (1986) 247.
137 D.E. Jänchen and H.J. Issaq, *J. Liq. Chromatogr.*, 11 (1988) 1941.

Chapter 4

Column liquid chromatography

H. POPPE

CONTENTS

4.1 INTRODUCTION

4.1.1 Liquid chromatography and high-performance liquid chromatography

By definition, liquid chromatography (LC) encompasses all chromatographic techniques in which the mobile phase is a liquid. Thus, thin-layer chromatography (TLC), size-exclusion chromatography (SEC), ion-exchange chromatography (IEC), to name only a few techniques, all belong to this class, independent of the nature of the stationary phase or the equipment used.

However, this chapter is devoted to column liquid chromatography (CLC). In this variety, chromatography is carried out in the *elution* mode; the sample constituents are visualized only after elution from the column, as they pass through the on-line coupled detector. CLC is now mainly carried out by the use of small particle sizes and consequentially application of high pressures, a technique known as high-performance (or high-pressure) liquid chromatography (HPLC) [1]. This chapter is focused on this technique and on developments that can be regarded as its offshoots.

It should be clear that HPLC defines the experimental technique, but does not say anything about the phase system. Indeed, all distribution mechanisms, e.g., size exclusion, reversed-phase and normal-phase adsorption, liquid/liquid partition, ion exchange (including the so-called ion chromatography) can be used in the HPLC mode.

4.1.2 Particle size

The essential role of small particles in HPLC has been amply discussed in Chapter 1. Also, the basic definitions of chromatography are given there. A brief recapitulation of the theory as it pertains to HPLC is given here.

The platenumber, N, of a column characterizes its general performance in obtaining separations, the efficiency. It is related to the experimentally observable σ_t (the standard deviation of the peak in time units) and t_R (the retention time) by (for isocratic elution)

$$N = (t_R/\sigma_t)^2 \tag{4.1}$$

The plateheight, H, a characteristic of the efficiency of the column construction independent of its length, is given by

$$H = L/N = L\,\sigma_t{}^2/t_R{}^2 \tag{4.2}$$

One can predict the values of H (and thus N) by using one of the plateheight expressions that result from the study of the chromatographic transport process in the column. These give the plateheight as a function of mobile-phase velocity, particle size, etc. There are many more or less similar expressions, depending on the model used and the phenomena taken into account. However, it is nearly always useful to switch to the dimensionless variables. These are

$$\text{Reduced plateheight, } h = H/d_p \tag{4.3}$$

$$\text{Reduced velocity, } v = \frac{ud_p}{D_m} \tag{4.4}$$

where d_p is the particle size, u is the linear velocity of the mobile phase, and D_m is the diffusion coefficient of the solute in the mobile phase.

The plateheight expression has simple forms like

$$h = \frac{2B}{v} + A + Cv \tag{4.5}$$

and when dimensionless variables are used, A, B, and C are dimensionless numerical constants (e.g., B = 0.75, A = 1, C = 0.03), which are in general fairly independent of the column characteristics, such as d_p, column diameter (d_c) etc. The function of Eqn. 4.5 implies that h has a minimum, usually in the range of 2-3, occurring at v values of 3-10. To the left of the minimum, i.e. at low velocities, the solute stays (too) long in the column and is given too much opportunity for longitudinal diffusion; to the right of the minimum the plateheight rises due to the C term in Eqn. 4.5, which represents the influence of slow equilibration of the solute, inter alia, between the two phases.

It should be noted that more refined plateheight equations are available (cf. Chapter 1), but as it is given here, it shows the essential traits. Clearly, one wants to have narrow peaks for high resolving power. The smaller the H value at a given length, the larger is N. It is therefore appropriate to decrease H by using small particle sizes, d_p (as $N = L/H = 1/h \times L/d_p$), and to attempt to work in the minimum in the h/v curve.

Obviously, this is limited by the sturdiness of the equipment; choosing smaller and smaller particle sizes leads to ever higher pressure (see Eqn. 1 of Chapter 1). Normally, one assumes that working above 30 to 40 megapascals (MPa), i.e. 300 to 400 times atmospheric pressure at the column inlet, is impractical.

The question of what particle size constitutes the best choice under such pressure-limited conditions has been dealt with by Knox and Saleem [2] in 1969. Readers interested in the theoretical derivations should consult that paper. Here, only the conclusions are given:

One starts with the assumption that ΔP is the maximum tolerable pressure drop, and that a platenumber N is necessary to perform the separation. The aim is to realize this in as short a time as possible. It is convenient to express this time as that of an unretained solute, t_o; the whole chromatogram is to take longer, but that involves a fixed factor. The expression one arrives at is

$$t_o = N^2 h^2 \phi \, \eta / \Delta P \qquad (4.6)$$

where ϕ is the shape factor for the pressure drop in the column (a constant with a value of 500-1000; cf. Chapter 1), and η is the mobile-phase viscosity, to which the pressure drop is also proportional.

This equation must be interpreted with some care. Assuming N, ϕ, and η to be constants and ΔP to have an upper limit, one cannot conclude for a given column that it is best to work at a minimum h value; if one would try to do so, neither N nor ΔP may have the required value. Rather, one should manipulate d_p, L, and v until one reaches the point where N and ΔP both have their required, or maximum value. This is accomplished according to the Knox-Saleem treatment when

$$d_p = (D_m \eta N \phi h \, v / \Delta P)^{1/2} \qquad (4.7)$$

Some illustrative values are given in Table 4.1.

The numbers in that table have some important bearings on practice:

(a) The most common HPLC columns, having roughly 5000 to 10 000 plates, should have a particle size in the order of 2-3 μm for the fastest analysis. The unretained solute would then be eluted after 10 sec. Most columns in use nowadays have 5-μm particles and are slower for that reason. Also, the packing efficiency is often less then optimal ($h > 2$). Optimum columns require a detector response faster than 0.1 sec and a volume

TABLE 4.1

UNRETAINED RETENTION TIMES AND PARTICLE SIZE FOR KINETICALLY OPTIMIZED CLC

Assumed parameters: $h_{min} = 2$; $\nu_{min} = 10$; $\Delta P = 40$ MPa; $\phi = 1000$; $\eta = 0.001$ kg m^{-1} sec^{-1}; $D_m = 10^{-9}$ m^2 sec^{-1}.

N	t_o (sec)	d_p (μm)	L (mm)	σ_t* (sec)	σ_V** (μl)
10^3	10^{-1}	0.7	1.4	$3\text{x}10^{-3}$	$6\text{x}10^{-1}$
10^4	10	2.3	4.6x10	$1\text{x}10^{-1}$	6 ←
10^5	10^3	7	$1.4\text{x}10^3$	3	$6\text{x}10^1$
10^6	10^5	23	$4.6\text{x}10^4$	$1\text{x}10^2$	$6\text{x}10^2$

* σ_t was found with $\sigma_t = t_o/\sqrt{N}$.

** σ_V was found from $\sigma_V = V_o/\sqrt{N}$, with V_o, the column deadvolume, $\pi/4\ d_c^2\ L\ \varepsilon_m$ and pertains to the unretained peak in a 4.6-mm ID column, with a porosity, ε_m, of 0.8.

standard deviation smaller than 6 μl.

(b) Moving from this optimized column (the one with the arrow) to less efficient but faster chromatography with $N = 10^3$ plates leads to very impractical dimensions and extremely small time and volume standard deviations. This area has not been explored very much experimentally; there seems to be very little interest in this sort of extremely fast analysis. However, it may become of importance when coupled-column techniques develop further: Analysis of fractions of the entire eluate of the first column on a second column may require extremely high speed on the latter.

(c) The attainment of higher resolution – desperately needed in many cases – is possible, but goes very much at the expense of analysis time. A ten-fold increase in resolution, e.g. in going from the second to the fourth row in Table 4.1, leads to an analysis time in excess of several days!

(d) Factor $h^2\phi$ plays a central role in Eqn. 4.6 and is indicative of how well the column performs in the time/resolution trade off. It has been dubbed "separation impedance", E, by Knox and Bristow [3]. The smaller E, the better is the performance of the column in generating high platenumbers at a given pressure drop. This concept was developed earlier in "capillary gas chromatography" by Golay [4], where it appears as the inverse, the "column performance factor". The same idea was also behind Halász' [5] proposal to calculate d_p (and with that, h) from the observed pressure drop. In that way the ϕ factor effect on E is incorporated in the h value.

(e) In the above, the column diameter is seen to play no role in the kinetic performance of the column. Indeed, this is true for all HPLC columns with diameters larger than ca. 250 μm. In such columns, with particle sizes in the range of 2-10 μm, the column aspect ratio, i.e. the ratio of column diameter to particle diameter, is larger than ca. 30. However, with smaller aspect ratios, present in some forms of micro-HPLC, the E values may be more favorable. Also, in open-tubular columns (Section 4.10) the E values are much smaller ($E=30$ under optimal conditions).

(f) Columns σ_t and σ_v of Table 4.1 will be referred to in Section 4.1.4.

4.1.3 Packing

A packing material for HPLC should fulfill the following requirements:

(a) It should implement a (phase) distribution system in accordance with the kind of separation aimed at; a completely inert packing, not sorbing or excluding any solutes, would not be of much use. Sorption may consist of distribution between two mutually immiscible liquid phases, the stationary phase being held in the pores of the packing material. Other distribution systems may involve the adsorption of solutes on a surface, which may be more polar (normal-phase chromatography) or less polar (reversed-phase chromatography, RPC) than the mobile phase. Other varieties are IEC (Chapter 5) and SEC (Chapter 6). In nearly all cases the use of a porous packing is necessary. The pore volume usually forms a significant part of the total column volume, often ca. 40%. The walls of the pores inside the particles constitute the larger part of the surface area of the material, usually in the range of 50-400 m^2 per ml of total column volume. However, with extremely small particles the outer surface itself provides a sufficiently large area.

(b) The packing material should be available in a well-defined particle size, with a narrow particle size distribution. Only with such batches can small E values be obtained.

(c) The particles should preferably be rigid and have sufficient mechanical strength so as to avoid collapse of the bed structure during packing and operation of the column.

(d) The chemical stability should be high; in particular, the material should be resistant toward chemical attack by the mobile phase.

Properties of packing materials are discussed in more detail in Section 4.9.

The packing process serves to arrange th particles in a bed of high regularity and stability. To obtain low plateheights it is essential that the bed be uniform across the width of the column. If not, the mobile phase (and the solutes!) will move faster where the permeability is greater and the bands will spread out considerably. Such effects can be visually observed in classical CLC experiments with colored solutes in glass columns.

It is generally accepted that the main cause of such radial differences in permeability is segregation of particles according to size during the packing process, e.g., larger particles arriving preferentially in the center, smaller ones near the wall. One way to avoid this is to start with a narrow distribution of particle sizes; the broader the size distribution, the more attention must be paid to the packing method.

However, the latter is always critical. Unfortunately, this is probably the least-understood aspect of HPLC. The numerous recipes that are in use are based on experience rather than theory. Many commercial firms market well-packed columns. It is therefore generally unwise for a laboratory to pack just one or a few columns, because it takes (apart from the cost of the equipment) quite a few fruitless attempts before a reasonable column is produced.

There are many ways to pack a column. Techniques such as sedimentation, dry packing with tamping, or just pouring the material into the column, are generally not successful if the particle size is below 20 μm. Slurry packing is virtually always used in HPLC. The material is suspended in a suitable liquid, and this slurry is pumped (indirectly; pushed forward by a "displacer' liquid, since one cannot have the slurry in the pump) into the column, normally at rather high flowrates and pressures. The bed is formed by (auto)filtration of the particles. Segregation of particles during this process is believed to be caused mainly by partial sedimentation, before or during packing, of the slurry: Larger particles sediment faster than smaller ones, and the resulting nonuniform slurry may lead to a nonuniform bed. Therefore, it is generally believed that the slurry should be reasonably stable and filtration should procede fast. Also, the slurry liquid should wet the surface of the particles without aggregation of particles.

Slurries have been stabilized in various ways:

(a) Balanced density. According to Majors [6] one chooses a slurry liquid with the same density as that of the particles.

(b) Surface-charge stabilization according to Kirkland [7]. By choosing a suitable slurry liquid composition (e.g., a high pH for silica) a surface charge builds up, and the slurry is stabilized by mutual repulsion.

(c) Viscosity stabilization. Sedimentation is slow in a viscous liquid. However, the filtration (packing) process itself is, of course, also slowed down by this high viscosity. Still, experience shows that this method usually works.

Nowadays most experimenters use a mixture of a lower alcohol and chloroform; probably several of the above effects make this method work.

Apart from the aspects mentioned above, a lot of other variables must be under control. Packing speed or (interrelated) packing pressure (possibly programed), chemical nature of the solvent, waiting time after preparation of the slurry and again after packing, final "compacting" of the bed with the slurry liquid or with some other liquid, etc., all have some influence.

As indicated, there is no proper theoretical understanding; for instance, there is not always an explanation why the chemical nature of the packing (silica as such, alkyl-modified silica, silica-based ion exchangers) has such a strong bearing on the conditions required in the packing process. Therefore, this section closes with a repetition of the advice not to undertake a column packing project unless there are very compelling reasons for it.

4.1.4 Ancillary equipment

As discussed in Section 4.1.2, the most commonly used HPLC column has a diameter of 4.6 mm and a length of about 100 mm, and it contains particles of about 5 μm. This "standard HPLC column" is taken as the starting point for the discussion of ancillary equipment.

Injector and detector should be of such a design that they do not impair the separation obtained in the column. As explained in Chapter 1, the peakwidth generated by the complete chromatographic system, consisting of injector, column, and detector, is related to the peakwidths generated by each part, as

$$\sigma_{V,tot}^2 = \sigma_{V,inj}^2 + \sigma_{V,col}^2 + \sigma_{V,det}^2 + \sigma_{V,con}^2 \tag{4.8}$$

where σ_V = volume standard deviation (e.g., expressed in μl or mm^3), σ_V^2 = volume variance (μl^2), tot = total, inj = injector, det = detector, col = column, con = connectors and terminators.

The resolution equation (cf. Chapter 1) takes only $\sigma_{V,col}$ into account, while the actually observed resolution is smaller to the extent that $\sigma_{V,tot}$ is larger than $\sigma_{V,col}$. In general, one wants to limit this loss in resolution to, say, 5%. Some arithmetic then shows that one must see to it that

$$\sigma_{V,det} < 0.3\,\sigma_{V,col} \quad \text{and} \quad \sigma_{V,inj} < 0.3\,\sigma_{V,col} \tag{4.9}$$

The volume standard deviations generated by injector and detector, required to be smaller by a factor of ca. 3, are not equal to the injected volume and the detection cell volume, respectively. However, when these parts are properly designed (absence of unnecessary or "dead" volumes), they are related to these geometric volumes. The best one can accomplish under ideal conditions is

$$\sigma_V = \frac{1}{\sqrt{12}} V = 0.29\,V \tag{4.10}$$

where V is the geometric volume of the detector cell or injector loop. These volumes, therefore, can at best (with the 5% loss in resolution criterion) be about equal to the column standard deviation.

It should be noted that the values for the latter quantity have been calculated here for an unretained component. For retained ones the values are larger by a factor of $(1 + k')$. Often one can accept losses in resolution for less-retained compounds, and in that case the volumes of detector and injector can be larger.

The limitation on these volumes is troublesome for two reasons: In the first place, the design and machining of injector and detector flow channels must be carried out with extreme care in order to avoid additional contributions to peakwidth. Fortunately, HPLC equipment with external contributions in the order of 10-20 μl is provided by instrument manufacturers, and this suffices for the less critical experiments.

The second reason is more fundamental: the volume limitations restrict the concentration sensitivity of the HPLC analysis in the following way: The concentration, $c_{i,e}$, of an eluted component, i, is equal to [8,9]

$$c_{i,e} = \frac{Q_{\text{inj}}}{\sqrt{2\pi} \cdot \sigma_V} = \frac{V_{\text{inj}} \cdot c_{i,s}}{\sqrt{2\pi} \cdot \sigma_V} \tag{4.11}$$

where $Q_{i,\text{inj}}$ is the injected amount of i; σ_V is the volume standard deviation of component i; V_{inj} is the injected volume of sample; and $c_{i,s}$ is the concentration of i in the injected sample solution.

It follows that the concentration at the column outlet is smaller than that in the sample by the dilution factor

$$\sqrt{2\pi} \cdot \sigma_V / V_{\text{inj}} \tag{4.12}$$

V_{inj} is limited because we do not want a loss in resolution. The minimum dilution factor is ca. 3. This can be much higher if the system is not optimized, or for more retained components for which σ_V is larger. As will be shown in Section 4.3, this problem can often be minimized by using on-column concentration techniques.

Apart from the chromatographic requirements mentioned, ancillary parts of the HPLC system must, of course, satisfy more general requirements imposed by the intended use as (routine) analytical devices: They must have a high precision and stability (injection volumes, signal response factors), must be easy to control and operate, and preferably be suitable for automatic operation.

4.1.5 Loadability

When the column diameter, d_c, is varied, while L, N, and H remain constant, the following relations hold: σ_V, $V_{\text{inj,max}}$, F_c, and $Q_{i,\text{max}}$ increase with d_c^2, while t_R, σ_t, and Δp remain constant.

The values of these parameters, for a 15-cm-long, 5-μm-particle column, at a velocity of 2 mm/sec with $h = 3$, are given in Table 4.2.

As explained in Chapter 1 and Section 4.1.4, there is a maximum volume that can be applied, given approximately by [8,9]

$$V_{\text{inj}} < \sigma_{V,\text{col}} \tag{4.13}$$

(see also Eqn. 4.10 and following text). The value of $\sigma_{V,\text{col}}$, with fixed column length, particle size, velocity, etc., is proportional to the square of the diameter of the column according to

$$\sigma_{V,\text{col}} = \varepsilon_m \cdot \pi/4 \cdot d_c^2 \cdot H \cdot \sqrt{N}\ (1+k') \tag{4.14}$$

where ε_m is the porosity of the column, the volume fraction occupied by the mobile phase.

Thus, as the H, N, and L $(=N \cdot H)$ values are more or less dictated by our desire to obtain good separations in reasonable time, the column diameter plays the key role in determining the volume scale of the experiment and of the ancillary devices. Often one

References on p. A221

TABLE 4.2

LOADABILITY OF A COLUMN, 15 cm LONG, 5-μm PACKING, FLOWRATE 2 mm/sec

For symbols, see text.

Column diameter (mm)	σ_V and $V_{inj,max}$ (unretained) (μl)	F_c (ml/min)	$Q_{i,max}$ (μg)	σ_t (sec)
10	100	7.5	250	0.8
4.6	20	1.6	50	0.8
3	8	0.7	20	0.8
2	4	0.3	10	0.8
1	1	0.07	2.5	0.8
0.5	0.2	0.02	0.6	0.8
0.25	0.06	0.005	0.15	0.8

would like to inject a large volume in order to make compounds detectable, or to collect significant amounts of material in a preparative mode. However, as resolution is the most precious asset in HPLC, one cannot afford to violate Eqn. 4.13. This violation can be avoided by choosing d_c in such a way that $\sigma_{V,col}$ is large enough. The same holds, mutatis mutandis, when the detection cell volume is the limiting factor; one can increase the volume scale of the column so that it is compatible with the size of the detector cell.

The amount that can be separated in one chromatogram is limited, as at high concentrations the distribution does not adhere to a linear isotherm (cf. Chapter 1). Recent work has led to a clear treatment [10-12]. It appears that a dimensionless load, W_{xn} ($2\cdot m$ in Refs. 10 and 12), can be defined that yields a good a priori estimate of extra peak broadening due to this kind of overload. The expression is not very easily derived for the general case, but for the case of Langmuir adsorption (and this applies well to normal- and reversed-phase adsorption systems as well as to ion exchange), the expression is relatively simple

$$W_{xn} = \frac{Q_i}{S_i} N \left(\frac{k'}{k'+1}\right)^2 = \frac{Q_i}{s_i} \left(\frac{k'}{k'+1}\right)^2 \tag{4.15}$$

where Q_i = the amount of i, S_i = the amount of i on the total surface of the column when this is saturated, and s_i = the same amount, but for the adsorbent in one plate, $s_i = S_i/N$.

It follows from both theory [10,12] and experiment [11] that the peakwidth, σ_Q, divided by the one observed at near-zero load, σ_0, is a universal decreasing function of W_{xn}, practically independent of adsorbent type, platenumber, diameter, etc.

$$\frac{\sigma_Q}{\sigma_o} = f(W_{xn}) \approx 1/\sqrt{(1 + W_{xn}/4)} \tag{4.16}$$

Defining a maximum acceptable load at the point in this function where $\sigma_Q/\sigma_o = 1.05$ (5% loss in resolution), one finds $W_{xn} = 2\text{-}4$. The fact that W_{xn} is a constant for, e.g., a 5% increase in peakwidth, leads to the important conclusion that loadability, $Q_{i,\max}$, is equal to (apart from a constant factor 2-4) the saturation amount in one plate. Thus, as packing density and surface area per gram do not vary too much when various column shapes are compared, it is proportional to the product $d_c^2 \cdot H \cdot q_s$, where q_s is the saturated amount per unit column volume. The increase of loadability width d_c^2 is, of course, most important, and this is quite obvious even without the above treatment. However, the occurrence of H in the expression is also significant: it shows that past and future attempts to increase kinetic performance (smaller H), e.g., by using smaller particle sizes, will lead to smaller and smaller loadabilities. It should be noted that the concentration at the column outlet, under such normalized overload conditions (fixed increase in peak width, say 5% as above), is constant, and determined by the nature of the phase system and solute, but not by column dimension and particle size. Finally, q_s is of great importance; the packing should have a reasonably high surface area (Section 4.9).

4.1.6 Types of LC columns; miniaturization

Table 4.3 gives an overview of commonly used column hardware. Some aspects of these columns will be treated in Section 4.9. Here it is appropriate to discuss the analytical advantages of miniaturization [13,14], i.e. columns with smaller diameters.

The amount of analyte, Q_i, leads to an eluate concentration $Q_i/(\sqrt{2\pi}\sigma_V)$. Clearly, the smaller peak volumes in small-diameter columns lead to much less dilution of the injected amount. As long as the detectors require no redesign because of the required volume standard deviation, this results in a very profitable increase in the signal-to-noise ratio. Microanalysis, a term used here to indicate situations where the amount of sample is limited, is therefore greatly improved by miniaturization of the column diameter (up to a certain point, vide infra).

In the opposite case the sample has a low analyte concentration, but large volumes of it are available. In this case, where it is better to speak about trace analysis, such an advantage is not found: Any analysis with a given column can be carried out with the same signal-to-noise ratio in a wider column, provided that the injection volume is adapted accordingly. However, it may still be preferable to use smaller-bore columns, because, for one thing, sampling and sample pretreatment may be easier on a smaller-volume scale.

Other advantages that have been mentioned for smaller-diameter columns are briefly discussed now, in the order of decreasing importance:

(a) Interfacing with sample pretreatment and especially detectors/spectrometers that are not capable of handling high flowrates. Special cases here are the coupling to GC detectors [14,15], such as the electron-capture detector (ECD) and the alkali flame-ionization detector (AFID), which allow the selective and sensitive detection of classes of

TABLE 4.3

TYPES OF LC COLUMNS

Diameter (mm)	Names used	Column hardware material	σ_V^{o}* (μl)
4.6	Normal/standard	stainless steel, occasionally glass	40
3-1	Small-bore/narrow-bore	stainless steel, glass	20-2
1-0.25	Microbore	Teflon, glass, fused silica, stainless steel	2-0.1
< 0.25	"Microcapillary"	fused silica, glass	< 0.1
Note:	Various column types for electrically driven chromatography, electro-kinetic micellar chromatography, capillary zone electrophoresis: mostly fused silica, 50-200 μm ID.		

* For $k' = 0$, calculated for $L = 1$ m, $d_p = 5\,\mu$m, $h = 2 \rightarrow N = 10^5$.

compounds. Also, coupling to mass spectrometry and Fourier transform infrared spectrometry often is more elegantly and effectively carried out on a smaller-volume scale.

(b) Savings in solvent, packing, and column hardware. It has been argued that only a small part of the total expense per analysis is connected with these items. Still, many analysts think this is a real advantage, also because of environmental and safety problems. All depends, of course, on the particular type of analysis. With very expensive solvent components and packings, e.g., in the area of enantiomer separations, this point may be taken very seriously. With ordinary packings and solvents the advantage will certainly stop to be relevant at a column diameter of 2-3 mm.

(c) When precious materials are being separated, e.g., valuable protein preparations [16], it may be very unwise to spread these out over large volumes and surfaces, with the increased risk of all kinds of unwanted effects, such as dilution, contamination, and irreversible adsorption.

(d) Higher platenumbers might be achievable [13,14,17]. As will be discussed in Section 4.9, small-diameter columns may be used to advantage for obtaining higher platenumbers.

4.2 SOLVENT DELIVERY SYSTEMS

4.2.1 Liquid-chromatography pumps

Although the use of electro-osmotic forces has been proposed for actuating the LC process (see Section 4.10), virtually all HPLC experiments at present depend on the induction of flow by means of a mechanical liquid pump. It is useful to have some understanding of their operating principles. For readers who find that convenient, we shall occasionally use analogies with electrical concepts.

The pressure over an element, i, such as a column or a connecting tube, is given by the expression

$$P = R_i \cdot F \quad (4.17)$$

where R_i is a resistance, analogous to that in Ohm's Law, and F is the flowrate. Equations for R_i can be found in the treatments in Chapter 1 on permeability of columns and tubes. It is important to note that R_i includes the viscosity and is therefore dependent on the composition of the liquid. Units for R_i may be, e.g., Pa/ml/min.

Another important aspect is the compressibility, χ, of liquids. The value of this is the fractional volume change when the pressure is increased by one unit, and its value is between $5 \cdot 10^{-10}$ Pa^{-1} (water) to $18 \cdot 10^{-10}$ Pa^{-1} for other liquids. Note that gases have a compressibility of 1/prevailing pressure. Supercritical fluids can have even larger compressibilities, so that the effects described in this section become much more relevant in SFC (Chapter 8). Although the small values for liquids may suggest that compressibility is unimportant in HPLC, under the highest pressure encountered (say, 40 MPa), the volume change is between 2% and 7%. Under many conditions this effect tends to influence accuracy and precision rather drastically.

An element, i, of volume V_i in the flow system acts as a reservoir of liquid when the pressure drops. Due to the expansion of the compressed liquid, it can deliver flow for some time. Upon pressure increase, the element could "absorb" flow. Thus, each such an element, like a capacitor in electric circuits, can bring about delays in pressure and flow. It is convenient to define the "capacity", C_i, (analogous to electrical capacity) of an element, i, by the absolute volume "absorption" when the pressure changes by one unit

$$C_i = V_i \cdot \chi \text{ (units, e.g., ml/Pa)}$$

Imagine what happens if such a capacity, under pressure, is connected to a resistance R_j, like a column, with the exit of the latter open to ambient pressure. The pressure will be released through the column. Solution of the associated differential equation shows that the pressure will drop according to an exponential function [18] of the time, t, with time constant τ

$$P = P(\text{start})\, e^{-t/\tau} \quad (4.18)$$

where τ can be calculated as the product

$$\tau = R_j \cdot C_i = R_j \cdot V_i \cdot \chi \quad (4.19)$$

Thus, it is relatively simple to get an idea of the size of these time delays. More accurate calculations would also require including the effect of mechanical deformations of the containers.

In electrical as well as in hydrodynamic circuits, there may be two ways of actuation:

(a) Current of flow source. An element (pump in our case) induces a given current or flow into the system, independent of what reactions (in the form of voltage or pressure increases) occur in the system. One may indicate that ideal (because that is what it is for LC) as an infinitely hard pump.

(b) Voltage or pressure source. The element (battery or pump) now imposes a voltage or pressure on a particular point in the circuit, no matter how the circuit reacts.

Fig. 4.1a shows a schematic drawing of a so-called reciprocating-piston LC pump. With an incompressible liquid and leak-free check valves, it would work as a flow source or hard pump; each stroke would force the volume displaced by the piston through the column. The flowrate is $n_s \cdot V_d$, where n_s is the stroke frequency and V_d is the displacement volume of the piston. Only the constancy of the electric motor could be a matter of concern as far as stability is concerned. However, the latter effect is ordinarily negligible. Problems arise primarily because the liquid is compressible and because the valves often leak.

Due to compressibility, a displacement (expressed as a volume) of the piston equal to $V_c \cdot \chi \cdot \Delta P$ is needed to reach the outlet pressure, ΔP, of the pump, equal to or sometimes higher than the operating pressure of the column. In this expression V_c is the

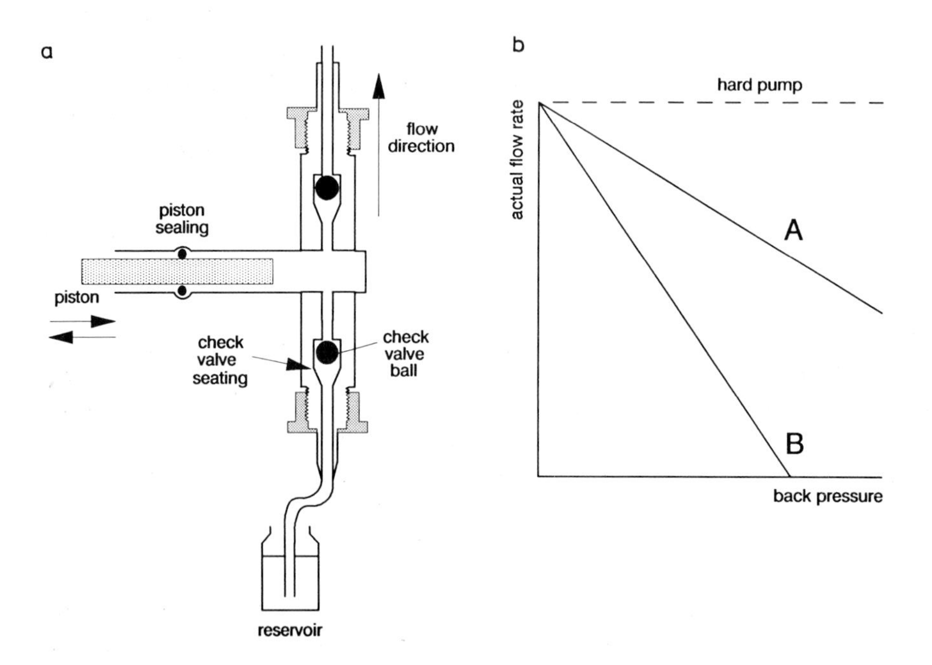

Fig. 4.1. Basic scheme of a reciprocating LC pump and working curve.

volume of the pumping chamber (again neglecting the mechanical deformation of the chamber), which is always larger than V_d. The higher the pressure, the stronger is this effect. The loss in flowrate is thus $n_s \cdot V_c \cdot \chi \cdot \Delta P$. One obtains a working curve, that is the flowrate plotted versus the pressure, as Line A in Fig. 4.1b. Curve B in this figure also takes the leakage into account. Assuming that this is a linear effect, one can represent that as a resistance, R_l, parallel to the pump circuit. The overall efficacy of pumping is equal to

$$1 - \left(\frac{V_c}{V_d} \chi - \frac{1}{R_l} \right) \frac{\Delta P}{F_i} \tag{4.20}$$

The steeper the working line, the softer is the pump; a hard pump has a nearly horizontal working curve. The flow can become zero at some value of ΔP; the pump would reach that pressure if one blocks the exit port. Small χ and ΔP values and small or no leakage would, of course, be favorable in this respect. More important for the design is that the ratio V_c/V_d be as low (as close to 1) as possible, because it so-to-speak multiplies the compressibility effect (see Eqn. 4.20).

The pump as such would enforce a pulsating flow, obviously an undesirable situation. To a first approximation the flow/time course would show a meander pattern. The pulse dampener serves to flatten out these pulses. It consists of a "capacity" that will change its volume, V_p, on changing pressure. It can be a rigid vessel, thanks to the compressibility of the liquid, but one can also use more effective devices, such as deformable stainless-steel containers; flat, spiralized tubes; or plastic tubes in a larger container with a suitable well-compressible liquid.

The capacity is again expressed in volume change per unit pressure, and it would be equal to $V_p \cdot \chi$ when only liquid compressibility is doing the job. Together with the flow resistance, R_c, of the column, this forms a damping combination, as discussed above. Its time constant, τ, equals

$$\tau = R_c \cdot C_p = \frac{\Delta P}{F_c} \cdot V_p \, \chi \tag{4.21}$$

where ΔP is now the pressure drop over the column. It may be clear that the larger τ is, i.e. the more liquid the damping container has "in store" and the less of it is used, the better is the smoothing action. In fact, it is the product $\tau \cdot n_s$ (dimensionless) which determines this. For a remaining "ripple" of 10% one would need $n_s \cdot \tau > 2$. Clearly, high pump frequencies, n_s, facilitate effective pump dampening. In general, one does not want to make V_p too large, because this leads to slow reactions to new pump settings. More important is the fact that the volumes needed to "prime" the pump with a new mobile-phase composition would be too large.

Before discussing the various types of pumps, it is necessary to mention one additional general point. The compressibility of the liquid causes the flowrates to vary over the length of the column. The question is what flow we want to keep constant. There is, unfortunately, no clear answer to that:

(a) Peak positions are determined by the column-averaged flowrate, and consequently one would like to eliminate the decompression for ca. 50%.

(b) Peak areas are, however, influenced by the value of the outlet flowrate. This would be an indication to keep the outlet flowrate constant.

Fortunately, in nearly all cases, the HPLC conditions (i.e. pressure) are sufficiently constant so that these effects can be included in the retention and peak intensity calibration.

If flow measurements could be carried out accurately and on-line, modern HPLC pumps would probably look quite different. In that case, a feedback mechanism, operating in the control software for the piston movement, could keep the flow precisely at the set value. Also, pressure and composition influences could be eliminated. Unfortunately, no such solvent-independent transducers are available.

Pressure, on the other hand, can be measured quite accurately on-line with pressure transducers. Many solvent delivery systems house such devices, not merely for displaying the prevailing pressure to the user. To an increasing extent, the pressure signal is used to actuate all kinds of feedback and feedforward mechanisms that control the piston movement. These will be mentioned in the next section.

4.2.2 Pumping systems

4.2.2.1 Requirements

The following requirements for LC pumps can be formulated:

(a) Ideally a constant flowrate or pressure should be generated. That amounts to infinitely "hard" or infinitely "soft" behavior, respectively. If this is impossible, one would at least want to have a constant "working curve", such as the one shown in Fig. 4.1.

(b) Flowrate or pressure should preferably be calibrated accurately.

(c) The change to another solvent should be fast and convenient. That means that volumes should be small, and unswept ("dead") volumes should not occur.

(d) Pulsation of flow should be kept to a minimum.

(e) The pump should be compatible with all solvents customary in HPLC.

(f) Flowrates should be adjustable between wide limits.

Common problems encountered during the use of LC pumps are briefly discussed now.

Air bubbles in the pump or detector can be an irritating problem, sometimes spoiling the reliability of the operation and of the data produced. Some patience at the startup of a new system is advisable. However, if the problem persists, the solvent in the reservoirs must be degased. This can be done by:

(a) Boiling (effective, inconvenient, may change solvent composition).

(b) Ultrasonic treatment (brings concentration of gases down to saturation level, not to zero. Combination with vacuum is more effective and has only minor disadvantages).

(c) Helium sparging (very effective, patented, has only a slight effect on composition, if carried out properly).

(d) Use of a flowthrough degaser, basically consisting of a tube that is permeable to gases but not to the solvent, surrounded by vacuum or helium.

All HPLC pumps, having moving parts such as check valves and pistons, are sensitive to dust and other particulate material in the liquid pumped. It is therefore wise to filter liquids before use and to insert a filter in the suction line in order to avoid malfunction and permanent damage. An often observed problem is that of suction tubes being too narrow or squeezed. The resulting underpressure in these tubes may lead to malfunction of the pump, inter alia because of bubble formation in the low-pressure part.

4.2.2.2 Fast reciprocating pumps

These pumps are of the basic type discussed as an example in Section 4.2.1. The pulse dampener is normally essential; it should, of course, be of the flowthrough type to allow effective priming. The smoothing action does deteriorate at lower flowrates, because n decreases and τ remains constant; controlling flowrates via stroke frequency poses problems when a large dynamic range is aimed at. Pulse dampening improves considerably when dual-piston pumps are used. Smoothing is much better, as there is no or only a short period of no flow delivery.

For obtaining a harder operation, sometimes feedforward software is included: At higher observed pressures the frequency is increased, in order to compensate for compressibility losses. Unfortunately, the compressibility is solvent-dependent, and the best one can do is use some average value for all solvents.

A variation on this type is the membrane reciprocating pump. Here, the piston displaces a hydraulic fluid, which is separated from the mobile phase by a membrane of stainless steel. Its main advantage is that the piston seal is not exposed to all kinds of fluids; its main drawback is the softer action, due to the additional compressibility in the hydraulic chamber.

4.2.2.3 Slow reciprocating pumps

These pumps are schematically the same as the fast ones. However, the displacement volume is larger, and the frequency lower for the same flowrate. The movement of the piston is linear with time during most of its travel in the pumping direction, leading to a constant flowrate. The interruption at the end of the stroke can be taken care of in various ways:

(a) fast refill when one piston is present,

(b) arranging for pump action of the alternate piston with dual-piston pumps,

(c) a combination of the two.

Such arrangements can be implemented by using software-controlled stepping motors, and/or by using special cam wheels for actuating the piston(s). However, this is more complicated than it looks, mainly because of the precompression needed at the beginning of every pump stroke. This would spoil a carefully designed timing sequence.

More sophisticated designs use the signal of a pressure transducer in a feedback

References on p. A221

mechanism to solve the interruption problem. As the set point for this feedback one can take the constant pressure observed during most of the pump stroke. In this way, a nearly perfectly "hard" pump can be produced.

4.2.2.4 Syringe pumps

In these pumps the displacement volume of the piston is sufficient for one or several chromatograms, and this eliminates all the worries about flow interruption and pulse dampening. Earlier designs, for standard HPLC (flowrate 0.5-3 ml/min) had displacement volumes of 250-500 ml. They declined gradually in popularity during 1975-1985. However, for small-bore columns and the associated low flowrates the reciprocating pumps were less successful, and a rebirth of the syringe-type pump took place.

These pumps may give the illusion that they are hard. They indeed are, statically: once the system is stabilized, they deliver a set flowrate (at inlet pressure!). However, due to the large capacity of the reservoir, the stabilization may take quite a while [18]. The time constant, τ, equals

$$\tau = V_r \cdot \chi \cdot R_C = V_r \cdot \chi \cdot \frac{\Delta P}{F_C} \tag{4.22}$$

where V_r is the chamber volume. V_r varies between a very low value and that of the freshly refilled pump (e.g., 250 ml). The value of τ may be even as large as 20 min (V_r = 250 ml, $\chi = 2 \cdot 10^{-9}$ Pa^{-1}, $\Delta P = 4 \cdot 10^{7}$ Pa, F_C = 1 ml/min). One could describe this as "dynamically soft". The problem is aggravated because the time constant does depend on the position of the piston; it is large at the beginning and low at the end of the pump cycle.

Good designs have facilities for dealing with this problem, such as precise pressure monitoring and fast precompression toward the expected (but unfortunately often unknown) pressure.

4.2.3 Gradient systems

4.2.3.1 General

Gradient elution, in which the composition varies according to a predetermined program during chromatography, is a must for many applications of HPLC. In the early days, physical tricks, such as stirred reservoirs, were used to obtain gradients. However, these lack the opportunity of tuning the gradient shape to the requirements. Nowadays, virtually all gradient systems make use of a computer-stored program, which controls the liquid streams that form the gradient.

The designs of gradient systems fall into two categories: low-pressure mixing, where the composition of the mixture is made up before it enters the high-pressure pump, and high-pressure mixing, where two (or more) programable HPLC pumps, each with their own solvent reservoirs, are used.

Low-pressure mixing is generally the less expensive solution. The mixture composition can be controlled in various ways, described in the following sections.

4.2.3.2 Time-proportioning systems

Fig. 4.2 shows how the average composition con be controlled by opening and closing of two valves in the suction part of the pump. However, for a smooth composition/time curve the pulsations in the composition must be dampened out. This is done by a mixing chamber, sometimes equipped with a magnetic stirrer. Such a chamber, of volume V_M, leads to a time constant τ_M of V_M/F_c. The pump chamber itself also contributes to V_M. This value should be large enough to cope with the fluctuations. Again, the product $n_M \cdot \tau_M$, where n_M is the frequency of the actuation of Valves A and B, determines the quality of the smoothing action. The value should be larger than, e.g., 16 for a 1% residual composition ripple. On the other hand, one does not want V_M to be too large, as it leads to a delay in the gradient (exactly equal to τ_M in the linear parts) and a distortion of the gradient shape in the nonlinear parts.

This is often an awkward compromise. It is, of course, easier to cope with, if the switching frequency of Valves A and B is high. However, even the fastest valves have switching times of some milliseconds. The time that a valve is open should, for reasonable accuracy, be at least 10-100 times this value, say at least 0.1 sec. Then, for metering a composition of 1% B in A, the cycle time cannot be shorter than 10 sec: The result is n_M = 0.1, and τ_M must be 160 sec, an almost unacceptable value.

The best mixing chamber volume does depend on the flowrate; for miniaturized columns it must be much smaller. Anyhow, in miniaturization it becomes increasingly difficult to generate reliable gradients.

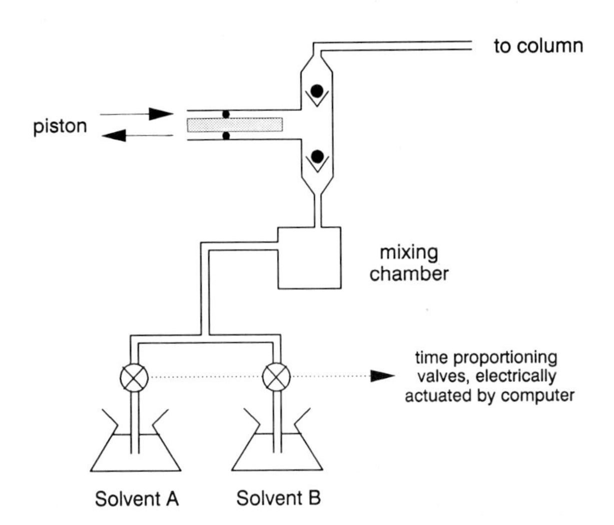

Fig. 4.2. Low-pressure gradient system with time-proportioning valves.

Another aspect of this type is the synchronization of pump stroke with the switching of Valves A and B. If this is not well arranged, very large, unpredictable deviations in the composition can occur. For instance, Valve B may open accidentally just at the time when no suction is taking place, and as a result no B is introduced into the pump at all.

4.2.3.3 Low-pressure metering

Another approach to precolumn gradient forming is the use of one or two (or even more, in the case of ternary and quaternary gradients) low-pressure metering pumps. In one design, even the total flowrate is controlled by these pumps; the high-pressure pump does not meter the flow, but merely augments the pressure.

Generally, however, there is one low-pressure pump less than the number of solvents. This circumvents synchronization problems. However, the pulsations in the flow induced by low- and high-pressure pumps still necessitate the use of a mixing chamber. The problems with the dynamic range (in flow and composition) are also not completely eliminated.

A practical problem encountered in the use of low-pressure mixing systems is that air bubbles, generated at the mixing point, enter the pump head. This is a result of the fact that sometimes mixing of two solvents, neither of which is saturated with air, can lead to a supersaturated mixture. Some low-pressure mixing systems have an air bubble release in the mixing chamber. Careful degasing may be necessary.

4.2.3.4 High-pressure mixing

The higher investment in high-pressure mixing systems gives good returns in the form of better gradient precision and a greater dynamic range.

The mixing vessel is still needed. For one thing, without it there might be a chance of insufficient radial mixing. Also, normally, pulsations in the flows of the two pumps would still induce fluctuations in composition with time. However, the size of the mixing vessel can often be much smaller than in the case of time-proportioning low-pressure mixing.

It should be noted that viscosity of the mobile phase may change drastically with composition, e.g. with methanol/water by a factor of 3. As a result, the column pressure will change. This serves as a warning not to rely too much on feedback and feedforward mechanisms in the main pumps. Also, in a gradient device composed of two syringe-type pumps, the compression in the two reservoirs may affect the gradient shape in a rather unpredictable way.

4.2.4 Thermostating

The distribution of solutes in the phase system generally depends on the temperature, the change being of the order of a few percent per degree centigrade. For this reason, it is good practice to control the temperature of HPLC experiments.

The first measure to take is to put the column in a thermostating mantle, through which water circulates, or in an airbath, similar to the ones used in GC. However, many users and even instrument designers are not aware of the fact that heat exchange with the column wall material is rather slow in the bed of standard HPLC columns. As a result, the temperature of the top part of the bed is determined by the temperature of the incoming mobile phase, rather than by the wall temperature, and the viscous heat dissipation in the column leads to an increase of the mean temperature in the bed.

These effects are accompanied by radial temperature gradients. These may be even more troublesome, as they lead to nonuniform migration of solutes and corresponding loss in platenumber. A detailed discussion of these aspects is to be found in Ref. 19. The main conclusions are briefly:

(a) Faster elution, larger diameters, and smaller particles amplify the effects.

(b) The most primitive way of operating the column, in still air, is best in terms of efficiency; the viscous heat dissipation then leads to a longitudinal gradient, but not to a radial one. Rigorous wall thermostating may improve reliability and reproducibility, but it can affect efficiency.

(c) Prethermostating of the mobile phase is always advantageous.

Prethermostating of the mobile phase should, of course, be applied after it has left the pump, because the latter generally exerts a poorly controlled thermal influence. A capillary in good thermal contact with the thermostating medium will serve the purpose. The diameter is immaterial as far as thermostating is concerned, but should be chosen small enough to avoid unwanted delays and gradient distortion. The required length (e.g., for obtaining a 100-fold reduction in the temperature difference) depends on the thermal conductivity of the solvents, its heat capacity, and the flowrate. For normal HPLC operation, a length of 10-20 cm is sufficient.

This section closes with the remark that all problems associated with slow heat transfer vanish when one uses smaller-bore columns. These can be effectively thermostated by immersion of the column in the thermostating medium.

4.3 SAMPLE HANDLING AND INTRODUCTION

4.3.1 Syringe injection

Injection with a syringe through an elastomer septum, as used in GC, is convenient because it allows the adjustment of the injected volume within a range of 0.1-1.0 times the maximum volume of the syringe. Below that first value the precision of injection generally becomes poor. However, two problems are associated with this technique. The first is that the tightness of the seal around the syringe needle at the high pressures used is often problematic. A 1- to 2-mm elastomer sheet takes care of the sealing; it is pierced at every injection by the needle, but mechanical pressure exerted by the screw cap should provide leak tightness. This system, quite successful in GC, is not entirely satisfactory in LC because of the higher pressures involved and because of possible degradation of the septum by the solvent (swelling, chemical attack).

A modification of the "on-flow" injection technique is "stopped flow". During the insertion of the needle, or even during the whole injection procedure, the flow is stopped. This allows the use of sealing rings or gaskets of less flexible but more stable materials than are used in the septa. In some systems the opening for the injection needle must be plugged by a screw cap or similar device before the flow is restarted. These systems introduce some uncertainty in the retention (volume, time) axis of the chromatogram and operate satisfactorily, but they are slightly more laborious in use. The other general drawback of manual syringe injection is the low precision in the volume injected. In GC this is often compensated for by the use of an internal standard, and for various reasons (among others because there is no good alternative) this is accepted in GC. However, in HPLC most of the work is carried out either with fully automated syringe injection, or with a valve.

4.3.2 Valve injection

Fig. 4.3 shows the basic features of an injection valve. The system allows switching the flow between two channels, one of which (the external loop in the figure) can be filled beforehand at ambient pressure with the sample solution. Usually, this is done by injecting 100-1000 μl by means of a syringe, the outlet of which is press-fitted to the "sample in" port of the valve.

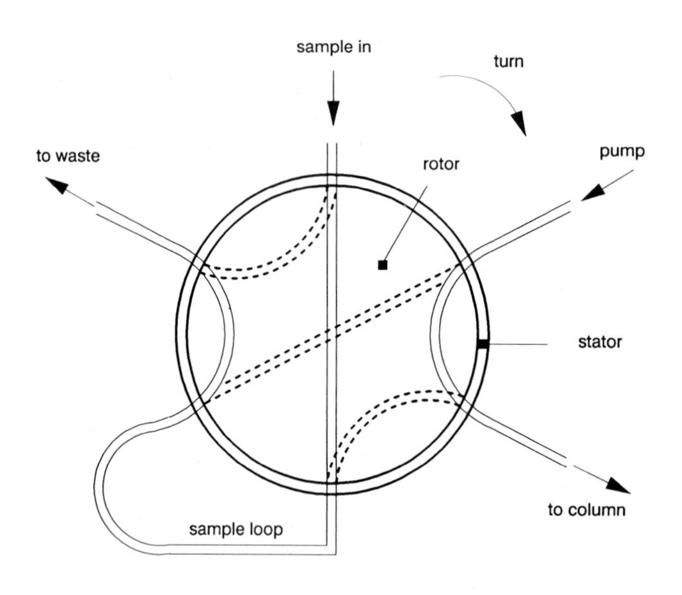

Fig. 4.3. Valve Injection with an external loop.

As mentioned, the main advantages of such systems are easy and trouble-free operation and high precision of injection volumes. In fact, the precision is probably so high that it is hardly ever the limiting factor in the whole system, and as a result, it is quite difficult to find a good estimate of its real value. The use of internal standards in HPLC should therefore be limited to those cases where quantitative errors originate in the sample pretreatment or in the detection process. However, under such conditions it is often much more difficult to find an internal standard that behaves exactly like the analytes.

When it comes to injecting smaller volumes, say 2 μl or less, e.g., in miniaturized chromatography, one usually resorts to another system, with an internal rather than an external loop. The connections (basically with the hardware as shown in Fig. 4.3) are made in such a way that the injected volume is that of the connecting channel in the rotor of the valve (internal loop). Depending on the design of the valve, this can be as small as 20 nl.

Two drawbacks remain. One of them is that a relatively large amount of sample is needed to flush the valve and/or loop properly with it (see discussion of dispersion phenomena at the end of this section). The other is that the injected volume is inflexible. These two drawbacks are circumvented in another design. Here, things are arranged in such a way that a metered amount of sample can be injected with a syringe into the loop while it is at ambient pressure, displacing the mobile phase. After injection, the syringe port is shut, and the valve is turned. By "backflushing", i.e. reversing the flow direction with respect to the filling operation, excessive dispersion of small samples is avoided. One might argue that such systems bring us back to the lack of precision of syringe injection and therefore are not of much use. However, here syringe injection is carried out under very favorable conditions: no backpressure and no need for speed. While precision is indeed less than that of the full-loop injection, the flexibility obtained, of importance especially in single-shot analyses and in the optimization stage of routine assays, is a great asset.

For all kinds of injections it is important to have some, at least qualitative, insight into the dispersion processes that may be in effect. A schematic diagram, as shown in Fig. 4.4, applying to valve injection, leads to some clarifying considerations on the basis of a simple model.

The block of sample ideally moves through the loop and the connecting tube into the column as an undistorted plug. However, dispersion will take place, and at the entrance to the column there will not be a rectangular plug, but a distorted one. To characterize the performance of the device in this respect one has to characterize the peakwidth at this point. The volume standard deviation, $\sigma_{V,\mathrm{col}}$, serves this purpose, although the only way to measure this for the highly non-gaussian peaks involved is that of the second moment (Chapter 1). Furthermore, it is good to normalize this value on the injected volume, V_{inj}. According to Karger et al. [9] we define

$$K = \sigma_{V,\mathrm{col}}/V_{\mathrm{inj}} \tag{4.23}$$

A large K means that the sample is extensively dispersed and diluted; small K values

References on p. A221

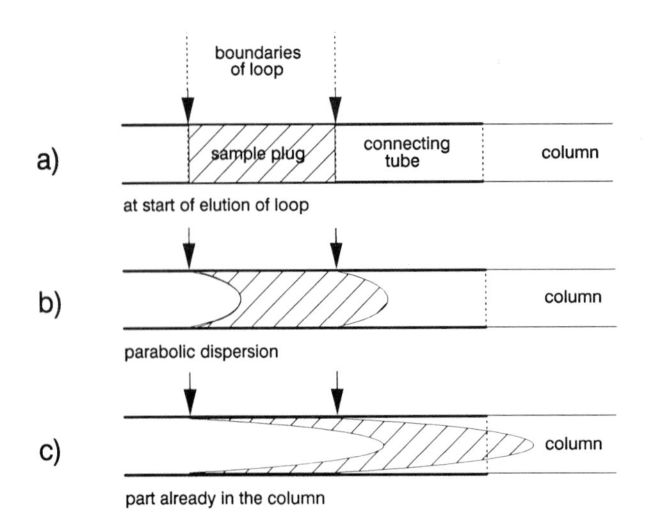

Fig. 4.4. Schematic view of the dispersion processes on injection.

indicate less influence of these effects. The smallest K value accessible is obtained when the plug is not distorted at all. The profile at the column entrance is a block of width (in volume – to obtain time measures one has to divide everything by F_c) V_{inj}, the standard deviation of which [20] is equal to $1/\sqrt{12}$ times the block width. Therefore, the smallest possible value of K is $1/\sqrt{12}$.

In Ref. 22 the value of K is also derived for a case where the loop behaves like a well-mixed vessel. In that case, the concentration profile at the outlet would be an exponential function declining by a factor, e, for each volume V_{inj}, passing through. The standard deviation would also be V_{inj}, and the K factor would be 1.

Unfortunately, these simple models bear little resemblance to reality, where flow proceeds through cylindrical channels with no mixing and little diffusion. Most of the dispersion is usually brought about by the laminar parabolic flow profile, which appears to prevail despite the presence of bends and diameter changes.

As depicted in Fig. 4.4, the velocity profile will disturb the front edge of the plug during passage through the connecting tube. For the trailing edge this also happens in the loop itself. When diffusion has no chance to average out the velocities of molecules, the front and tail will be shaped into rotation paraboloids, and the dispersion will be rather high. For instance, even when the size of the connecting tube is negligible, the volume needed to sweep out all the sample material from the loop is many times the volume of the latter. A more detailed description of these phenomena in Ref. 21 leads to rather complicated expressions.

The dispersion can be reduced considerably by the action of radial diffusion, the theory of which has been described in Chapter 1 under the treatment of capillary chromatography. A rough indication of the conditions under which this occurs is useful: The

Fourier number $D_m \cdot t/d_c^2$, where t is the mean residence time, is indicative of the number of times a molecule samples the various streamlines in the channel. This expression can be rearranged to $V_{inj} \cdot D_m/F_c/d_c^2$. If the number is high, e.g. 100, each molecule moves with approximately the average velocity. Thus, the smaller the diameter of the tubes, the less is the influence of the parabolic dispersion. Therefore, if the best compromise between large injection volume and small input peakwidth (small K factor) is needed, it is wise to work with connecting tubes and loops as narrow as possible. Coiling the tubes may also be useful, as radial homogenization is improved as a result of secondary flow.

4.3.3 Sample pretreatment

Like other analytical techniques, when applied to actual analytical problems, HPLC must be combined with one or several sample pretreatment techniques, also called "preconcentration" or "cleanup" methods, in order to:

(a) transfer the analytes to the chromatographic environment (solvent),

(b) free the analytes of interfering constituents (which may be present in huge amounts) to enhance selectivity,

(c) concentrate the analytes to enhance sensitivity.

The techniques involved include liquid/liquid extraction, liquid extraction of solid samples, liquid/solid adsorption, crystallization, dialysis, distillation, and electrophoresis. Some of them are so closely related to their use in HPLC and were developed in this context, that a few words should be devoted to them in this chapter.

The original pretreatment step in chromatography was liquid/liquid extraction. The choice of the extracting solvent determines the sensitivity and selectivity. The method is well documented, not too difficult to optimize, but laborious in routine operation and not easy to automate.

The principles of liquid/solid extraction will be discussed below for the case of RPC, but similar techniques can be used with other phase systems. The first experiments [22] were carried out with the column adsorbent itself. When the sample is in water solution, an extremely weak eluent in RPLC, it is possible to inject very large volumes of samples without causing the peakbroadening that is normally associated with large injection volumes. This is due to the strong adsorption of the analytes on the top part of the column during injection. Since water is a weak eluent, the analytes wait there until the mobile phase is pumped into the column, and then they are released from the packing as a narrow plug. Very large concentration factors, 100-1000, can easily be obtained in this way.

However, it was soon realized that this type of sample handling is not selective. Since the complete sample enters the column, the result may be disastrous: deterioration, even clogging, of the column, and overcrowded chromatograms. Pretreatment was therefore carried out separate [23] from the main HPLC column (now often referred to as the "analytical column").

The basic scheme of such a system is shown in Fig. 4.5. First, the precolumn is loaded with the sample, the volume of which can be as much as 100 to 10 000 times the

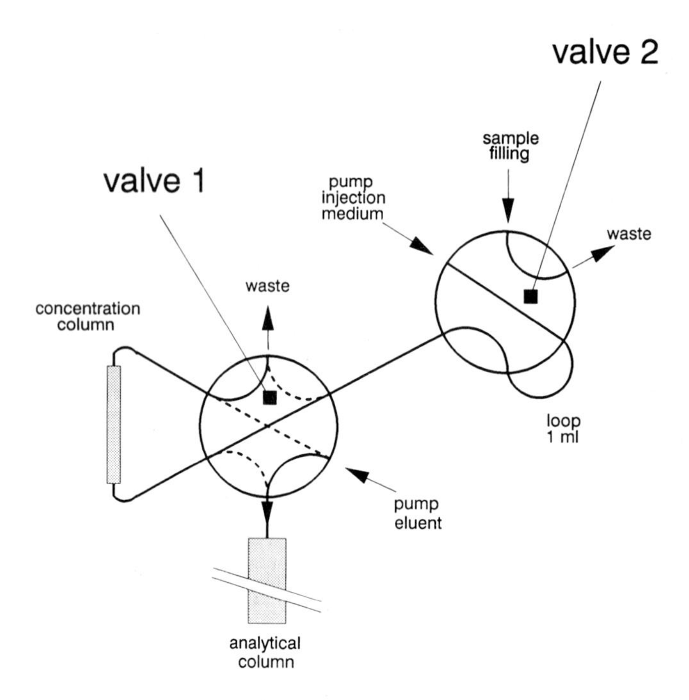

Fig. 4.5. Basic precolumn technique for HPLC.

volume that could be applied by direct injection. Next, the precolumn may be flushed with water (or some other selected solvent mixture) to remove any unwanted material present in the mobile phase of the precolumn. Finally, the valve is rotated, and the analytes are desorbed and swept into the main column by the mobile phase.

The advantages of these techniques, which have become known under the names of precolumn techniques, preconcentration columns, solid-phase extraction, and column switching, are numerous and significant:

(a) Very large concentration factors can be obtained, especially for aqueous samples. This is the result of the steep dependence of log k' on the organic modifier concentration (Chapter 1).

(b) The pretreatment enhances selectivity, prolongs column life, and cleans up the chromatograms. Unretained substances do not enter the main column at all, nor do compounds that are irreversibly retained. Also, by properly selecting the chemical nature of the sorbent in the precolumn and choosing one that is different from that in the main column it is possible to reject unwanted components, so that the analyte peaks stand out on a clear background.

(c) The technique is easily automated; the loading of samples, flushing, and switching to the main eluent can be under computer control with electrically controlled valves and a suitable sample changer.

Precolumn techniques have been under continuous study during the last 10 years.

Especially Frei and Brinkman and their associates have been very active in this field. An overview of their work can be found in Ref. 24. A brief survey of its main aspects should be given here.

The breakthrough volume, i.e. the volume of sample that can be passed through before the analyte emerges at the exit of the column, has been studied. Breakthrough can be caused either by column overload (saturation of the surface, most likely by extraneous matrix constituents) or by regular linear elution of the analytes by the sample solvent (that would also occur in the absence of matrix constituents). It has been found that breakthrough is a rare problem; it occurs only in the most critical situations, e.g., when analyzing polar solutes while aiming at high concentration factors.

The scale on which precolumns can be operated well is continuously decreasing. It turned out that the ratio of volumes of precolumn to main column can often be made very small, in the range of 1:1000 or less. Good interfacing with 1-mm- or even 250-μm-ID columns is now also possible, with suitable miniaturized versions.

Dispersion in precolumns is not so troublesome, especially because the volume ratio can often be chosen to be very low. A low platenumber in the precolumn then will not matter, as it results in only a small volume standard deviation, added to the overall variance. Using "backflush" when desorbing improves this further. Larger particles, offering convenient packing and low backpressure, thus can be used in most cases.

It is very rewarding to explore selective sorbents, and to define and construct the chemical surface that is most selective (in the context of the matrix and the detection method used) for the analytes. Apart from adjusting the polarity, one can resort to ion exchange, complex formation with metal atoms [25] on the surface, and even to immobilized antibodies [26,27].

Automation of such systems has been studied, and commercial systems are available. It is now practicable to go to more complicated schemes, with two or more columns, having complementary selectivities. In multicolumn chromatography, a fraction of a chromatogram from one column, often popularly called a "heart cut", is injected into another column, having a different selectivity, in the hope that overlapping peaks in the first chromatogram will be resolved in the second one. The precolumn technique is already a rough heart cut, because neither unretained nor very strongly retained components enter the second (main) column. Therefore, the basic tenet of multicolumn operation (Section 4.8.4), that the columns must have different selectivities, applies equally well to the precolumn techniques.

Another way to apply precolumn techniques is off-line, in a hand-operated low-pressure system. In this case, the columns are cheap, "disposable", and used for one or only a few analyses. This technique is also rather popular, and the precolumns are marketed by several suppliers under puzzling names. Often they are small plastic tubes, "cartridges". Some of these can be press-fitted to syringes and injection valves. Much of what has been said about the high-pressure on-line modifications also applies to this technique.

A new method for on-line coupling to HPLC is zone electrophoresis. A separation of low-molecular-weight ionized materials, e.g., from proteins with lower electrophoretic mobility, can be obtained by low-resolution capillary zone electrophoresis [28]. Sample size

References on p. A221

and duration of the experiment are the limiting factors. Within the time of one chromatogram the automated system, coupled on-line to the HPLC system, can process about 10 μl of sample.

4.3.4 Automatic sampling systems

The widespread use of HPLC in routine analyses has led to the development of numerous automatic sampling systems. These generally incorporate a sample tray, onto which 10-100 vials, containing samples and standards, can be loaded. These vials are often screw-caped with a septum, in order to avoid evaporation and contamination. The mechanical design depends on the type of injection system used. Simple valve injection is most frequently used. The transport of sample from the vials to the loop is actuated by vacuum or an overpressure on the vial. There are also automatic instruments where syringe injection is used. Full automation of this process eliminates many of the disadvantages of this injection technique.

The whole operation can be controlled by a microprocessor. In most cases, the operation is fully time-based, but some designs allow feedback from the shape of the chromatogram, e.g., starting with a new cycle when a given peak has been observed. This requires that also data acquisition, including sample identification and data reduction, be coupled with it, and this is indeed possible with most commercial designs. Once it is decided to perform HPLC in unattended operation, other problems surface: safety aspects, check on empty solvent reservoirs, temperature cycles during the 24-h operation, etc.

4.4 COLUMNS AND CONNECTORS

4.4.1 Materials

The most important variable in HPLC is the mobile phase. Therefore, it is imperative to select hardware (instrument parts, column, and tubing materials) that does not restrict this free choice.

Stainless steel is the most important material for HPLC hardware. It is resistant to most mobile phases. Its high chemical stability is due to the presence of an invisible oxide layer. This layer may be attacked by strongly reducing and metal-complexing agents, such as formic acid, chloride ions, citrate ions, etc., and such chemicals must be used with some care. This layer explains why a treatment with 15% nitric acid, sometimes used when electrochemical detection is applied, decreases corrosion of stainless steel.

In some instruments stainless steel (mostly of Type 316) is replaced by other metals or alloys with alleged superior properties, such as titanium. Often, these are more difficult to machine, and the equipment is correspondingly more expensive. One reason given for the use of such materials is the fact that small amounts of metals, such as Fe, Ni, and Cr, are released from stainless steel. These could under certain circumstances denature proteins chromatographed in such equipment. The real value of such expensive "biocompatible" equipment is still being debated.

Next in importance are materials for use in tubes, gaskets, valve rotors or stators, and other parts that must be more or less deformable. Teflon, for instance, is resistant to virtually all chemicals. Mechanically, it has the drawback that at high pressures it is irreversibly deformed (creeping, flowing). Newer materials, such a PEEK and KAPTON do not have this disadvantage. Glass is used as column tubing in some designs, especially in cartridge-type columns. It is a fine, inert material, that can withstand surprisingly high pressures, when properly used.

Increasing use is made of fused-silica capillaries for connectors and reactor tubes, small-diameter columns, and as flowcells for optical detection. The material is available with outer diameters of a few tenths of a mm, and inner diameters ranging from 1 to 600 μm. Mechanical stability is obtained by a coating, usually of polyimide. For optical experiments this needs to be removed, e.g., by heating in a flame. However, the bare silica tube is as breakable as glass. Still, the easy manipulation, good transparency, and chemical inertness of fused-silica tubing are very attractive.

4.4.2 Connecting tubes

Connecting tubes are normally made of stainless steel, with an outer diameter of 1/16 in., while the inner diameter usually is between 0.1 and 0.4 mm. When volumetric dispersion is small ($\sigma_V < 10\ \mu l$) the choice of length and diameter may become critical. The Aris-Taylor equation (Chapter 1) gives guidelines for this choice.

Connecting tubes are generally fitted with coupling unions of the "zero-deadvolume" type, contributing indeed only in the order of 900 nl [29] to the volume variance. These constitute one little device without which modern HPLC experiments would be virtually impossible. Unfortunately, life is not as easy as it could (and should) be: different manufacturers sell many types of these connectors, with different dimensions and cone angles of the ferrule and seating. The same holds for the connecting ports of valves, column terminators, and detectors. Using the wrong ferrule may result in irreparable damage to, e.g., an injection valve.

4.4.3 Column tubes and terminators

Column terminators are usually rather complex in design, and this is no surprise considering the various requirements they should meet:

(a) absence of significant contribution to the volume variance, i.e. absence of dead, unswept volumes;

(b) presence of a frit or screen that keeps the packing in position; usually this is impermeable to particles of $>2\ \mu$m;

(c) pressure resistance;

(d) reasonably uniform distribution of solvent flow and injected solutes over the cross section of the packed bed.

Various designs marketed by a number of firms may be satisfactory. A matter of concern sometimes is the position of the ferrules on the column tube and the connecting line. When a connection is changed, a deviation in this position may cause appreciable peak broadening because of an increased deadvolume.

To keep the packing in place, the choice is between frits and screens, both of stainless steel. Frits normally have a higher flow resistance. Also, sometimes adverse affects are blamed on the high surface area of frits. On the other hand, screens require a stronger supporting construction, and are difficult to obtain for the smallest particle sizes used. Some terminators have special provisions to ensure that the mobile phase and the solute are distributed well over the cross section of the packing. Normally, radial channels are machined into the terminator, at the seating of the screen or frit. Some firms market cartridge columns, allowing the tubing with the packing to be exchanged, without the need to buy new terminators.

4.5 DETECTION

4.5.1 Introduction

Detectors serve the purpose of translating concentration changes in the column effluent into electrically observable signals. From the chromatographic point of view detectors are regarded as accessories to the main part of the chromatograph, the column. However, from the more general analytical point of view they are more essential than the column. Indeed there are a number of HPLC systems where the information from the detector is much more important than the separation in the column. In such cases, the latter may serve only to separate the analyte from interfering material, e.g., proteins may be removed from a drug, which is then determined at high selectivity and low detection limit by, e.g., a high-resolution spectroscopic technique. Such situations occur with increasing frequency, as HPLC becomes a sample processing technique in such areas as spectrometry (especially MS), immunochemical analysis, and enzymatic analysis.

Classical development and study of HPLC detectors started from a vantage point quite different from that in which the detector is considered an aid for monitoring the column separation process. As a result, the trend in the chromatographic literature is to formulate requirements resulting from the nature of the chromatographic column and the processes occurring in it. Our discussion will follow that route, but occasionally attention will be paid to the alternative approach, where the detector is taken as the starting point and where requirements for the column rather than for the detector are formulated.

4.5.2 Characteristics

The performance and usefulness of detectors can be assessed by means of the following check list:

(a) *Contribution to peakwidth,* conveniently expressed as volume, $\sigma_{v,d}$, or time, $\sigma_{t,d}$, standard deviation. Sometimes it is also adequate to use the time constant, $\tau_{t,d}$, or even the "volume constant" $\tau_{v,d}$, i.e. in those cases where the response of the detector is a pure exponential curve.

(b) The *detection limit* of the detector, preferably expressed as the concentration in the detector when the signal equals two or three times the noise level.

(c) The upper limit of the linear range; the *linear dynamic range* is defined as the ratio of this upper limit to the detection limit.

(d) The selectivity of the response; when nearly all solutes are detectable, one speaks about *universal* response, if only classes of solutes can be detected the response is selective.

(e) *Compatibility* with various mobile phases.

A few additional remarks should be added. The contributions to peakwidth are due due to:

(a) cell volume (sweeping takes a time proportional to V_{cell}/F_c),

(b) electrical or electronic effects, and

(c) dispersion in connector tubes that are part of the detector.

Considering the dependence on flowrate, and expressing peak dispersion in time units, the first effect (a) is roughly inversely proportional to flowrate, the last (c) is constant, and the second (b) is somewhere in between [30]. In miniaturized chromatography the cell volume effect therefore often predominates.

In discussions on quantitation we need the concept of calibration factors, although this in itself does not tell us anything about the quality of the detector. This quantity, S_i, is alternatively denoted as response factor or sensitivity (the latter term is often also used to indicate detection limit, in contradiction to IUPAC recommendations). S_i is the ratio of signal (change) to the concentration (change) *in the detector*. Ideally it should be a constant. If it is flowrate-independent, one speaks of a *concentration-sensitive* device. If it is proportional to flowrate, the response is clearly proportional to the mass flow of solute, and one speaks of *mass flow sensitivity*. Some devices, e.g., the electrochemical detector, exhibit an intermediate behavior.

Another calculation unit is the *noise*, strictly speaking all variations in output not carrying the desired information. Among the random noise, which is of greatest interest, three types can be distinguished:

(a) high-frequency noise, which can be filtered out easily, either in the analog circuit or by suitable algorithms in the data reduction step;

(b) mid-frequency noise, which is of the greatest importance, because it interferes with area and height determinations;

(c) low-frequency noise, which is often indistinguishable from drift and is also readily eliminated by suitable integration devices.

In the following it will be assumed that the (a) and (c) types of noise are negligible or taken care of in the data reduction. The intensity of mid-frequency noise can be expressed as the standard deviation (the statistical concept, not the peakbroadening concept), σ_d, around a best linear fit through a time record; this is approximately equal to one fifth the "peak-to-peak" noise in that record.

The detection limit of the device is not strictly independent of the time during which one can measure; a very-long-lasting peak can be observed better, because more intensive

References on p. A221

damping of the noise can be applied. The theory of this is complicated [31]. However, the error made in neglecting this is not too serious. The detection limit, c_i, is thus roughly equal to $k \cdot \sigma_d/S_i$, where $k=2$ or 3.

The upper limit of the linear range depends, of course, on the criterion (percentage deviation from the extrapolated line) applied. Normally, one accepts 5% as such.

The calibration over the full range of accessible concentrations is often presented as a log/log plot of signal vs. concentration. Such plots may conceal severe nonlinearity, as an intrinsically nonlinear response may give a straight line, with only the slope deviating from 1. Basically, it is better to represent the behavior over a large concentration range as a plot of the calibration factor, S_i, vs. log concentration. In the ideal case, this would be a straight horizontal line.

Normally, the "signal" is quantitated as the output voltage that is presented to a recorder or integrator, and this fact again stresses that the values of the calibration factor and of the noise alone are unimportant; they can be made larger or smaller by means of electric circuits. However, in quite a number of cases it is possible to express "signal" in terms of a suitable physical quantity; e.g., in absorbance detectors one can use absorbance units; in refractive index detection, the refractive index itself. In that case, it is possible to state sensitivity and noise in those terms. This is highly preferable, as it allows comparison of different detection schemes and instruments. In these cases it is also quite useful for commercial instruments to give outputs in volts, which are calibrated in, e.g., absorbance units.

The *determination of* S_i can be performed statically or dynamically. In static measurements one switches the mobile-phase stream from one (zero) concentration of i to another, and observes the (change in) signal. Switching can be conveniently performed by means of an injection valve, equipped with a large loop, usually with no column in the system. In dynamic measurements the peak area, A_i, is observed as a function of the amount injected, with or without insertion of a column. The value of S_i then follows from

$$S_i = A_i \cdot F_c / Q_i \qquad (4.24)$$

where Q_i is the amount injected and F_c is the flowrate.

It should be noted that Eqn. 4.24 is also quite useful for other purposes: it allows calculation of the expected peak area when the calibration factor S_i is known, e.g., from a knowledge of the molar absorptivity of the species. Too little use is made of such equations to predict the performance of detection schemes.

4.5.3 Optical detectors

4.5.3.1 UV/visible absorption

The vast majority of HPLC analyses are carried out with the UV/visible absorption detector. It measures absorption of radiation in the wavelength range between 190-800 nm according to the Lambert-Beer law. The usual photometric scheme involves a monochromatic radiation source (either a line emitter or a continuous source with monochromator or filter), an absorption cell, a photoreceptor, and a data handling unit that, among others, does the log conversion for obtaining a signal linear in concentration.

The part most characteristic of a HPLC detector is the cell. In the common form it is cylindrical, with a length of 6-10 mm, a diameter of ca. 1 mm, and a volume between 5 and 8 μl. The windows are made of 0.5- to 1-mm quartz discs, press-fitted between gaskets. Inlet and outlet tubing are usually of stainless steel, 0.15-0.25 mm in ID. Present technology of instrument makers allows manufacture of cells with virtually no unnecessary dispersion effects.

The molar absorptivity, ε_i, of the analyte, i, (liter mol^{-1} cm^{-1}) determines the calibration factor. For instance, for a compound with an ε_i of 10^4 (a factor 10 times smaller than the maximum that can be expected theoretically) and with a molecular weight of 200, the calibration factor with a 1-cm cell would be S_i = 10 000 absorption units (AU) mol^{-1} liter, or S_i = 0.05 AU/ppm. The noise level is often as small as $5 \cdot 10^{-5}$ AU, so that a detection limit of $5 \cdot 10^{-5}/0.050 = 0.001$ ppm is not uncommon. With a peak standard deviation of 200 μl this corresponds to 0.2 ng injected.

The variations in UV/visible detectors are very large, mainly because of the variations in the wavelength choice. There are:

(a) Single-wavelength instruments. The wavelength is set by the nature of the light source (often a mercury lamp with 254 nm as the most frequently used line) or of the filter.

(b) Multiple-wavelength devices. In these one choose among a number of wavelengths, often those of mercury, cadmium, or zinc discharge lamps.

(c) Continuously variable wavelength devices. The usual combination of tungsten ("halogen type") and deuterium lamps is used as a light source to cover the spectrum from 190 to 650 nm (or a smaller portion with only one source). A monochromator selects the desired wavelength.

(d) Programable variable-wavelength instruments. These have the additional option to change the wavelength during chromatography, according to a program, so that each peak can be measured at optimum wavelength. With some of these instruments there is the additional option of scanning the full spectrum. However, in most cases the scan does not proceed fast enough to catch the spectrum "on the fly", and the flow must be stopped in order to avoid deformation of the spectrum.

(e) Photodiode-array instruments. In these the absorbance is measured simultaneously at 200 to 500 wavelengths, equidistant over the spectral range. The diode array that makes this possible consists of a corresponding number of photoreceptors that independently integrate the impinging light intensities. Every (fraction of a) second the diodes are interrogated by a computer, and the intensity data are stored after conversion to absorbances.

The optical arrangement is such that polychromatic light falls through the cell and next passes through a grating or prism. There is no exit slit in this "monochromator"; instead, the diode array is positioned in this plane. The possibilities and pitfalls in the application of such a device, which belongs to the class of HPLC-coupled spectrometers, are discussed in Section 4.8.

The linear range of UV/visible detectors goes to about 1 or 2 absorbance units;

References on p. A221

beyond that, stray light and spectral bandwidth will impair the linearity. This corresponds to a linear dynamic range of 10^4 to 10^5.

UV detection is selective. It only responds to compounds having chromophores. It is by far the most widely used detector. The mobile phase should, of course, be transparent in the region of interest. This limits the choice of solvents and mobile-phase additives, but, in general, this causes no problems, as one can compose mobile phases of all polarities and buffers, etc., required in HPLC, by mixing the admissible solvents.

4.5.3.2 Miniaturization of absorption detectors

The widespread use of this type of detector, and the general interest in miniaturized chromatography, has led to many attempts to scale down the volume of such detectors. Unfortunately, this has not been very successful. Obviously, the cell volume should not be decreased by shortening the path *length,* as the signal would decrease according to Lambert's law. However, decreasing the cell *diameter* is likewise unfavorable. As has been shown by theoretical treatment [32,33], the amount of light passing through the cell decreases rather rapidly (fourth power of volume), and the resulting small photon fluxes cannot be measured without strong statistical fluctuations, manifested as increased noise. Some UV/visible instruments for micro-LC are available, but their performance in terms of concentration detection limit is sometimes disappointing.

In some new forms of LC *on-column* UV detection is used. Nearly always fused silica is the material for the column. Its protective coating (often polyimide) must be removed at the measuring position in order to obtain UV transparency. On-column detection presents optically an extreme degree of miniaturization. In the first place, the path length is very small, with concomitant decrease in signal-to-noise ratio. In the second place, vignetting of the beam by means of diaphragms is usually needed in order to ascertain that a large fraction of the radiation is actually passing through the liquid, rather than through the wall material or even around the tube. Vignetting might also be needed for avoiding peak-broadening (especially in capillary zone electrophoresis, where the peaks are very narrow). For instance, if the beam strikes the column over 1 mm of its length, detection would contribute to the peak width corresponding to 1 $mm^2/12$ in the position variance. The vignetting results in a decreased photon flux, and this may lead to an additional loss in signal-to-noise ratio.

4.5.3.3 Fluorescence

Some organic molecules, when absorbing UV/visible (excitation) radiation, re-emit part of the energy in random directions (emission), and at a somewhat longer wavelength. The measurement of the emitted light provides a useful detection method. As only a few percent of all compounds exhibit fluorescence, the method is very selective. Also, the method generally has a low detection limit. Very small light intensities (some tens of photons per sec) can be measured very well. In fact, this is not the main limitation, but rather the noise introduced by extraneous radiation, such as Raleigh and Raman scattering and reflection. Still, depending on the compound, the detection limit can be a factor of 10 to 1000 lower than that in absorption detection, with similar linear dynamic ranges.

A proper implementation of the technique requires:

(a) A strong radiation source. In commercial equipment high-pressure mercury lamps or (pulsed) xenon arcs are used, and in research equipment lasers. Normally, the desired wavelength range is isolated with filters or a monochromator.

(b) A cell, with UV-transparent windows, with as low a scattering as possible.

(c) A filter or monochromator for isolating the desired emission wavelength range, and especially, to prevent the excitation wavelength from entering the photodetection path.

(d) A very sensitive photodetection device; this is usually a photomultiplier.

The optical diversity of fluorescence devices is even larger than that of absorption detectors, as there are two optical paths. Thus, there are: a dual filter, a filter monochromator, a monochromator filter, and a dual monochromator system. Also, some instruments are equipped for programing the wavelength(s). This is more rewarding than in UV, as the choice of the two wavelength settings can be more critical in fluorescence. The fluorescence detector is used not only for the analysis of compounds with native fluorescence, such as polyaromatic hydrocarbons, and for many drugs, but also in combination with pre- and postcolumn derivatization methods, leading to fluorescent derivatives (Section 4.6). As it is very selective, it is eminently suited to the trace analysis of such compounds. The selectivity also implies that the application range is not very wide. Mobile-phase selection is the same as for absorption detection, but solvent purity requirements are sometimes more stringent.

4.5.3.4 Refractive index

This was one of the first detection principles applied in HPLC. The refractive index (RI) of the eluate can be measured in various ways:

(a) Measuring the diffraction of a light beam passing through a prism-shaped cell, containing the eluate. This is called the diffraction-type RI detector.

(b) Measuring the intensity of the reflected beam at the interface between the liquid and, e.g., a glass block. When approaching the critical angle of incidence, this intensity depends very strongly on the refractive index of the liquid. This is called the Fresnel-type RI detector.

(c) Using interference. A light beam is split; one half passes through a sample cell, while the other passes through a reference cell. On recombination of the beams, the path length difference in wavelength determines the type of interference (constructive or destructive). As this is a function of the RI in the paths, a sensitive measurement can be derived.

A reference cell is also used in the diffraction and Fresnel-type RI detectors. This, inter alia, is necessary to compensate for the strong temperature dependence of refractive indices.

Volume standard deviations, being in the range of 5-10 μl, are satisfactory for normal HPLC columns, but too large for small-bore columns. The noise level is of the order of

10^{-7} RI units. As the calibration factor (change of RI per volume fraction of solute) is generally smaller than 0.1, detection limits are usually above 1 ppm. This is the reason why RI is only applied when other detection principles fail.

RI is one of the few options for universal detection in HPLC. Only a few compounds in a given mobile phase will not respond, because the difference in RI happens to be zero. However, this problem can always be solved by changing the mobile phase. The linear range is good; in most cases chromatography will become nonlinear before the detector does.

4.5.3.5 Other optical detectors

Optical rotation can be exploited and is very interesting because of the selectivity, especially in biochemical and food-related applications. However, the optical rotation effect is small, and the detection limits are disappointing. Only with expenditure of extreme instrumental efforts have good results been obtained in research situations [34], but the required pathlength still mandates cell volumes that are too large. A similar situation exists in circular dichroism measurements.

The technique of low-angle laser light scattering (LALLS) is of more widespread use. It is applied in SEC analysis of polymers. With a laser as a light source, the scattering of macromolecules can be measured at low angles. This makes it easy to interpret the signal in terms of molecular size of the eluted polymers. By combination with a "concentration" detector (UV or IR) it is then possible to "double-calibrate" molecular weights: once by the chromatographic retention and once by the LALLS signal. The technique, for which equipment is commercially available, has found rather widespread use in polymer laboratories. Further details can be found in Refs. 35-37.

4.5.4 Electrochemical detectors

Electrochemical methods are of great importance in HPLC detection [38]. The most important type is the amperometric detector. In this detector, the mobile phase flows along a surface of an electrode. The latter can have various shapes: flat or cylindrical, or it can even be porous. The electrode is at a fixed potential with respect to the solution (measured by means of a suitable reference electrode). Solutes of proper structure can undergo electrochemical oxidations or reductions. For instance, catechol can be oxidized to a quinone. The released electrons are carried away by the electrode, and the resulting current is a good measure of the concentration of catechol.

In most systems, the rate of the electrochemical reaction is mass-transfer limited; the concentration of the catechol at the electrode surface is zero, because every molecule arriving there reacts immediately, and diffusion towards the electrode supplies new molecules to be oxidized. Meanwhile, the flow sweeps away many other molecules, and only a small fraction (0.1 to 10%) of them actually react before leaving the cell.

The electrode potential is the most important parameter in amperometric detection. It can be varied between a negative limit of – 0.2 to – 1.0 V and a positive limit of 0.5 to

2.0 V, very much dependent on the nature of the electrode and of the mobile phase. Outside these limits, the solvent or the electrode material will react and lead to excessive currents. Positive potentials promote the oxidation of analytes, while negative potentials may lead to reductions. Each analyte that responds at all (and only a few compounds respond within the accessible potential range) has a characteristic potential (halfwave potential), above which (in the case of oxidation) or below which (in the case of reduction) the reaction proceeds. Thus, by choosing the potential one can control response and selectivity.

When flowrates are low or the surface area of the electrode is large (e.g., for porous electrodes) it might be possible that no molecule "escapes conversion". One then speaks about coulometric detection. One attractive feature of this is that the amount of analyte can be stoichiometrically calculated from the current integral. Apart from that, it is not much different from the amperometric version.

Such detectors can easily fulfill the requirements with respect to contribution to peak-width (Section 4.5.2). Even with ultraminiaturized systems, chromatograms without external peakbroadening can be obtained [39,40]. Another attractive feature is the low detection limit, which, in a standard HPLC operation, can be as low as 0.01 ppb, corresponding to a 10-pg injection with a 200-μl peak standard deviation.

The selectivity of detection and the concomitant limitations in scope are as extreme as is the case of fluorescence detection. The system is in widespread use for the determination of catecholamines and similar biogenic amines, for which it is the detector with the best performance/price ratio.

The choice of solvents is limited to those having sufficient electrical conductivity. As this implies a certain polarity, the choice of the phase system is definitely more restricted than for, e.g., UV detection. The use of derivatization schemes is less widespread than for fluorescence detection, probably because the operation of the detector requires somewhat more skill and experience. Also, derivatives with electrochemically active tags (mostly oxidizable) tend to be less stable than fluorescent derivatives. The performance of amperometric detectors, in particular in terms of sensitivity and selectivity, can be improved under some conditions by the application of pulse techniques in the electrochemical measurement. This is a rather contradictory field, for which the reader should consult other sources [41,42].

The electrochemical detector next in importance is based on conductivity measurement. The electrical conductivity is a bulk property to which all ions in solution contribute. It can be measured with high precision, in cells of suitably small dimensions. These consist of a flow-through cell with two electrodes, which are, e.g., part of a Wheatstone bridge circuit. The latter is energized by an AC source in order to avoid the effect of electrode polarization. Very low concentrations of one electrolyte in a nonconducting solvent can be determined very well; detection limits can be as low as 10^{-8} *M*. However, nearly all mobile phases used to separate ions (in IEC and ion-pair chromatography) have a substantial conductivity, which means that analytes have to be measured "on top of" this background. Under such conditions, detection limits are impaired, even with the best stabilized devices.

References on p. A221

One way to circumvent this deterioration of detection limits is to remove the eluted salt in IEC. This can be done with membrane suppressors, or by means of a second (reactor) column that, e.g., exchanges sodium ions for hydrogen ions. If the counterion is that of a very weak acid, e.g., carbonate, the released hydrogen ions recombine with this, and the residual conductivity is very low. As a result, analytes can be measured under more favorable conditions. This patented concept has led to the development of a sub-branch of HPLC, commercialized under the name "Ion Chromatography" (Chapter 5).

Other electrochemical devices, such as the potentiometric detector, e.g., based on ion-selective electrode technology, are still in the research stage and not yet commercially available. Some of these are suitable for extreme miniaturization and on-column detection in very narrow columns.

4.5.5 Other detection principles

The situation in HPLC detection has always been unsatisfactory: No good universal detector with low detection limit is available, while selective detectors, like fluorescence and electrochemical detectors have a selectivity that is not easily predictable and can hardly be manipulated. There have been many attempts to find new principles that could lead to better performing devices. As yet, however, the situation has not changed too much.

In one concept, known as the light-scattering or "mass" detector, the mobile phase is nebulized. The solvent evaporates, and the analytes can be measured by the light scattering of droplets or particles. The detection limit is not very low, and the calibration curve is nonlinear.

Transport detectors have also been devised and commercialized. They similarly exploit the expected volatility difference between mobile phase and analytes. The eluate is deposited on a carrier (belt, wire, or disk), which is in continuous motion. The carrier and the deposits pass through an oven, where the mobile phase evaporates. Next, the solutes are measured, e.g., by means of a flame-ionization detector, after being released from the carrier in some way or other. This has not led to a real breakthrough either. Problems encountered are: Mechanical complexity, memory effects, and wetting problems with resulting imprecision and high detection limits.

Other approaches, mainly investigated in the context of miniaturized chromatography, are based on the use of GC detectors, which are generally considered to be superior. With the nitrogen/phosphorus and the electron-capture detector good results have been obtained [14,15] by evaporation of the eluate and feeding all of the resulting gas phase to the devices. Depending on the solvent, one must go to smaller-diameter columns; with RPC eluents 1 mm is indicated.

4.6 REACTIONS IN HPLC

4.6.1 Precolumn derivatization

Many analytes do not have an absorbing chromophoric group, nor do they lend themselves to other sensitive detection principles. However, they may be capable of reacting with a suitable reagent to form compounds that are readily detectable by, e.g., a fluorescence detector. This section deals with such reactions, carried out before separation. These reactions should fulfill the following requirements:

(a) The identity of the analyte should be preserved; reactions in which a large part of the molecule is degraded, e.g., via oxidation or cleavage reactions, are of no use. Thus, these reactions are always derivatizations.

(b) The derivatives must be readily detectable, preferably with high selectivity; the more sensitive and uncommon (e.g., fluorescence at long wavelenghts) the response is, the better.

(c) The products should be easy to separate.

(d) One rather than several products should be formed.

(e) Products should preferably be formed in high yield. In any case, the yield should not depend on the matrix composition.

Most precolumn derivatization reactions involve the "tagging" of such functional groups as hydroxyl, amino, or carboxyl. The tag then consists of a fluorescent moiety, with a reactive binding group. In the reagent a suitable leaving group, such as chloride, is displaced by the analyte functional group, e.g., by nucleophilic attack on an amine.

The reaction conditions play a decisive role in this kind of derivatization. Especially the solvent in which the sample is dissolved may be of importance. If this is water, many reactions with, e.g., alcohol groups cannot be carried out, as the water will destroy the reagent. It is then necessary to transfer the sample first to a nonprotic solvent, either by solvent extraction or by solid-phase extraction (Section 4.3.3). Good overviews of various reactions are given in Refs. 43-45.

Precolumn derivatization can be carried out in various ways. Most simply it is performed in small vials, which can be put in a heating block or rack in a thermostated bath if controlled or elevated temperature is required. These racks can be inserted into automatic sample changers afterwards.

This off-line operation has the important drawback that the time between the formation of the derivatives and chromatography is not constant; derivatives that are not perfectly stable, such as the *o*-phthalaldehyde (OPA) derivatives of amines, can decompose to varying extents. An on-line method is sometimes preferable, particularly when the conversion is time-dependent. This can be done in two ways.

In the first, the derivatization is carried out in the injection process [46]. In the commercialized version, an injection syringe draws up the sample solution and then the reagents in a tube. Next, the liquids are moved forward and backward in the tube to accomplish mixing. Finally, the mixture is injected into the column.

Another approach to on-line derivatization is the use of solid-phase reagents [47-49]. In this procedure, the "tag" is bound to a support, from which it can exchange with the functional group of the analyte. All one needs to do is to push the sample solution into a column filled with such material, wait some minutes, and then push the liquid further into the injection valve, or directly into the column.

References on p. A221

It is clear that neither version allows the use of very slow reactions, which may require either long reaction times or elevated temperatures.

4.6.2 Postcolumn reaction

In postcolumn derivatization, or more generally, reaction detection (PCRD) the effluent from the column is treated with a reagent in an on-line mode, either by mixing it with a reagent solution in a T-piece, or by introducing it into a reactor with solid-phase reagent. PCRD has a number of advantages compared to precolumn derivatization:

(a) The reaction proceeds in a diluted and purified solution of the analyte. Good reproducibility of the reaction and the absence of side reactions can be expected, whereas in the precolumn mode the various matrix components may give rise to undesirable chemical effects.

(b) It is not necessary for the reaction to lead to only one, chemically well-defined compound. Any conversion to, e.g., a fluorescent material, whatever the complexity of its composition, is acceptable, as long as it is reproducible.

(c) Reactions that give rise to unstable derivatives can still be used, because there is a short and constant time interval between the formation of the derivative and the measurement, whereas in the precolumn mode the derivative must be stable at least for the duration of the chromatographic procedure. Also, in the off-line precolumn mode varying time delays would ruin the reproducibility.

(d) In development work it is often found convenient that the separation is not affected by the choice of the reagent for the derivatization, whereas in the precolumn mode the separation must be redesigned each time one tries another derivative.

However, the precolumn mode may be preferable for other reasons:

(a) A reagent that gives a response in the detector can be used, as it is separated from the derivatives of the analytes by HPLC, whereas in the postcolumn mode such a reagent cannot be used.

(b) Very slow reactions can be used, whereas in the postcolumn mode the maximum reaction time is ca. 30 min.

The technical design of the postcolumn reactor has been studied rather extensively by various groups. An excellent overview is given in Ref. 50. From this review the following main points emerge:

The reactor, or delay element, should stall the analytes long enough to allow them to react. The delay time, t_r, needed depends on the reaction rate. The dispersion in the reactor, on the other hand, should not be so great that chromatographic resolution is lost. Thus, the reactor should have a large t_r, but a small time standard deviation $\sigma_{t,r}$. The problem of designing such a reactor is largely similar to that of designing a good separation column; note that $(t_r/\sigma_{t,r})^2$ is the platenumber of the reactor.

There are three types of such reactors: open-tubular reactors, packed-bed reactors, and open-tubular reactors with segmented flow. The optimization of each of these has been studied extensively.

Open tubular reactors should be coiled, because that diminishes dispersion. Still, with reaction times in excess of 1 min the dispersion becomes too large. Packed-bed reactors allow the use of longer reaction times, albeit at the expense of higher pressures, experimental difficulties, and additional cost of pumps. For very slow reactions the segmented-flow reactors are most suitable. In these, the reaction mixture is transported without any appreciable dispersion by segmenting it with bubbles of either a gas or another, immiscible liquid. The added complication is that at the end of the reactor the segmentation gas or liquid must be removed in a "debubbler", which, firstly, may give rise to experimental difficulties and, secondly, tends to introduce appreciable dispersion.

Some varieties of reactors, by their very nature, offer additional advantages in certain situations. The packed-bed reactor is a very elegant expedient, if it can be packed with the reagent or a catalyst, either because this is solid, or it can be converted to a solid by appropriate chemical bonding to a solid support. This area is reviewed in Ref. 51.

The segmented-flow reactor lends itself to on-line extraction techniques. If the products but not the reagent can be extracted into the segmenting liquid, even detectable reagents can be used. One example involves the formation of ion pairs with a fluorescent anion by positively charged amines, eluted in an aqueous phase. The anion is only extracted by the organic phase if an organic cation is present. The organic phase therefore displays no background [52].

A special case is that of the chemiluminescence detection, a technique that exploits the light emission accompanying some chemical reactions [53]. The detector can be very simple. The detection cell of a fluorescence detector and associated photodetection arrangement is normally used. The excitation optics and even the emission filter or monochromator are not needed. With the peroxydioxalate reaction solutes not involved in the reaction (e.g., polyaromatic hydrocarbons) also give a response, via intermolecular energy transfer.

A nearly ubiquitous problem in PCRD with liquid reagents is the pulsation of the two liquid streams. High-pressure pumps, used with the high-pressure-packed-bed reactors, as well as the peristaltic pumps used in open-tubular and segmented-flow reactors, introduce pulsations in the reagent stream, while the HPLC flow itself may also be pulsating. These pulsations result in fluctuations in the mixing ratio. As there is nearly always some background signal in at least one of the liquids, these fluctuations lead to baseline instabilities and noise. This, and the desire to simplify equipment, may be the reason why there is much interest in "nonflow" reagents. Solid-phase reagents were already mentioned; another example of "nonflow" reactions is photochemical reaction, described in Ref. 54 and reviewed in Ref. 55.

The use of enzymes in PCRD deserves special mention. Because these biocatalysts are enormously selective, there is much interest and activity in this field. Many enzyme-catalyzed reactions lead to the formation or consumption of NAD, which can be easily monitored by UV, fluorescence, or electrochemically. Other selectively detectable species may also be involved in detection by enzyme reaction. Here too, there is a trend towards the use of solid-phase reactors: Coupling of enzymes to a support in a packed column or to the wall of a tubular reactor simplifies the equipment and may reduce the cost of expensive enzymes. The possibilities for variations, e.g., with tandem enzyme reactions

References on p. A221

and reactors, are virtually endless. The applications are, of course, mainly limited to the analysis of naturally occurring compounds, although some other examples, involving enzyme-blocking effects, exist. This field is reviewed in Ref. 56.

4.6.3 Indirect detection

Although indirect detection does not necessarily involve a reaction in the column, this may sometimes be the case, and therefore this variety of detection is treated here. In indirect detection a component giving a response in the detector, the marker or probe [57,58], is added to the mobile phase. Eluted nondetectable analytes can be monitored by means of the changes in the marker concentration. In order for such changes to occur, it is necessary that the analyte influence the distribution behavior of the marker. This can be the result of:

(a) competition in the phase distribution of the two compounds, e.g., in adsorption or ion-exchange systems;

(b) reactions that are coupled to the distribution equilibrium for both compounds, e.g., in proteolytic equilibria, or in ion-pairing systems.

The technique has found its widest application in the detection of ions, e.g., in "ion chromatography" as an alternative to conductometric detection [59,60].

The theory of indirect detection is rather complex, and interested readers should consult the original literature [61,62]. From that literature the following rough guidelines can be derived:

(a) The response is determined exclusively by the distribution behavior of marker and analyte.

(b) Apart from the analyte responses, there are usually one or more "system peaks" in the chromatogram. These are generated by the injection, but are not related to a specific solute. The more complicated the mobile phase, the more system peaks can be expected.

(c) The response may be positive or negative. Often the sign depends on whether the solute capacity factor is smaller or larger than that of a system peak.

(d) The response can be larger than stoichiometric; i.e. the change in the concentration of the marker may be larger than that of the analyte. This is especially likely to occur when the capacity factor of the analyte approaches that of (one of) the system peak(s).

(e) The analytical utility depends very much on the stability of the detector used. As the peaks "ride on" the background of the marker, it is necessary to use a detection system that allows monitoring relatively small concentration changes ("dynamic reserve") [63].

In most instances indirect detection has been implemented by UV detection, using marker substances like naphthalenesulfonic acid. However, the use of fluorescence detection has also been demonstrated [63]. The technique may be especially useful for the analysis of otherwise undetectable solutes, occurring in relatively high concentrations. For trace analysis, extensive sample cleanup may be necessary, as the dynamic range of the system is not large [62].

4.7 SPECTROMETERS

4.7.1 General aspects

The interpretation of chromatographic peaks is often difficult. Homogeneity of peaks may be doubtful, and their identity may be mistaken, and often there is no clue to their identity. One approach to solving such problems is that of coupled-column chromatography. This strategy will be treated in Section 4.8.4. However, the additional information needed can also be obtained by spectrometrically analyzing the eluate. Techniques like MS, IR or UV/visible absorption, and NMR spectrometry are extremely well suited to answer such questions.

The approach may be implemented off-line or on-line. In the first case, a fraction of the eluate, corresponding to a peak, is investigated by one of the spectrometric techniques. One advantage is that the spectrometric analysis is not forced into a time frame dictated by the HPLC separation. Also, the off-line approach allows optimizing the spectrometric analysis independently, e.g., using special introduction techniques or changing the medium (solvent, gas phase, or adsorbed state) in which the analytes occur. In cases where HPLC as well as spectrometry must be used to the limits of their capabilities, the off-line method is probably the method of choice. As there are not many special techniques or methods involved, except that of transferring the analytes to the appropriate medium, it is not necessary to go into more detail on off-line methods.

However, in general, off-line methods are rather awkward and time-consuming, especially when information is required on not just one peak, but throughout the chromatogram. That means that many fractions would have to be processed. An on-line, "on-the-fly" method, of which GC/MS is an example, is then indicated. The problems encountered in LC are briefly summarized:

(a) *Interfacing*. Spectrometry is often not compatible with the physical state of the analytes as they are eluted from the column. They must be transferred to a suitable phase system, e.g., vacuum in MS or adsorbed on KBr in FTIR. As this must be done on the fly on a 1-sec time scale, with negligible bandbroadening, this may pose many problems.

(b) *Detection or identification limit*. In not a few cases (see below) the amount needed for detection, or for obtaining an interpretable spectrum (minimum identifiable quantity, MIQ), is large compared to what is usually delivered by a HPLC separation. For instance, proton NMR requires many micrograms.

(c) *Speed of spectral scanning*. With some spectrometers the acquisition of a full spectrum in less than one second may not be possible without serious compromises.

(d) *Storing data*. Data streams in excess of 100-500 numbers per second may occur. Fortunately, modern, cheap computer technology (even PC-type equipment) allows "dumping" such amounts of data.

(e) *Data presentation*. The user must be able to access the data in an ergonomically acceptable way. This has been accomplished by means of "3-D" representations of the signal intensity, plotted as a function of retention time and the spectrometric coordinate (e.g., wavelength), by presenting cross sections (signal vs. retention time for a particular

References on p. A221

wavelength, or signal vs. wavelength for a given retention time), and by "contour plots", that describe, in a retention time/wavelength plane, the signal intensity coded in colors and "iso-signal lines", as in altitude maps.

(f) *Data handling*. Still, on the basis of the acquired raw data it is necessary to answer such questions as mentioned above: Is the peak homogeneous? What is its identity? Can conclusions about the structure be drawn? Programs that perform these tasks are still a subject of continuous study, especially in chemometrics [64].

4.7.2 Various spectrometers

UV spectrometry has now become commonplace, and the technical aspects are more or less standardized. The solid-state diode-array multichannel photoreceptor allows acquisition of spectra many times per second. Cell volumes are in the range of 5-10 μl.

The information supplied by a UV spectrum is limited. Still, the analytical utility of the HPLC/UV combination is large, first because of the low detection limits, and, second, because of the quantitative reliability of the absorption profiles. With the latter, all sorts of linear data-reduction techniques become applicable [64-66]. These, inter alia, allow "deconvolution" of peaks, i.e. elution curves of unresolved components can be constructed on the basis of the three-dimensional data structure. Without any resolution of components this requires the a priori knowledge of the spectrum of each component. However, it has now been amply shown that even without this prior knowledge, but with marginal resolution, the spectra and elution profiles can still be obtained [66]. This is of great value for "peak tracking" in automated phase-system optimization [64] (Section 4.8) and validation of chromatographic results. However, for direct quantitative analysis calibration would still be needed.

The suitability of UV spectra for identification and library search techniques, especially in relation to the dependence of the spectra on the mobile-phase composition, are a matter of concern [67-71]. Chemometric techniques can be expected to bring further improvements in this area.

In *mass spectrometry* the interfacing problem dominates research activities. A large number of solutions to the problem of transferring the solute from the liquid phase to the high-vacuum MS environment have been proposed. This leads to a rather confused situation with respect to important points, such as detection limit, contribution to peak-broadening, and type of spectra.

Many interfacing solutions preclude the exploitation of fragmentation in the mass spectrometer, and the sole occurrence of the (pseudo)molecular ion detracts very much from the information potential of the method. The increased use of the thermospray interface reinforces this. Efforts to improve this situation, either the "smart way" by adapting the interfacing, or the "brute force" approach of using LC/MS/MS are being studied intensively.

The strength of MS also resides in the low MIQ values. This has allowed it to be coupled to high-resolution capillary GC, a technique generally applicable only to samples of one microgram or less, while still maintaining a reasonable dynamic range. Although such low- and sub-picogram sensitivities cannot normally be obtained in HPLC coupling, MS is still the most sensitive kind of spectrometry that can be considered for HPLC coupling. Good overviews of HPLC/MS can be found in Refs. 72 and 73, while the new, promising technique of electrospray ionization is discussed in Refs. 74 and 75.

UV and MS are the techniques that are accessible to routine analysis and which are commercially supported on a reasonable scale. The remaining techniques are still in the research stage.

Fluorescence detectors hold the second place in popularity in HPLC, and a multiwavelength instrument (excitation or emission, or both) would seem to be a very valuable asset. However, the scope would be small, as relatively few compounds fluorescence. Also, the fluorescence spectra obtained after pre- and postcolumn reactions generally do not reveal as much about the identity (let alone structure) of the analytes as do native fluorescence or UV spectra. The conclusion from the above is that the future for the HPLC/fluorescence spectrometry combination is rather dim.

Fourier transform infrared spectrometers are now available commercially for GC/FTIR coupling. The spectra are highly informative for structural analysis. The information is complementary to that obtained by MS, especially in the determination of geometric isomers, and is therefore warmly welcomed by many scientists. Sensitivity is the main problem in the development of the technique; the MIQ values, although continuously improving, are of marginal analytical utility. In IR the measuring time significantly influences the S/N ratio; the larger the number of scans that can be added, the higher is the attainable S/N ratio. Such a situation is not favorable for on-line coupling, as there is too little time available for the acquisition of repeated spectra.

Another vexing problem mentioned in the literature on HPLC/FTIR coupling is that of canceling the absorption by the solvent. Normal LC solvents, with the exception of carbon tetrachloride and some other nonpolar chlorinated alkanes, leave only relatively small "windows" open for observing the analytes in the infrared.

Early experiments on HPLC/FTIR have therefore been carried out with such mobile phases in normal-phase adsorption systems, contrary to the prevailing trend in HPLC towards RPC systems with more polar and aqueous mobile phases [76]. Subsequently, a number of solvent elimination techniques have been investigated. These are reviewed by Griffith and Conroy [77].

Nuclear magnetic resonance in combination with HPLC has been studied less intensively than FTIR, but still a number of interesting papers have appeared on this subject. The problems are similar, sensitivity being the most critical one, even more so than in FTIR. Proton NMR is best in that respect, but there the universally present solvent protons generate signals that not only obscure interesting areas of the spectrum, but also tend to saturate the measuring system.

Because of the limited mass sensitivity, it seems expedient to use rather large columns in order to transport as much material as possible to the spectrometer. The work of Albert et al. [78] is an example of the use of semipreparative HPLC columns. In contrast to FTIR, for NMR the balance between off-line and on-line coupling undoubtedly is more in favor of the former. The high investment in a NMR instrument reinforces this point.

References on p. A221

Other techniques are also being considered for HPLC coupling. Supersonic jet spectrometry, fast-scanning or multi-electrode electro-chemical amperometry, and circular dichroism have all been tried for this purpose. However, their analytical utility is generally unlikely to match that of HPLC/UV and HPLC/MS in the near future.

4.8 OPTIMIZATION

4.8.1 Introduction

Optimization of a HPLC system for a particular sample is often necessary merely for obtaining the required separation. In other cases the separation is satisfactory, but faster analysis or greater sensitivity may be required. Optimization of LC may be rather complicated, as many variables play a role. An inventory of variables leads to an impressive list: nature of the stationary and mobile phases, particle size, column length and width, mobile-phase velocity and temperature, etc. The first two items alone may again involve a range of adjustable parameters. When such aspects as detection limit and column loadability are of added interest, the detector choice, possible reaction types, and sample pretreatment lead to additional variables. It is useful to approach the optimization in steps. This is normally done in the following iterative sequence:

(1) The phase system is chosen, e.g., a RPC or IEC packing, together with the constituents of the mobile phase.

(2) Variables, such as content of modifier or buffer concentration and pH, are adjusted in order to obtain a reasonable ($k' <$ ca. 20) retention for the last-eluted peak.

(3) With this, the capacity factors, k_i', of components and the relative retentions, $\alpha_{i,j}$, of successive pairs of components, i and j, are fixed. After measuring these, the number of theoretical plates, N_{req}, required to obtain the separation, can be calculated according to the inverse resolution equation

$$N_{req} = 16 \cdot R_{ij}{}^2 \cdot \left(\frac{k_i' + 1}{k_i}\right)^2 \cdot \frac{1}{(\alpha_{ij} - 1)^2} \tag{4.26}$$

where R_{ij} is the required resolution in the chromatogram. In the general case, where n components must be separated, there are $n - 1$ successive pairs, ij, for which the resolution may be critical. With a given phase system, one of these determines the required platenumber. This obviously leads to an excess of resolution for other pairs. When the value of N_{req} is not acceptable, i.e. normally $>$5000-10 000, it is necessary to return to Stage 2 and to modify the phase system, or even to Stage 1 and to change to a different chromatographic principle, e.g., from IEC to RPC.

(4) Other conditions are examined. For instance, if detection is inadequate and this cannot be remedied easily, e.g., by the choice of column dimensions or detection parameters, it is again necessary to return to Stage 2 or 1.

(5) The required platenumber is implemented in the most appropriate way, given the

available pressure, particles sizes, and column lengths. Depending on the importance of fast analyses, it may be necessary to go to the smallest possible particles and highest pressures. In some cases one may find out at this stage that the process is too slow, and that another phase system is necessary.

Stages 3, 4, and 5 in the above scheme are the most transparent: The effects of variables like particle size, column length, linear velocity, and capacity factors, resulting from Stage 2, on the chromatogram and resolutions are readily and accurately predictable. The same holds, e.g., for the detectability.

However, the changes in the phase system, accomplished in Stages 1 and 2, generally produce poorly predictable effects on the capacity factors. It should be kept in mind that extreme accuracy in capacity factors is required to assess the value of a phase system. In a pair of solutes, *i* and *j*, a value of $\alpha_{i\,j}$ for the ratio of the k' values of 1.06 is adequate for separation, while with $\alpha_{i\,j} = 1.00$ separation is impossible.

It is this stage of the optimization that leads to the least satisfactory approach, which often consists of a trial-and-error strategy, where various combinations of mobile and stationary phase are tested in succession.

4.8.2 Computer-aided phase system optimization

There have been numerous attempts to develop more systematic procedures of optimization. These are excellently reviewed in Schoenmakers' book [79]. All of them make use of computers to help in filing, arranging, and manipulating the chromatographic data. A short description of the merits of these systems is appropriate, but necessitates recapitulation, however brief, of the formal optimization process.

Optimization of a number of variables in an experimental situation requires the exact definition of the parameter that is to be as large (or as small) as possible. This quantity, known as the "objective function", in the following designated by *Y*, is not easy to define in HPLC optimization. Clearly, resolution is the first candidate. However, for *n* components there are $n - 1$ resolutions in the chromatogram. A consecutive sum of these is useless. For instance, a system could be rated highly when one or more resolutions are (close to) zero but are compensated by very large other resolutions. This problem can be solved by taking some more or less complicated nonadditive functions of the resolutions. The simplest is: *Y* is equal to the smallest resolution, as was already assumed above. Other functions may, e.g., add up valley/peak ratios (a defined function of resolution and peak intensity ratios). This likewise avoids laying too much weight on large resolutions.

A second problem is that not all resolutions are of interest; only the analytes of importance ought to be well separated from other peaks. It might even happen that one is to determine, say, five compounds, but that two of these are of utmost importance. This then should be expressed by using weight factors in assembling *Y*.

Next, the time required by the chromatogram should be taken in to account. In general, the resolution equation, as well as the Knox equation (Eqn. 4.6) predict that longer analysis times allow better resolutions. When just counting some measure of resolution, any formal optimization system will tend to choose conditions leading to a long analysis time. The way in which the duration of the chromatogram is to be accounted for in order to avoid this is complicated. A good discussion is to be found in Ref. 79.

We assume from here on that some function, Y, has been determined, which can be derived from an experimental or computer-predicted chromatogram. As an example, one may think of

$$Y = R_{smallest} \cdot t_{last}{}^{p} \tag{4.27}$$

where $R_{smallest}$ is the smallest resolution occurring in the chromatogram, t_{last} is the retention time of the last-eluted component, and p is some (negative) exponent, e.g., $-1/4$ when in the pressure-limited case [79].

Fig. 4.6 illustrates all this for the simple case of a two-component mixture, with one mobile-phase modifier concentration that can be varied. For more variables, such plots are, of course, more-dimensional.

Assuming that one starts to resort to such a scheme at Stage 2, one must then scan the various compositions of the phase system and find the maximum in Y. The strategies for this are broadly classified below:

(a) *Experiments.* This naturally requires many experiments. With 2 continuously variable parameters, such as the content of two organic modifiers in RPC, each given only 10 values, it is necessary to evaluate 100 chromatograms.

(b) *Models.* The dependence of all capacity factors can be modeled on the mobile-phase composition [80]. This may be successful in many cases, e.g., dependencies on pH, organic modifier concentration, and ion concentrations in ion-pair and IEC. In those cases a few measurements of capacity factors suffice for a numerical prediction of all values at intermediate compositions. Thus, with 3 to 20 determinations of capacity factors it is possible to map the retention over the entire range of variables. The search for the highest Y value can next be performed entirely by a computer. Such systems can be supplemented in such a way that the complete chromatogram is constructed and other aspects of the separation, e.g., detection limits and interference can also be judged from the predicted chromatogram [81].

(c) *Simplex.* In the simplex method [82-84] the chromatogram is initially evaluated at three compositions (for two variables, in general one more than the number of variables). On the basis of the observed Y values the simplex algorithm indicates another composition, where possibly a better Y value is to be found. After the Y value of this chromatogram has been determined experimentally, further rules indicate how to search for (further) improvement. Such a "hill-climbing" strategy has the large advantage that it is not dependent on any model for the retention, or whatever other rules may apply to the system under study. However, its behavior may be similar to that of a blindfolded hiker trying to find the highest point in a mountainous area by always walking uphill. He finds a top, but there may be much higher summits close by. He could only reach the other summits by going through a valley first. The method is said to lead to a local optimum, rather than a global one. Especially when peaks change their elution order, there are such local optima: at the

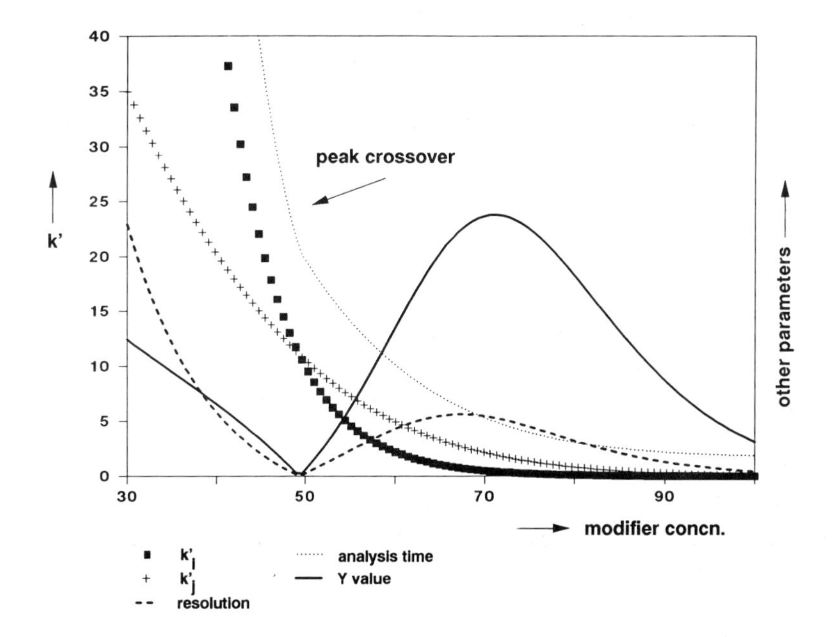

Fig. 4.6. Capacity factors (+, Component 1; ■, Component 2), resolution (— — —), analysis time (● ● ●), and objective function *Y* (———) as a function of mobile-phase modifier concentration.

intermediate value(s) of the parameter(s) the value of *Y* is very low (a valley in the *Y* map), and it will be impossible for the simplex to pass this region.

(d) Another *sequential approach* [85] assumes that between two sets of conditions the retention varies linearly. On the basis of a few experiments it is then possible to predict a better set. This is evaluated experimentally, and on the basis of this result further improvement can be predicted for yet another set. This procedure is halfway between the full mapping and simplex approach. Although it is possible to pass through a *Y* valley, there is still the risk that a local optimum may be found.

The field of computer-aided optimization of phase systems is still under development. A few central topics in such studies are:

(a) Optimization with a sample containing (an unknown number of) unknown components. Retention mapping is then very difficult [86]; often it is even impossible to keep track of the peak identity when conditions are changed, unless coupled spectrometric detectors are used [87]. Otherwise, one must use the simplex or similar method.

(b) On paper, it is nearly always possible to vary many parameters, but in practice one cannot vary more than two or three. Which ones are likely to be the most rewarding is thus an important question.

(c) Even when the most promising variables have been identified, it is necessary to fix the limits of the useful range in which they should be varied [88].

(d) Further complications arise when only a few components are to be quantitated.

4.8.3 Gradient elution

The theory of gradient elution has been treated in Chapter 1, mainly in terms of the linear solvent strength (LSS) gradient, where the capacity factors decrease exponentially with time. The general strategy for obtaining the best time/resolution combination with a given low and high solvent strength combination is pretty clear there. However, when sample constituents of diverse chemical structure occur, it may be desirable to find the best solvent pair for improving resolution of particular pairs in the chromatogram. Thus, one may wish to decide, e.g., whether it is better to change pH or counterion concentration, or to increase elution strength with, e.g., the organic modifier content.

Such questions lead to the same kind of problems as were outlined in the previous section, but at a higher level of complexity. This is even more so as gradient systems often make it possible to generate ternary or even quaternary mixtures. This allows adjustment of the selectivity of the solvent for individual pairs during chromatography, as each pair has its own time period in which it effectively migrates; the selectivity of the mobile phase in that period should be fine-tuned to that pair. A rather complicated example of such a scheme is given in Ref. 89.

4.8.4 Column switching

There remain cases in which a satisfactory separation cannot be obtained. For sample mixtures with more than 10-20 detectable components this is not surprising: As shown in a number of papers [90,91], it is very unlikely that such a separation can be obtained with platenumbers in the range of 10^4. Although the peak capacity may then be something like 50, the capacity factors will never form a series of such regularity that the chromatogram is filled up uniformly. Also, there are many cases where the number of components simply exceeds the peak capacity.

A *column switching system,* also called *multi-column operation,* (or, in the jargon, "multi-dimensional" chromatography), resolves these problems. An unresolved peak cluster from one column is transferred, preferably in an on-line mode, to a second column. The phase system in the second column can be optimized for just the separation of this cluster. The technique is also referred to (cruelly) as "heart cutting". A review on this technique is to be found in Ref. 92. Here, the main aspects are briefly summarized:

(a) A central point in the discussion is the relation of capacity factors in the first and the second column. There should be as little relation between them as possible ("uncorrelated" or "orthogonal", in the jargon). When that is the case, a peak cluster in Column 1 will spread out in Column 2.

(b) The on-line experimental accomplishment in HPLC is hampered by the additivity of peak variances. Peaks from Column 1 may be too broad to serve as injection plugs in Column 2. In many cases, therefore, one needs an intermediate refocussing step at the inlet of Column 2.

(c) The other, related, troublesome point is the compatibility of the mobile phases used for two columns. Mobile Phase 1 may not be totally suitable to be injected into Mobile

Phase 2, either because it is immiscible, or because its eluotropic strength is too high.

(d) When only a few compounds are to be quantitated, the time requirement of the method may be acceptable; many parts of the first chromatogram can be discarded, and only a few fractions need to be analyzed in Column 2. However, a more or less comprehensive analysis of the sample would require a number of second chromatograms larger than the peak capacity of Column 1. With the latter being in the order of 50, the time requirement would increase ca. 100-fold, as compared to single-column chromatography.

Despite these difficulties, it is believed that for difficult separation problems column-switching techniques are the best approach, the main reason being that there hardly are any alternatives. As mentioned in Section 4.4, some sample pretreatment techniques with precolumns may also be regarded as column-switching methods.

4.9 PACKING MATERIALS

4.9.1 Supports

A packing material serves, in the first place, to form the chromatographic bed; it may or may not be involved in the distribution process. In liquid/liquid chromatography (LLC) it is not; the stationary phase is a liquid that resides in the pores of the packing. In that case the packing is just the *support*.

On the other hand, when adsorption or ion exchange controls the distribution process, the surface or the functional groups of the packing are essential; the packing is support and stationary phase at the same time. In many cases it consists of a base, skeleton material that may have been chemically modified. The packing is then sometimes called a "phase", an inappropriate abbreviation for "stationary phase". It is better to stick to the neutral word packing, column material, or sorbent, or a specific word like adsorbent or ion exchanger, if the predominant mechanism of sorption is known.

Porous silica is the most widely used support for HPLC packings. Therefore, its properties will be considered first, while other materials are discussed briefly in Section 4.9.4.

Silica consists mainly of SiO_2, where the silicon atoms are bridged by oxygen atoms, each silicon atom being at the center of a tetrahedron. On the surface, however, one remaining valency is normally occupied by an OH group, referred to as a silanol group. Silanol groups, several varieties of which are known, are very polar and rather reactive, acting as an acid. They can be removed by heating to about 700°C to form Si–O–Si, siloxane groups. This product is no longer useful for chromatography. Drying of silica, e.g., preceding derivatization reactions, should take place at 200-250°C.

Porous silica is a material of remarkable mechanical and chemical properties [94]. A large part of the material, in fact some 50-70% by volume, consists of pores. The skeleton material between the pores is often only ca. 1 nm thick. It can nevertheless withstand the shear forces exerted by the pressure-driven solvent flow in the columns. Another remarkable fact is the openness of the pore structure. The fast mass transfer in the pore space that can be observed chromatographically (the C term is only increased by factor of 1.5-2

References on p. A221

due to the tortuosity of the pore space [94-96] indicates that all pores are well interconnected.

Yet another advantageous property of silica is that it can be prepared with quite a narrow particle size distribution. Originally, mostly irregular particles of silica were used; narrow size distribution was obtained by various methods of classifying (sieving, air classifying, elutriation) techniques. However, spherical silica is now being used more and more. It is produced in very narrow size distributions from the outset, so that little or no classifying is necessary. The spherical shape is often considered more favorable for chromatography, possibly because spherical material packs more uniformly.

Still another favorable feature of silica is its chemical inertness. It does not dissolve and is not attacked by a large range of mobile phases useful for HPLC. However, its chemical inertness is not completely satisfactory, and that is the main reason why much effort is devoted to the development of other materials (Section 4.9.4). It suffers from two deficiencies:

(a) In basic solutions its skeleton is hydrolyzed; the material just slowly dissolves. This constitutes a serious limitation on the free choice of mobile phase, especially in ion-pair and ion-exchange systems. For that reason, use of pH values above 8 is generally considered impractical.

(b) The silanol groups are very polar. This may be considered to be an advantage in normal-phase chromatography, where the adsorption on the silanol-covered surface is the basic distribution mechanism. Even here, however, the extremely high polarity, and especially the nonuniformity of the adsorptive strength of the surface, leads to difficulties. These were already assessed experimentally in one of the first papers on modern LC separations [97].

In RPC, the silanols remaining after modification with nonpolar groups or moieties of medium polarity are held responsible for many undesirable effects in this type of separations. These will be discussed in more detail in Section 4.9.3.

In the study of packing materials a number of nonchromatographic techniques are used for the characterization of the porous materials. These are now briefly reviewed.

The *specific surface area* is the area of all pore walls per unit weight, usually expressed in m^2/g. It may vary between 10 and 500 m^2/g. Determinations are nearly always carried out by the Brunauer-Emmet-Teller (BET) method. This consists of measuring the amount of nitrogen sorbed as a function of (the log of) the nitrogen pressure at the boiling point of nitrogen. In the resulting plot, the point where the surface is just covered by a monolayer of nitrogen molecules can nearly always be clearly distinguished. By assigning a conventional, accepted value to the surface area occupied by one molecule (e.g., 0.17 nm^2) one can calculate the area of the sample.

Another property is the *specific pore volume*, expressed in ml/g, and usually in the range of 0.4-1.2 ml/g. In most cases this value is determined by mercury porosimetry: The sample is evacuated, and mercury is added to it at increasing pressures. Initially, the mercury cannot enter the pores, due to surface tension. With increasing pressure, smaller and smaller pores become accessible. In this way, it is possible to prepare a plot of *pore size distribution,* which indicates what percentage of the pore volume corresponds to a given range of pore diameters. Of course, the pore diameters reported are understood to be equivalent diameters; the pores are not cylindrical, but the experimental results are treated as if they were. Limitations of the technique are observed at the small pore sizes, diameters < 5 nm requiring extremely high pressures. For these pores additional information can be obtained from the BET method. Surface area, pore volume, and mean pore diameter are interrelated. At a given pore volume the surface area is greater when the pore diameter is smaller. A discussion of these relations is to be found in Refs. 93 and 99.

The choice of the pore diameter is an important topic. For various reasons it is advantageous to choose the smallest possible pore diameter: a large surface area results in a high sample capacity, in the direct adsorption mode as well as in cases where some interacting groups are attached to the surface. Also, in LLC the stability of the stationary phase is much greater when smaller pores are used. However, the pore diameter should be large compared to the dimension of the solute molecules. Otherwise, a large part of the surface will be inaccessible to the solute. Even before that would happen, diffusion within the pore space will be hampered, and the resulting slow mass transfer will lead to low efficiency. With fragile large molecules also undesirable changes in conformation may occur when they are forced into pores that are too narrow.

Thus, for low-molecular-weights solutes (<500 D), one uses materials with a specific surface area of 200-400 m^2/g and a pore diameter of 5-10 nm. For increasingly larger molecules one is forced to use wider-pore materials, with surface areas that are much smaller (10-50 m^2/g). Eventually, one may decide to refrain from the use of porous materials and rely on the adsorption on the outer surface [99] of the particles. It is instructive to note that at a 1-μm particle size the area of the outer surface, normalized to the mobile-phase volume in the column, is about the same as that in a column packed with a 5-m^2/g porous material of larger size. An alternative is to use pellicular packings, where a thin layer of porous material is coated on an otherwise impervious particle [100].

In view of the above, it is fortunate that silica technology allows the preparation of particles of widely different pore diameters: the full range of 6 nm to 0.4 μm is available. Very-wide-pore materials are necessary for the separation of the largest molecules, such as proteins of 10^6 D.

For SEC the pore structure plays an even more important role, as the separation is directly dependent on it. The mean pore diameter should be adapted to the range of molecular sizes to be separated. For the highest selectivity in a given MW range, the pore size distribution should be narrow. On the other hand, when a large range of molecular sizes is to be analyzed in one chromatogram, it may be better to use a broad pore size distribution, or even to mix packings having different pore sizes.

4.9.2 Adsorption and partition

Adsorption is generally understood to be the process in which a component is enriched at the interface between two phases. In partition, the component dissolves in two bulk phases. Thus, it must be dispersed in three dimensions, and molecules must move freely within the bulk of the phase. However, because of the finite size of molecules, in

References on p. A221

adsorption the enriched layer must also have a finite volume (layer thickness), and therefore a distinction between the two modes is not straightforward. A more precise definition is: Partition occurs when the equilibrium relation between the concentrations in the two phases does not depend on the volume of each of these. In particular, in HPLC the equilibrium should not depend on the thickness of the stationary-phase layer.

The most popular form of HPLC, RPC on an alkyl-modified silica packing, was originally believed to be best classified as partition; the cloud of alkyl groups at the surface was thought to be equivalent to a alkane solvent, with the sole advantageous difference that it could not be stripped off. However, it is now clear from various studies that this is too simple a picture of RP distribution:

(a) Correlation of retention data with those for L/L systems, consisting of aqueous solutions and alkanes, shows that the selectivities are quite different.

(b) Increasing the chainlength of the alkyl groups on the surface does not lead to a proportional increase in capacity factors, as would be expected for partition.

(c) The nature of the base silica often – to our regret, because this is the main cause of the problems with reproducibility of RP packings – exerts a strong influence on the relative retention.

Most investigators now believe that in typical RP systems the solute is at least partially enriched at either the boundary between the alkyl cloud and the support, or at the boundary of the alkyl cloud with the mobile phase. This would explain the fact that a plateau in the *k*′ value is reached for many systems on increasing the length of the alkyl groups on the surface.

4.9.3 Surface modification

The large variety of useful phase systems in HPLC is for the most part due to the ability of the silica surface to be modified chemically in so many ways. The fact that so many of these modifications have indeed been made, studied, and marketed reflects the need for packings that are suitable for the large variety of compounds separated by HPLC. In this section, the chemical basis of modifications is discussed first. The second part is devoted to the relation between the nature of the sample and the choice of the modification.

The silanol groups at the silica surface are the anchoring groups for most modifications. Nearly all surface-modifying moieties are attached to these by reactions such as

$$\text{Si–OH} \quad + \quad \text{Me–}\underset{\text{Cl}}{\overset{\text{Me}}{\text{Si}}}\text{–C}_{18}\text{H}_{37} \longrightarrow \text{Si–O–}\underset{\text{Me}}{\overset{\text{Me}}{\text{Si}}}\text{–C}_{18}\text{H}_{37}$$

silanol   dimethyloctadecylchlorosilane   "octadecylsilica"

in which the Cl atom in the silane is displaced by the O atom of the silanol. Variations on this principle are:

(a) the use of leaving groups other than Cl atoms, in order to increase the rate and yield of the reaction, or in order to obtain byproducts that are less undesirable than HCl, and

(b) the use of reagents with two or more leaving groups (so called polyfunctional reagents), e.g., dimethoxysilanes or triethoxysilanes, for the same purpose, and in the hope that more than one silanol will bind to one silane molecule. However, it is unlikely that 3 silanols will ever do so.

It is often held that monofunctional reagents, such as the one given above, lead more easily to a reproducible material. With polyfunctional reagents and traces of water present during the reaction, polymerization reactions may occur, leading to a much higher carbon content. Such reactions are expected to be less controllable. Still, sometimes such packings are preferred, because they show special selectivities for certain groups of compounds.

The ultimate purpose of the modification is to alter the interactive nature of the surface, from polar to nonpolar or less polar. The very high polarity of the support is therefore definitely a disadvantage; it "shines through". Even a few remaining sites, having strong interactions with polar groups on solutes, may give rise to, e.g., asymmetrical peaks. One therefore seeks to hide the original polarity as much as possible. Nevertheless, it is found that especially basic compounds, such as amines, often still interact strongly with the support.

In this context, it should be noted that because of stereochemical effects not all silanols can be converted. Even with the small-sized trimethylchlorosilane, only about half of the silanols react, the remaining ones being inaccessible to the reagent. The longer and bulkier the groups on the reagent, the stronger is this effect. On the other hand, when the reagent does not "see" the silanols any more, it is likely, and indeed often found, that the solutes do neither; with an exhaustively modified packing the above-mentioned undesirable effects are significantly lessened. There is now overwhelming evidence that the original surface and the silanols are rather effectively shielded by the alkyl cloud when it is as dense as possible. Also, proper pretreatment of the support before bonding is important.

It follows that the surface coverage is an important parameter. It is usually expressed in μmol per square meter (of the base silica!). The density of silanols is roughly 8 μmol/m^2. Division by the (determined or estimated) amount of silanols in the same units gives the percent coverage. As mentioned, this can be no higher than ca. 50%. These numbers are, of course, also related to the carbon content of the packing, which is easily determined by elemental analysis.

In the production of many commercial packings an additional treatment, "endcapping" is applied. This involves a second reaction with, e.g., trimethylchlorosilane, that because of its small size is better able to reach the least accessible silanols.

A large variety of other more or less complicated functionalities than just alkyl groups can be attached to the silica surface. This can be done basically in two ways:

(a) Preparing the complete chemical moiety, with a spacer connected to a silane with

References on p. A221

leaving groups and allowing this product to react with the silica.

(b) Modifying the silica first with, e.g., an aminosilane compound or an epoxysilane compound and next building the required moiety on the surface by using the synthetic possibilities of the amino or epoxy group.

Definition of the chemical structure and the completeness of conversion are important problems in these techniques. Method (a) has the advantage that standard spectrometric methods can be applied during the entire synthesis of the moiety to follow yield and purity. The final bonding step may be difficult and slow, though. In Method (b) one does not have the latter problem, but it is much more difficult to verify the chemical structure of the product.

A brief survey of the modifications used, in the context of their fields of application, is given below:

Moderately polar adsorbents are prepared by using cyano-, nitro-, or diol-substituted alkylsilanes, instead of alkylsilanes. When used with the customary mobile phases, consisting of water and an organic solvent, retention may be much smaller, and the selectivity may be drastically different from that obtained with alkylsilicas. Especially, the methylene increment (the slope in the plot of log k' vs. number of methylene groups in the solute) is generally smaller than in RPC. Thus, there is less spacing between homologs in the chromatogram, while the separation according to types of functional groups is as large or larger.

These packings are already so polar that with sufficiently nonpolar mobile phases the separation is of the normal-phase type; the methylene increment is negative. Thus, diol packings may be used as an alternative to bare silica, with the advantage that the polar interactions are less extreme, extending the application range to rather polar compounds. Also, the surface is energetically more uniform, resulting in more symmetrical peaks.

Ion exchangers are based on such groups as primary, secondary, or tertiary amino groups for anion exchange, and sulfonic acid or carboxylic acid groups for cation exchange. They are widely used in the analysis of ionic compounds, including, e.g., amino acids, peptides, nucleobases, nucleosides, and proteins. Neutral compound are sometimes also separated on ion exchangers of this type, e.g., sugars on aminoalkyl packings.

For relatively small analytes, IEC on such packings, which are not always very stable or reproducible, finds a strong competitor in *ion-pairing systems* [101]. These are normally implemented on regular RPC packings. The mobile phase contains an "ion-pair former", salts like sodium dodecyl sulfate or butyl trimethylammonium bromide, which have more or less long nonpolar chains. The organic ions should have a charge opposite to that of the analytes. In a RPC system, these long-chain ions form pairs with the analytes, which are preferentially adsorbed.

With relatively short-chain ion-pair formers, the retention of analytes increases with the concentration of the reagent, as would be expected from the mass action law, which pushes the equilibrium towards the formation of ion pairs. However, with longer chains, e.g., with sodium dodecyl sulfate, such an effect is weak or not observed at all. The reason is that such reagents react in the same way with their own counterions, in this case sodium, and cover the surface entirely with the resulting ion pairs. They are held so strongly at even very low ion-pairing reagent concentrations that the packing may be considered as converted to an ion exchanger for all practical purposes.

In ion-pair systems, the variation of type and concentration of the ion-pair former (and pH, as generally with ionic solutes) gives ample opportunity to adjust the selectivity of the separation to the sample at hand, with the same column. This may explain the rather great popularity of such systems, despite the rather complicated chemistry behind them.

Attachment of *chiral selectors* is an important application of surface modification techniques. The selectivity toward enantiomers may be based on various types of interactions: ion pairing, ion exchange, metal-ion complexation, but also polar interactions with ester- or amide-type moieties. The field of enantiomer HPLC separations is too large and diverse to be covered here; interested readers should consult a review article [102,103].

New approaches to the modification of packings based on silica address the following problems:

(a) In view of the difficulties with the surface reactions with the silanols (which, by the way, also confine the application of these modification techniques to silica), it may be worthwhile to avoid chemical bonding, but instead to polymerize a material with suitable stationary-phase properties within the pores of the support, where it is further held mechanically. Such "polymer-coated packings" appear to offer a number of advantages, but it is still too early to estimate the role they will play.

(b) In protein separations (Chapter 14) it is difficult to control interactions with the surface of the support. Untoward interactions that are not of much concern in work with small molecules, can give rise to excessive retention, peak tailing, irreversible adsorption, etc. in the case of these molecular giants. Also, denaturation of proteins is an important problem. If interactions, such as those with silanols or with nonpolar groups, are too strong, they may in themselves lead to these undesired effects. In other cases, the required high elution strength of the mobile phase can only be obtained by resorting to mixtures (e.g., high methanol content in RPC) which will lead to denaturation. These applications would therefore benefit much from the development of very hydrophilic packings, in which interactions are small on a per-atom basis. Such packings should contain, e.g., polyol or sugar groups.

4.9.4 Alternatives to silica

The chemical properties of silica, i.e. the susceptibility to basic hydrolysis and the high polarity of the surface, form a stumbling block in the development of many applications. Much research effort has been and is being devoted to the development and study of other supports. However, improvements are often obtained only at the expense of other essential qualities of a packing material. The alternative materials can be divided into inorganic and organic types.

The chromatographic properties of alumina (the γ-modification) have been reviewed [104,105]. Contrary to silica, which is acidic, it has an amphoteric character; it can interact with acids as well as bases, and even serve as an ion exchanger. Its greatest advantage over silica is that it is stable throughout a wide pH range. Alumina does not contain

References on p. A221

well-defined functional groups, such as silanols, that could serve as starting points for stable stoichiometric modifications. It appears that modification of its properties could be made by, e.g., polymer coating.

Oxides of other metals are presently also investigated for their suitability as HPLC packings [106,107].

Porous graphitic carbon has been introduced as an alternative to alkyl-modified reversed-phase packings [108]. It is stable over the entire pH range of 0-14. Capacity factors are generally much larger on carbon, especially when the analyte contains aromatic groups. As a result, much less polar mobile phases must be used. Although the present manufacturing process leads to a rather uniform surface, it is still more difficult to obtain linear isotherms in RPC with this material.

Another material of interesting properties is hydroxyapatite [109]. It appears to be quite suitable for the separation of proteins and has moderate ion-exchange properties.

Organic materials are generally polymers. Possibilities of varying the interactive properties of these materials are endless, and from this standpoint these packings are much more attractive than, e.g., silica. However, it is not easy to find synthetic routes for all these materials that also meet the requirements regarding pore structure and mechanical stability.

A material with a long history is styrene/divinylbenzene polymer. It forms the support for sulfonic and amine-type ion exchangers, which have been in use (e.g., in amino acid analyzers) on a massive scale. The wide experience with such ion exchangers may serve to illustrate some of the important points in the use of organic supports:

(a) The material may swell when in contact with solvents. The problem is that the swelling is dependent on the composition of the solvent; changes in ionic strength or organic modifier content change the degree of swelling. A fine column, packed with such materials may be destroyed on solvent change.

(b) The particles are not as rigid as silica. Under the influence of the viscous stress exerted by the mobile-phase flow the bed may compact (locally), with the result that pressure increases and efficiency decreases. Some materials can be used only up to a certain inlet pressure, depending on the dimensions of the column.

Both phenomena are strongly dependent on the divinylbenzene/styrene ratio, the so-called crosslinking percentage. With 4% crosslinking the swelling and compaction are intolerable, with 12% virtually negligible. However, the 12% material is hardly suitable for chromatography, because the diffusion through the highly crosslinked polymer becomes prohibitively slow. This problem can be partly circumvented by synthesizing the material in such a way that pores many nanometers in diameter are formed, in addition to the pores of molecular dimensions that are always present in the polymers ("macroreticular resins"). However, this may again affect the mechanical stability.

Such compromises are typical of the field of organic packing materials. The chemistry behind the various solutions presented is often obscured because of the confidentiality of proprietary processes. Even the nature of the organic polymer used is often not disclosed on presentation of the packings, neither in the commercial nor in the scientific literature.

Methacrylate is often a constituent of the polymers used. In combination with other monomers, it yields materials of suitable hydrophilicity, such as that required for protein separations When this is successfully done, the support can next be modified, e.g., by adding ion exchange groups, chiral selectors, or antibodies (in the case of affinity chromatography) to obtain the selectivity desired. Some materials are sold with only the suitable anchoring groups, allowing the user to attach the required selector to the packing conveniently.

It should be added that at the time of this writing many research groups in commercial manufacturing and outside are very active in exploring the possibilities of a great variety of polymer systems and production processes. For a review on packing materials the reader is referred to Ref. 110.

4.10 SOME NEW FORMS OF LIQUID CHROMATOGRAPHY

4.10.1 Introduction

HPLC has found numerous, if not innumerable, applications and has been the key element in the solution of very important analytical problems, ever since it first became available for analyses (ca 1970). Still, despite the general appreciation of its performance, it leaves much to be desired. Two points immediately come to mind when a comparison with contemporary (capillary) gas chromatography is made:

(a) The resolving power and especially the peak capacity of HPLC is too low. Gas chromatography easily generates 300 peaks in one 30-min chromatogram, while in HPLC this is only possible when analysis times of more than a day can be accepted.

(b) Another weakness of HPLC is unfavorable detection limits. GC with the universal flame-ionization detector has detection limits of ca. 10-100 pg per peak for all carbon-containing compounds. In HPLC 1-10 ng is the detection limit for a compound with good UV absorption. While it could be argued in the early days of HPLC that the disadvantage is virtually absent when (the more relevant) injected concentrations are considered, as larger sample volumes can be used in HPLC, this is no longer true. In the first place, injection techniques in GC permit injection of 10-100 μl or more of solutions. In the second place, the increased efficiencies obtained over the years in HPLC have made it necessary to limit injected volumes to 10-30 μl or less.

The two limitations of HPLC mentioned above have led to much activity to develop miniaturized LC in various forms, following the pioneering work by Scott and Kucera [13]. It was and is believed that miniaturization would allow better access to higher plate-numbers and improved mass detection limits.

Present approaches to improve the situation are (apart from the small-bore columns treated in Section 4.1.3):

(a) use of micropacked or "microcapillary" chromatography;

(b) chromatography in open channels, especially open-tubular LC (OTLC);

(c) electrically driven chromatography (EDC).

All of these approaches are only possible with an appreciable degree of miniaturization. Detection is the critical problem here. However, the extent of miniaturization required

References on p. A221

will differ; it is greatest for OTLC, (detection volume $<$ 1 nl), and most modest for the "microcapillary" mode (depending on the variety, 1 to 100 nl). The prospects for reaching these limits looked rather dim for quite a while. However, after 10 years of detector development, promoted to no small extent by interest in capillary zone electrophoresis (CZE), it can be stated that now a number of detection schemes easily fulfill this requirement, although none of them are commercially available and some have significant disadvantages.

As was indicated in Section 4.5.2, these schemes include laser-excited fluorescence detection, electrochemical detection, mass spectrometry, and UV absorption, albeit sometimes at the expense of decreased concentration sensitivity. Unfortunately, the two first methods are of very limited scope, therefore, there is much interest in the development of derivatization reaction techniques that can enlarge their scope. Mass spectrometry is universal, but expensive, and the interfacing problem has not been solved satisfactorily. The electro-spray interface, recently applied in CZE by various groups, appears to be most promising [74,75]. In it, liquids are nebulized by application of a strong electric field at the exit of a tube. After evaporation of the solvent, solutes appear to generate (pseudo)molecular ions – without the application of any additional ionization method – often with multiple charges. Liquid flows that can be handled by the method are compatible with those in miniaturized LC forms and CZE. The method holds great promise for high sensitivity, quantitation, and structural characterization of high-molecular-weight species, such as biopolymers.

Another problem is the performance of the pump. It becomes increasingly difficult to meter the flowrate accurately when it decreases to a few microliters per minute. Although commercial instrumentation is continuously being improved, in several cases one still has to abandon flowrate metering in favor of constant-pressure operation.

4.10.2 Capillary chromatography

The separation impedance, E, expresses the tradeoff between analysis time and efficiency (Section 4.1.2, Eqn. 4.6, and Table 4.1). It is at least ca. 4000 for regular HPLC columns. A substantial reduction of its value can be obtained by using columns with a small ratio ($<$ ca. 30) of column to particle diameter, the aspect ratio. Most columns, announced as "microcapillary" (suggesting the obsolescence of macrocapillaries?) fall into this class. Such columns are prepared in two ways:

(a) In the normal packing process, a fused-silica tube is the column. Earlier, glass and Teflon tubing was also used for columns.

(b) Drawn, packed capillaries are prepared by drawing glass capillaries from "preform" tubes, filled with packing, in a glass-drawing machine. The result is a packing structure of often very low density, which is stabilized by partial fusion of packing and glass wall. The improved permeability of such columns offers some promises for faster and better separations ($E = h^2\phi$ is smaller, up to a factor of 3-10) [111].

Whatever method is used, the concept implies that the column diameter and with that the volume scale of LC is fixed by the choice of particle size. These columns can only be constructed in diameters of 50-250 μm or less. The peak standard deviations are in the nanoliter range.

The improvements in E are believed to result from smaller values for h as well as for ϕ: With the small diameters the radial equilibrium is fast, so that flow nonuniformities do not contribute to h. The less dense packing gives lower pressure drops, reflected in smaller ϕ values. Altogether, the gain in E values, which can be considered as a gain in analysis time for a given platenumber, can amount to a factor of 3-10 [111-114].

However, the method is primarily intended to increase the efficiency. The accessible platenumber in a given time (Eqn. 4.6) improves only with the square root of E, and the resolution with the square root of that, altogether leading to resolutions that are 1.3 to 1.7 times greater. As one normally desires more substantial improvements, one works in general under conditions where analysis time is also longer that in ordinary HPLC, so that greater improvement in resolution is obtained. Many rather impressive separations of complicated samples have been demonstrated following this approach [111,114].

The efficacy of the concept depends very much on the platenumber aimed at. With low platenumbers, the optimal particle size may be <3 μm (Eqn. 4.7); the diameter should then be ca. 50 μm. This requires extreme miniaturization of detection, in roughly the same degree as required for OTLC. On the other hand, with a million plates as the goal, one would have an optimal d_p of some 20 μm, and a 200-μm column with these properties could be constructed. This is much more reasonable from the point of view of detection.

4.10.3 Open-channel systems

Theoretically, much greater improvements in E could be obtained, if it where possible to exploit the concept of OTLC. The open-tubular mode, now becoming the standard in GC, may lead to values of separation impedance of <30. The gain of a factor of 100 would reflect itself in an improvement of 10 in platenumber, or 3 in resolution. This would constitute an enormous boost in the potential of HPLC.

In 1979 Knox and Gilbert [115] theoretically assessed the possibilities of following this route. They first applied the same principle as that which led the way from classical LC to HPLC: adjustment of the capillary diameter (instead of the particle size a packed column) until a minimum in separation impedance (i.e. a minimum in h) is obtained, while the sturdiness of the equipment is exploited. Unfortunately, an equation analogous to Eqn. 4.7 predicts that this is to occur only at very small tube diameter; e.g., for conditions such as given in Table 4.1, a tube diameter of 0.3 μm is the result.

Such dimensions would lead to excessively small peak volume standard deviations, in the order of 0.1 pl (10^{-12} l). It was regarded as unrealistic to assume that in the (then) foreseeable future detection systems could be developed with a contribution to peakwidth (e.g., cell or measurement volume) as small as that. Therefore, Knox and Gilbert [115] reversed their theoretical approach and calculated the separation performance (e.g., time required for a given platenumber) that could be obtained at various values of the detectors' peakwidth contribution.

References on p. A221

TABLE 4.4

KINETIC PERFORMANCE OF OPTIMIZED OPEN-TUBULAR LIQUID CHROMATOGRAPHY, WITH DETECTOR VOLUME LIMITATION

Conditions: $D_m = 1 \cdot 10^{-9}$ m^2/s, $\eta = 0.001$ kg m^{-1} sec^{-1}, $\Delta P = 2 \cdot 10^7$ Pa. Assumed h/ν dependence: $h = 2/\nu + 0.08\nu$.

Volume standard deviation	N	d_c (nl)	L (μm)	Impedance E (mm)	t_0 (sec)	t_{packed} (sec)
10^{-2}	10^{+4}	2.7	$1.7 \cdot 10^{+2}$	$1.2 \cdot 10^{+3}$	6.1	$2.0 \cdot 10^{+1}$
10^{-1}		4.9	$5.3 \cdot 10^{+2}$	$3.8 \cdot 10^{+3}$	$1.9 \cdot 10^{+1}$	
10^{0}		8.7	$1.7 \cdot 10^{+3}$	$1.2 \cdot 10^{+4}$	$6.0 \cdot 10^{+1}$	
10^{+1}		15	$5.3 \cdot 10^{+3}$	$3.8 \cdot 10^{+4}$	$1.9 \cdot 10^{+2}$	
10^{+2}		27	$1.7 \cdot 10^{+4}$	$1.2 \cdot 10^{+5}$	$6.0 \cdot 10^{+2}$	
10^{-2}	10^{+5}	2.7	$5.5 \cdot 10^{+2}$	$1.3 \cdot 10^{+2}$	$6.6 \cdot 10^{+1}$	$2.0 \cdot 10^{+3}$
10^{-1}		4.9	$1.7 \cdot 10^{+3}$	$3.9 \cdot 10^{+2}$	$2.0 \cdot 10^{+2}$	
10^{0}		8.7	$5.4 \cdot 10^{+3}$	$1.2 \cdot 10^{+3}$	$6.1 \cdot 10^{+2}$	
10^{+1}		15	$1.7 \cdot 10^{+4}$	$3.8 \cdot 10^{+3}$	$1.9 \cdot 10^{+3}$	
10^{+2}		27	$5.3 \cdot 10^{+4}$	$1.2 \cdot 10^{+4}$	$6.0 \cdot 10^{+3}$	
10^{-2}	10^{+6}	2.4	$2.1 \cdot 10^{+3}$	$2.4 \cdot 10^{+1}$	$1.2 \cdot 10^{+3}$	$2.0 \cdot 10^{+5}$
10^{-1}		4.7	$5.8 \cdot 10^{+3}$	$4.9 \cdot 10^{+1}$	$2.5 \cdot 10^{+3}$	
10^{0}		8.6	$1.7 \cdot 10^{+4}$	$1.3 \cdot 10^{+2}$	$6.6 \cdot 10^{+3}$	
10^{+1}		15	$5.4 \cdot 10^{+4}$	$3.9 \cdot 10^{+2}$	$2.0 \cdot 10^{+4}$	
10^{+2}		27	$1.7 \cdot 10^{+5}$	$1.2 \cdot 10^{+3}$	$6.1 \cdot 10^{+4}$	

Some results of such a calculation are briefly summarized in Table 4.4, which shows in the last column for comparison the times required when a packed column with optimized particle size is used. As can be seen, it does not make much sense to go to OTLC for low platenumbers. The packed column can do the job faster in most cases, unless an extreme degree of miniaturization can be effected.

The future of OTLC therefore clearly depends on its ability to yield high platenumbers. Still, it can be seen that even for 10^5 plates, one needs a detector volume of 10 nl or less in order to outperform the packed column in speed. Apart from the availability of suitable detectors, further development of OTLC will depend very much on the ability to prepare 2- to 10-μm columns with suitable stationary-phase wall coates. The simplest approach to that, starting with the easily handled fused-silica tubes, is to apply silica modifications to the inner surface. The monolayer of, e.g., octadecyl moieties thus obtained can bring about the required retention in the RP mode. However, the phase ratio (being the ratio of surface area to volume in this case) is inconveniently low; a 10-μm tube leads to 0.4 m^2/ml, while in packed columns one usually has 100- to 1000-fold higher values. Although enough retention can be obtained by choosing a mobile phase of lesser strength, this is not a very promising solution. In the first place, one soon runs into problems with solubility in the mobile phase. In the second place, the concentrations in the stationary layer already become very large when mobile-phase concentrations are barely detectable; with such a small phase ratio, the distribution constant must indeed be very large for a reasonable k' value. For these two reasons, it may be expected that the dynamic range of such a chromatographic method would be extremely small.

Attempts to apply thicker layers, with higher sample capacity and retention on 2- to 10-μm tubes follow various approaches:

(a) Use of GC techniques to immobilize silicone materials, after static or dynamic coating [116]. It has been shown that the usefulness of such materials can be improved by having them swell by sorption of an alkane from the mobile phase [117].

(b) Application of porous silica to the wall [118] by coating a precursor, such as polyethoxysiloxane on the wall and hydrolyzing it to produce porous silica in situ. This then can be used (i) as such, in normal-phase chromatography, (ii) in the RP mode after derivatization to alkyl-bonded silica, and (iii) as a support for a genuine liquid/liquid system with a low-molecular-weight stationary liquid phase.

(c) Polymerization of acrylates has also been shown to lead to suitable phases for OTLC [119].

The severe detection problems in OTLC have led various researchers to consider physical arrangements that would combine the kinetic advantage of a small separation impedance with a larger volume scale. Meyer at al. [120] studied the parallel operation of a bundle of 100-1000 capillaries. Giddings et al. [121] developed a rectangular channel or slit for the same purpose. The channel was 50 μm high, but had a width of a few centimeters. Stationary phase was applied to the horizontal wall. The channel height determines the separation impedance, which is as favorable as for tubes. However, the volume scale is larger in proportion to the channel width.

Such designs suffer from one important drawback (apart from complexity): There is no equilibrium across the width of the channel to wipe out any nonuniformities in migration rates. Therefore, the solutes must move in all parts of the system at nearly exactly the same rate. For instance, when the velocities, v, of the mobile phase or the migration rates, $u = v/(1+k')$, were to differ by only 1%, the resulting peak would have a width of 1% of the retention time. After conversion of the width to a standard deviation, (factor $\sqrt{12}$), this leads to a platenumber of 12 000. This would be the maximum attainable, if other, genuine dispersion mechanisms were negligible.

As OTLC is likely to be of interest because of much higher platenumbers, it follows that extreme specifications are to be set for the equality of flowrates and retention throughout the systems. Channel dimensions and degree of stationary-phase loading must be held constant in all the 1000 capillaries and over the entire width of the channel to well within 1%. It is doubtful whether this can ever be accomplished.

References on p. A221

It must be noted, however, that the situation may improve when electrically induced, rather than pressure-induced flow is applied. Electrically induced flow (Section 4.10.5) is hardly dependent on channel dimensions. Still, the stationary-phase coating should be uniform.

Another attempt [122] has been made to solve the problem of nonuniformity of the flow and retention in a slit system. In the arrangement chosen the slit was formed between two concentric cylinders, with the inner, free-floating cylinder rotating. The thickness of the annular slit thus automatically becomes uniform. The system is in all likelihood too complicated to be of any practical value.

4.10.4 Capillary zone electrophoresis

The main aspects of this technique are covered in Chapter 11. However, it is worth treating this complicated subject in outline form again here as an introduction to Section 4.10.5.

An electric field, E (10 to 100 kV/m), applied to a liquid with ionic conductivity leads to electrophoresis of ions. Each ion is moving with a velocity u_i relative to the liquid. This velocity is proportional to the field, E, with proportionality constant μ_i, the electrophoretic mobility

$$u_i = E \cdot \mu_i \qquad (4.28)$$

To a first approximation, the value of μ_i for each ion can be considered to be a constant, although it depends on the nature and the ionic composition of the solution. When sample ions are migrating at relatively low concentrations and in narrow zones in a background carrier electrolyte, the technique is called zone electrophoresis. Under these conditions, the sample ions behave independent of each other, inter alia because the local field does not depend on local composition.

Since the pioneering work by Everaerts and coworkers [123] and Jorgenson and Lucacs [124] an instrumental version of zone electrophoresis is known. Instead of observing the position of analytes on a medium like a gel plate, the passage of analytes as a function of time is monitored in a flow-through detection mode.

This process, known as capillary zone electrophoresis (CZE), is implemented in a fused-silica tube, which is typically 50 cm long with an inner diameter of 25-100 μm. Detection is often carried out on-column, using UV absorption (Section 4.6.3.2) or fluorescence, but electrochemical, postcolumn reaction detectors, and MS are also applied. Injections are performed either by electromigration, which means that the electrophoresis is used to transfer material from a sample vial to the tube, or by pushing the sample liquid into the tube by applying a slight pressure difference across the tube.

An important phenomenon in CZE and EDC (Section 4.10.5) is that of electro-osmosis: The liquid as a whole moves under the influence of the applied field. This is caused by the excess charge normally acquired by a liquid when it is in contact with the wall material. In most cases the charge is positive. With silica tubes, and above pH 2, the positive charge is probably due to the partial dissociation of the acidic silanols. Thus, the liquid moves in the same direction as positive ions.

The electro-osmotic flow differs from pressure-induced flow in one important aspect: the velocity profile across the tube diameter is flat, rather than parabolic. This can be explained by the fact that the excess charge resides in a thin (0.01- to 1-μm) layer, close to the wall, the so-called diffuse electrical double layer. The electric force has a grip on only this part of the liquid. However, the part within this annulus is carried along passively, as it were, as no forces counteract its movement.

The electro-osmotic velocity, u_{eo}, is proportional to the field, according to

$$u_{eo} = E \cdot \mu_{eo} \tag{4.29}$$

where μ_{eo} is the electro-osmotic mobility. Its value is dependent on the chemical nature of the wall surface and the liquid, but does not depend on the dimensions of the channel (at least under usual conditions [125]).

The resulting "plug flow" in the tube has some aspects desirable for chromatographic separations, which will be discussed in Section 4.10.5. For CZE it implies that (in general) the overall movement of positive ions is increased, that of negatives ions is decreased, and neutral species will also move. Fortunately, the additional movement does not lead to peak dispersion, as there is plug flow.

The peakbroadening in CZE systems can be very small, and thus highly efficient separations (10^5 to $>10^7$ plates) can be obtained. This results from the fact that, under proper conditions, there is only one peakbroadening mechanism; that of longitudinal molecular diffusion. The peakwidth variance, σ_z^2, induced by this mechanism equals $\sigma_z^2 = 2 \cdot D \cdot t$, where t is the residence time and D is the diffusion coefficient of the solute. By working at high field strengths, resulting in small t values, this term can be made very small. This state of affairs, i.e. that faster operation and shorter "columns" lead to better separations, is still astonishing to many chromatographers.

Indeed, the high separation efficiency of CZE as compared to other versions of electrophoresis is due, to a large extent, to the intensity of the applied field, at least for low-MW analytes. For large molecules, such as proteins, the small values of D allow low-field-strength experiments without much peakbroadening.

The desire to use strong fields has led to some experimental consequences which must be discussed before turning to the chromatographic significance in Section 4.10.5. The carrier solution must have a minimum electrolytic conductivity for various reasons. In the first place, it is required to have a constant field across the tube, that is not influenced by the presence of analyte ions. The latter may therefore not contribute significantly to the conductivity of the liquid (E = current density/conductivity). The carrier conductivity should be 100 to 1000 times larger than that expected from the solutes, in the center of their zones. Otherwise, distorted, triangular peaks will result. A second reason is that a pH buffer is generally needed in order to stabilize the proteolytic equilibria, and with that the charges of analytes. Also, this buffer must have a large concentration relative to that of the solutes. Finally, the electro-osmotic flow is only plug-shaped provided the electrical double

References on p. A221

layer thickness is small compared to the channel dimension. At very low ionic strengths the double layer is larger; at 10^{-5} mol/l it is 0.1 μm. This is unimportant for CZE, but it is relevant for the EDC techniques covered in Section 4.10.5.

Normally, therefore, carrier solutions are 10^{-4} to 10^{-2} mol/l in electrolytes. The resulting current, in combination with the high field, represents an appreciable energy. The dissipated heat must be transported to the surrounding thermostating medium, first through the carrier liquid itself, next through the tube material, and finally through the stagnant layer of thermostating liquid or gas.

Insufficient heat transfer in the carrier itself may lead to a loss of efficiency: thermal nonuniformity gives nonuniformity of migration velocities and corresponding peak-broadening. These nonuniformities cannot be wiped out by radial diffusion, as (even for low-MW solutes) this happens only fast enough in tubes of <10 μm. Too slow a heat transfer in the other parts does not necessarily impair efficiency, but it does endanger reproducibility and accuracy of peak positions. The thermal management of CZE and EDC systems is therefore a central problem in the technique, which has been amply discussed in two papers (125,126]. The following conclusions can be drawn from these:

(a) Higher fields and higher carrier concentrations require the use of narrower and narrower tubes.

(b) The heat transfer through the silica wall does not pose a major problem.

(c) Heat transfer in (the stagnant layer of) a thermostating gas may cause problems.

The choice of the dimensions of CZE separation tubes, possibly made intuitively originally, can be justified rather precisely on the basis of these considerations. With available equipment and avoiding special high-voltage experimental techniques, $6 \cdot 10^4$ V is about the upper limit for the total potential drop across the tube. For efficient and fast separations, the length should be $<$ 0.3-1 m. The carrier concentration is fixed by the aim to elute substantial concentrations of the solute, allowing for the factor 100-1000 needed to stabilize the conductivity. Given these conditions, the heat dissipation per unit volume can be calculated. This, in turn, dictates the maximum diameter of the separation tube.

The resulting dimensions are typically 0.5 m length and 50 μm ID. There is a continuous trend to even smaller dimensions, but the values given already represent an appreciable degree of miniaturization; peak standard deviations are in the order of 2 nl. This is not much different from the miniaturization required for successful OTLC. Although detection is presently still a stumbling block in the development of these techniques, it must also be said that the increased mass sensitivity in such miniaturized systems (which exists despite the loss in concentration sensitivity) is probably of great importance in all kinds of biochemical research. The most extreme example of this was demonstrated by Kennedy and Jorgenson [127], who analyzed the contents of single cells by means of capillary separation techniques. Also, the intense interest in CZE has led to vigorous efforts to find better detection systems; the most prominent result is probably the advent of the electrospray/mass spectrometry coupling [74,75].

4.10.5 Electrically driven chromatography

A number of chromatographic separation techniques that exploit electrophoretic and electro-osmotic phenomena have been proposed recently. Experimentally, they closely resemble the CZE technique. The optimization in terms of system dimensions is very similar if not the same, as are the problems with respect to miniaturized detection.

In a couple of remarkably innovative papers, Terabe et al. [129,130] introduced "micellar electrochromatography" (MEC). In a CZE system they added micelle-forming surfactants in the carrier. The micelles move in the CZE system under the influence of their charge. However, these micelles have the ability to sorb foreign species, including neutral ones. Thus, injected analytes can distribute themselves between the bulk part of the carrier and the micelles.

This could be thought of as a chromatographic system in which both phases move. Neutral solutes with no "retention" (i.e. those which do not distribute themselves at all into the micelles) will emerge at the position corresponding to the electro-osmotic flow, while analytes with high retention will be eluted at the micelle position. The net migration velocity of most neutral analytes will lie between these limits. The technique possesses the following important advantages:

The high efficiency of CZE is preserved. Apparently, the mass transfer between the phases is fast enough to avoid any significant C-term effects. It is logical to assume that this is because the micellar phase is much more dispersed than traditional HPLC packings. Micelles are huge agglomerates on the molecular scale, but they are still very small, compared to 3- to 5-μm particles.

Another advantage of MEC is that all analytes (with the possible exception of ions with the "wrong" charge, electrophoretically moving against the electro-osmotic flow) are eluted. As in TLC, the "general elution problem" (Chapter 1) is nonexistent. The full k' range is projected onto the elution time scale.

Inspired by this work, Knox and Grant [125] proposed another variety of EDC. They argued that instead of micelles one could also use a solid colloid material, such as silica-based particles, below 1 μm in size. Such solid particles acquire a surface charge (the counterpart of the excess charge of the liquid) and, as a result, they move electrophoretically as do the micelles. Like these, they could serve as a moving "stationary" phase. Apparently, this idea has not been tested experimentally.

This has been done with yet another variety of EDC. In this case, the traditional concept of a column, with the particles fixed in position, is again used. However, rather then inducing flow by means of pressure, the electro-osmotic flow is applied. This concept was already put forward long before the inception of CZE [131], but the idea remained dormant until the experimental progress achieved in CZE made it practical to implement it.

It may seem at first sight that the method offers little advantages over capillary LC. However, there are two significant advantages: The first is that better h values can be obtained with the same particle size. The plug flow also prevails in the interparticle channels in a packed column. More important, probably, is the fact that the velocity does not depend on the width of the channel. Thus, dispersion due to nonuniform flow within and

References on p. A221

between channels, generally combined under the "A term", can be expected to be much lower. Indeed, Knox and Grant [125,126] showed experimentally a large improvement in h values.

Another, probably more decisive, advantage is that particles of smaller size can be employed. In normal HPLC optimum d_p values are in the range of 3-20 μm, depending on the platenumber and the available pressure. In EDC there is no objection to the use of much smaller particles. The lower limit has not yet been assessed experimentally, but it can be expected to be far into the submicrometer range. The limit is, as far as we can see, determined by the effect of "double-layer overlap" and has been calculated [125].

Summarizing this section, the interaction of electrophoretic and chromatographic concepts and the mutual borrowing of experimental techniques in the two fields leads to very interesting developments, to which vigorous research efforts are being devoted. The key objectives are high separation efficiency, application to high-molecular-weight solutes, miniaturization, and high mass sensitivity.

4.11 PREPARATIVE LIQUID CHROMATOGRAPHY

4.11.1 Introduction

The most common application of HPLC is in analysis, where producing information on the sample composition is the primary goal. However, there is an increasing need for preparative LC, where producing pure substances is the objective. Standard analytical HPLC columns may furnish ca. 100 μg (Section 4.1.4 and Eqn. 4.17). Such amounts may already be sufficient, for, e.g., biochemical and biological experiments with modern techniques.

To obtain more material the experiment can be repeated and corresponding fractions pooled. To scale up the column, the diameter may be made larger and the flowrates higher (Section 4.1.4). Even larger amounts may exceed the range in which standard HPLC equipment (pumps, injectors, detectors) can be used, and specialized equipment must be obtained. At the point the operation becomes quite expensive, in terms of investment and operation, and it becomes essential to evaluate the relations between the chromatographic conditions, operating mode, etc., and the economics of the process. That is, the cost per unit weight of the product, taking into account prices of packing, hardware, (regeneration of) solvent, etc., must be minimized. Solving such problems is the province of chemical engineers rather than of chromatographers. Readers interested in this field will find more information in an excellent review by Guiochon and Katti [132].

The following subsections will summarize the optimization of a preparative separation. In such an optimization three main parameters of a preparative separation should be considered:

(a) the yield, i.e. the fraction of the compound of interest that is eventually isolated in pure form;

(b) the purity, i.e. the fraction of the compound of interest in the isolate;

(c) the throughput, i.e. the amount of the compound of interest purified per unit time.

High throughputs can be obtained either on a small, fast column by repeated injections, or by one injection into a large, slow column.

4.11.2 Choice of the phase system

This choice is the most critical step, maybe even more so in preparative than in analytical HPLC. The main requirement is that there be as large a selectivity as possible for critical pairs (involving the compound of interest) in the chromatogram. The smaller the selectivity, the more theoretical plates are needed. Apart from the question whether these could be generated at all, the operation of large-platenumber columns always takes a long time, and throughput would be small.

Solvent economy is important. High concentrations and a small capacity factor of the last peak are obviously favorable in this respect, and this again shows the importance of the choice of the phase system. Experience teaches that the larger the scale, the higher is the proportion of the solvent cost in the total expense. Solvent prices (including environmental and safety measures that have to be taken) as well as possibilities for solvent recovery become decisive factors under such conditions. As volumes increase, mixed and high-boiling solvents become increasingly less attractive. In large-scale production, according to general opinion, one must first decide on the solvent and then find the stationary phase. This is quite opposite to analytical practice.

Phase systems that are useful in analytical operation therefore can be very disappointing for preparative work. A second difficulty may be that the solubility in the mobile phase can be too small, a third that the removal of the mobile-phase constituents is troublesome.

4.11.3 Elimination of excess resolution

Once the phase system has been decided upon, the available pilot chromatogram may look like Fig. 4.7a. In order to obtain larger amounts of pure substance in one chromatogram, either the injected volume or concentration, or both, could be increased. These are the first measures one would take in scaling up. The limitations of this approach have been discussed in Chapter 1 and in Section 4.2.1.4. Large volumes worsen the resolution because of the width of the injection plug; high concentrations do so because the distribution between phases becomes nonlinear, also leading to broadened (and asymmetrical) peaks. The latter effect is designated as thermodynamic broadening [10-12], in contrast to the kinetic broadening associated with platenumber and plateheight. The overload is increased until yield or purity become unacceptably low.

It is generally agreed now that the product of volume and concentration of the sample, the total mass, is limited by relations such as Eqn. 4.16. That is, for a given amount of sample it does not matter what the injection volume is, as long as it does not exceed the σ_V of the column.

As shown in Fig. 4.7, the acceptable extent of overloading depends very much on the initial resolution in the chromatogram. For instance, if this is already critically low at the

References on p. A221

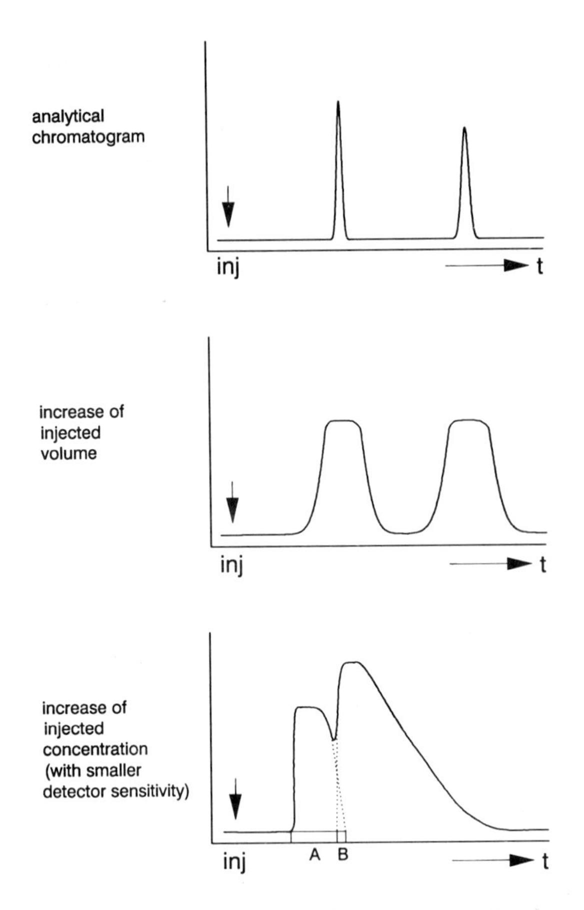

Fig. 4.7. Preparative separation developed from an analytical HPLC experiment. Separation of a binary mixture. Top: under analytical conditions; low concentrations, small injection volume. Middle: improved production by injection of a larger volume, such that resolution is still acceptable. Bottom: even larger production by increasing the amount such that thermodynamic broadening just leaves an acceptable resolution. Outlet concentrations are much higher in the figure at the bottom. A: collection interval for first component; B: mixed fraction, which may be reinjected if yield is important.

outset, there is "no room" for increasing the amount collected; not even a part of the resolution could then be sacrificed without impairing the purity or yield. It has been suggested, therefore, in work dealing with just volume overload, that the initial platenumber should be several times the value required for analytical work. However, the situation is more complicated, as may be clear from the following.

4.11.4 Kinetic or thermodynamic peakbroadening

A very efficient column would allow an appreciable overload. However, as the time required to elute a column increases rapidly with the platenumber aimed at, this may not be the best choice; it might be better to start with fewer plates and, e.g., obtain the same throughput by repeated injections. This issue is not fully resolved yet. It is partly obfuscated by the aspect of partial peak collection mentioned below. Still, there are arguments, well expressed in Ref. 133, that in optimized preparative HPLC both causes of peakbroadening contribute about equally to the peakwidths. If no significant thermodynamic broadening is visible, the throughput (and solvent economy!) could easily be increased by injecting larger amounts. On the other hand, if kinetic broadening mechanisms (plateheight) area virtually negligible, the column could be redesigned (e.g., by using bigger particles) for faster elution.

4.11.5 Partial peak collection

Elementary considerations of preparative LC assume that the compound of interest should occur in a peak that is well separated from all others. However, it could be much more favorable to overload more drastically. As shown in Fig. 4.7, one could nevertheless collect a pure substance; the collection boundaries ("A") have to be chosen on the basis of a target compromise between yield and purity. If the yield is critical, one could recirculate Fraction "B", possibly after reconcentration.

4.11.6 Displacement chromatography

The displacement mode (Chapter 1) is very suitable for preparative purposes. There are no gaps of "empty" mobile phase emerging, so solvent economy and production rate normally are high. Although the applicability of the method is dependent on the occurrence of competition in the distribution process, the method appears to work in many cases. This nearly forgotten mode of chromatography has been applied and studied intensively with modern HPLC equipment in recent years [134,135].

REFERENCES

1 J.J. Kirkland and L.R. Snyder, *Introduction to Modern Liquid Chromatography,* Wiley-Interscience, New York, 1979.
2 J.H. Knox and M. Saleem, *J. Chromatogr. Sci.,* 7 (1969) 614.
3 J.H. Knox and P. Bristow, *Chromatographia,* 10 (1977) 279.
4 M.J.E. Golay, in D.H. Desty (Editor), *Gas Chromatography 1958,* Butterworths, London, 1958, p. 36.
5 R. Endele, I. Halasz and K. Unger, *J. Chromatogr.,* 99 (1974) 377.
6 R.E. Majors, *Anal. Chem.,* 44 (1972) 1722.
7 J.J. Kirkland, *J. Chromatogr. Sci.,* 10 (1972) 593.
8 J.F.K. Huber, J.A.R.J. Hulsman and C.A.M. Meijers, *J. Chromatogr.,* 62 (1971) 79.
9 B.L. Karger, M. Martin and G. Guiochon, *Anal. Chem.,* 46 (1974) 1640.

10 H. Poppe and J.C. Kraak, *J. Chromatogr.*, 255 (1983) 395.
11 J.A. Eble, R.L. Groab, P.E. Antle and L.R. Snyder, *J. Chromatogr.*, 384 (1987) 25.
12 S. Golshan-Shirazi and G. Guiochon, *J. Chromatogr.*, 506 (1990) 495.
13 R.P.W. Scott and P. Kucera, *J. Chromatogr.*, 169 (1978) 51.
14 F.A. Maris, R.J. van Delft, R.W. Frei, R.B. Geerding and U.A.Th. Brinkman, *Anal. Chem.*, 58 (1986) 1634.
15 J.C. Gluckman and M. Novotny, *J. Chromatogr.*, 333 (1985) 291.
16 E.C. Nice, C.J. Lloyd and A.W. Burgess, *J. Chromatogr.*, 296 (1984) 153.
17 G. Guiochon, *J. Chromatogr.*, 185 (1979) 3.
18 M. Martin, G. Blu, C. Eon and G. Guiochon, *J. Chromatogr.*, 112 (1975) 399.
19 H. Poppe and J.C. Kraak, *J. Chromatogr.*, 282 (1983) 399.
20 J.C. Sternberg, in J. Giddings (Editor), *Advances in Chromatography*, Vol. 2, Dekker, New York, 1966.
21 J.M. Reijn, W.E. van der Linden and H. Poppe, *Anal. Chim. Acta*, 126 (1981) 1.
22 J.N. Little and G.J. Fallick, *J. Chromatogr.*, 112 (1975) 389.
23 J. Lankelma and H. Poppe, *J. Chromatogr.*, 149 (1976) 587.
24 R.W. Frei and K. Zech (Editors), *Selective Sample Handling and Detection in High-Performance Liquid Chromatography*, Journal of Chromatography Library Series, Vol. 39A, Elsevier, Amsterdam, 1988.
25 M.W.F. Nielen, H.E. van Ingen, A.J. Valk, R.W. Frei and U.A.Th. Brinkman, *J. Liq. Chromatogr.*, 10 (1987) 617.
26 C. van Water, D. Tebbal and N. Haagsma, *J. Chromatogr.*, 478 (1989) 205.
27 K.S. Boos, B. Wilmers, R. Sauerbrey and E. Schlimme, *Chromatographia*, 24 (1987) 363.
28 W.Th. Kok, *Chromatographia*, 24 (1987) 442.
29 R.P.W. Scott and P. Kucera, *J. Chromatogr. Sci.*, 18 (1980) 49.
30 H. Poppe, *Anal. Chim. Acta*, 114 (1980) 59.
31 H.C. Smit and H.L. Walg, *Chromatographia*, 8 (1975) 311.
32 W. Baumann, *Fresenius Z. Anal. Chem.*, 284 (1977) 31.
33 H. Poppe, *Anal. Chim. Acta*, 145 (1983) 17.
34 D.R. Bobbit and E.S. Yeung, *Anal. Chem.*, 56 (1984) 1577.
35 M. Martin, *Chromatographia*, 15 (1982) 426.
36 H.G. Barth, W.E. Barber, C.H. Lochmüller, R.E. Majors and F.E. Regnier, *Anal. Chem.*, 58 (1986) 223R; H.G. Barth, W.E. Barber, C.H. Lochmüller, R.E. Majors and F.E. Regnier, *Anal. Chem.*, 60 (1988) 387R.
37 H.A. Stuting, I.S. Krull, R. Mhatre, S.C. Krzysko and H.G. Barth, *LC-GC Mag.*, 2 (1989) 34.
38 A. Krstulovic, H. Colin and G. Guiochon, in J.C. Giddings (Editor), *Advances in Chromatography*, Vol. 24, Dekker, New York, pp. 83 ff.
39 R.L. St. Claire III and J.W. Jorgenson, *J. Chromatogr. Sci.*, 23 (1985) 186.
40 L.A. Knecht, E.J. Guthrie and J.W. Jorgenson, *Anal. Chem.*, 56 (1984) 475.
41 D.C. Johnson and W.R. LaCourt, *Anal. Chem.*, 62 (1990) 589A.
42 J.G. White, R.L. St. Claire III and J.W. Jorgenson, *Anal. Chem.*, 58 (1986) 293.
43 U.A.Th. Brinkman, *Chromatographia*, 24 (1987) 190.
44 R.W. Frei and J.F. Lawrence, *Chemical Derivatization in Liquid Chromatography*, Elsevier, Amsterdam, 1976.
45 R.W. Frei and J.F. Lawrence (Editors), *Chemical Derivatization in Liquid Chromatography*, Vol. 2, Plenum Press, New York, 1982.
46 R. Schuster, *J. Chromatogr.*, 431 (1988) 271.
47 T.Y. Chou, C.X. Gao, S.T. Golan, I.S. Krull, C. Dorschel and B.A. Bidlingmeyer, *J. Chromatogr.*, 454 (1988) 169.
48 C.X. Gao, T.S Chou, S.T. Golgan, I.S. Krull, C. Dorschel and B.A. Bidlingmeyer, *J. Chromatogr. Sci.*, 26 (1988) 501.
49 T.Y. Chou, S.T. Colgan, I.S. Krull, C. Dorschel and B.A. Bidlingmeyer, *J. Chromatogr. Sci.*, 26 (1988) 445.
50 B. Lillig and H. Engelhardt, in I.S. Krull (Editor), *Reaction Detection in Liquid Chromatography*, Dekker, New York, 1986, p. 1.

51 S.T. Colgan and I.S. Krull, in I.S. Krull (Editor), *Reaction Detection in Liquid Chromatography,* Dekker, New York, 1986, p. 227.
52 R.W. Frei, J.F. Lawrence, U.A.Th. Brinkman and I. Honigsberg, *J. High Resolut. Chromatogr. Chromatogr. Commun.,* 1 (1979) 11.
53 J.W. Birks (Editor), *Chemiluminescence and Photochemical Reaction Detection in Chromatography,* VCH Publishers, New York, 1989.
54 M. Nieder and H. Jaeger, *J. Chromatogr.,* 413 (1987) 207.
55 I.S. Krull and W.R. LaCourse, in I.S. Krull (Editor), *Reaction Detection in Liquid Chromatography,* Dekker, New York, 1986, p. 303.
56 L.D. Bowers, in I.S. Krull (Editor), *Reaction Detection in Liquid Chromatography,* Dekker, New York, 1986, p. 195.
57 H. Small and T.E. Miller Jr., *Anal. Chem.,* 54 (1982) 462.
58 J. Crommen, G. Schill, D. Westerlund and L. Hackzell, *Chromatographia,* 24 (1987) 252.
59 W. Buchberger and G. Rieger, *J. Chromatogr.,* 482 (1989) 407.
60 T. Okada and T. Kuwanoto, *Anal. Chem.,* 55 (1983) 1001.
61 J. Crommen, G. Schill and P. Herne, *Chromatographia,* 25 (1988) 397.
62 H. Poppe, *J. Chromatogr.,* 506 (1990) 45.
63 W.D. Pfeffer, T. Takeuchi and E.S. Yeung, *Chromatographia,* 24 (1987) 123.
64 B.G.M. Vandeginste, G. Kateman, J.K. Strasters, H.A.H. Billiet and L. de Galan, *Chromatographia,* 24 (1987) 127.
65 J.K. Strasters, H.A.H. Billiet, L. de Galan, B.G. M. Vandeginste and G. Kateman, *Anal. Chem.,* 60 (1988) 2745.
66 A.F. Fell, T.P. Bridge and M.H. Williams, *J. Pharm. Biomed. Anal.,* 6 (1988) 555.
67 H.J. Boessenkool, P. Cley, C.E. Goewie, H.H. van den Broek and H.A. van't Klooster, *Mikrochim. Acta,* 2 (1987) 75.
68 S. Ebel, *Fresenius Z. Anal. Chem.,* 327 (1987) 794.
69 D.W. Hill, D.W. Kelley and K.J. Langner, *Anal. Chem.,* 59 (1987) 350.
70 D.W. Hill, *J. Liq. Chromatogr.,* 10 (1987) 377.
71 A.F. Fell, B. Clark and H.P. Scott, *J. Chromatogr.,* 316 (1984) 423.
72 W.M.A. Niessen, *Chromatographia,* 21 (1986) 277; 21 (1986) 342.
73 K. Biemann and S.A. Martin, *Mass Spectrom. Rev.,* 6 (1987) 1.
74 E.D. Lee, W. Mück, J.D. Henion and Th.R. Covey, *J. Chromatogr.,* 458 (1988) 313.
75 R.D. Smith, J.A. Loo, C.J. Barinaga, C.G. Edmonds and H.R. Udseth, *J. Chromatogr.,* 480 (1989) 211.
76 J.W. Helgeth and L.R. Taylor, *Anal. Chem.,* 59 (1987) 295.
77 P.R. Griffith and C.M. Conroy, in J.C. Giddings, E. Grushka, J. Cazes and P.R. Brown (Editors), *Advances in Chromatography,* Vol. 25, Dekker, New York, 1987, p. 105.
78 K. Albert, M. Nieder, E. Bayer and M. Spraul, *J. Chromatogr.,* 346 (1985) 17.
79 P.J. Schoenmakers, *Optimization of Chromatographic Selectivity, a Guide to Method Development,* Journal of Chromatography Library, Vol. 35, Elsevier, Amsterdam, 1986.
80 J.L. Glajch, J.J. Kirkland, K.M. Squire and J.M. Minor, *J. Chromatogr.,* 199 (1980) 57.
81 L.R. Snyder, J.W. Dolan and D.C. Lommen, *J. Chromatogr.,* 485 (1989) 65 and 90.
82 S.N. Deming, J.C. Bowers and K.D. Bower, in J.C. Giddings, E. Grushka, J. Cazes and Ph. Brown (Editors), *Advances in Chromatography,* Vol. 24, Dekker, New York, 1984, p. 35.
83 J.C. Berridge, *J. Chromatogr.,* 485 (1989) 3.
84 S.N. Deming, J.M. Palasota, J. Lee and L. Sun, *J. Chromatogr.,* 485 (1989) 15.
85 L. de Galan and H.A.H. Billiet, in J.C. Giddings, E. Grushka, J. Cazes and P.R. Brown (Editors), *Advances in Chromatography,* Vol. 25, Dekker, New York, 1986, p. 63.
86 M. Otto, W. Wegscheider and E.P. Lankmayr, *Anal. Chem.,* 60 (1988) 517.
87 J.K. Strasters, H.A.H. Billiet, L. de Galan, B.G.M. Vandeginste and G. Kateman, *Anal. Chem.,* 60 (1988) 2745.
88 L. de Galan, D.P. Herman and H.A.H. Billiet, *Chromatographia,* 24 (1987) 108.

89 J.L. Glajch and J.J. Kirkland, *Anal. Chem.,* 54 (1982) 2593.
90 J.M. Davis and J.C. Giddings, *Anal. Chem.,* 57 (1985) 2168.
91 M. Martin, D.P. Herman and G. Guiochon, *Anal. Chem.,* 58 (1986) 2200.
92 M.M. Bushey and J.W. Jorgenson, *Anal. Chem.,* 62 (1990) 161.
93 K.K. Unger, *Porous Silica, its Properties and Use as Support in Column Liquid Chromatography,* Elsevier, Amsterdam, in preparation.
94 L.R. Snyder, J.L. Glajch and J.J. Kirkland, *J. Chromatogr.,* 218 (1981) 299.
95 J.H. Knox and H.P. Scott, *J. Chromatogr.,* 282 (1983) 297.
96 J.P. Crombeen, H. Poppe and J.C. Kraak, *Chromatographia,* 22 (1987) 319.
97 L.R. Snyder, *J. Chromatogr.,* 5 (1961) 468.
98 E. van Kreveld and N. van den Hoed, *J. Chromatogr.,* 149 (1978) 71.
99 H. Giesche, K.K. Unger, U. Esser, B. Eray, U. Truedinger and J.N. Kinkel, *J. Chromatogr.,* 465 (1989) 39.
100 Y-F. Maa and Cs. Horváth, *J. Chromatogr.,* 445 (1988) 71.
101 R.K. Gilpin, S.S. Yang and G. Werner, *J. Chromatogr. Sci.,* 26 (1988) 388.
102 W. Lindner, *Chromatographia,* 24 (1987) 97.
103 I.W. Wainer and M.C. Alembik, *J. Chromatogr. Sci.,* 40 (1988) 355.
104 H. Engelhardt and H. Elgass, in Cs. Horvath (Editor), *High Performance Liquid Chromatography, Advances and Perspectives,* Vol. 2, Academic Press, New York, 1980, p. 57.
105 C.J.M.C. Laurent, H.A.H. Billiet and L. de Galan, *Chromatographia,* 17 (1983) 394. See also *Chromatographia,* 18 (1984) 47, *J. Chromatogr.,* 285 (1983) 161, and *J. Chromatogr.,* 287 (1984) 45 for the separation of proteins.
106 M.P. Rigney, T.P. Weber and P.W. Carra, *J. Chromatogr.,* 484 (1989) 273.
107 M.P. Rigney, E.F. Funkenbusch and P.W. Carr, *J. Chromatogr.,* 499 (1990) 291.
108 J.H. Knox, B. Kaur and G.R Millward, *J. Chromatogr.,* 352 (1986) 3.
109 Y. Yamakawa, K. Miyasaka, T. Ishikawa, Y. Yamada and T. Okuyama, *J. Chromatogr.,* 506 (1990) 319.
110 R.E. Majors, *LC-GC,* 7 (1989) 212 and 304.
111 J.W. Jorgenson and E.J. Guthrie, *J. Chromatogr.,* 255 (1983) 335.
112 M. Novotny, *J. High Resolut. Chromatogr. Chromatogr. Commun.,* 10 (1987) 248.
113 C. Borra, M. Soon and M. Novotny, *J. Chromatogr.,* 385 (1987) 75.
114 R.T. Kennedy and J.W. Jorgenson, *Anal. Chem.,* 6 (1989) 1128.
115 J.H. Knox and M.T. Gilbert, *J. Chromatogr.,* 186 (1979) 405.
116 O. van Berkel, H. Poppe and J.C. Kraak, *Chromatographia,* 24 (1987) 739.
117 S. Folestad, B. Josefsson and M. Larsson, *J. Chromatogr.,* 391 (1987) 347.
118 P.P.H. Tock C. Boshoven, H. Poppe, J.C. Kraak and K.K. Unger, *J. Chromatogr.,* 447 (1989) 95.
119 S. Eguchi, P.P.H. Tock, J.G. Kloosterboer, C.P.G. Zeger, P.J. Schoenmakers, J.C. Kraak and H. Poppe, *J. Chromatogr.,* 516 (1990) 301.
120 R.F. Meyer, P.B. Champlin and R.A. Hartwick, *J. Chromatogr.,* 211 (1983) 433.
121 J.C. Giddings, J.P. Chang, M.N. Myers, J.M. Davis and K.D. Caldwell, *J. Chromatogr.,* 255 (1983) 359.
122 H. Poppe, J.C. Kraak, S.J.J. van der Linde and N. van Vught, *Chromatographia,* 20 (1985) 618.
123 F.E.P. Mikkers, F.M. Everaerts and Th.P.E.M. Verbruggen, *J. Chromatogr.,* 169 (1979) 11.
124 J.W. Jorgenson and K.D. Lucacs, *J. Chromatogr.,* 218 (1981) 209.
125 J.H. Knox and I.H. Grant, *Chromatographia,* 24 (1987) 135.
126 J.H. Knox, *Chromatographia,* 26 (1988) 329.
127 R.T Kennedy and J.W. Jorgenson, *Anal. Chem.,* 61 (1989) 436.
128 C.G. Edmonds, J.A. Loo, C.J. Barinaga, H.R. Udseth and R.D. Smith, *J. Chromatogr.,* 474 (1989) 21.
129 S. Terabe, K. Otsuka, K. Ichikawa, A. Tsuchiya and T. Ando, *Anal. Chem.,* 56 (1984) 111.
130 S. Terabe, K. Otsuka and T. Ando, *Anal. Chem.,* 57 (1985) 834.
131 V. Pretorius, B.J. Hopkins and J.D. Schieke, *J. Chromatogr.,* 99 (1974) 23.

132 G. Guiochon and A. Katti, *Chromatographia,* 24 (1987) 165.
133 J.H. Knox and H.M. Pyper, *J. Chromatogr.,* 363 (1986) 1.
134 Cs. Horváth, J. Frenz and Z. El-Rassi, *J. Chromatogr.,* 255 (1983) 273.
135 A.W. Liao, Z. El-Rassi, D.M. LeMaster and Cs. Horváth, *Chromatographia,* 24 (1987) 881.

Chapter 5

Ion-exchange chromatography

H.F. WALTON

CONTENTS

5.1 INTRODUCTION

Ion exchange, in the sense of this chapter, is the reversible interchange of ions of like charge between a solution and a solid, insoluble material in contact with it. It proceeds by equivalents; as one chemical equivalent of one kind of ion enters the solid, one chemical equivalent of another kind must leave. Secondary reactions can occur, like the combination of hydrogen ions with anions of weak acids to form uncharged species, and metal ions can enter the solid exchanger in the form of coordination complexes, but always electrical neutrality must be maintained. The structure of the solid is not altered during the exchange, save for swelling and shrinking associated with solvent uptake and release, and these changes, too, are reversible.

All ion exchanges are reversible, but the equilibrium distributions vary widely, and it is these variations that make ion-exchange chromatography possible. Primarily, ion exchange is regulated by electrostatic interactions between the ions being exchanged, the mobile ions, and the fixed ions, ions that are attached covalently to the molecular lattice of the ion exchanger. Other kinds of interactions may occur, however. Most solid ion exchangers used in chromatography are organic polymers, and as such, they are like organic solvents and interact with organic ions and even uncharged organic molecules. This interaction is often called "reversed-phase" interaction. Thus, ion-exchange chromatography has a very wide field of application, ranging from the simplest inorganic cations and anions to the very large ions of proteins and nucleic acids. Carbohydrates can be separated by ion exchange, as can the optical isomers of amino acids.

Ion exchange was first recognized in soils, where it is responsible for the retention of fertilizers and plant nutrients. In industry, the earliest and still the most important use of ion exchange is water conditioning. Hydrometallurgy is another very important application. However, this book is concerned with chromatography. The emphasis of this chapter will be on high-resolution, high-performance liquid chromatography that uses ion exchange or ion-exchanging materials.

5.2 ION EXCHANGERS

5.2.1 General considerations

For a solid substance to exchange ions with a solution, it must have ions of its own, and these ions must be able to move freely in and out of the molecular structure. The exchange can occur at the surface, and this is the case with clay minerals and with porous silica, but in general the solid must have an open, permeable structure. It must also carry ionic, electrically charged groups that are anchored to the solid structure. These are called the fixed ions. Balancing the charges of the fixed ions are the mobile ions or exchangeable ions, also called counterions. These are the ions that take part in ion exchange.

Ion exchangers can be inorganic or organic. They can have organic structures grafted on to an inorganic core, generally porous silica. An interesting class of inorganic ion exchangers is the artificial zeolites or "molecular sieves"; their crystal structure has large cavities and channels in which the mobile ions reside. They are very important as catalysts and in special separation processes, but they have no applications to chromatography, so we shall not describe them here. For chromatography, the most important ion exchangers by far are those made from organic polymers.

5.2.2 Polymeric exchangers

5.2.2.1 Synthesis

The first synthetic organic ion exchangers were condensation products similar to Bakelite, the early plastic made from phenol and formaldehyde. Materials like Bakelite were called "synthetic resins", and the ion exchangers made from them by Adams and Holmes in 1935 [1] were called "ion-exchange resins". The name has stuck, and persists to this day.

Condensation products are not very stable, and their structures are ill-defined. A great step forward was made in 1944, when D'Alelio patented ion-exchanging polymers based on styrene and crosslinked with divinylbenzene [2]. Most polymers used today for chromatography are of this type. They are made by the process of bead polymerization, in which the two liquid monomers, styrene and divinylbenzene, are mixed together and stirred as droplets in a bath of hot water, carrying a little detergent. As polymerization proceeds, the liquid drops become solid, transparent beads. The general chemical structure of the ion exchangers produced from the polymer beads is

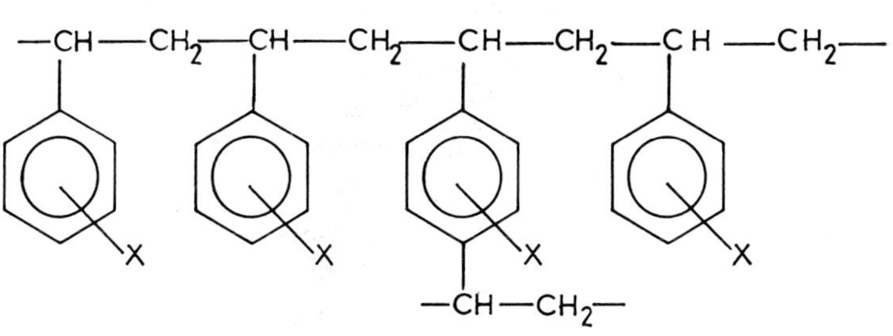

—CH—CH_2—CH—CH_2—CH—CH_2—CH—CH_2—

X X X X

—CH—CH_2—

5.2.2.2 Functional groups

In this formula, X represents an ionic functional group. The commonest functional groups are -SO_3H and -CH_2NR_3Cl. The group -SO_3H is introduced by treating crosslinked

References on p. A263

polystyrene beads with fuming sulfuric acid. It is strongly ionized; $-SO_3^-$ is the fixed ion, H^+ the mobile ion. Because the mobile ions are cations and can be exchanged with other cations, sulfonated polystyrene is called a cation exchanger. A polymer carrying the groups $-CH_2NR_3Cl$ is called an anion exchanger, because the fixed charges are positive, and it is the anions, carrying negative charges, that can be exchanged for other ions from the solution. The anion-exchanging groups are introduced by a two-stage reaction. First, the styrene/divinylbenzene polymer is treated with chloromethyl ether, which adds the side chain, $-CH_2Cl$; then this chloromethylated product is treated with a tertiary amine, generally trimethylamine, $N(CH_3)_3$. The product has a quaternary ammonium ion as the fixed ion, chloride ion as the mobile ion. Chloride can be exchanged with other negative ions, including the hydroxide ion, and the quaternary hydroxide is an extremely strong base.

Other functional groups can be introduced, if desired. A common anion exchanger has the fixed ion $-CH_2N(CH_3)_2C_2H_4OH^+$. This group has a stronger attraction for the hydroxide ion than the one just mentioned, that is to say, its hydroxide is a weaker base, and this is an advantage in chromatography, if the hydroxide ion is used as the eluent. The hydroxide is nevertheless a quaternary amine and must be classed as a "strong" base, in contrast to primary, secondary, and tertiary amines.

Starting with chloromethylated polystyrene, carrying the group $-CH_2Cl$, many functional groups can be attached. One of them is the iminodiacetate group, $-CH_2N(CH_2COOH)_2$. This binds heavy metal ions by chelation. Chelating resins are not much used in chromatography, but they are very useful for retaining and preconcentrating traces of heavy metals from water, including sea water [3]. Chelating exchangers that deserve special mention – and scores of such exchangers have been made [4] – are those with 8-hydroxyquinoline functionality [5] and the "Srafion" resins, made from polystyrene by attaching the sulfoguanidine group, $-CH_2S(:NH)NH_2$ [6,7]. These resins have a strong affinity for gold and the platinum metals and are used in hydrometallurgy. The phosphonic acid group, $-PO_3H_2$, has a certain chelating character and is a moderately weak acid. General methods for attaching chelating groups to polystyrene are reviewed by Griessbach and Lieser [8].

Exchangers that are weak acids or weak bases have special applications in biochemical analysis. Their capacity for exchanging ions other than H^+ and OH^- can be adjusted by regulating the pH. Weakly acidic and weakly basic exchangers can be made from polystyrene, but it is much more common to use different types of polymer matrix. A weakly acidic cation exchanger with functional -COOH groups is made by polymerizing methyl methacrylate, plus divinylbenzene or divinyl malonate as a crosslinking agent, then hydrolyzing the ester groups. Acrylic resins are softer than those made from polystyrene, and thus less suited to chromatography, but an important class of acrylic polymers is the "Spherons" [9-11]. They are made by copolymerizing ethylene glycol mono- and dimethacrylates. The basal unit of this polymer has the structure

```
       |
      CH2
       |
CH3 —— C —— COOCH2CH2OH
       |                      |
      CH2                    CH2
       |                      |
CH3 —— C —— COOCH2CH2 —— OOC — CH3
       |                      |
      CH2                    CH2
       |                      |
```

The polymer is used "as is" for size-exclusion chromatography, and the alcohol groups, $-CH_2OH$, may be oxidized to -COOH. Other functional ionic groups can be added. The "Spherons" are moderately rigid and are used for high-performance chromatography of proteins and biological macromolecules.

Weakly basic resins are made from polyethyleneimine, $-(CH_2CH_2NH)_n-$, by crosslinking with epichlorhydrin. A special class of ion exchangers that are hydrophilic and adapted to biomedical analysis are the "Sephadex" exchangers [12]. They are made by crosslinking dextran with epichlorhydrin and then introducing functional groups. One of the most useful is "DEAE-Sephadex", where DEAE means diethylaminoethyl. It is weakly basic. Its functional group is: $-O-CH_2CH_2N(C_2H_5)_2$. The parent base, 2-diethylaminoethanol, has (in its protonated form) $pK_a = 9.80$ at 25°C and zero ionic strength.

Care is needed in using ionization constants of small molecules to predict the combining power of ion exchangers that incorporate these molecules, like DEAE in the example just cited. The ratio of ionized to nonionized functional groups in an exchanger depends not only on the pH of the surrounding solution, but also on all exchangeable ions present, and on the spacing of ionic charges in the exchanger network (analogous to a polyelectrolyte chain). The general effect is that the proportion of ionized groups changes more gradually with pH and over a wider range of pH than would be the case if the parent small molecules were titrated in aqueous solution.

5.2.2.3 Crosslinking

The formula sketched above (Section 5.2.2.1) shows three styrene units and one unit of divinylbenzene. The divinylbenzene unit binds two styrene chains together. This binding is called crosslinking. One would call this polymer "25% crosslinked" because the monomer mix contained 25 mole per cent of divinylbenzene (DVB). The degree of crosslinking is a very important quantity, for it determines the degree to which the ion-exchanging polymer swells when placed in water, hence the concentration of ions in the swollen exchanger and the speed with which ions can diffuse in and out. It also influences the ion-exchange selectivity. The greater the degree of crosslinking, the greater the differences in the strengths of binding of different mobile ions.

High crosslinking is preferred for the greater ionic capacity, greater selectivity, greater rigidity, and greater resistance to pressure in a column. Low crosslinking gives faster exchange, especially with large ions, but at the cost of softness and possible collapse under pressure. A comprise must be made between high and low crosslinking. The best degree of crosslinking for most purposes in 8%, that is, 8 moles of divinylbenzene to 92 moles of styrene in the monomer mix.

As a practical matter, the monomers used to make commercial ion-exchange resins are never pure. Divinylbenzene is a mixture of *ortho-*, *meta-*, and *para*-isomers and contains up to 50% of ethylvinylbenzene; however, the advertised or nominal crosslinking always expresses the actual divinylbenzene content. It is easy to see why commercial ion-exchange resins, even some of the resins used for chromatography, are not uniform between different manufacturers and different batches. The special grades of ion-exchange resin used for high-performance chromatography are, however, made with pure monomers.

5.2.2.4 Particle size

An important consideration is particle size. Resins used for high-performance chromatography must have small and uniform particle size, generally 10 μm and less. The advertised particle sizes are those of the resins swollen in water, not the dry resins.

5.2.3 Macroporous polymers

The word "macroporous" means "having large pores". Macroporous ion exchangers (also called "macroreticular") have the form of beads that under a high-powered microscope look rough and opaque, not smooth and transparent as do the "gel-type" polymers we have been describing. The beads that one sees are aggregates of much smaller particles, tens of nanometers in diameter, between which are pores or channels that allow easy penetration of all but the largest molecules and ions. The beads are easily crushed by mortar and pestle, but they are sufficiently rigid to be packed into a chromatographic column, where they have the great advantage that volume changes are minimal; changes in ionic concentration, nature of the exchangeable ions and even in the nature of the surrounding solvent have little effect on the volume. Mass transfer is rapid between macroporous ion exchangers and the solution, and this makes for good chromatography. A problem in the past has been nonuniformity of exchange sites, which caused bad tailing in chromatographic bands. Modern macroporous exchangers do not seem to have this defect.

Much research is being done today on macroporous sorbents and ion exchangers, most of it in industrial laboratories, to produce materials that are mechanically stable and give minimum volume changes and narrow chromatographic peaks. It is important to realize that there is not just one type of macroporous structure; there are all degrees of "macroporosity". Macroporous polymers might better be called "solvent-modified" polymers. They are made by mixing the monomers with a solvent that is chemically inert and is recovered afterwards without change. The solvent may be a poorer solvent for the polymer than for the monomer, or it may be a better solvent for the polymer.

When the monomers are mixed in the presence of a catalyst, polymerization begins at a number of nuclei. Polymer chains start to form and spread outwards from the nuclei. In a liquid mixture of styrene and DVB, styrene combines somewhat faster with DVB than it does with itself [13], so that in the later stages of polymerization there is less crosslinking agent. The linear polymer chains reach out until they meet chains formed from another nucleus, and the nuclei are joined. If, now, a solvent is present that is a good solvent for the polymer, toluene for example, the chains will be more or less straight and will soon connect the nuclei, giving an extended network. If, however, the solvent is a poor solvent for the polymer, e.g., hexane, the growing chains will coil back on themselves and intertwine with each other, forming internal crosslinks. Only a few will join with the chains from other nuclei, and the end product will be the loose aggregate of tiny sub-micrometer particles that was described above. Another way to describe the process is to say that the polymer precipitates as it is formed.

The formation and properties of solvent-modified styrene/DVB polymers is described in a classic series of papers by Millar and coworkers [14,15]. It is clear that many variations are possible. A long and definitive review of macroporous polymers and ion exchangers made from them is given by Seidl et al. [16].

5.2.4 Silica-based ion exchangers

Porous silica is itself a fine example of a macroporous, solvent-modified inorganic polymer. Its production is a fine art; pore size and surface area can be varied at will; the greater the pore size, the smaller the surface area. The overall, macro-scale particle diameter can also be controlled. Porous silica is an ion exchanger because of the surface silanol groups, -SiOH. Alkali and alkaline-earth cations may be separated by chromatography on porous silica, with aqueous lithium salts as eluents. The order of elution is the same as that for a sulfonated polystyrene resin. Generally the ion-exchanging properties of porous silica are unwanted, a nuisance that is suppressed by "end-capping", reaction of the silanol groups with trimethylchlorosilane to convert -SiOH to $-SiOSi(CH_3)_3$. This is the final step in preparing alkyl-bonded phases for reversed-phase chromatography.

Carbon chains carrying ionic functional groups can be bonded to porous silica, creating an important class of ion exchangers for use in high-performance chromatography. They have been reviewed by Unger [17]. Cation and anion exchangers have been prepared, with strong-acid, weak-acid, strong-base, and weak-base functional groups. Some of the structures are shown in Fig. 5.1. The last examples in this figure are the so-called "Glycophases", developed for the chromatography of proteins. The sidechain is hydrophilic and can be made as long as one wishes. The others are the bases for commercial ion-exchanging silicas. Their capacities are quite high, some 0.5 to 1.5 meq/g, depending greatly on the conditions of synthesis, the surface area of the porous silica, and the extent of surface coverage. For ion chromatography these capacities may be inconveniently high (see below).

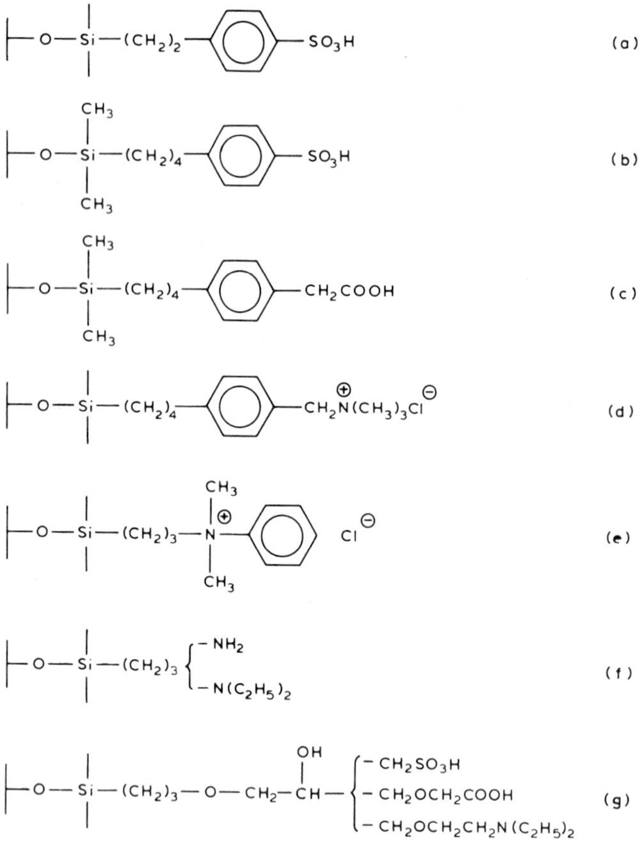

$$\vdash O-Si-(CH_2)_2-C_6H_4-SO_3H \quad (a)$$

$$\vdash O-Si(CH_3)_2-(CH_2)_4-C_6H_4-SO_3H \quad (b)$$

$$\vdash O-Si(CH_3)_2-(CH_2)_4-C_6H_4-CH_2COOH \quad (c)$$

$$\vdash O-Si-(CH_2)_4-C_6H_4-CH_2\overset{\oplus}{N}(CH_3)_3Cl^{\ominus} \quad (d)$$

$$\vdash O-Si-(CH_2)_3-\overset{\oplus}{N}(CH_3)_2-C_6H_5 \;\; Cl^{\ominus} \quad (e)$$

$$\vdash O-Si-(CH_2)_3\begin{cases}-NH_2\\-N(C_2H_5)_2\end{cases} \quad (f)$$

$$\vdash O-Si-(CH_2)_3-O-CH_2-CH(OH)-\begin{cases}-CH_2SO_3H\\-CH_2OCH_2COOH\\-CH_2OCH_2CH_2N(C_2H_5)_2\end{cases} \quad (g)$$

Fig. 5.1. Examples of silica-based ion exchangers. Vertical line at left represents the silica support. References: (a) 18, 19; (b) 20, 21, 22; (c) 20; (d) 23, 24; (e) 18, 19; (f) 24; (g) 25. Group (g) shows examples of Glycophases; see text.

One can also attach chelating groups to silica. Among the many groups that have been attached to porous silica are 8-hydroxyquinoline [26], dithiocarbamate, and 1,3-diketone [27]. These materials are used to recover trace metals from water.

The advantages of silica-based ion exchangers in chromatography are their mechanical rigidity and very fast mass transfer. The main disadvantage is their limited pH range. Above pH 7-8 the materials dissolve, because of the acidity of the silanol groups; below pH 2-3, silicon/carbon bonds are hydrolyzed, and the organic chains separate from the silica surface.

5.2.5 Ion exchangers for chromatography; surface-modified polymers

For high-performance chromatography it is essential to have fast mass transfer between the mobile and the stationary phases. The big handicap of liquid chromatography, compared with gas chromatography, is the slowness of diffusion in the liquid phase. Diffusion rates in a liquid are typically one ten-thousandth of those in a gas. Hence the need for small particles in liquid chromatography, and with small particles comes the need for high pressures. Diffusion rates of ions in a gel-type polystyrene-based ion exchanger are at most one-tenth of those in water [28]. Not only do the exchanging ions have to move through a medium that is obstructed by polymer chains, they must also jump from one fixed ion to another, from one potential-energy well to another over an activation-energy barrier. The barrier is greater, the higher the charge of the mobile ion; divalent cations diffuse one-tenth as fast as univalent cations.

Thus, it is highly necessary for efficient ion-exchange chromatography to make the distances over which the ions must diffuse as short as possible. One way to accomplish this is to use small resin beads, 10 μm or less in diameter, and this can be done. Another way is to confine the ion-exchanging region to a thin film or layer on the surface of a larger, chemically inert bead. In 1967, in a paper that may mark the beginning of modern high-performance liquid chromatography, Horváth et al. [29] described "pellicular resins". They took glass beads, 30-40 μm in diameter, roughened their surface with HF, and dipped them in a liquid styrene/DVB mix, which was allowed to polymerize, covering the glass bead with a film of polymer that was later sulfonated. Diffusion paths were short, yet the glass beads were large enough to give little resistance to flow. With these coated beads they performed fast, efficient chromatography of nucleosides and nucleic acid bases. But these "pellicular resins" had two disadvantages: their ion-exchange capacity was very low, and the polymer film was brittle and flaked away from the glass support.

In 1975, a landmark paper by Small et al. [30] described two kinds of ion exchanger in which the active stationary phase was confined to a thin layer on the surface, not of glass, but of a polymer bead. To make a cation exchanger, uniform beads of styrene/DVB polymer, some 20 μm in diameter, were dipped briefly in hot sulfuric acid, then removed and washed. A layer of sulfonated polymer, 1-2 μm thick, was formed that was integrally and chemically bonded to an inert core.

To make a surface-functional anion exchanger, it was difficult to form a layer of quaternary ammonium ions directly on the polymer surface, so Small et al. took a different path. They first sulfonated the polystyrene beads by a very brief immersion in sulfuric acid, so that the ion-exchanging layer was very thin, and then they brought the beads into contact with a suspension of very small particles of a conventional strong-base anion exchanger. The negative sulfonate ions attracted the positive quaternary ammonium ions and bound the resin particles very strongly to the surface of the polymer bead. Only a small fraction of the quaternary ammonium ions on the resin bind the particles to the bead; the rest make it function as an anion exchanger. In the early work, the dispersion of strong-base resin was made by grinding resin beads. Later it was made as a latex by "emulsion polymerization"; styrene/DVB monomer mix was dispersed in a hot, relatively concen-

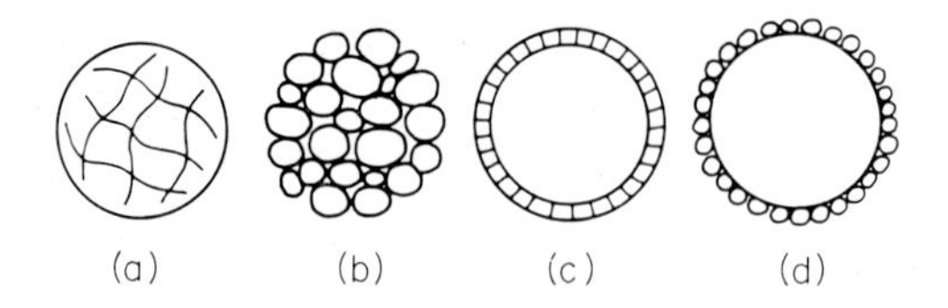

Fig. 5.2. Ion exchangers for high-performance chromatography. (a) Gel-type; (b) macroporous; (c) surface-functional; (d) latex-coated. Diameters are approximately 10 μm. From Ref. 108, by permission of CRC Press, Inc., Boca Raton, FL, USA.

trated solution of a surface-active agent, above its critical micelle concentration, and a water-soluble chain initiator, like hydrogen peroxide, was used. The ion-exchanging groups were introduced after polymerization or before.

Latex-coated polymer beads are used increasingly, both for anion and for cation exchange. Mass transfer in the latex particles is very fast, and by choosing the ratio of diameters of the latex particle and the bead support, the ion-exchange capacity can be controlled and can be made quite high, if desired. A new development is to coat macroporous beads with latex particles that adhere to the external surface but are too large to enter the pores. The coated beads can function efficiently as ion exchangers and also as reversed-phase supports. The reversed-phase or hydrophobic character of polystyrene-based ion exchangers (Section 5.1), which is always present, is exaggerated in these new exchangers. These beads tolerate small concentrations of acetonitrile in the mobile phase without swelling appreciably [31].

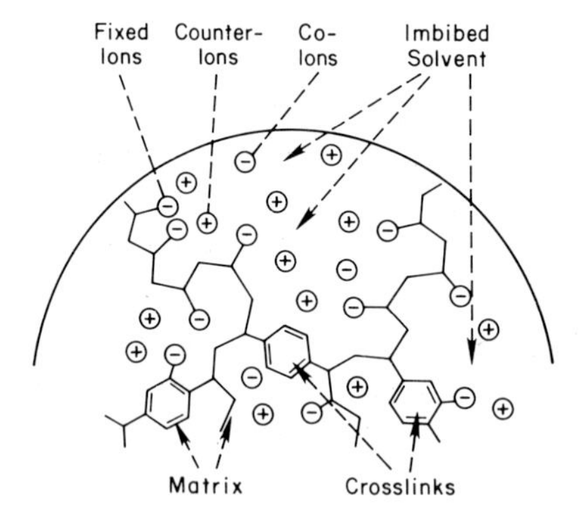

Fig. 5.3. Gel-type, polystyrene-based cation-exchange resin. Zigzag lines represent the carbon chain of the polystyrene "backbone"; hexagons represent divinylbenzene cross-links; negative charges, the fixed sulfonate ions; positive charges, exchangeable ions or counterions; isolated negative charges, co-ions. Penetrating the entire network is the imbibed solvent, generally water.

One can, of course, take a macroporous styrene/DVB polymer and place ionic functional groups on its inner surfaces. This was done by Gjerde et al. [32]. They used the standard two-stage chloromethylation/amination sequence that was described above to make and anion exchanger, with the important feature that chloromethyl ether was produced in situ from paraformaldehyde and HCl. Thus, the handling of chloromethyl ether, which is highly carcinogenic, was avoided. The proportion of ionic groups, and with it the ion-exchange capacity, could be controlled over a wide range.

Fig. 5.2 shows schematically the comparison between gel-type, macroporous, and two types of surface-active ion exchangers. Fig. 5.3 shows the internal structure of a gel-type cation-exchange resin, with its polymer chains (the aromatic rings of these are omitted for clarity), fixed ions, counterions and co-ions, and above all, the imbibed water.

5.3 SOME PROPERTIES OF ION-EXCHANGE RESINS

5.3.1 Gel-type resins

We have used the term "gel-type resins" to distinguish the homogeneous ion-exchanging polymers, described in Section 5.2.2, from the surface-functional, macroporous, and composite ion exchangers just described. Placed in water, a "gel-type" ion-exchange resin bead takes up water and swells to an extent that depends on the crosslinking. The interior of the swollen bead resembles a drop of concentrated electrolyte solution, and its physical properties can be modeled on those of a solution of toluenesulfonic acid. Much elegant physical chemistry has been done with gel-type polystyrene-based ion-exchange resins, mainly the strong-acid cation exchangers, because these materials are more uniform and easily characterized than quaternary-base anion exchangers. In this chapter we shall note a few properties that are of interest in chromatography.

5.3.2 Internal pressure

As the resin bead takes up water and swells, the crosslinks hold it in check. The internal pressure at equilibrium may be estimated by calculating the osmotic pressure of an ideal solution having the same molal ionic concentration as the solution inside the resin. An 8% crosslinked sulfonated polystyrene resin, carrying exchangeable sodium ions, has an internal concentration ca. 6 molal, that is, 6 mols of sodium ions and 6 mols of fixed sulfonate ions per kg of water. The osmotic pressure of an ideal 6 molal solution of one kind of ion at room temperature is about 150 atm. A more rigorous calculation [33] of the internal pressure is based on the measured vapor pressure of the swollen resin and uses the thermodynamic relation

$$\frac{\partial \bar{G}}{\partial P} = \bar{V}$$

References on p. A263

where G, P, and V are free energy, pressure, and volume, respectively, and the bars indicate partial molal quantities. This calculation also gives an internal pressure around 150 atmospheres. The important point is that this pressure is very high, and that, as a practical consequence, one should never pack a column with a dry gel-type resin, but always mix the resin with water first and pack the column with the slurry.

The degree to which a resin swells depends on the hydration of the ions in the resin and on their charge. A lithium-loaded cation-exchange resin swells more than a sodium-loaded resin, but a calcium-loaded resin swells less than the same resin loaded with sodium or potassium ions, in spite of the fact that Ca^{2+} is more hydrated than Na^+. A likely explanation is the "electrostatic crosslinking", caused by the interaction of the doubly charged calcium ion with two singly charged fixed ions on two different polymer chains.

In the normal course of chromatography, as one kind of counterion is exchanged for another kind, and eluents of different concentrations are passed successively through the column, the beads of a gel-type exchanger expand and contract. As long as these volume changes are small, a well-packed column can accommodate them, for the beads are elastic and can undergo slight deformation at their points of contact; but large volume changes ruin the performance of the chromatographic column. This is a serious disadvantage of gel-type resins in high-performance chromatography, especially with resins of low crosslinking, and it is a reason why current practice favors surface-functional and macroporous resins.

5.3.3 The Donnan equilibrium; co-ion uptake

When an ion-exchange resin is in equilibrium with a solution of a salt, the chemical potential of the dissolved salt, and its activity, must be the same inside and outside the resin bead. Consider a gel-type cation-exchange resin in contact with a dilute NaCl solution. At equilibrium, the activity of NaCl must be the same within the resin as outside. The activity of the salt is the product of the activities of the two ions.

$$a_{NaCl} = a_{Na} \times a_{Cl}$$

Equating activities with concentrations, and using bars to indicate the resin phase,

$$[\overline{Na}^+][\overline{Cl}^-] = [Na^+][Cl^-] = c^2 \qquad (5.1)$$

where c is the molal concentration of the salt outside the resin. (In the corresponding equation for a uni-divalent salt, like $CaCl_2$, the chloride concentration is squared.)

Inside the resin the concentration of the counterions, Na^+ in this case, is very high. Therefore, the concentration of chloride ions, which we call the co-ions, must be very small. A quick calculation shows that if the internal resin solution is 6 molal and the external NaCl concentration is 0.1 molal, the internal Cl^- concentration is 0.0016 molal. In other words, the chloride ions are virtually excluded from the interior of the resin. However, as the external salt concentration is increased, the internal co-ion concentration rises rapidly. A tenfold increase in the external salt concentration causes a ca. 100-fold increase in the internal concentration of a 1:1 electrolyte.

The Donnan equilibrium has several important consequences for ion-exchange chromatography. If we want to know the void volume of a column of cation-exchange resin, for example, we can inject a small amount of a dilute solution of sodium nitrate. The nitrate ions are excluded from the resin virtually completely, and they may be detected at the column exit by their ultraviolet absorbance. The method of ion-exclusion chromatography, to be described below, depends on Donnan exclusion of co-ions, and so does the operation of ion-exchanging membranes and the "membrane suppressor" used in ion chromatography.

5.3.4 Ion-exchange capacity

The "capacity" of an ion exchanger is the quantity of exchangeable ions per unit quantity of exchanger. It is usually expressed in gram-equivalents of ions per kilogram of exchanger, or more conveniently, milliequivalents per gram (meq/g) on a dry basis. For exchangers that are strong acids or strong bases, fully ionized under all conditions, there is no difficulty in expressing the capacity, nor in measuring it experimentally. The capacity of a strong-acid cation-exchange resin is found by converting it to the hydrogen form by washing it with excess of a dilute strong acid, followed by pure water, then drying in air, putting a weighed portion into a flask with a salt solution, and titrating with standard sodium hydroxide solution, using an indicator or a pH electrode. A second, weighed portion is placed in an oven and dried to remove moisture, then weighed again. For exchangers that are weak acids or bases, the effective ion-exchange capacity (for exchange of ions other than H^+ and OH^-) depends on pH and on the nature of the displacing ions. One must be careful in measuring or expressing the capacity of most strong-base anion exchangers, for they contain not only quaternary ammonium fixed ions, as described above, but also a surprisingly large proportion, up to 50%, of weakly basic tertiary amine groups, The titration with standard acid of a "strong-base" anion-exchange resin taken out of a bottle, particularly one that has been stored for a long time, shows two inflection points, the first due to neutralization of the strongly basic groups, the second due to the neutralization of weakly basic groups.

For ion exchangers in columns it is useful to express the capacity in meq/ml of the packed bed. The total capacity of a column of strong-acid cation-exchange resin is found by first passing an excess of acid through the column to replace all cations by hydrogen ions, then washing it with pure water, then passing through a dilute solution of NaCl or KCl, rinsing the column, and titrating the acid in the effluent.

Gel-type, 8% crosslinked, strong-acid sulfonated polystyrene cation exchangers have capacities of ca. 5 meq/g on a dry basis (i.e. the equivalent weight of the repeating polymer unit is ca. 200) and in a column, ca. 2 meq/ml of packed bed. For 8% crosslinked strong-base anion exchangers the capacities are about half this.

References on p. A263

5.3.5 Equilibrium

A typical ion-exchange reaction between two positive ions, A and B, of equal charge can be written thus

$$A^+ + BRes = ARes + B^+$$

where Res means the fixed negative ion of the resin. The equilibrium constant of this reaction is

$$\frac{[ARes][B^+]}{[BRes][A^+]} \times \frac{\bar{\gamma}_A \gamma_B}{\bar{\gamma}_B \gamma_A} = K \qquad (5.2)$$

The quantities γ are activity coefficients; barred quantities refer to the resin phase. The ratio of the activity coefficients of the two ions, A^+ and B^+, equals $\gamma^2_{AY}/\gamma^2_{BY}$, where Y is a univalent anion, like Cl^-. The activity coefficients of the salts AY and BY in solutions can be measured experimentally, and in dilute solutions they are nearly equal and cancel. Exchanges between ions of different charge are more complex, and activity coefficients do not cancel. It is more realistic in that case to quote a concentration quotient, Q, in which concentrations only are shown.

For uni-univalent exchanges in strong-acid and strong-base resins the quotient Q is nearly independent of the ratio of concentrations of A and B. In general, it increases as the proportion of ions B increases; i.e., the more B is present, the harder it is to put more B into the resin. The effect increases with crosslinking. The dependence of Q on ionic concentrations was studied by Reichenberg [34] and Bonner and Smith [35]. In analytical chromatography, ions of one kind, the analyte ions, are present in small concentrations compared with the ions of the eluent, and Q may be considered as a true constant.

In chromatography, the distribution ratio, D, defined as

$$D = \frac{\text{conc. of analyte ions in exchanger}}{\text{conc. of analyte ions in solution}}$$

is more useful than the equilibrium constant. It depends on the concentration of the eluting ions. For a uni-univalent exchange, say an analyte ion A^+ against H^+ as the displacing agent,

$$D = \frac{\text{conc. of A in exchanger}}{\text{conc. of A in solution}} = \frac{Q}{[H^+]} \qquad (5.3)$$

Units are equivalents of cation per equivalent of resin and equivalents of cation per equivalent of solution. Since under the conditions of analytical chromatography the proportion of ions A is always small, the concentration of hydrogen ions in the resin may be considered unity. In preparative chromatography, however, this simplification is not justified.

For exchanges between ions of equal charge, *D* does not depend on the total concentration of the solution; i.e., if water is added to an equilibrium mixture of resin and solution, the proportion of ions in each phase does not change. This is not true if the exchanging ions have different charges. For the reaction

$$Ca^{2+} + 2NaRes = 2Na^{2+} + CaRes_2$$

diluting the solution moves calcium ions into the resin; making the solution more concentrated shifts calcium ions out of the resin. This effect is exploited to advantage in water softening. In chromatography, where one eluent ion displaces one analyte ion, the distribution ratio, *D*, is inversely proportional to the eluent concentration, but where it takes two eluent ions to displace one analyte ion, *D* is inversely proportional to the square of the eluent concentration. In general, divalent cations are held much more strongly by an exchanger than univalent cations, and are eluted much later. By using a more concentrated eluent both ions are eluted sooner, and the two peaks are brought nearer together. (However, raising the eluent concentration raises problems of detection in "ion chromatography"; see below.)

5.3.6 Elution volume

It remains to relate the ion-exchange equilibrium quotient and the distribution ratio to the elution volume in chromatography. Let us start with the familiar equation relating the elution volume, V_e, with the interstitial or void volume, V_i, and the column distribution ratio or capacity factor, k'.

$$V_e = V_i + k' V_i \tag{5.4}$$

The quantity k' is the amount of solute in the stationary phase divided by the amount of solute in the mobile phase in any segment of the column. We evaluate it from the equilibrium quotient, *Q*, and the volumes of the stationary and mobile phases in the column. Let these volumes be V_S and V_i respectively. In a column packed with gel-type resin, V_S is the volume of the resin beads. In a column packed with a surface-active or a composite ion exchanger, V_S is the volume of the active stationary phase. It does not include inert material, like the polystyrene core of a surface-sulfonated or latex-coated exchanger. Let the ionic concentration in the active stationary phase be C_S meq/ml, and let us assume that activities in this phase are proportional to concentrations.

First, suppose that a small amount of a singly charged cation, A, is displaced by a large excess of hydrogen ions, H. (Ionic charges are omitted for simplicity.) The equilibrium quotient for the ion-exchange reaction is

$$Q = \frac{[\bar{A}][H]}{[A][\bar{H}]}$$

The column distribution ratio is

$$k' = \frac{[\bar{A}]V_S}{[A]V_i} = \frac{QC_SV_S}{[H]V_i}$$

Under the conditions of chromatography, $[\bar{H}] = C_S$. The elution volume is

$$V_e = V_i + \frac{QC_SV_S}{[H]} = V_i + \frac{QC}{[H]} \quad (5.5)$$

where C is the total ionic capacity of the column. It is important to note that when this capacity is lowered, the concentration of the eluent (here, hydrogen ions) must be lowered too if the elution volume is to remain constant.

Making a similar calculation for the displacement of calcium ions by hydrogen ions, where one Ca is displaced by two H, the elution volume is

$$V_e = V_i + \frac{QCC_S}{[H]^2} \quad (5.6)$$

In a composite ion exchanger the capacity of the stationary phase can in principle be varied in two ways, one changing V_S, the volume occupied by the active material, the other by changing C_S, the concentration of active sites (fixed ions) within the active phase. The (corrected) elution volume increases as the square of the concentration of active sites, but only as the first power of the volume of active phase. It is hard to test these relations experimentally, for it is hard to vary the concentration of active sites without changing the structure of an exchanger. However, the volume of the active phase may be changed by varying the depth of sulfonation in surface-sulfonated polystyrene. There is strong evidence that there is a sharp boundary in such materials between the active surface shell and the inactive core [36], and it has been shown that the distribution coefficients of alkali-metal cations in such materials are directly proportional to ion-exchange capacity; furthermore, the points for totally sulfonated crosslinked polystyrene fall on the same line in the graph of distribution coefficient versus capacity as do the points for partially sulfonated polystyrene, indicating that the sulfonated layer has the same properties as fully sulfonated gel-type exchangers [37].

5.3.7 Ion-exchange selectivity

The chromatographer is interested in relative distribution ratios, that is, ion-exchange selectivities. Table 5.1 shows selectivities of common cations and anions in 8% crosslinked strong-acid and strong-base exchangers. The difference between "Type 1" and

TABLE 5.1

ION-EXCHANGE SELECTIVITIES

Shown are relative values for 8% crosslinked strong-acid and strong-base resins. "Type I" anion exchangers have functional groups $-N(CH_3)_3^+$, "Type II" have $-N(CH_3)_2CH_2CH_2OH^+$. Adapted by permission from Table 7, Price List K, Bio-Rad Laboratories.

Cations		Anions		
			Type I	Type II
H^+	1.0	OH^-	1.0	1.0
Li^+	0.85	F^-	1.6	0.3
Na^+	1.5	Cl^-	22	2.3
NH_4^+	1.95	Br_-	50	6
K^+	2.5	I^-	175	17
Rb^+	2.6			
Cs^+	2.7	ClO_3^-	74	12
Ag^+	7.6	BrO_3^-	27	3
		IO_3^-	5.5	0.5
Mg^{2+}	2.5	NO_2^-	24	3
Ca^{2+}	3.9	NO_3^-	65	8
Sr^{2+}	5.0	HCO_3^-	6	1.2
Ba^{2+}	8.7	HSO_3^-	27	3
		HSO_4^-	85	15
Fe^{2+}	2.5	$H_2PO_4^-$	5	0.5
Zn^{2+}	2.7			
Co^{2+}	2.8	Acetate	3.2	0.5
Cu^{2+}	2.9	Phenate	110	27
Ni^{2+}	3.0	Salicylate	450	65
Pb^{2+}	7.5	Benzenesulfonate	500	75

"Type 2" anion exchangers is that "Type 1" has fixed ions $-CH_2N(CH_3)_3^+$, while "Type 2" has $-CH_2N(CH_3)_2C_2H_4OH^+$. "Type 2" holds hydroxide ions more strongly than "Type 1", but the hydroxide is still a strong base. We can make certain generalizations about selectivity orders:

For the cations of the first two groups of the periodic table the orders of strengths of binding to the exchanger are:

$$Li^+ < H^+ < Na^+ < K^+ < Rb^+ < Cs^+ \text{ and } Mg^{2+} < Ca^{2+} < Sr^{2+} < Ba^{2+}$$

These are the orders of ionic hydration. The more strongly hydrated the ion is, the more it stays out of the resin; the more it is attracted to the aqueous phase, where there is more water.

References on p. A263

For the simple halide ions the order is

$F^- < Cl^- < Br^- < I^-$

An elution order found in ion chromatography, which corresponds to the affinity sequence, is

$F^- <$ formate $< BrO_3^- < Cl^- < NO_2^- < Br^- < NO_3^- < SO_4^{2-} < S_2O_3^{2-}$

Here, the selectivity depends to some extent on the fixed ions, $-NR_3^+$. The size and hydrophobicity of the R groups affect the selectivity [38].

With anions, hydration cannot be very important, for physical evidence shows that singly charged anions are hardly hydrated at all. A more important effect is that of ionic size. To dissolve in water, a large molecule or large ion must break bonds between water molecules. The energy spent in breaking hydrogen bonds is released when the ion moves into the resin. In support of this idea, we cite one regularity, a rule that always holds: In a series of oxy-anions of a given element in different states of oxidation but having the same charge, the larger the anion the more strongly bound it is by anion exchangers. Thus, NO_3^- is more strongly held than NO_2^-, $H_2AsO_4^-$ than $H_2AsO_3^-$, SO_4^{2-} than SO_3^{2-}, and the chlorine oxy-anions are bound in the order $ClO_4^- > ClO_3^- > ClO_2^- > ClO^-$. The oxy-anions of phosphorus show the same tendency; the larger the anion, the stronger the retention, provided the ionic charges are the same. Anions of organic acids behave similarly; methacrylate is held more strongly than acrylate, fumarate (*trans*-) is held more strongly than maleate (*cis*-). Fumarate has no more carbon atoms than maleate, but its effective volume in water is greater.

As with cations, the greater the charge on an anion, the stronger the binding, at least in the concentration range used in chromatography. However, a high charge/radius ratio favors ionic hydration, and exceptions to this statement can be found.

These selectivity orders hold for ion exchangers having sulfonate or quaternary ammonium functional groups. These fixed ions are large and have low field strength. Smaller fixed ions having higher field strength, like carboxylate ion, may show different selectivity orders. The effect of field strength of fixed ions on selectivity is the subject of an important theory by Eisenman [39].

5.4 ION CHROMATOGRAPHY

5.4.1 High-performance ion-exchange chromatography; detection by electrical conductivity

Early attempts at chromatography of ions on ion-exchange resins used gravity flow in glass columns open to the atmosphere. Fractions of the effluent were collected and individually analyzed. Analysis times were of the order of hours, and resolution was poor.

Nevertheless some useful separations were accomplished, notably the separation of lanthanide ions by Boyd and Ketelle [40] and the separation of amino acids by Moore and Stein [41].

"High performance" requires high resolution, high speed, and continuous detection. These conditions were met by Small et al. and described in their 1975 paper [30], which is generally considered to mark the birth of "ion chromatography". High speed and high resolution were achieved by the use of surface-functional exchangers. Continuous detection of ionic species was accomplished through electrical conductivity.

All ions conduct an electric current, and measurement of electrical conductivity was the obvious way to detect and measure ionic species as they emerge from the column. The problem was that the eluents used, e.g., 0.7 *M* HCl, had a very high conductivity of their own. The changes of conductivity that occurred as analyte ions left the column were too small to be detected. However, eluents as concentrated as 0.7 *M* hydrochloric acid were needed, because the exchangers in the columns were gel-type resins having a high capacity, some 5 meq/g for cation exchangers. The surface-functional resins introduced by Small et al. had a much lower capacity, 0.005 meq/g or less. Therefore, the eluents had to be much more dilute (see Eqn. 5.5). Their electrical conductivity was correspondingly smaller. There was now a better chance of measuring the changes of conductivity that occurred as analyte ions were eluted.

TABLE 5.2

EQUIVALENT CONDUCTANCES, $\lambda°$, AT 25°C

Units, ohm^{-1} cm^2 $equiv^{-1}$.

H^+	350	OH^-	198
Li^+	39	F^-	54
Na^+	50	Cl^-	76
K^+	74	NO_3^-	71
Mg^{2+}	53	HCO_3^-	45
Ca^{2+}	60	Acetate	41
Cu^{2+}	55	Benzoate	32
NEt_4^+	33	SO_4^{2-}	80

Table 5.2 gives equivalent conductances of several cations and anions. The most highly conducting cation, by far, is the hydrogen ion. Suppose that sodium ions (equivalent conductance 50) are displaced from a column by hydrogen ions (equivalent conductance 350) in a solution of hydrochloric acid. The Cl^- concentration remains constant during elution; therefore, the sum of the equivalent concentrations of Na^+ and Cl^- must be constant too. (Remember that ion exchange proceeds by equivalents.) When the Na^+ concentration increases, the H^+ concentration falls, and the conductivity there-

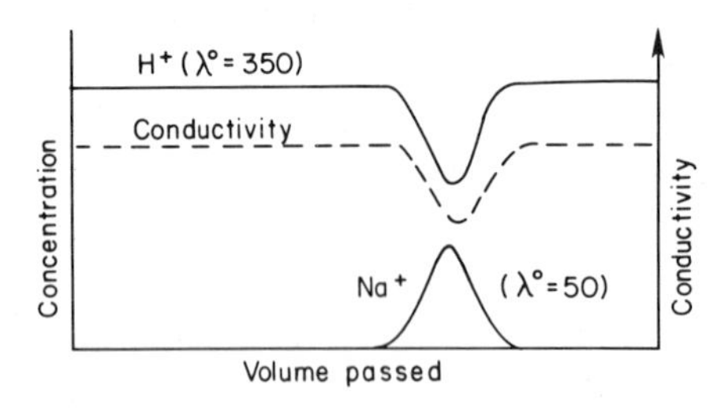

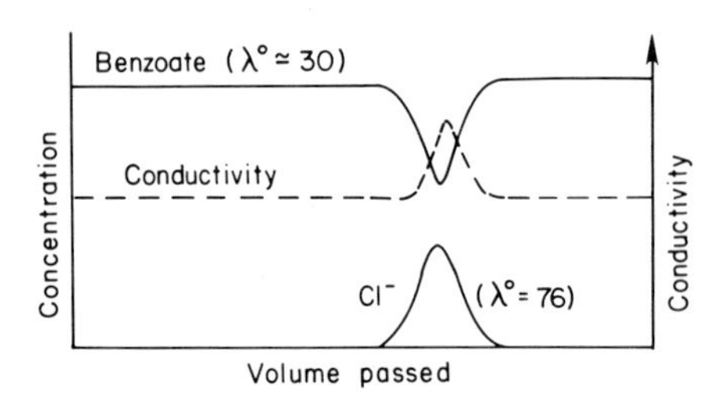

Fig. 5.4. Concentrations and ionic conductivities at the column exit during ion-exchange elution. Reproduced by permission from H.F. Walton, Ion-Exchange Chromatography, ACS Audio Course C-93; American Chemical Society, Washington, DC, USA.

fore falls (see Fig. 5.4). In anion exchange, if the relatively fast-moving sulfate ions were displaced by slow-moving benzoate ions, the conductivity would rise as the sulfate ions emerged.

There are two ways to use electrical conductivity to detect ionic species in chromatography. One is to measure the conductivity of the effluent and look for changes, such as those shown in Fig. 5.4, keeping the background conductivity as low as possible by using dilute eluents, and choosing the eluents to give maximum conductivity changes. This method is called nonsuppressed or single-column ion chromatography. The other way is to remove selectively the excess eluent from the solution as it leaves the column, and then measure the conductivity. This method is called suppressed ion chromatography. The first method is simpler and requires less equipment; the second promises greater sensitivity. Small et al. [30] chose the second way.

5.4.2 Suppressed ion chromatography

In this method, an eluent must be chosen that can be selectively removed. For the chromatography of cations this can be hydrochloric or nitric acid, ca. 0.005 *M*. These eluents are strong enough to elute singly charged cations, but not doubly charged ions, like those of the alkaline earths. To elute alkaline-earth cations, a good eluent is "DAP", 2,3-diaminopropionic acid [42]. The ionization constants (p*K* values) of this acid are 1.3, 6.7, 9.4; used with 0.05 *M* HCl, about half the acid is in the doubly charged, fully protonated form, and this is a strong displacing agent. The membrane suppressor (see below) converts the left-over DAP to its uncharged dipolar-ion form.

For anion chromatography, NaOH or KOH is a logical choice, but these eluents are seldom used, because the hydroxide ion is a weak displacing agent (see Table 5.1). More commonly, a solution is used that is 0.0024 *M* in carbonate, 0.003 *M* in bicarbonate. The divalent carbonate ion is a stronger displacing agent than singly charged bicarbonate. Clearly, the eluent strength can be regulated by changing the ratio of concentrations of these two ions. Mixtures of carbonate and hydroxide ions are sometimes used. Suppose phosphate ions are to be eluted; they can be $H_2PO_4^-$, HPO_4^{2-}, or PO_4^{3-}, the proportions depending on pH. The more highly charged is the anion, the more strongly it is bound. The order of elution is phosphate, bromide, nitrate, sulfate in the carbonate/bicarbonate eluent cited above, but bromide, nitrate, sulfate, phosphate in an eluent containing carbonate and hydroxide.

In the original scheme of Small et al. the excess eluent was removed by a column of high-capacity gel-type exchanger that followed the column of surface-functional, low-capacity exchanger, in which the separation was performed. In the separation of cations, the eluent was dilute HCl. The second column, which was called the suppressor column, was packed with a strong-base anion-exchange resin in its hydroxide form. In this column an eluted salt, such as NaCl, was converted to its corresponding base, here NaOH, which

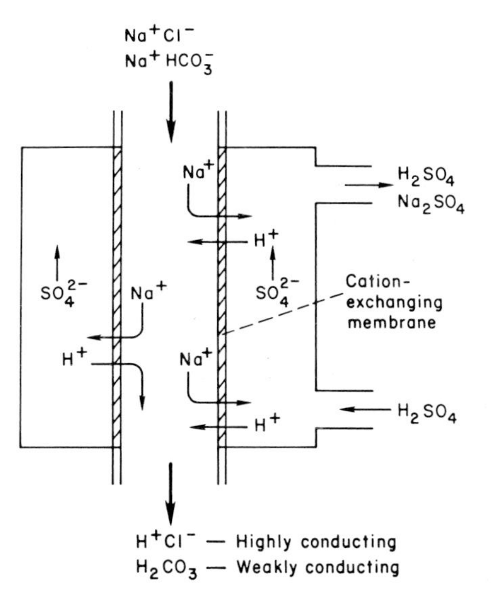

Fig. 5.5. Membrane suppressor for ion chromatography. Chloride ions are eluted from an anion-exchange separator column by a solution of sodium bicarbonate. The solution passes between two cation-exchange membranes that allow cations to pass but not anions. Leaving the suppressor, the solution enters the conductivity cell. The internal volume of the channel between the membranes is 40 μl. The drawing is not to scale. From Ref. 108, by permission of CRC Press, Inc., Boca Raton, FL, USA.

References on p. A263

has a very high electrical conductivity. The excess of HCl from the cation separation was absorbed by the strong-base resin, and its hydrogen ions were converted to water. In the separation of anions the separator column was packed with a strong-acid, high-capacity cation-exchange resin in its hydrogen form. An eluted salt, like NaCl, was converted to its corresponding acid, here HCl, which again has a very high electrical conductivity. The excess of carbonate/bicarbonate eluent was converted to carbonic acid, the conductivity of which, while not zero, is very low.

The suppressor columns had two disadvantages: they caused additional band spreading, and they eventually became exhausted and had to be regenerated. Today, the suppressor columns have been replaced by the "micro membrane suppressor" [43]. This device uses ion-exchange membranes. Its principle is shown in Fig. 5.5. The column effluent flows between two very thin menbranes that are high-capacity ion exchangers. Because of the Donnan equilibrium (Section 5.3.3), a cation-exchanging membrane that has fixed negative ions excludes negative ions in the solution but allows positive ions to pass, in the sense that they can change places with other positive ions, maintaining electrical neutrality. On the outside of the two membranes flows a regenerant solution which is either a strong acid or a strong base.

In Fig. 5.5 the elution of chloride ions from an anion-exchanging separator column is shown. The eluent normally used contains sodium carbonate and bicarbonate; here, for simplicity, only bicarbonate is shown. Chloride ions come out of the column as sodium chloride, accompanied by excess sodium bicarbonate. The membranes allow sodium ions to pass and be replaced by hydrogen ions from the regenerant solution, which is a dilute solution of sulfuric acid. The chloride ions cannot pass, save for the very small amounts permitted by the Donnan equilibrium. The excess bicarbonate ions are converted to carbonic acid. The solution that leaves the membrane suppressor and enters the conductivity cell contains the chloride ions (the analyte) in the form of hydrochloric acid, plus carbonic acid which gives a low background conductivity. Typically, 9 common anions are eluted in 10 min with good peak separation.

For cation chromatography the membranes are anion-exchanging and the regenerant solution that flows on the outside is a solution of tetramethylammonium hydroxide. Sodium hydroxide would do, but sodium ions leak across the membranes more rapidly than the larger tetramethylammonium ions.

Donnan exclusion is more effective, the higher the concentration of fixed ions in the membranes. Membranes of high ionic concentration allow the use of fairly concentrated eluents, up to 0.1 *M*. A great virtue of the membrane suppressor is that it permits the use of concentration gradients in elution.

5.4.3 Nonsuppressed ion chromatography

In nonsuppressed ion chromatography the conductivity cell is placed directly at the column exit. Differences in conductivity are measured between the solutions entering and leaving the column. Reliance is placed on a very sensitive and stable conductivity detector, and eluents are chosen that will give maximum changes in conductivity when peaks

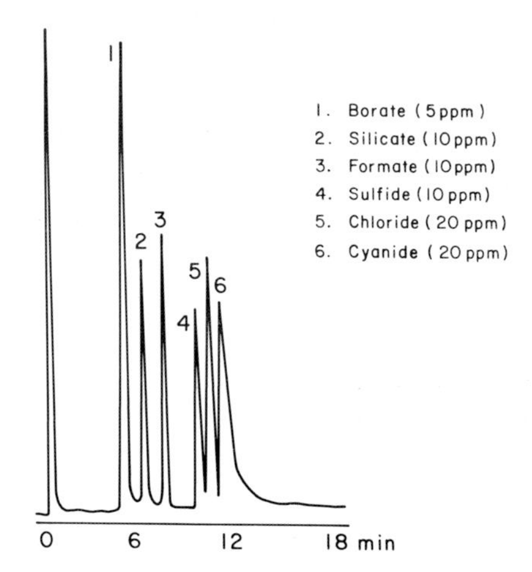

Fig. 5.6. Nonsuppressed chromatography of weak-acid anions. The eluent is 4 m*M* NaOH/0.1 m*M* sodium benzoate. Sample volume, 100 μl. Ordinates show decrease in conductivity. By courtesy of Wescan Instruments, Inc.

appear. Eluents are made as dilute as possible to minimize the background. Concentrations of 0.001 *M* and less are commonly used.

For chromatography of the alkali metal ions, a typical eluent is 0.001 *M* nitric acid. To elute doubly charged ions, a good choice is a salt of ethylenediamine, $H_2NCH_2CH_2NH_2$ or "en". A 0.001 *M* solution of $enH_2(NO_3)_2$ (pH 6.1) separates calcium and magnesium ions from one another and from the alkali metal ions, which are eluted as a single peak near the void volume. The conductivity drops when Ca^{2+} or Mg^{2+} replaces the faster-moving ion enH_2^{2+} [44].

For anion chromatography the most common eluents are sodium benzoate and sodium phthalate. The anions of these salts are slow-moving, and elution of other anions is marked by a rise in conductivity. Phthalate is a stronger eluent than benzoate, so that lower concentrations must be used; moreover, the doubly charged phthalate ion, P^{2-}, is bound more strongly than singly charged HP^-. The proportion of these two ions, and hence the eluent strength, can be regulated by regulating the pH. In their first studies Gjerde et al. [32] used 0.2 m*M* benzoate and 0.1 *M* phthalate, the phthalate being pH 6.25 to produce the doubly charged ion. (The p*K* values of phthalic acid are 2.9 and 5.5.) Another popular eluent is a solution of sodium borate (borax) and sodium gluconate, each 0.0013 *M* [45]. Gluconic acid has hydroxyl groups that coordinate with borate ions and make boric acid a stronger acid.

The main advantage of nonsuppressed ion chromatography over suppressed ion chromatography is, of course, instrumental simplicity. Other advantages are the use of eluents in the neutral pH range (which makes it possible to use silica-based exchangers) and the ability to measure ions of very weak acids, such as bicarbonate, silicate, and borate. These can be detected and measured with 0.001 *M* KOH as eluent [46]. They

References on p. A263

cannot be detected in suppressed ion chromatography because the suppressor would convert them into their weakly ionized, nonconducting acids. Fig. 5.6 shows a chromatogram of weak-acid anions.

Both modes of ion chromatography are fast and sensitive. Analysis times less than 1 min can be obtained if desired; nonsuppressed chromatography has the advantage in this respect. The sample volumes are generally 50 to 100 μl; the concentrations are in the low parts-per-million range; detection limits are tens of ng and less.

5.4.4 Methods of detection other than conductivity

The distinction between suppressed and nonsuppressed chromatography disappears if a detection method other than electrical conductivity is used. Let us, therefore, look at other means of detection.

One that is nearly universal is indirect photometric detection [47]. The eluent ion absorbs light, while the analyte ions do not. To detect simple inorganic anions a solution of sodium phthalate is used. Phthalate ions absorb in the ultraviolet. When a nonabsorbing ion, like chloride or sulfate, is eluted, the concentration of phthalate ions falls, and so does the absorbance. For the chromatography of cations a light-absorbing displacing ion is used. This can be Cu(II), Ce(III), or benzyltrimethylammonium ion [48]. Cerium(III) is fluorescent as well as absorbing in the ultraviolet, so that a decrease of fluorescence marks the elution of other cations; this effect was used to measure sodium, ammonium, and potassium ions in urine. The eluent was 10^{-5} *M* $Ce(ClO_4)_3$ [49].

Indirect photometric detection is very sensitive, but it requires the measurement of a small signal against a high background, and it is easily affected by stray light. It cannot be used with gradient elution.

Electrochemical or amperometric detection [50] is used for ions that can be electrochemically oxidized or reduced. The eluate flows between a counter-electrode and a working electrode of carbon, platinum or gold, sometimes silver, which is maintained at a constant potential with respect to a reference electrode. Most often, this detector is used with oxidizable organic ions, such as those of phenols or catecholamines [51]. These simple anions can be detected by oxidation at a silver anode [50]: Cl^-, Br^-, I^-, HS^-, CN^-, SCN^-, $S_2O_3^{2-}$.

Applied to organic ions, the amperometric detector has the drawback that the anode becomes contaminated by intermediate oxidation products and must be cleaned periodically. To avoid this difficulty, Hughes and Johnson [52] introduced "pulsed amperometric detection". A potential is applied for a brief time to a small gold or platinum electrode that is sufficiently positive to oxidize the compound sought, and the current is read. Then the potential is raised to a high value so that oxygen is evolved and carbon compounds sticking to the electrode are burned off. Meanwhile, a layer of metal oxide is formed. Next, the potential is brought to a negative value and held there until the oxide is completely reduced back to metal. The cycle is repeated several times a second. A typical cycle would be

0.2 V for 0.06 sec (detection, current measurement)
0.6 V for 0.06 sec (cleaning, burning off)
– 0.8 V for 0.24 sec (reduction of metal oxide)

These potentials are referred to a Ag/AgCl/1 *M* KCl electrode. Different potentials are used for different analytes and different electrode materials.

These times and potentials are used for the detection of sugars, which is a very important application of ion chromatography. Sugars are very weak acids, having pK_a values near 12. They are separated and measured by anion chromatography, with 0.01 *M* KOH as eluent [53]. Sorbitol, arabinose, glucose, fructose, lactose, and sucrose are eluted in this order, and oligomers are separated, larger molecules being retained longer. Many other kinds of organic compounds can be measured by pulsed amperometry, including amino acids.

5.4.5 Postcolumn derivatization; transition metals, lanthanides

Transition metals and heavy metals are separated and measured by ion chromatography. The selectivity of ion exchange is not very good; ion exchange is therefore supplemented by complex-ion formation in solution. Eluents containing oxalate, citrate, and pyridine-2,6-dicarboxylate have been used, with a column of cation exchanger. The anions form negatively charged complexes with transition metals and so pull the metal ions off the exchanger, or reduce their retention. Selectivity is achieved through differences of stability of the complex ions.

To detect the metal ions as they emerge from the column the effluent is mixed with a reagent that produces colored compounds. Useful reagents are PAR [4-(2-pyridylazo)resorcinol] and Arsenazo I (see formulae below, left and right respectively)

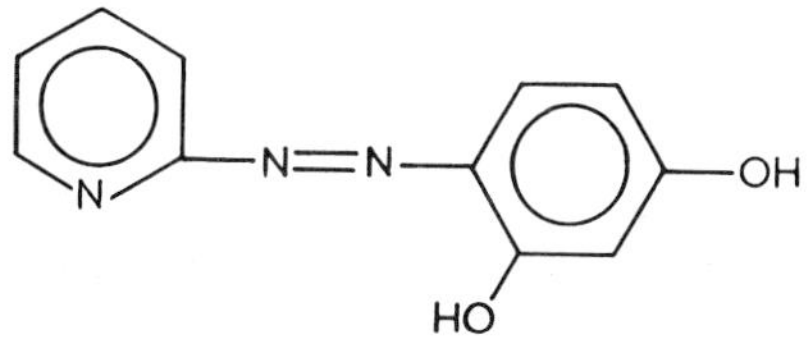

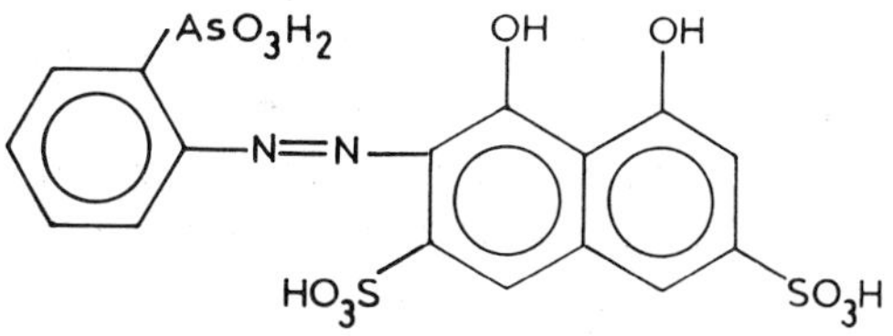

The detection limit for transition metals with PAR is 0.1-1 ng. The volume of sample is generally 0.05-0.1 ml. Very dilute samples, like drinking water, may be preconcentrated by

References on p. A263

passing a large volume, 1 to 50 ml or more, through a small column of high-capacity cation-exchange resin, then stripping the retained ions with a small volume of appropriate eluent. Cassidy and Elchuk [54] used a 3 x 3-mm resin cartridge to retain the heavy metals, then turned a valve to bring the cartridge in series with the analytical column and into the stream of eluent, which was citrate of pH 4.8, starting at 0.08 *M* and rising in a linear gradient to 0.20 *M*. In this manner, Cassidy and Elchuk were able to measure cobalt ions at the ng/l level in reactor-loop water. Clearly, detection by postcolumn derivatization is compatible with gradient elution.

Gradient elution with 2-methyllactic acid (α-hydroxyisobutyric acid) was used to separate lanthanide ions in $<$ 30 min, a remarkable feat [55]. Detection was by postcolumn reaction with Arsenazo I. The lanthanides were eluted in the order of their atomic numbers, with the smallest ion, Lu, eluted first and the largest, La, last. The smallest ion forms the most stable anionic complex, as expected from Coulomb's law.

It is appropriate to compare ion chromatography with atomic absorption spectrometry for analysis of transition metals. Flame atomic absorption is more sensitive by one or two orders of magnitude; of course, the sensitivity of atomic absorption varies greatly from one element to another. Ion chromatography has the advantage of being a multi-element technique. Measuring several elements at one time, it reveals elements one did not expect to see, and it provides speciation, i.e., it distinguishes different compounds of the same metal and different oxidation states, e.g., Fe(II) from Fe(III), and Cr(III) from Cr(IV). Moreover, ion chromatography is directly compatible with preconcentration or trace enrichment procedures.

5.4.6 Scope of ion chromatography

When it was first used, the term "ion chromatography" meant high-performance chromatography of ionic species on columns of surface-functional, low-capacity exchangers with detection by electrical conductivity. The term still implies high performance: continuous detection and automatic recording, high resolution, and high speed. But high performance can be achieved with other means of detection than conductivity, and low capacity is not essential; in fact, exchangers of high capacity are often necessary. An early study of Elchuk and Cassidy [55] on the separation of the lanthanides compared a column of 13-μm gel-type cation-exchange resin with columns of bonded cation-exchanging silicas, 5 and 10 μm in diameter, and found it to give just as good a resolution with only a small sacrifice in speed. Newer macroporous or solvent-modified polymers combine high ion-exchange capacity with high speed and resolution [56].

5.4.7 Applications of ion chromatography

The range of applications of ion chromatography is so great that it would be impossible to list them all. Reference is made to several books that stress applications [57-61] and to Chapter 12 of this book. Chapter 4 of Ref. 60 is especially recommended, also Ref. 109. Environmental analysis, fertilizers, steam power plants, electroplating baths, foods, and pharmaceuticals are among the many fields in which ion chromatography is used.

The first applications of ion chromatography were to simple inorganic anions and cations, especially anions, for these were difficult to determine by other means. Metallic ions could be measured very easily by atomic absorption spectroscopy, but anions, such as nitrite, nitrate, sulfite, sulfate, and the myriad oxy-anions of phosphorus were very difficult to detect and measure. For these anions, especially sulfate, ion chromatography answered a great need. Ion chromatography also made it easy to determine important nonmetallic cations, such as ammonium and substituted ammonium ions – like the alkanolamines used in soap formulations.

A common use of ion chromatography is water analysis. No sample preparation is necessary, except, often, dilution with distilled or deionized water. The usual "anion test solution" is 4 mg/l in Cl^-, 20 mg/l in NO_3^-, 25 mg/l in SO_4^{2-}; using 50 μl for injection, these concentrations give good-sized peaks. Many natural and municipal water supplies have higher concentrations than these. Note that bicarbonate ions are not observed in suppressed ion chromatography. They can be observed and measured by nonsuppressed chromatography, though the peak is eluted early. Bicarbonate may also be determined by ion exclusion (see below).

The cations commonly found in water are Na^+, Mg^{2+}, and Ca^{2+}. It is difficult to measure all these in one isocratic run, because the divalent ions are held much more strongly than the univalent ions. Various gradient-elution and column-switching methods have been proposed [42].

When it is necessary to measure small concentrations of minor constituents in the presence of large concentrations of a major constituent, preliminary sample preparation is necessary. The major constituent must be removed, at least partially, before ion chromatography is performed. A common way to do this with cations is to pass the solution through a short column of iminodiacetate chelating resin (Chelex-100), which retains transition metals and heavy metals much more strongly than it does Mg, Ca, or Na. Heavy metals can be recovered from sea water by this resin [3,62]. The retained metals are stripped from the column by a dilute acid, and then analyzed by ion chromatography.

A preliminary separation may sometimes be avoided if the major constituent is eluted after the minor ones. The minor ions are pushed ahead and are eluted as distinct peaks, with the detector at high sensitivity, before the sudden sharp rise in conductivity caused by the major constituent. However, when the major constituent is eluted first, the minor ones appear as small peaks on a large, sloping "tail", and quantitation is more difficult. In cases like this, column switching diverts most of the major constituent to waste and reduces the background against which the analyte peaks are measured.

For the analysis of extremely dilute solutions, preconcentration is necessary. This may be accomplished on a separate, small column or on the column that will be used for the chromatographic separation. A classic example of preconcentration is the measurement of common ions in the meltwater from Antarctic snow and ice [63]. Samples of 5-10 ml were passed directly into the analytical columns. Because the background was nearly pure water, the ions were held at the entrance to the columns and did not start to move

References on p. A263

until the eluent was passed through. Two separator columns were used, one for cations (Na^+, NH_4^+) and one for anions (Cl^-, NO_3^-, SO_4^{2-}); they were followed by suppressor columns. The concentrations measured ranged from 20 μg/l (Na^+) to 120 μg/l (SO_4^{2-}). The authors noted: "The amount of meltwater required is about an order of magnitude less than for neutron activation and two orders less than for atomic absorption."

5.5 ION-EXCLUSION CHROMATOGRAPHY

Ions having the same charge sign as the fixed ions of an exchanger are excluded from the exchanger, save for the small amounts permitted by the Donnan equilibrium (Section 5.3.3). However, uncharged molecules are free to enter the swollen resin.

Suppose we have a column of strong-acid, high-capacity gel-type cation exchanger, carrying mobile hydrogen ions, with a dilute strong acid as the mobile phase. Let us inject a mixture of two acids, one strong and completely ionized, the other weak and largely nonionized. The strong acid, excluded from the exchanger, is eluted at the void volume; the weak acid enters the exchanger and is eluted near the total water volume (as it might be measured by injecting D_2O and noting the dip in refractive index when this is eluted). Acids of intermediate strength that are partly ionized are eluted at intermediate volumes in the order of their acid strengths. This effect was used years ago to separate acetic acid from HCl. On the analytical scale, Turkelson and Richards [64] used ion exclusion to separate acids formed in the Krebs cycle. They used a column of 4% crosslinked sulfonated polystyrene and eluted with 0.001 *M* HCl. The acids came out nearly, but not quite, in the inverse order of their ionization constants, with the strongest, *cis*-aconitic, emerging first. A similar study was made by Woo and Benson [65], who used a proprietary gel-type ion-exchanging polymer that would stand high pressures in a closed column. Their eluent was 0.005 *M* sulfuric acid; eleven acids from oxalic to butyric were eluted almost in the order of their pK_a values. Ion exclusion was the main mechanism of separation, though "hydrophobic interaction", the attraction between the carbon chains of the solute molecules and the polymer matrix of the exchanger, played a part too. The longer the alkyl chain of an ion or molecule, the more strongly the substance is retained.

Among the many recent applications of ion-exclusion chromatography are the determination of fluoride ions [66] and bicarbonate ions [67]. Each is the anion of a weak acid and is converted to its parent acid, wholly or partly, by lowering the pH. The fluoride ion is weakly held by anion exchange and is eluted so near the void volume that its peak cannot be distinguished from the "water dip". (This is the drop in conductivity that accompanies the pulse of pure water from the injected sample as this pulse travels down the column, leaving the ionic species behind.) But the pK_a of HF is 3.45, and bringing the pH below this value reduces exclusion and increases retention on a cation-exchange resin. To measure bicarbonate ions in blood serum, it suffices to inject the sample into a strong-acid cation-exchange column, with pure water as the eluent. Bicarbonate ions are converted to carbonic acid, which enters the swollen resin and is retained; chloride ions remain ionic and are excluded from the resin, emerging at the void volume.

An afterthought concerns the use of pure water as an eluent for weak acids. As the peak advances and the concentration of acid in the mobile phase rises, the proportion of undissociated acid rises, and retention (on a reversed-phase column) increases. The result is an unsymmetrical chromatographic peak that shows "fronting". This effect can be avoided by adding acid or a buffer to the mobile phase, thus stabilizing the ratio of undissociated acid to anions, and with it the partition ratio.

5.6 ION-PAIR CHROMATOGRAPHY

Ion pairs are formed in water by the association of positive and negative ions. Evidence for their formation comes from electrical conductivity and light absorption. They are best formed between large organic ions of single charge. Ion pairs can be extracted from water by organic solvents, preferably polar solvents, like long-chain alcohols or chloroform, and they can be sorbed by hydrophobic sorbents, including alkyl-bonded silica and porous polymers, hence their importance in chromatography. The formation of ion pairs is best understood by invoking the hydrogen-bonded structure of water, in the same way that anion-exchange selectivity is understood (Section 5.3.7).

Ion-pair chromatography (also called ion-interaction chromatography) was first seen as an alternative to ion exchange. In the early and very important studies of Knox and colleagues [68-70] the authors wished to analyze mixtures of dyes and dyestuff intermediates that were aromatic sulfonic acids. Anion-exchange chromatography was the natural choice. However, even on high-performance surface-functional exchangers, the separation was inefficient and gave broad peaks with bad tailing. Knox et al. used a standard reversed-phase column of C_{18}-silica and, as the mobile phase, a dilute solution of a cationic surfactant, cetyltrimethylammonium bromide. Detection was by visible or UV absorbance. The negative analyte ions now emerged from the column in combination with the positive quaternary ammonium ions, and the peaks were very sharp. To analyze mixtures of positive ions, like those of catecholamines, sodium dodecyl sulfate was added to the mobile phase. In both cases, a solvent modifier was added, some 25% of methanol or 2-propanol.

The technique can be used to analyze inorganic ions, particularly large ions, like polyphosphate and perchlorate, but, in general, ion-pair chromatography is best suited to large organic ions, like those of alkaloids, peptides, antibiotics, and surfactants. These are ions that themselves could be used as pairing ions in the mobile phase.

The first interpretation of ion-pair chromatography was that the surfactant ions associated with the analyte ions in the mobile phase to form ion pairs of differing stability, which then distributed themselves to varying degrees between the mobile and the stationary phase. However, this idea failed to explain certain facts, one of which was that the eluent had to be passed through the column for a considerable time before the retention volumes became constant. It was also found that the nature and concentration of the counterion of the surface-active ion-pairing agent, like sodium ions in sodium dodecyl sulfate, influenced retention. If the sodium ion concentration was increased by adding a

References on p. A263

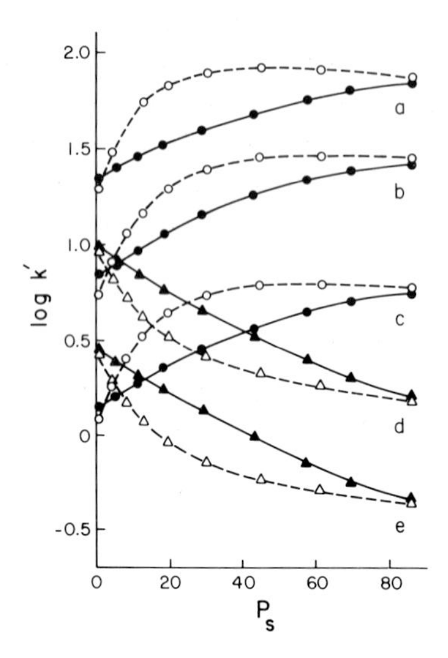

Fig. 5.7. Ion-pair chromatography of large anions and cations. Stationary phase, octadecyl-bonded silica, 5 μm; mobile phase, 0.025 *M* H_3PO_4/0.025 *M* NaH_2PO_4 plus added sodium butyl sulfonate and/or sodium bromide. Abscissae, P_s, are pairing-ion concentrations in the stationary phase, units μmol/gram. Ordinates, k', are column capacity factors (as logarithms). Solutes: (a) morphine; (b) tyrosine; (c) adrenaline; (d) *p*-toluenesulfonic acid; (e) benzenesulfonic acid. Note that the first three species retained are cations, the last two anions. (From Ref. 74, with permission.)

simple sodium salt, the retention of positive analyte ions was decreased, suggesting an ion-exchange displacement. Kissinger [71] proposed that the ion-pairing agent, the so-called "hetaeron", was adsorbed by the octadecyl chains of the bonded-silica stationary phase, along with its counterions, thus converting the bonded phase into an ion exchanger. The analyte ions were then separated by ion exchange. In contrast to conventional ion exchange, the ionic capacity of the stationary phase is not constant. An extensive treatment of the equilibria involved, along with supporting experimental data, is given in a paper by Knox and Hartwick [72].

A series of papers by Bartha, Vigh et al. [73-75] emphasizes the importance of the concentration of *sorbed* surfactant in the stationary phase, as well as concentrations of counterions and organic modifiers. Fig. 5.7 shows the effect of these factors on the retention of large organic ions. Note that all these ions are retained in the absence of surfactant. Adding the surfactant increases the retention of ions of opposite charge to the surfactant ions, and decreases the retention of ions having the same charge. These findings are consistent with a postulated ion-exchange model.

If the pairing agent is very hydrophobic and very strongly adsorbed, one can indeed use the ion-exchange model. This is true of the work of Cassidy and Elchuk [76] on the separation of lanthanide and transition-metal ions with "dynamically coated" and "permanently coated" ion exchangers. They used columns of bonded silica and also PRP-1 (a porous styrene/divinylbenzene polymer) and coated them with either sodium *n*-octane sulfonate or sodium C_{20} sulfonate by pumping solutions of these salts in methanol or acetonitrile through the columns, and then displacing the excess by water. The C_{20} sulfonate stuck so fast to the column that it remained permanently in place until it was washed out with an organic solvent. The C_8 sulfonate was not so strongly bound and had to be kept in place by making the eluent 0.005 *M* in sodium octane sulfonate. To separate lanthanide ions, they used α-hydroxyisobutyric acid in a concentration gradient from 0.05 *M* to 0.4 *M*. Very sharp peaks were obtained. The 14 lanthanides plus Y, U, and Th were separated in 16 min [79]; Th was eluted after Yb, U after Tm, Y after Ho. A great advantage of the "coating" technique is that the concentration of fixed ions and the ion-exchange capacity of the column can be changed at will and quite rapidly. The same coating technique was used for the chromatography of anions. A porous polymer was coated with cetylpyridinium ions to make an anion exchanger [80].

5.7 LIGAND-EXCHANGE CHROMATOGRAPHY

5.7.1 General

Ligand-exchange chromatography depends on the exchange of electron-donor ligands around a central metal ion. The exchange can occur in the stationary or the mobile phase. In its original form, introduced by Helfferich in 1961 [79], ligand-exchange chromatography was performed on a column of cation-exchange resin, loaded with ions of copper or nickel. The ligands that were exchanged were uncharged ligands, ammonia and amines. A mixture of amines could be separated by introducing a small amount of the mixture at the column inlet, then passing aqueous ammonia through the column. The amine that was most weakly coordinated to the metal ions was displaced first and emerged first from the column, accompanied by excess ammonia. The metal ions did not move, but remained on the column. Of course, they were displaced to a small extent by ion exchange with the ammonium ions that exist in an aqueous ammonia solution. However, loss of metal ions could be minimized by choosing an exchanger that held the metal ions strongly. Helfferich used a polyacrylate resin with functional carboxyl groups. Alternatively, the metal loss could be compensated by adding metal ions in low concentration to the ammonia influent.

The method is very versatile. One can choose the metal ions, the exchanger, and the displacing ligands. The ligands to be exchanged need not be uncharged; they can be singly charged, like ions of amino acids. More possibilities are opened by exchanging the ligands in the mobile phase. In this case the stationary phase need not be an ion exchanger. It can be a bonded silica reversed-phase packing that retains uncharged metal/ligand complexes but not ionic species. The metal ion chosen is usually Cu(II), because this ion forms the most stable complexes. The displacing ligand is generally ammonia.

References on p. A263

5.7.2 Optical isomers of amino acids

The most dramatic application of ligand-exchange chromatography is the separation of optical isomers of amino acids, developed primarily by Davankov et al. [80,81]. Davankov and colleagues first synthesized a solvent-modified, "macronet isoporous" styrene polymer with long and rigid crosslinks. By chloromethylation they attached to this polymer the naturally occurring *levo*-form of the amino acid proline, making a chiral ion exchanger that had the following structure

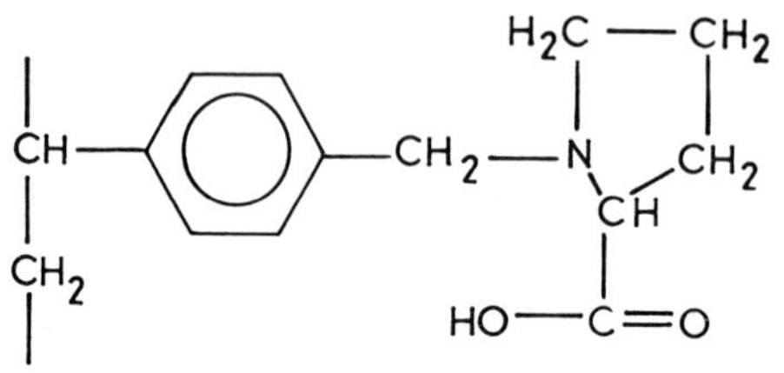

H_2C — CH_2
CH — C_6H_4 — CH_2 — N CH_2
CH
CH_2
HO — C=O

Through a column packed with this resin they passed a copper/ammonia salt solution. Copper ions were introduced up to a maximum of one Cu to two proline units. Each Cu bridged two polymer chains. In an early experiment [82] 0.5 g of racemic DL-proline was introduced into a 0.9 x 50 cm column and was washed down with water. Optically pure L-proline appeared at the column exit. When no more proline appeared, 1 *M* ammonia was passed through. Now, optically pure D-proline appeared. Sorption and displacement followed these equations

$$\text{RPro-Cu-ProR} + \text{Pro}^- = \text{RPro-Cu-Pro} + \text{RPro}^-$$
$$\text{RPro-Cu-Pro} + 2NH_3 = \text{RPro-Cu}(NH_3)_2^+ + \text{Pro}^-$$
$$\text{RPro-Cu}(NH_3)_2^+ + \text{RPro}^- = \text{RPro-Cu-ProR} + 2NH_3$$

In the second equation, it is understood that water molecules can take the place of ammonia molecules. Water can displace L-proline, which is weakly bound; the stronger base, ammonia, is required to displace D-proline.

Of course, any α-amino acid can take the place of proline in the solution, and all amino acids can be resolved into their D- and L-forms with varying efficiencies. Separation factors range from near unity (no separation) to as high as 7. Other cyclic amino acids can be substituted for L-proline as the fixed ligand. Extensive tables of separation factors have been published [83]. Except for histidine, all the L-isomers are eluted first; i.e., in the exchanger the LL-combination is less stable than the DL-combination. Experiments in aqueous solution with the model compound *N*-benzylproline show unequivocally that the L-Cu-D complex is more stable than the L-Cu-L complex by some 6 kJ per mole, which corresponds to a separation factor of 10 [86].

Chromatography on chiral resins is inefficient, because mass transfer is slow. (Nevertheless, the capacity is high, and the separation is excellent for the preparative scale.) Davankov achieved faster mass transfer and much sharper chromatographic peaks by taking a C_{18}-bonded silica and impregnating it with a long-chain *N*-alkyl proline (see Section 5.6), then adding Cu(II) [85]. The mobile phase was 15% methanol, 10^{-4} *M* in copper(II) acetate. Gubitz [86] obtained similar results by attaching L-proline to porous silica.

All these methods put the chiral resolving agent, L-proline, in the stationary phase. A different approach is to add L-proline to the mobile phase, that is, to use a chiral eluent. The stationary phase is a standard reversed-phase packing, C_8- or C_{18}-silica; the mobile phase contains proline and Cu(II) acetate in the mole ratio 2:1. Here, the D-form of the other amino acid is eluted first, indicating again that the DL-combination is more stable in solution than the LL-combination. Separation factors are only 1.5 or less, but the peaks are sharp, and resolution is excellent. Analysis times are $<$ 30 min [87].

Among the first researchers to use a chiral eluent were Hare and Gil-Av [88]. They used as the stationary phase a gel-type ion-exchange resin. The mobile phase was 0.004 *M* Cu(II)/0.008 *M* proline/0.05 *M* sodium acetate (pH 5.5). With L-proline in the mobile phase the L-isomers were eluted first. It appeared that the resin bound the 1:1 complexes, CuL^+, but not the uncharged 1:2 complexes, CuL_2. Analysis times were up to 90 minutes, and separation factors were very high.

In all these analyses the amino acids were detected by postcolumn reaction with *o*-phthalaldehyde and ethanethiol, which react with primary amines to give strongly fluorescent products [89]. Proline and other secondary amines do not respond, nor does ammonia.

5.7.3 Carbohydrates

Another application of ligand-exchange chromatography, one that is used on the manufacturing scale, is the separation of sugars and carbohydrates on columns of cation-exchange resins, loaded with calcium ions. The complexes are very weak and the eluent is pure water. It must be noted that ligand exchange is not the only mechanism. Hydrogen bonding and interaction with the water in the resin are important too, as is size exclusion [90]. Nevertheless, it is indisputable that sugars that have favorable orientations of their hydroxyl groups, like ribose in its 6-membered ring form, which has the axial/equatorial/axial configuration, form coordination complexes with calcium ions that cause them to be bound to the resin [91]. Water is the displacing ligand in their chromatography. Sugar alcohols, like mannitol and sorbitol, form relatively strong complexes with Ca(II) and are bound correspondingly strongly. But, in contrast to the Cu/amino acid complexes, having formation constants of the order 10^8 l/mol (for the acid anions), the formation constant of the Ca/sorbitol complex is about 1 l/mol [90].

References on p. A263

Chromatography on calcium-loaded resins with water as eluent and refractive index detection is routinely used in the analysis of corn syrup and other sugar products [92]. An alternative method that is gaining acceptance is anion exchange with strongly alkaline eluents and pulsed amperometric detection (Section 5.4.4). The first method to use ion exchange for sugar analysis was with a strong-base anion-exchange resin in the borate form. Borate ions form cyclic esters with *cis*-diols, like sugar and sugar alcohols, and thus bind sugars to the resin. Recently this method has been adapted to automated analysis of sugars [93]. It could, perhaps, be classified as "ligand exchange". A general account of ligand-exchange chromatography is given in Ref. 80.

5.8 SPECIAL APPLICATIONS

Applications of ion-exchange chromatography to specific compound classes are described in other chapters of this book, with full references. Here we offer a few general remarks on methodology.

5.8.1 Amino acids

Underivatized amino acid mixtures are analyzed by cation-exchange chromatography, and detected by postcolumn derivatization with a color-producing reagent (ninhydrin) or one that produces fluorescence (*o*-phthalaldehyde plus ethanethiol). Amino acids are amphoteric ions that can either gain protons to form cations or lose protons to form anions. Cations are formed in the low-pH region. To separate amino acids, a pH gradient is used, either continuous or in steps, starting at pH 3.25 or thereabouts and moving upwards. The pH affects the ionic charge and the proportion of the amino acid in the cationic form. The salt concentration also affects the elution, because the cations of the salt, usually Na^+ or Li^+, compete with amino acid cations for the exchanger sites. The exchangers used are high-capacity gel-type resins in particle sizes below 10 μm.

The first amino acids to be eluted are those having two carboxyl groups, the dibasic amino acids, glutamic and aspartic. The last to be eluted are those having two amino groups, the dinitrogen amino acids, lysine, arginine, histidine, and tryptophan. These are held so strongly that it is customary to analyze them on a separate column, using a second sample and strong eluent. In between are the neutral amino acids. The order of elution of these acids shows clearly the effects of hydrophobic interaction, hydrogen bonding, and the attraction of aromatic rings (in tyrosine and phenylalanine) to the matrix of the polystyrene resin.

Amino acid analysis by ion exchange has been perfected to a fine art, in which small effects of pH, buffer composition, and temperature are manipulated to give adequate peak separations throughout the elution sequence. The mixtures found in physiological fluids are more complex and present a greater challenge than the mixtures in protein hydrolyzates.

5.8.2 Proteins

Protein molecules, like those of the amino acids of which they are made, can exist as cations or anions. They can be analyzed by anion-exchange or cation-exchange chromatography. The binding to ion exchangers depends on pH and on the concentration of competing ions, i.e., on the salt concentration. Generally they are eluted from ion exchangers at constant pH with a salt gradient, say from zero to 0.6 *M* NaCl. Elution gives very sharp peaks, because the attachment of a large protein molecule to an ion exchanger takes place at several sites at once. Thus, the distribution ratio depends on a high power of the concentration of the competing ions [94].

Another feature of the interaction between proteins and ion exchangers is that the pH of zero charge, as indicated by ion exchange, is not necessarily the same as the pH of zero charge shown by electrophoresis, the isoelectric point. This fact indicates that the distribution of ionic charges over the surface of a large protein molecule is not uniform, but that the region involved in binding may have a different population of carboxyl and amino groups from the molecule as a whole. It is the overall balance between positive and negative ions that determines the electrophoretic mobility. In other words, electrophoresis probes the entire surface of the protein macromolecule, whereas chromatography "sees" only a portion of it [95].

Finally, and most importantly, special ion exchangers are used for the chromatography of proteins. These exchangers have hydrophilic character, so as not to denature the proteins, and they have their ionic groups at the end of long molecular chains, so that they can make contact with the ionic sites of the proteins. Both silica-based and polymer-based exchangers are used. Fig. 5.1 shows examples of such exchangers. Weakly basic and weakly acidic exchangers are often preferred, because controlling the pH gives better control of elution; see Section 5.2.2.2 for a discussion of the ionization of weak-acid and weak-base exchangers.

5.8.3 Anion exchange of metal chloride complexes

In previous editions of this book, much space has been devoted to the separation of metal ions by anion exchange in HCl solutions. Interest in such separations has receded with the development of ion chromatography, but there are still circumstances under which these separations are useful, such as the measurement of low concentrations of certain metals in the presence of large concentrations of others, the collection of traces from matrices such as sea water, and post-irradiation separations in activation analysis.

Most metal ions form anionic complexes with chloride ions. These complexes vary enormously in stability and in their strength of binding by anion exchangers. At one extreme are the ions of alkali and alkaline-earth metals, Y, and the lanthanides; these form no chloride complexes at all in aqueous solution. At the other extreme are the triply charged ions of Fe, Ga, In, Tl, and Au, which form singly charged anions MCl_4^-, which in turn are held very strongly by anion exchange. (This behavior could be expected from the

References on p. A263

ideas advanced in Section 5.3.7.) The distribution ratio between the resin and the solution depends greatly on the HCl concentration, and can range from near zero up to 10^6.

Thus, many separation schemes have been devised in which one takes a short open glass column of strong-base anion-exchange resin, washed with HCl, and places a small amount of a mixture of metal salts on it. One washes the metal salts down with HCl of various concentrations, starting with a high concentration, say 9 *M* or 6 *M*, and proceeding downward. The metal that is most weakly held comes out first, and the metal that is held most strongly emerges last. The individual metals are recovered as solutions in aq. HCl. Excess acid is removed by evaporation, and the amount of metal is measured by any method of choice, the most accurate, in general, being EDTA titration.

The strengths of binding and the elution orders can be drastically altered by the use of nonaqueous solvents. For example, Fe(III) is sorbed by a strong-base anion-exchange resin from aq. 9 *M* HCl with distribution ratio 10^4, and Ni(II) is not sorbed at all. From 1 *M* HCl in 80% aq. acetone, Ni is sorbed significantly, while Fe(III) is not held at all. The likely interpretation is that in acetone and other electron-donor solvents Fe(III) forms uncharged ion pairs, $SH^+FeCl_4^-$ where S is the solvent molecule, while electrostatic attractions in general are strengthened by the lowering of the dielectric constant, thus stabilizing $NiCl_4^{2-}$.

The first systematic study of anion exchange of metal-chloride complexes was made by Kraus and Moore [96]. In 1955 Kraus and Nelson [97] published a summary account that is a classic. It is often cited and is reproduced verbatim in a collection of papers on ion-exchange chromatography [98]. A feature of this paper is a chart in the form of the periodic table that has graphs of distribution ratios (Section 5.3.5) vs. HCl concentration for each element that is retained. Later these authors measured distributions of metals between HBr and HF [99] and cation and anion exchangers. Korkisch and coworkers [100-102] explored the effects of nonaqueous solvents on anion and cation exchange of metal complexes and devised numerous methods for isolating trace metals from complex matrices. He developed "combined ion exchange and solvent extraction (CIESE)" [103]. Strelow et al. [104-106] applied ion exchange to the complete analysis of rocks and minerals as well as to the isolation of trace metal impurities from highly pure materials.

These separation schemes are long and tedious, yet they provide definite, unequivocal analyses that can be used to check and calibrate faster physical methods. It is hard to adapt them to high-performance chromatography, because they use corrosive eluents and require frequent eluent changes.

5.9 CONCLUSION

Only a few leading references are quoted in this chapter. It remains to cite books on chemical analysis that describe ion exchange in a general way [107,108] as well as a recent, highly authoritative book on ion chromatography [109] that includes concise accounts of the principles of ion exchange and the chromatographic process and Ref. 110, a comprehensive treatise that lists many applications.

REFERENCES

1 B.A. Adams and E.L. Holmes, *J. Soc. Chem. Ind. (London),* 54 (1935) 1T.
2 G.F. D'Alelio, U.S. Patents 2,366,007 and 2,366,008 (1944).
3 R.E. Sturgeon, S.S. Berman, J.A.H. Desaulniers, A.P. Mykytiuk, J.W. McLaren and D.S. Russell, *Anal. Chem.,* 52 (1980) 1585.
4 S.K. Sahni and J. Reedijk, *Coord. Chem. Rev.,* 59 (198) 1.
5 F. Vernon and H. Eccles, *Anal. Chim. Acta,* 63 (1973) 403.
6 G. Koster and G. Schmuckler, *Anal. Chim. Acta,* 38 (1967) 179.
7 A. Warshawsky, M.M.B. Fieberg, P. Mihalik, T.G. Murphy and Y.B. Rao, *Sep. Purif. Methods,* 9 (1980) 209.
8 M. Griessbach and K.H. Lieser, *Fresenius Z. Anal. Chem.,* 302 (1950) 109, 191, 184.
9 O. Mikes, P. Strop and J. Coupek, *J. Chromatogr.,* 153 (1978) 23.
10 O. Mikes, P. Strop, J. Sbrozek and J. Coupek, *J. Chromatogr.,* 180 (1979) 17.
11 O. Mikes, P. Strop, M. Smrz and J. Coupek, *J. Chromatogr.,* 192 (1980) 159.
12 J.C. Janson, *Chromatographia,* 23 (1987) 361.
13 R.H. Wiley and E.E. Sale, *J. Polym. Sci.,* 42 (1960) 479, 491.
14 J.R. Millar, D.G. Smith, W.E. Marr and T.R.E. Kressman, *J. Chem. Soc. (London),* (1963) 218, 2779; (1964) 2740.
15 J.R. Millar, D.G. Smith and T.R.E. Kressman, *J. Chem. Soc. (London),* (1965) 304.
16 J. Seidl, J. Malinsky, J. Dusek and W. Heitz, *Adv. Polym. Sci.,* 5 (1967) 113.
17 K.K. Unger, *Porous Silica,* Elsevier, Amsterdam, 1979.
18 G.B. Cox, *J. Chromatogr. Sci.,* 15 (1977) 385.
19 G.B. Cox, C.R. Loscombe, M.J. Slucutt, K. Sugden and J.A. Upfield, *J. Chromatogr.,* 117 (1976) 269.
20 P.A. Asmus, C.E. Low and M. Novotny, *J. Chromatogr.,* 123 (1976) 109.
21 M. Caude and R. Rosset, *J. Chromatogr. Sci.,* 15 (1977) 405.
22 J.P. Lefevre, A. Divry, M. Caude and R. Rosset, *Analusis,* 3 (1975) 533.
23 P.A. Asmus, C.E. Low and M. Novotny, *J. Chromatogr.,* 119 (1976) 25.
24 P. Gareil, A. Heriter, M. Caude and R. Rosset, *Analusis,* 4 (1976) 71.
25 S.H. Chang, R. Noel and F.E. Regnier, *J. Chromatogr.,* 120 (1976) 321; 125 (1976) 103.
26 J.R. Jezorek and H. Freiser, *Anal. Chem.,* 51 (1979) 366, 373.
27 F.K. Chow and E. Grushka, *Anal. Chem.,* 50 (1978) 1346.
28 F.G. Helfferich, *Ion Exchange,* McGraw-Hill, New York, 1962, Chap. 6.
29 C. Horváth, B. Preiss and S.R. Lipsky, *Anal. Chem.,* 39 (1967) 1422.
30 H. Small, T.S. Stevens and W.C. Bauman, *Anal. Chem.,* 47 (1975) 1801.
31 M.E. Potts, S.S. Heberling, J.R. Stillian, C.A. Pohl, V.E. Summerfeldt, V.B. Barreto and C.R. Deveza, paper presented at Pittsburgh Conference, Atlanta, GA, 1989.
32 D.T. Gjerde, J.S. Fritz and G. Schmuckler, *J. Chromatogr.,* 186 (1979) 509.
33 F.G. Helfferich, *Ion Exchange,* McGraw-Hill, New York, 1962, Chap. 5.
34 D. Reichenberg, *Ion Exchange: A Series of Advances,* Dekker, New York, 1966, Vol. 1, Chap. 7.
35 O.D. Bonner and L.L. Smith, *J. Phys. Chem.,* 61 (1975) 326.
36 G. Schmuckler and S. Goldstein, *Ion Exchange and Solvent Extraction,* Dekker, New York, 1977, Vol. 7, Chap. 1.
37 P. Hajos and J. Inczedy, *Ion Exchange Technology,* Ellis Horwood, Chichester, 1984, p. 450.
38 R.E. Barron and J.S. Fritz, *J. Chromatogr.,* 316 (1984) 201.
39 G. Eisenman, *Advances in Analytical Chemistry and Instrumentation,* Wiley, New York, 1965, Vol. 4, p. 213.
40 B.H. Ketelle and G.E. Boyd, *J. Am. Chem. Soc.,* 69 (1947) 2800.
41 S. Moore and W.H. Stein, *J. Biol. Chem.,* 192 (1951) 663.
42 R.D. Rocklin and D.L. Campbell, paper No. 1018 presented at the Pittsburgh Conference on Analytical Chemistry, New Orleans, LA, 1988.
43 J. Stillian, *LC Magazine, Eugene, Oregon,* 3 (1985) 802.

44 J.S. Fritz, D.T. Gjerde and R.M. Becker, *Anal. Chem.*, 52 (1980) 1519.
45 G. Schmuckler, A.L. Jagoe, J.E. Girard and P.E. Buell, *J. Chromatogr.*, 356 (1986) 413.
46 T. Okada and T. Kuwamoto, *Anal. Chem.*, 57 (1985) 258, 829.
47 H. Small and T.E. Miller, *Anal. Chem.*, 54 (1982) 462.
48 T.A. Walker, *J. Liq. Chromatogr.*, 11 (1988) 1513.
49 J.H. Sherman and N.D. Danielson, *Anal. Chem.*, 59 (1987) 490.
50 R.D. Rocklin and E.L. Johnson, *Anal. Chem.*, 55 (1983) 4.
51 D.A. Roston and P.T. Kissinger, *Anal. Chem.*, 54 (1982) 429.
52 S. Hughes and D.C. Johnson, *Anal. Chim. Acta*, 132 (1981) 11.
53 R.D. Rocklin and C.A. Pohl, *J. Liq. Chromatogr.*, 6 (1983) 1577.
54 R.M. Cassidy and S. Elchuk, *J. Chromatogr. Sci.*, 18 (1980) 217.
55 S. Elchuk and R.M. Cassidy, *Anal. Chem.*, 51 (1979) 1434.
56 D.J. Woo and J. R. Benson, *J. Chromatogr. Sci.*, 22 (1984) 386.
57 D.T. Gjerde and J.S. Fritz, *Ion Chromatography*, Huethig, Heidelberg, 2nd Edn., 1987.
58 F.C. Smith, Jr., and R.C. Chang, *The Practice of Ion Chromatography*, Wiley, New York, 1983.
59 J. Weiss, *Handbook of Ion Chromatography*, translated by E.L. Johnson, Dionex Corp., Sunnyvale, CA, 1986. (For an update see J. Weiss, *Fresenius Z. Anal. Chem.*, 327 (1987) 451.)
60 R.E. Smith, *Ion Chromatography Applications*, CRC Press, Boca Raton, FL, 1988.
61 O. Shpigun and Yu.A. Zolotov, *Ion Chromatography in Water Analysis*, Halsted Press (Wiley), New York, 1988.
62 T.M. Florence and G.E. Batley, *Talanta*, 23 (1976) 179.
63 M. Legrand, M. DeAngelis and R.J. Delmas, *Anal. Chim. Acta*, 156 (1984) 181.
64 V.T. Turkelson and M. Richards, *Anal. Chem.*, 50 (1978) 1420.
65 D.J. Woo and J.R. Benson, *LC Magazine, Eugene, Oregon*, 1 (1983) 238.
66 R.E. Hannah, *J. Chromatogr. Sci.*, 24 (1986) 336.
67 J.R. Kreling and J. DeZwaan, *Anal. Chem.*, 58 (1986) 3028.
68 J.H. Knox and G.R. Laird, *J. Chromatogr.*, 122 (1976) 17.
69 J.H. Knox and J. Jurand, *J. Chromatogr.*, 125 (1976) 89.
70 J.H. Knox and J. Jurand, *J. Chromatogr.*, 149 (1978) 297.
71 P.T. Kissinger, *Anal. Chem.*, 49 (1977) 883.
72 J.H. Knox and R.A. Hartwick, *J. Chromatogr.*, 204 (1981) 3.
73 A. Bartha and Gy. Vigh, *J. Chromatogr.*, 260 (1983) 337; 265 (1983) 171.
74 A. Bartha, H.A.H. Billiet, L. de Galan and Gy. Vigh, *J. Chromatogr.*, 291 (1984) 91; 303 (1984) 29.
75 A. Bartha and Gy. Vigh, *Chromatographia*, 20 (1985) 587.
76 R.M. Cassidy and S. Elchuk, *Anal. Chem.*, 54 (1982) 1558.
77 D.J. Barkley, M. Blanchette, R.M. Cassidy and S. Elchuk, *Anal. Chem.*, 58 (1986) 2222.
78 R.M. Cassidy and S. Elchuk, *J. Chromatogr.*, 262 (1983) 311.
79 F. Helfferich, *Nature (London)*, 189 (1961) 1001.
80 V.A. Davankov, J.D. Navratil and H.F. Walton, *Ligand Exchange Chromatography*, CRC Press, Boca Raton, FL, 1988.
81 V.A. Davankov, A.A. Kurganov and A.S. Bochkov, *Advances in Chromatography*, Dekker, New York, 1984, Vol. 22, Chap. 3.
82 V.A. Davankov and S.V. Rogozhin, *Dokl. Akad. Nauk SSSR, Chemistry Proceedings*, 193 (1971) 94; English translation, Consultants Bureau, p. 460.
83 V.A. Davankov and Yu.A. Zolotarev, *J. Chromatogr.*, 155 (1978) 285, 295, 303.
84 A.A. Kurganov, L.Ya. Zhuchova and V.A. Davankov, *J. Inorg. Nucl. Chem.*, 40 (1978) 1081.
85 V.A. Davankov, A.S. Bochkov, A.A. Kurganov, P. Roumeliotis and K.K. Unger, *Chromatographia*, 13 (1980) 677.
86 G. Gubitz, *J. Liq. Chromatogr.*, 9 (1986) 519.

87 E. Oelrich, H. Preusch and E. Wilhelm, *J. High Resolut. Chromatogr. Chromatogr. Commun.,* 3 (1980) 269.
88 P.E. Hare and E. Gil-Av, *Science,* 204 (1979) 1226.
89 J.R. Bensen and P.W. Hare, *Proc. Nat. Acad. Sci. U.S.,* 72 (1975) 619.
90 H.F. Walton, *J. Chromatogr.,* 332 (1985) 203.
91 R.W. Goulding, *J. Chromatogr.,* 103 (1975) 229.
92 L.E. Fitt, W. Hassler and D.E. Just, *J. Chromatogr.,* 187 (1980) 381.
93 S. Honda, M. Takahashi, K. Kakehi and S. Ganno, *Anal. Biochem.,* 113 (1981) 140.
94 W. Kopaciewicz, M.A. Rounds, J. Fausnach and F.E. Regnier, *J. Chromatogr.,* 266 (1983) 3.
95 P.G. Righetti, *Isoelectric Focusing: Theory, Methodology and Applications,* Elsevier, Amsterdam, 1983.
96 K.A. Kraus and G.E. Moore, *J. Am. Chem. Soc.,* 75 (1953) 1460.
97 K.A. Kraus and F. Nelson, *Proc. 1st UN Conf. on Peaceful Uses of Atomic Energy,* 7 (1955) 113.
98 H.F. Walton (Editor), *Ion-Exchange Chromatography. Benchmark Papers in Analytical Chemistry,* Vol. 1, Dowden, Hutchinson and Ross, Stroudsburg, PA, 1976.
99 F. Nelson, R.M. Rush and K.A. Kraus, *J. Am. Chem. Soc.,* 82 (1960) 339.
100 I. Hazan and J. Korkisch, *Anal. Chim. Acta,* 32 (1965) 46.
101 W. Koch and J. Korkisch, *Mikrochim. Acta,* (1973) 245.
102 J. Korkisch, *CRC Handbook of Ion-Exchange Resins,* Vols. I-VI, CRC Press, Boca Raton, FL, 1989.
103 J. Korkisch, *Separ. Sci. Technol.,* 1 (1966) 159.
104 F.W.E. Strelow, in J.A. Marinsky and Y. Marcus (Editors), *Application of Ion Exchange to Element Separation and Analysis, Ion Exchange and Solvent Extraction,* Vol. 5, Dekker, New York, 1973, Chap. 2.
105 F.W.E. Strelow, C.J. Liebenberg and A.H. Victor, *Anal. Chem.,* 46 (1974) 1409.
106 F.W.E. Strelow, *Anal. Chim. Acta,* 183 (1986) 307.
107 J.X. Khym, *Analytical Ion-Exchange Procedures in Chemistry and Biology,* Prentice-Hall, Englewood Cliffs, NJ, 1974.
108 H.F. Walton and R.D. Rocklin, *Ion Exchange in Analytical Chemistry,* CRC Press, Boca Raton, FL, 1990.
109 H. Small, *Ion Chromatography,* Plenum Press, New York, 1989.
110 P.R. Haddad and P.E. Jackson, *Ion Chromatography,* Elsevier, Amsterdam, 1990.

Chapter 6

Size-exclusion chromatography

LARS HAGEL and JAN-CHRISTER JANSON

CONTENTS

6.1 INTRODUCTION

Size-exclusion chromatography (SEC) is defined as the differential elution of solutes, from a bed of a porous chromatographic medium, caused by different degrees of steric exclusion of the solutes from the porevolume due to the differential molecular size of solutes. Thus, in SEC, solutes are eluted strictly according to decreasing molecular size, and the maximum available volume for separation is equal to the total porevolume of the packing medium.

Although size-exclusion effects were already recognized for zeolites, charcoal, and ion-exchange resins in the Forties, and were discussed as the basis for possible separations of biological molecules [1], it was not until the middle of the Fifties that such phenomena were occasionally applied to biochemical separations [2,3]. More systematical studies, which led to commercially available materials, were performed by Porath and Flodin in 1959 (Sephadex®) [4], Hjertén and Mosbach in 1962 (BioGel P) [5], and Hjertén in 1962 (Sepharose®) [6]. A historical review of the development of Sephadex was recently given by Janson [7]. Some important contributions to the development of media, especially designed for aqueous SEC, are listed in Table 6.1.

The need for a good method of obtaining molecular weight distributions (MWDs) of synthetic polymers, soluble in organic solvents, prompted Moore [22,23] to develop

TABLE 6.1

HISTORICAL REVIEW OF SEC GEL MATERIALS DESIGNED FOR BIOPOLYMERS

Year	Chemical composition	Ref.
1955	Starch	2
1956	Starch	3
1959	Crosslinked dextran	4
1961	Agar	8
1962	Crosslinked polyacrylamide	5
1962	Agarose	6
1965	Controlled-porosity glass	9
1966	Agarose/polyacrylamide	10
1971	Crosslinked agarose	11
1973	Hydroxyethyl methacrylate	12
1974	Poly(acryloyl morpholine)	13
1976	Glycerylpropyl silica	14
1976	Poly(*N,N*-bisacrylamide)/dextran	15
1978	Hydrophilic vinyl polymer	16
1978	Hydrophilic bonded porous silica	17
1978	Regenerated cellulose	18
1979	Poly(acryloylamino-hydroxymethyl-propanediol)	19
1984	Heavily crosslinked agarose	20
1988	Crosslinked agarose/dextran	21

beads of lightly crosslinked polystyrene of suitable poresizes in the early Sixties, thus extending SEC to nonaqueous solvents.

For the past three decades, SEC has been applied to numerous tasks, such as determination of molecular weight (MW) and MWDs, buffer exchange, studies of solute/solute interactions, concentration of solutes, determination of solute diffusivity and solute shape, determination of pore dimensions and porevolume and, of course, to the fractionation and purification of solutes.

The considerable popularity of SEC in biochemical purification methods can be ascribed to two characteristics of this technique. Firstly, its very high recovery, which, even for sensitive proteins, often approaches 100% on gels consisting of natural polymers, and secondly, its ability to remove undesirable aggregates (dimers, oligomers, etc.) from protein products. The latter application has made SEC the premier "final polishing" step, in industrial-scale purification processes. The major alternative sieving techniques are ultrafiltration and dialysis, both of which utilize membranes of different shapes and configurations. It is fair to say that membrane techniques in general are superior to SEC when it comes to handling large volumes and when the solutes differ greatly, i.e. a decade, in molecular weight. However, whenever total, or close to total, removal of contaminant species is required and feasible on the basis of size- and shape-dependent separation, SEC is clearly the method of choice.

The major application area of nonaqueous SEC is the characterization of polymer composition (prepolymers in resins, polymer additives, etc.), and polymer reactions, together with semipreparative fractionations of polymers (Section 6.4.1).

SEC has been explored from several different perspectives, and this unfortunately led to ambiguous nomenclature (e.g., gel filtration, molecular sieve chromatography, gel chromatography, gel-permeation chromatography, etc.). In the pioneering work by Porath and Flodin the technique was termed gel filtration, as proposed by Arne Tiselius [24]. The application of SEC to separations in nonaqueous solvents was called gel-permeation chromatography [23]. These designations are still sometimes used to differentiate between the application of SEC to separations in aqueous and nonaqueous solutions. However, as the term size-exclusion chromatography is more descriptive and universal it can be recommended for general purposes.

The trend in chromatographic media has been towards smaller particle size to increase the separation efficiency, both for analytical and preparative purposes. This has primarily been utilized for a substantial decrease in the separation time from, originally, 24 h to less than an hour for fast SEC. This has led to a very rapid adoption of the HPLC mode of SEC. Very fast SEC has been suggested for on-line process control with response times of a few seconds. Another trend is the tailoring of media to specific application areas (e.g., separation of IgG from transferrin in the purification of monoclonal antibodies), where an optimum selectivity rather than a minimum particle size is desired. A trend towards miniaturization of chromatographic systems for analytical SEC is also noticeable, and microcolumn SEC has been applied in studies of very small amounts of substances at low solvent consumptions.

References on p. A304

The major applications of SEC are now focused in four areas: determination of MWDs, group separations (e.g., DNA from restriction enzymes), buffer exchange (e.g., desalting), and fractionation of solutes (e.g., monomers from oligomers).

During the last decade our understanding of the separation process in SEC has increased considerably, yet some fundamental questions remain to be answered, e.g., what physico-chemical property of solutes is the decisive parameter in separations by SEC, is there an optimal pore structure of SEC media, etc. Nevertheless, it is now possible to predict the results of SEC separations fairly accurately.

The aim of this chapter is to provide practical guidelines for optimization of the separation system for various modes of SEC and to illustrate the prediction of results on the basis of recent advances in the theory of SEC for simulation purposes. Present applications of major interest will also be treated, with special reference to aqueous SEC of macromolecules.

6.2 THEORETICAL ASPECTS

As opposed to other liquid chromatographic techniques, the separation in SEC is thermodynamically driven by entropy only [25] and may be pictured by imagining the reduced internal volume of the porous medium that will be available for penetration as solutes become progressively larger. Solutes will thus be eluted according to decreasing molecular size, and furthermore, the conformational loss in entropy predicts that the distribution coefficient in SEC, K_D, will vary linearly with solute size in a double logarithmic plot [26]. The distribution coefficient can be related to the standard free energy change, $\Delta G°$ for the transfer of solute molecules from the mobile phase to the stagnant liquid phase within the pores according to [27]

$$K_D = \exp(-\Delta G°/\mathcal{R}T) = \exp(-\Delta H°/\mathcal{R}T + \Delta S°/\mathcal{R}) \tag{6.1}$$

where K_D is the relative concentration of solutes in the stagnant phase as compared to that of the mobile phase, at equilibrium, $\mathcal{R}$ is the gas constant, T is the temperature, and $\Delta H°$ and $\Delta S°$ are, respectively, the standard enthalpy and entropy differences between solutes in the two phases. It is generally assumed that no change in enthalpy takes place in ideal SEC, and therefore, K_D will only be influenced by entropy changes as solutes move from one phase to the other (i.e. $K_D = \exp(\Delta S°/\mathcal{R})$. Thus, the temperature should generally not influence the retention volume in SEC. However, increases in temperature may induce increases in the hydrodynamic radius of polymers and in this way influence the elution volume. This was the explanation offered for the slight, i.e. 2-6%, decrease in the elution volume observed for dextran when the temperature was increased from ambient to 60°C [28]. However, large influences of the temperature on the elution volume are indicative of enthalpy effects caused by solute/matrix interactions [29].

6.2.1 Inherent features and limitations of size-exclusion chromatography

From the theory of SEC outlined above, it is obvious that this chromatographic technique is very useful for the determination of solute size as well as for the separation of solutes according to size, provided that no enthalpy effects are present. Furthermore, operational conditions, such as mobile-phase composition, temperature, flow velocity, etc., will not influence the retention volume in ideal SEC as long as these parameters do not influence the size of the solutes or the pore structure of the separating medium. However, this also means that the selectivity and the size separation range are not adjustable but are solely related to the poresize distribution of the SEC medium. Therefore, media of different poresizes, covering different fractionation ranges, are often supplied (Section 6.3.2).

Due to the absence of enthalpy effects, the distribution coefficient will vary between 0 (for molecules that are excluded from the stagnant liquid phase, i.e. for which $\Delta S^\circ \ll 0$) and 1 (for molecules that may permeate the total stagnant liquid phase, i.e. for which $\Delta S^\circ = 0$). This range corresponds to the breakthrough or void volume, V_o, and the total liquid volume, V_t, of the chromatographic bed. The effective separation volume in SEC is thus limited by the porevolume of the chromatographic medium used (see Fig. 6.1).

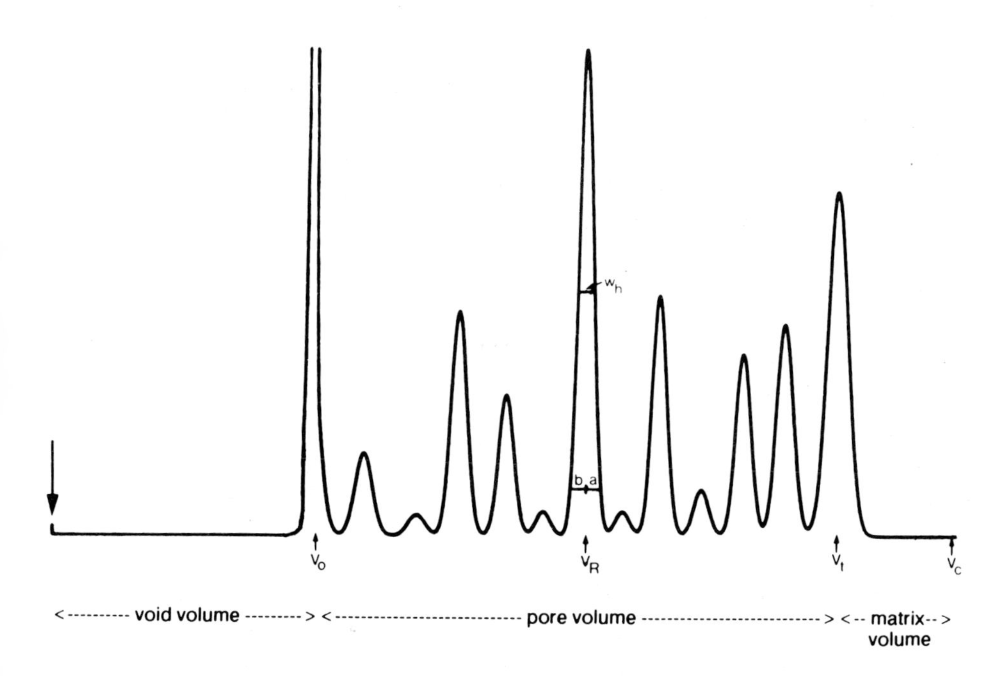

Fig. 6.1. Separation volume in SEC. Solutes are ideally eluted between the void volume, V_o, and the total liquid volume of the column V_t. In adsorptive modes of chromatography (i.e. RPC, IEC, etc.) solutes are eluted after V_t (V_t corresponds to $k' = 0$). The retention, or elution volume of the solute, V_R, the peakwidth at half peakheight, W_h, and a and b, the partial peakwidths at 10% peakheight, are used for calculation of the column performance (Section 6.3.3). The figure also illustrates the limited peak capacity of SEC, i.e., $n_{1.5} \approx 13$ (cf. Eqn. 6.3).

6.2.1.1 Characterization of size

The separation principle of SEC makes it invaluable for the characterization of the size of single molecules and size distribution of polymers. Since the separation is regulated by the size of the solvated solute, solutes of similar molecular weight but different shape will be eluted at different retention volumes, rod-shaped molecules being eluted earlier than flexible-coil solutes, which, in turn, will be eluted earlier than spherically shaped molecules [30]. This phenomenon may be illustrated by the behavior of double-stranded DNA fragments of increasing length (see Fig. 6.2). Up to ca. 18 basepairs the shape of the molecule is globular, and the elution behavior is similar to that of globular proteins. Larger fragments behave as stiff rods and are eluted much earlier than proteins of corresponding molecular weights [31]. As the length of the molecule increases, the flexibility also increases, and above ca. 150-200 basepairs the molecule will behave more like a flexible coil [31,32]. Thus, when determining the molecular weights of unknown solutes, it is very important to realize the impact of solute shape (which may be affected by chainlength, degree of branching, tertiary structure, etc.) on the elution volume. This problem can be obviated by using the column only as a separating medium and determining the absolute molecular weight of the eluites by on-line low-angle laser light scattering, provided that the refractive index increment of the solute is known [33]. It may in this context be noted that proteins are, with regard to shape, a very inhomogeneous class of substances. Thus, ferritin has been characterized as a perfect sphere [34], fibrous proteins have the shape of rigid rods, and globular proteins are more or less cigar-shaped. Still, many proteins fall on a common calibration curve [35], and SEC has proven to be a reliable technique for the estimation of molecular weights of, especially, globular proteins [36]. The shape of proteins may be normalized to random coils after reduction of S-S bridges and treatment with denaturating agents, such as 6 *M* guanidine hydrochloride [37]. SEC in this mode

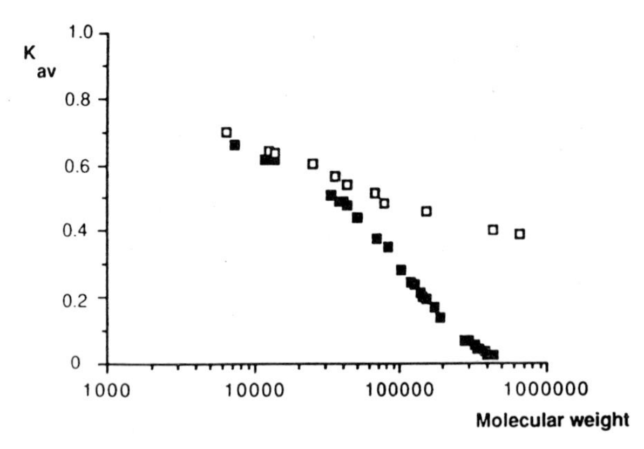

Fig. 6.2. Selectivity curves obtained from SEC of DNA restriction fragments (■) and standard proteins (□) on a 10.6 x 1-cm-ID Superose® 6 column with 0.02 *M* Tris-HCl (pH 7.6), containing 0.15 *M* NaCl. The DNA fragments were derived from digestion of pBR322 with HaeIII. (Reproduced from Ref. 31, with permission.)

has been shown to be an alternative to SDS/electrophoresis for determination of protein size [38].

The assay of MWDs of natural as well as synthetic polymers was among the first applications described for SEC (Section 6.4.1). The elution curve of a sample offers qualitative information about the MWD, which may be used for the purpose of comparison between samples. The MWD of the sample and various estimates of the MWD [39], e.g., the weight-average molecular weight, M_w, the number-average molecular weight, M_n, the centrifugation-average molecular weight, M_z, and the polydispersity of the distribution, M_w/M_n, may be calculated from the elution curve, provided that the column is properly calibrated [40]. It is important to realize that none of these estimates is equivalent to M_p, the molecular weight that corresponds to the elution peakmaximum of a polydisperse sample (for which $M_w/M_n > 1$). The estimates are related in the following way: $M_n \leq M_p \leq M_w \leq M_z$ [40]. A rough approximation of M_p is given by $M_p \approx \sqrt{M_w \cdot M_n}$ [41].

The MWD and estimates thus obtained will be influenced by the column zonebroadening. A mathematical correction of the MWD for the zonebroadening was described by Tung [42] and has been further refined [43-45]. However, sometimes the correction procedure may introduce erroneous MWD estimates [46,47]. Fortunately, correction for the zonebroadening is less important when the MWD of the sample is broad [47] and when the relative zonebroadening is kept small [48]. By using low velocities, i.e., ca. 15, 5, and 2 cm/h for 10-μm, 30-μm, and 100-μm particle-sized media, packed in 25-, 50-, and 100-cm-long columns, respectively, the accuracy in MW estimates of a polymer with $M_w \approx 10^5$ will, theoretically, be better than 99.5% without any correction procedure [49,50].

Calibration of the column is preferentially made with substances which are homologous to the sample to be studied. If such reference substances are not available, the column may be calibrated with the aid of well-characterized secondary references (e.g., globular proteins or fractions of polystyrene, dextran, polyethylene glycol, pullulan, etc.). Calibration according to the concept of hydrodynamic volume, $[\eta] \cdot M$, where $[\eta]$ is the intrinsic viscosity of the solute, was proposed by Benoit and co-workers [51]. This so-called universal calibration has been found to be applicable to flexible and spherical solutes in good solvents [40,51-56], but not to rod-like molecules [32]. Hamielec and Ouano [57] pointed out that calibration according to the hydrodynamic volume of polydisperse samples should be made using data of $[\eta] \cdot M_n$. The molecular weight of a chromatographed flexible molecule of known intrinsic viscosity may be calculated from $M_1 = M_2 \cdot [\eta_2]/[\eta_1]$, using a universal calibration made with well-characterized samples. A similar, though somewhat more elaborate approach involving an iterative correction of the established universal calibration curve may be used for polydisperse samples [58].

6.2.1.2 Nonadsorptive mode

Separation according to size does not involve any manipulation of solute surface properties or solute/sorbent interaction. It is therefore a very gentle separation technique, which is complementary to adsorption chromatography and offers a wide choice of pH,

References on p. A304

ionic strength, additives, and, in most cases, solvent polarity. However, some constraints may be due to the need to eliminate undesirable solute/sorbent interactions.

Often, a neutral buffer with an ionic strength of 0.1-0.2 is recommended to prevent ionic interactions between solutes and the gel medium in aqueous SEC. Hydrophobic interactions may be suppressed by the addition of organic modifiers or sodium dodecyl sulfate to the mobile phase [59]. Dubin and Principi [60] used simplex optimization for the selection of the mobile-phase composition. They found that Superose® 6 exhibited "ideal" SEC for globular proteins in a 0.38 *M* sodium chloride/phosphate buffer of pH 5.5. The result was also used for selecting the mobile-phase composition for the study of MWD of cationic polymers [61]. In some cases the shape of the solutes may be influenced by the ionic strength or the pH of the eluent or additives used, e.g., as a result of molecular expansion of charged polymers, association of proteins, or interactions of additives with proteins.

The choice of solvents used for nonaqueous SEC may influence the expansion of polymers as well as sorptive effects. Expansion of the polymer coil is largest in a good solvent and smallest in a theta solvent. However, sorptive effects may be obtained also when using a nontheta solvent, as found in the abnormal behavior of polystyrene on a polystyrene/divinylbenzene (PS/DVB) packing with toluene as eluent [54]. As opposed to tetrahydrofuran, which functions well for SEC of polystyrene, toluene was found to be unsuitable as a solvent for PS/DVB [62].

6.2.1.3 Separating power

Since the solutes are eluted according to the fraction of the porevolume that is available, the total separation volume in SEC is determined solely by the porevolume, V_p, of the packing medium, as illustrated in Fig. 6.1. The retention volume, V_R, is given by

$$V_R = K_D \cdot V_p + V_o \tag{6.2}$$

where K_D is the distribution coefficient, i.e. the available pore fraction (which ranges from 0 to 1), and V_o is the extra-particle void volume. The relative porevolume of packing media varies between 52% and 97%, the lower value being typical for silica-based materials and the larger value for semi-rigid polymer-based media [63]. The limited separation volume in SEC reduces the peak capacity to [50]

$$n_{R_s} = 1 + V_p/V_t \cdot (N/16)^{1/2}/R_s \tag{6.3}$$

where n_{R_s} is the number of peaks separated with a resolution factor of R_s, V_t is the total liquid volume of the column, i.e., $V_p + V_o$, and N is the maximum number of theoretical plates of the column. The peak capacity of SEC columns is often cited as $n = 1 + 0.2 \cdot N^{1/2}$. However, a prerequisite for this equation is that N have the same value for all solutes [64]. Since this is not the case in SEC of macromolecules, Giddings [64] suggested that an average plate count should be used in the calculation. Eqn. 6.3 is based on the assumption that a maximum number of peaks are produced at optimum flow velocity (Section 6.2.2.2) and has the advantage of including the porevolume in the calculation of peak capacity. However, both equations show that the peak capacity of a SEC column is very small as compared to other LC techniques. A high-performance SEC column will not be able to separate more than ca. 13 peaks, if complete separation between peaks (i.e. $R_S = 1.5$) is to be achieved. As can be calculated, there is no dramatic gain in maximum peak capacity when going from 100-μm to 10-μm media, partly because of the smaller porevolume and also the shorter column lengths used for the latter media [50]. Thus, when using columns for high-performance SEC there is a gain not in separating power but in separating speed (Section 6.3.1.1).

6.2.1.4 Solute diffusivity

As the separation process of SEC proceeds by molecular diffusion of solutes, i.e., into the pores of the medium, equilibration with the sterically available porevolume and then diffusion back again to the extra-particle mobile phase, diffusion distances and diffusion times are important parameters. These are especially critical for the separation of macromolecules, which diffuse 10-100 times more slowly than small inorganic solutes. Thus, while the diffusivity of solutes generally is high enough for equilibrium to be reached between the extra-particle volume and the porevolume, it will still affect the broadening of solute zones upon their passage through the bed (Section 6.2.2.2).

6.2.2 Influence of critical operational parameters

Since the zonebroadening in size-exclusion chromatography is diffusion-controlled, parameters influencing the total diffusion time (e.g., flow velocity, column length, solute diffusivity) or total diffusion distance (e.g., particle-size and poresize distribution of packing material) will be critical. The contribution of different parameters to the separation efficiency may be described by the equation for resolution, which for SEC can be transformed to [63]

$$R_S = (1/4)\log(M_1/M_2)\,[b/(V_o/V_p + K_D)]^{1/2}\,(L/H)^{1/2} \qquad (6.4)$$

where R_S is the resolution of two solutes of molecular weights M_1 and M_2, eluted from a packing material with a slope of the selectivity curve equal to $-b$, i.e., $b = -\,d(K_D)/d(\log M)$, with a void volume of V_o and a pore volume of V_p. K_D is the average distribution coefficient of the sample, L is the column length, and H the average plateheight of the solutes.

6.2.2.1 Matrix selectivity

The slope of the selectivity curve reflects the poresize distribution (PSD) of the material (Section 6.2.3.1). As pointed out by Knox and Scott [65], the selectivity curve will always be more shallow than the integral curve for the PSD. A large slope yields a large selectivity (i.e., peak-to-peak distance) and a small slope yields a large working range of the column. The most favorable separation condition is obtained by selecting a medium where the solutes are eluted at K_D equal to or larger than 0.2 [63]. In some situations, e.g., desalting or group fractionation, a medium which selectively excludes or includes the component of interest is the best choice.

6.2.2.2 Bandbroadening

The plateheight is a measure of the column zonebroadening (Section 1.8), which may be described by the general formula of Van Deemter [66], i.e. $H = A + B/u + Cu$, which for SEC of macromolecules is expressed by [67]

$$H = 2\beta d_p + 2[0.6D_m + \lambda D_m(1/R - 1)]/u + R(1 - R)d_p^2 u/(30\,\lambda\, D_m) \tag{6.5}$$

where d_p is the mean particle size, β is a packing-dependent variable (sometimes denoted as λ), D_m is the diffusivity of the solute in the mobile phase, λ is the obstruction factor to diffusion in the porous network (sometimes denoted as γ_S), R is the relative zone velocity, i.e. V_o/V_R, and u is the interstitial velocity of the mobile phase. For small solutes, diffusion in the mobile phase is sufficiently large to reduce the influence from the dispersion of solutes due to multiple flowpaths (i.e., the first term in Eqn. 6.5) [68]. This results in a more complex *A* term in the Van Deemter equation, as discussed in Chapter 1.

A characteristic of the Van Deemter curve is the minimum zonebroadening, H_{min}, obtained at the optimum velocity, u_{opt}, given by

$$u_{opt} = (D_m/d_p)\lambda(60\{2/[3\lambda\, R(1 - R)] + 1/R^2\})^{1/2} \tag{6.6}$$

which in turn yields the expression of H_{min}

$$H_{min} = d_p\,(2\beta + 2\{2/30\,[0.6\,R(1 - R)]/\lambda + (1 - R)^2\}^{1/2}) \tag{6.7}$$

Thus, while the minimum zonebroadening is related to the particle size of the medium, the mobile-phase velocity for this minimum is directly proportional to the ratio between the solute diffusivity and the particle size. The influence of the diffusivity and eluent velocity on the plateheight is illustrated in Fig. 6.3. Separations of high-molecular-weight solutes must be performed at a relatively low flow velocity to avoid excessive bandbroadening. The figure also shows that the minimum of the curve flattens out for small solutes with large diffusion coefficients, thus offering a larger velocity range for minimal zonebroadening than the range available for larger solutes.

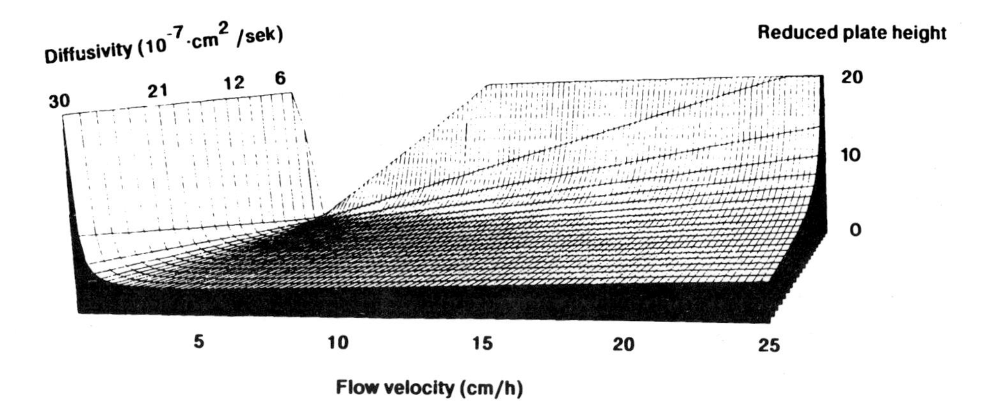

Fig. 6.3. Column zonebroadening (i.e. reduced plateheight) as a function of flow velocity for solutes of different diffusivities. Diffusivities range from $30 \cdot 10^{-7}$ to $1.5 \cdot 10^{-7}$ cm^2/sec, which corresponds to the molecular weight range $1 \cdot 10^3$ to $6 \cdot 10^6$ for peptides and proteins. Calculated from Eqn. 6.5 assuming a particle size of 33 μm.

diffusion coefficients, thus offering a larger velocity range for minimal zonebroadening than the range available for larger solutes.

One obvious way to speed up the separation is to increase the solute diffusivity to obtain a larger u_{opt}. The diffusion coefficient is to a large extent dependent on the viscosity of the solution. For proteins it may be estimated from [69,63]

$$D_m \approx 8.89 \cdot 10^{-8} \, (T/\eta) \, M^{-1/3} \tag{6.8}$$

where D_m is the diffusion coefficient, in cm^2/sec, T is the absolute temperature, and η the viscosity in cP. The viscosity itself is very temperature-dependent. For water, it decreases by a factor of 2 as the temperature increases from 4°C to 30°C. This fact must be taken into account when chromatography is to be carried out at reduced temperatures (Section 6.3.4). For polymers, the diffusion coefficient may be calculated from [50,70]

$$D_m \approx 1.36 \cdot 10^{-7} \, (T/\eta) \cdot \left(K \cdot M_v{}^{-(1+a)}\right)^{-1/3} \tag{6.9}$$

where M_v is the viscosity-average molecular weight and K and a are the Mark-Houwink constants for the polymer. The values of these constants are dependent upon the type of polymer, the solvent, and the temperature. Typical values for dextran in an aqueous solvent are [71] $K = 0.223$ ml/g and $a = 0.43$ and for polystyrene in THF [72], $K = 0.016$ ml/g and $a = 0.70$.

6.2.2.3 Sample volume

The parameters discussed so far have been focused on the zonebroadening in the column. However, in some situations the peak dispersion from other parts of the system must also be considered. Since the sample is not enriched on the column packing (as opposed to, e.g., in ion-exchange or reversed-phase chromatography), the influence of sample volume and dispersion in the precolumn parts of the system require special considerations in SEC. The latter effect may be made negligible by avoiding excessive deadspaces in tubings, connectors, and sample applicators. The optimal sample volume in preparative chromatography is a compromise between throughput and purity, whereas the critical sample volume in analytical chromatography will be determined by the column efficiency. The contribution of the injected sample volume, V_{inj}, to the peakwidth may be described by [73]

$$\sigma^2_{inj} \approx V^2_{inj}/K_{inj} \quad (6.10)$$

where K_{inj} is related to the shape of the injected plug and thus to the design of the injector. The optimal value of K_{inj} is equal to 12 (i.e., the variance of a square-wave distribution), while, in practice, values around 5 are more common for small sample volumes [74]. When very large sample volumes are injected, the elution profile may no longer be approximated by the Gaussian shape typical for column zonebroadening but will progressively adopt the square-wave shape of the injected plug. The elution volume of solutes will, of course, increase with the sample volume, unless the injection point is set at half the injected volume (i.e., by subtracting $V_{inj}/2$ from V_R).

In analytical SEC, and when the optimal performance of preparative-scale columns is tested, the sample volume should be kept very small, i.e., typically 0.2% of the bed volume or even smaller for microparticulate materials. Otherwise, the sample volume will contribute significantly to the total peakwidth [74]. Extra-column broadening of these small sample volumes must, of course, be avoided by reduction of tubing diameters and lengths and by using a detector with a small cellvolume (with some instruments the heat exchanger needs to be disconnected to minimize deadspaces).

6.2.2.4 Sample concentration

The concentration effect in SEC is generally very small, though effects have been reported in high-precision SEC [75]. Variations in the elution volume and peakwidth with changes in the sample concentration have been attributed to polymer expansion, viscosity phenomena in the extra-particle volume, and secondary exclusion effects [76]. Concentration effects will be more pronounced for flexible polymers of high molecular weight and when columns packed with media of small particle size are used [75]. As a general rule, for proteins the concentration of sample solutions should not exceed 70 mg/ml. At this concentration, overload effects have been noted [63]. This is not surprising, since the distance between neighboring molecules in a solution of, e.g., 50 mg/ml bovine serum

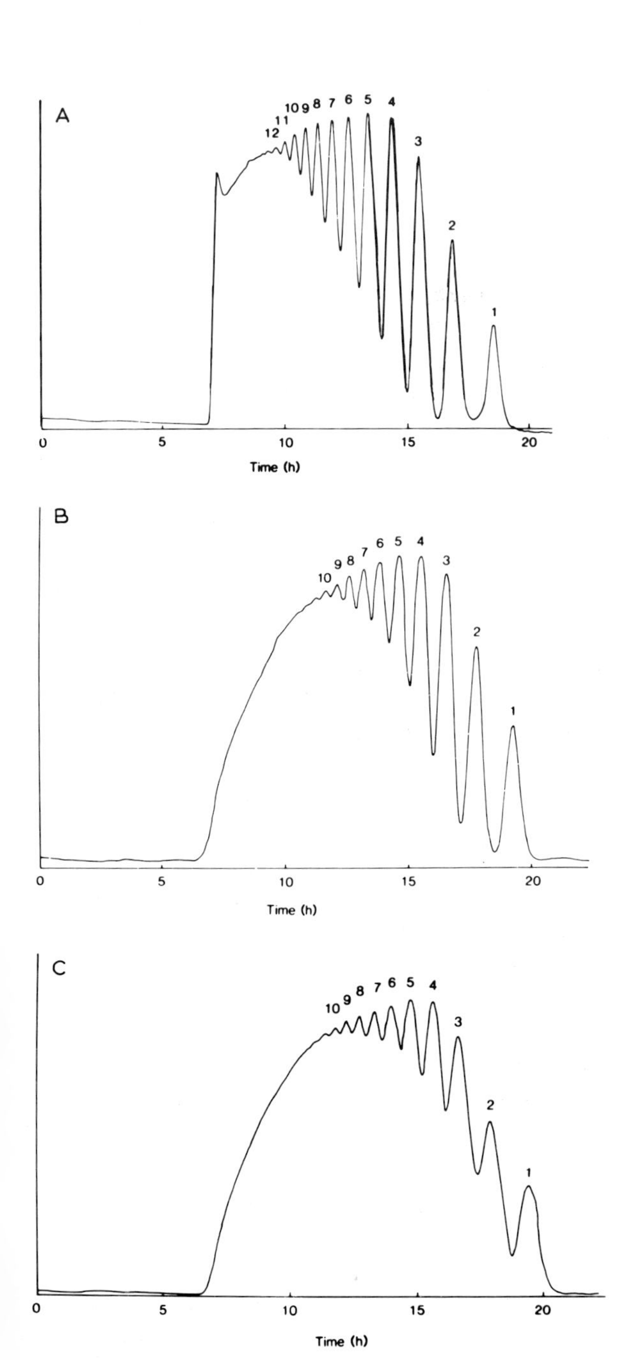

Fig. 6.4. Fractionation of glucose oligomers. (A) Two 90 x 6.5-cm-ID columns of Sephadex® G-25 SF; (B) as in (A), followed by a 50 x 6.5-cm-ID column of Sephadex G-50, Special Grade; (C) as in (B) but with the Sephadex G-50, Special Grade, column placed first. Sample: 10 g of dextran mixture ($M_W \approx 1100$) in 50 ml. (By courtesy of T. Andersson, Pharmacia.)

References on p. A304

albumin would be of the same magnitude as the size of the molecule itself. In the fractionation of polymers the concentration effect has been found to be significantly larger, and concentrations of less than 10 mg/ml have been recommended for obtaining fractions of polystyrene of low polydispersity [77]. However, it is possible to operate at somewhat higher concentrations without loss of resolution by performing the fractionation in a theta solvent [78]. Overload effects of dextran solutions have been found for sample concentrations corresponding to a relative viscosity of 1.5 [79]. The hydrodynamic expansion, and thus the viscosity of solutions, varies significantly among different types of molecules. Using the relative viscosity of 1.5 as a tentative limit for viscosity overload effects, the maximal sample concentrations that may be applied in SEC are: 70 mg/ml of a globular protein, 5 mg/ml of dextran of M_w ca. 10^6, 10 mg/ml of dextran of M_w ca. $2 \cdot 10^5$, and 5 mg/ml of polystyrene of M_w ca. $2 \cdot 10^5$, provided a good solvent is used. These figures may have to be reduced by a factor of 2 for high-precision analytical work. Sample concentrations of 30 mg/ml of proteins and 6 mg/ml of tRNA have been found to be the maximum applicable in high-resolution SEC [80,81]. A recent study showed that effects on contraction of macromolecular coils and on viscosity-induced distortion was evident even at polymer concentrations of a few mg/ml [82]. This may be due to a decrease of the hydrodynamic volume of the flexible macromolecules as the concentration increases. Upon passage through the column, the solute concentration will decrease and, thus, the size of the solutes will continuously increase, resulting in a reduced fractionation efficiency. However, in the fractionation of small-molecular-weight solutes the concentration effect is still small, as illustrated by Fig. 6.4, where 10 g of glucose oligomers was successfully separated on Sephadex G-25. It is also seen that overload effects are reduced by arranging the columns in the order of increasing poresize of the media (Figs. 6.4B and C).

6.2.2.5 Processing speed

Processing large volumes at a defined resolution in a fixed time may be accomplished either by dividing the total sample into a large number of portions and chromatographing each of them very fast, or chromatographing a smaller number of portions with larger volumes at lower speed. In the first case, the limiting factor for the resolution is the column zonebroadening (i.e. the C term in the Van Deemter equation), and in the latter case, the resolution will be limited by the large peakwidths due to the large volume applied to the column. The optimal conditions for processing V_{feed} ml sample per hour will correspond to the injected volume [83]

$$V_{inj,opt} \approx (V_{feed} \cdot K_{inj} \cdot V_c \cdot V_p \cdot d_p{}^2/15 \cdot D_m)^{1/3} \tag{6.11}$$

and the corresponding nominal velocity is given by

$$u_{nom} \approx V_{feed} \cdot L/V_{inj,opt} \tag{6.12}$$

The nominal flow velocity is given by the volumetric flowrate divided by the column cross-sectional area. These equations are useful for selecting optimum conditions and for scaling up separation schemes. A change in any of the parameters will, of course, affect not only the optimum injection volume but also the resolution, since the nominal flow velocity is altered. The effect on resolution may be calculated from the change in flow velocity by using Eqns. 6.12, 6.5, and 6.4 and the relationship $u = u_{nom}V_c/V_o$. Theoretical predictions from these equations were found to be in accordance with experimental results [83].

6.2.3 Important properties of packing materials

6.2.3.1 Particle size and porevolume

Two important parameters of packing materials are given in Eqn. 6.11, viz. the porevolume, V_p, and the particle size, d_p. The porevolume determines the maximum separating volume, i.e. peak-to-peak volume, and the particle size influences the column zone-

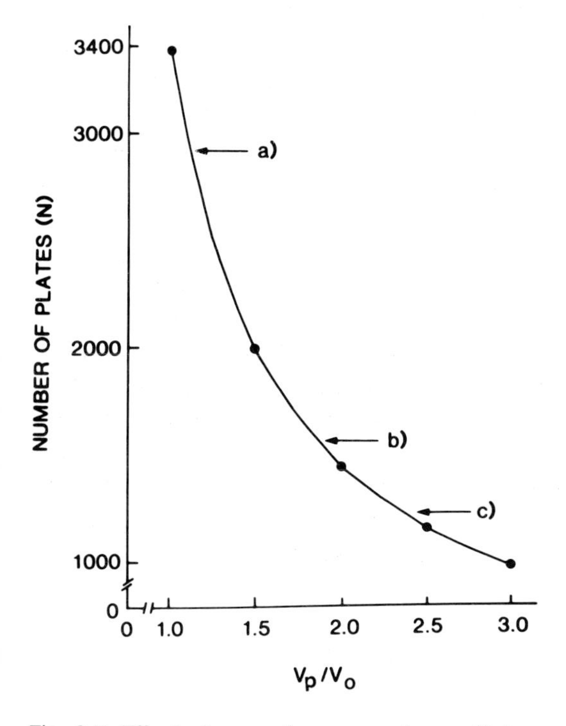

Fig. 6.5. Effect of porevolume on column efficiency required to give a resolution factor of 1.0 for two samples, eluted at K_D = 0.5 and 0.4, respectively. Typical values of permeabilities, V_p/V_o, for different media, as given in Ref. 63, are denoted by arrows: (a) rigid medium; (b) semirigid medium; (c) conventional medium. (Adapted from Ref. 84, with permission.)

References on p. A304

broadening, i.e., the peakwidth. As indicated in Eqn. 6.4, it is possible to compensate for a low porevolume with a small particle size (yielding a high efficiency) and vice versa, as illustrated in Fig. 6.5. In essence, this is another way of keeping the peak capacity constant (cf. Eqn. 6.3), the trade-off is separation time. A third important parameter is the selectivity, which is a function of the poresize distribution (PSD) of the packing material. Due to the nature of SEC, the selectivity curve of a medium will have a more shallow slope than the PSD. Thus, a single-poresize packing medium will fractionate flexible molecules on the basis of size, over a molecular weight range of ca. 2 orders of magnitude [30,85]. The influence of low selectivity may be counteracted by a large porevolume.

The impact of the pore structure of chromatographic materials on the separation efficiency has yet to be elucidated. However, preliminary results in affinity chromatography suggest that open-pore structures, as in agarose, favor the adsorption/desorption process [86]. The pore structure will also influence the obstruction to diffusion in the porous network [87,88] and may in this way play a significant role in the broadening of high-molecular-weight solutes in SEC.

6.2.3.2 Chemical and physical stability

The chemical and physical stability of the packing medium is very important. The limitations of pH stability of some of the established media are well known, e.g., silica-based media dissolve at pH above 8, and polymeric media may be susceptible to either acidic or basic hydrolysis at extreme pH values. Thus, in the preparation of therapeutic drugs, trace amounts of leakage products from the chromatographic column may need to be assayed [89]. Leakage of matrix fragments may occur as a result of harsh treatments, such as washing procedures. While the leakage products may be washed away by several column volumes of eluent, such a treatment may create adsorptive groups on the matrix which will affect the separation. Safe procedures for cleaning the packing material are therefore very important, especially in process SEC. High physical stability of the media allows a high flow velocity, to be used without compression of the bed. However, in conventional SEC of high-molecular-weight solutes this property is of limited importance, since the slow mass transfer imposes a practical upper limit on the flow velocity for a critical separation (see Fig. 6.3). As the particle size is decreased, the applicable eluent velocities are increased and the relative rigidity of the matrix needs to be enhanced. This will in most cases result in an increase in the matrix volume, and thus, the pore volume will subsequently be reduced. A rigid matrix is, of course, desirable when the column is to be quickly flushed. Rigid materials, such as silica, will also be unaffected by variations in temperature or solvent composition. The chemical and physical stability of the media will also influence the expected cycle lifetime of the chromatographic bed and thus the economy of the separation step (Section 6.3.1).

6.2.4 Prediction of separation results

Owing to the development of chromatographic theory, supported by practical experience, over the last twenty years it is now possible to use computer simulation for optimizing the separation. The availability of commercial software for this purpose is growing steadily [90,91]. The separation of proteins, with the aid of SEC, may be readily predicted from some basic relationships [63,92]. Some of the parameters that need to be known are related to characteristics of the packing material, i.e., the particle size (distribution), the pore fraction and the selectivity (or PSD), whereas other parameters are unique for the specific experimental situation, such as the column dimension, the void fraction, the flow velocity, and the size and concentration of the solutes.

The selectivity, i.e., K_D as a function of solute size, is a parameter which is preferably calibrated with the class of samples to be separated. For globular proteins and flexible macromolecules, the hydrodynamic volume has proven to be a common size-exclusion-determining variable (Section 6.2.1.1). Thus, protein separations may be simulated by using data obtained with other molecules, e.g., dextran fractions, on the basis of their hydrodynamic volume. However, this is only useful for a rough approximation, which requires subsequent fine-tuning by using experimental data. This approach was used for obtaining accurate molecular weight estimates of polycarbonates chromatographed on a column calibrated with narrow fractions of polystyrene [58]. Empirical relationships between calibration curves for solutes of different shapes can also be used. In this way, it was possible to determine fragment lengths of rod-shaped double-stranded DNA from a calibration with globular protein standards [31].

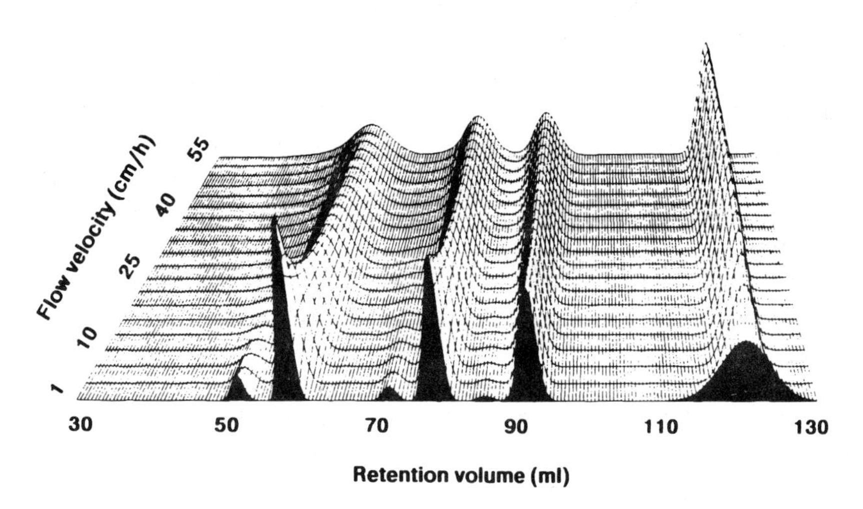

Fig. 6.6. Prediction of the influence of flow velocity on the separation of a protein mixture. The separation of thyroglobulin, bovine serum albumin, myoglobin, and cytidine on Superose® 6 Prep. Grade, was simulated for flow velocities from 1 to 60 cm/h. Other conditions chosen in accordance with Fig. 6 in Ref. 93.

References on p. A304

The utilization of Eqns. 6.2 and 6.5 for the prediction of the separation of a protein mixture is illustrated in Fig. 6.6. Comparison with the corresponding experimental results published earlier [93] indicates that this simple procedure may be very useful for studies of the influence of different parameters on the separation results [92].

6.3 PRACTICAL ASPECTS

6.3.1 Modes of size-exclusion chromatography

In the Introduction we specified that SEC is performed with beds of porous media. This is to distinguish SEC from other types of liquid chromatography where steric exclusion is utilized, e.g. hydrodynamic chromatography, field-flow fractionation, and secondary exclusion in the extra-particle volume. The latter effect becomes noticeable when the solute size is very large and the particle size small [94]. The effect is negligible for solutes of sizes less than 1% of the particle size [95].

In liquid chromatography distinctions are often made with reference to pressure, performance, speed, resolution, and scale. For SEC, the peak capacities vary very little between different types of systems (Section 6.2.1.3). The real differences seem to be in the separation time and sample capacity. Thus, the distinction in LC with respect to performance is inapplicable to SEC. The term high-performance liquid chromatography refers to columns with platenumbers of 2000-20 000 [96]. Most columns for SEC (e.g., with $d_p \leq 100\ \mu m$ or $L \geq 50$ cm) have maximum platenumbers exceeding 2000. Yet, there is a need to distinguish between some fundamentally different application areas of SEC. SEC methods can be classified with respect to the aim of the separation, i.e., production of information or production of substances. The information-producing category may be further subdivided into fast SEC, where speed is the most important feature, standard SEC, where flexibility and cost are important, and microcolumn SEC, where the amount of sample is critical.

6.3.1.1 Fast size-exclusion chromatography

In fast SEC the objective is to obtain the desired information as fast as possible; cost is of no major concern. Therefore, packings of small beads, typically having particle sizes between 5 μm and 13 μm, are used exclusively. Most of these materials are available only in the form of prepacked columns (Tables 6.2 and 6.3). In order to cope with the high backpressures generated by the narrow void channels between the small particles at the high flow velocities used for fast analysis, only high- or medium-pressure systems (up to 250 and 50 bars, respectively) are suitable for fast SEC. The separation time depends, of course, on the required resolution. Calculations of practical flow velocities for optimal zonewidths of medium-sized proteins indicate separation times between 10 and 60 min for particles 5 and 13 μm in diameter [63]. In extreme situations the separation time may be substantially reduced, but at the expense of resolution. Thus, fast SEC has recently been proposed for a 3-sec on-line process analysis of a high-molecular-weight polymer [97].

TABLE 6.2

CHARACTERISTICS OF SOME MEDIA FOR SEC OF BIOPOLYMERS

Data stated as given by the manufacturers.

Medium	Matrix	Particle size (μm)	Approx. fractionation range for proteins (kD)	Supplier
Media for fast SEC				
Zorbax GF 250	Silica	4	Not stated	Du Pont
Zorbax GF 450	Silica	6	Not stated	Du Pont
G2000SWXL	Silica	5	$5 \cdot 10^3$ - $150 \cdot 10^3$	Toyo Soda
G3000SWXL	Silica	5	$10 \cdot 10^3$ - $500 \cdot 10^3$	Toyo Soda
G4000SWXL	Silica	7	$20 \cdot 10^3$ - $10 \cdot 10^6$	Toyo Soda
Superose 12	Agarose	10	$1 \cdot 10^3$ - $2 \cdot 10^6$	Pharmacia
Superose 6	Agarose	13	$5 \cdot 10^3$ - $40 \cdot 10^6$	Pharmacia
Protein-Pak 60	Silica	10	$1 \cdot 10^3$ - $20 \cdot 10^3$	Waters
Protein-Pak 125	Silica	10	$2 \cdot 10^3$ - $80 \cdot 10^3$	Waters
Synchropak GPC 100	Silica	10	$3 \cdot 10^3$ - $630 \cdot 10^3$	SynChrom
TSK SW 2000	Silica	10	$5 \cdot 10^2$ - $60 \cdot 10^3$	Toyo Soda
TSK SW 3000	Silica	10	$1 \cdot 10^3$ - $300 \cdot 10^3$	Toyo Soda
TSK SW 4000	Silica	13	$5 \cdot 10^3$ - $1 \cdot 10^6$	Toyo Soda
Media for standard and process SEC				
Superose 12 Prep. Grade	Agarose	34	$1 \cdot 10^3$ - $2 \cdot 10^6$	Pharmacia
Superose 6 Prep. Grade	Agarose	34	$5 \cdot 10^3$ - $40 \cdot 10^6$	Pharmacia
Superdex™ 75 Prep. Grade	Agarose/dextran	34	$3 \cdot 10^3$ - $70 \cdot 10^3$	Pharmacia
Superdex 200 Prep. Grade	Agarose/dextran	34	$10 \cdot 10^3$ - $100 \cdot 10^3$	Pharmacia
Toyopearl				
HW-40S	Polymeric*	25-40	$1 \cdot 10^2$ - $10 \cdot 10^3$	Toyo Soda
HW-50S	Polymeric*	25-40	$5 \cdot 10^2$ - $80 \cdot 10^3$	Toyo Soda
HW-55S	Polymeric*	25-40	$1 \cdot 10^3$ - $700 \cdot 10^3$	Toyo Soda

(Continued on p. A286)

TABLE 6.2 (continued)

Medium	Matrix	Particle size (μm)	Approx. fractionation range for proteins (kD)	Supplier
HW-65S	Polymeric*	25-40	$50 \cdot 10^3$ - $5 \cdot 10^6$	Toyo Soda
HW-75S	Polymeric*	25-40	$5 \cdot 10^5$ - $50 \cdot 10^6$	Toyo Soda
Sephacryl®				
S-100 HR	Dextran/bis-acrylamide	47	$1 \cdot 10^3$ - $100 \cdot 10^3$	Pharmacia
S-200 HR	Dextran/bis-acrylamide	47	$5 \cdot 10^3$ - $250 \cdot 10^3$	Pharmacia
S-300 HR	Dextran/bis-acrylamide	47	$10 \cdot 10^3$ - $1.5 \cdot 10^6$	Pharmacia
S-400 HR	Dextran/bis-acrylamide	47	$20 \cdot 10^3$ - $8 \cdot 10^6$	Pharmacia
S-500 HR	Dextran/bis-acrylamide	47	$20 \cdot 10^3$ - $3 \cdot 10^9$	Pharmacia

* Copolymer of oligoethyleneglycol, glycidylmethacrylate, and pentaerythroldimethacrylate.

However, excessive eluent velocities can lead to frictional heating effects or shear degradation of polymers [98]. Barth and Carlin [99] concluded in a review of polymer shear degradation that the flowrate should be less than 1 ml/min when chromatographing molecules of M_w ca. 10^6 on a 4-mm-ID column, packed with 10-μm material. The utility of materials of particle size less than 5 μm for chromatography of high-molecular-weight polymers is therefore questionable [99]. Fast separations of polymers can be achieved by using moderate flowrates and short columns.

6.3.1.2 Standard size-exclusion chromatography

Standard SEC encompasses most of the usual laboratory applications. The major features of standard SEC are the large variety of media of different selectivity and matrix properties and the great flexibility of gel/column combinations due to the availability of media in bulk and moderate cost, as compared to fast SEC. Yet, most separations achievable with standard SEC are comparable to those of fast SEC, though at a considerably longer separation time (Section 6.3.4.2). In some cases, separations by standard SEC cannot be accomplished with fast SEC, simply because packings possessing suitable selectivity properties are not available. Most media for standard SEC have a mean particle size in the range of 30-150 μm [63]. Optimum eluent velocities for protein separation range from 2 to 20 cm/h, and the pressure drop over the packed columns is very

TABLE 6.3

CHARACTERISTICS OF SOME MEDIA FOR SEC OF SYNTHETIC POLYMERS

Data stated as given by the suppliers.

Medium	Matrix	Particle size (μm)	Separation range of medium family* (MW)	Supplier
Media for nonaqueous SEC				
Zorbax PSM	Silica	6	$1 \cdot 10^2$ - $2 \cdot 10^6$ (polystyrene)	Du Pont
μStyragel	Styrene/ divinylbenzene	10	50 - $1 \cdot 10^6$ (polystyrene)	Waters
Ultrastyragel	Styrene/ divinylbenzene	7	50 - $1 \cdot 10^6$ (polystyrene)	Waters
LiChrogel PS	Styrene/ divinylbenzene	5, 10	$1 \cdot 10^2$ - $1 \cdot 10^7$ (polystyrene)	E. Merck
Shodex A	Polystyrene	10	10 - $2 \cdot 10^7$ (polymer not stated)	Alltech
PLGel	Styrene/ divinylbenzene	5, 10	$1 \cdot 10^2$ - $1 \cdot 10^7$ (polystyrene)	Polymer Lab.
Media for aqueous SEC				
TSK PW	Methacrylate	10-17	$1 \cdot 10^2$ - $1 \cdot 10^6$ (dextran)	Toyo Soda
TSK PW_{XL}	Methacrylate	5	$1 \cdot 10^2$ - $1 \cdot 10^6$ (dextran)	Toyo Soda
PL GFC	Not stated	10	$1 \cdot 10^2$ - $1 \cdot 10^7$ (dextran)	Polymer Lab.
Ultrahydrogel	Methacrylate	6-13	$1 \cdot 10^2$ - $7 \cdot 10^6$ (polyethylene oxide)	Waters
Shodex OHpak B	Methacrylate	Not stated	$1 \cdot 10^2$ - $1 \cdot 10^7$ (dextran)	Alltech
Asahipak GS	Vinyl alcohol	9	$1 \cdot 10^2$ - $1 \cdot 10^7$ (pullulan)	Asahi
μBondagel	Silica	10	$2 \cdot 10^3$ - $2 \cdot 10^6$ (polypropylene glycol)	Waters

(Continued on p. A288)

TABLE 6.3 (continued)

Medium	Matrix	Particle size (μm)	Separation range of medium family* (MW)	Supplier
MCI GEL CQP	Methacrylate	10	$10 - 1 \cdot 10^6$ (polyethylene glycol)	Mitsubishi

* Each family of gels is comprised of 3-8 gels of different poresizes, ranging from 100 Å to 4000 Å. The separation range stated indicates the working range of the gel family, not of individual gels.

small. Therefore, inexpensive low-pressure systems (up to 5 bar) may be employed in standard chromatography. Separation times for this type of media will typically be 3-24 h. The lower figure is typical for 30-μm media. Recently, several media of small particle size (in this context, small means 30-50 μm) have been introduced, which are usable in standard SEC (Table 6.2). Improvement in column design and packing procedures enables the preparation of very cost-efficient columns for SEC [83]. However, for some applications, the separating power of the traditional media, such as Sephadex and BioGel, are still unsurpassed (Fig. 6.4). Heitz [100] used soft gels of poly(vinyl acetate), possessing large pore volumes, packed in long columns, for ultimate resolution of oligomers.

6.3.1.3 Microcolumn size-exclusion chromatography

In microcolumn SEC, columns of very small inner diameter (typically 0.25 mm) are used. Their major advantages are small sample loads, low solvent consumption, and high platecounts due to the use of long columns, prepared from fused-silica capillaries. However, rather sophisticated equipment is needed for on-column injection of samples of only a few hundred nanoliters, for pumping eluents at tens of microliters per minute, and for detector cell volumes of 1 μl or less. Microcolumns are more difficult to pack than traditional columns, and reduced plateheights of 5 have been reported [101]. Microparticulate media [101,102], ordinary fast SEC media [103,104], and traditional "soft" media have been used as packing materials [105]. Detection of polymers in femtomole quantities [102] and hundredfold reduction in sample volumes necessary for frontal analysis [105] have been objectives for miniaturization. Since eluent flow velocities for microcolumn SEC are the same as for fast SEC, there is no gain in separation time with this mode. Proteins are separated typically in 1 h on a 1 m x 0.26-mm-ID column, packed with TSK G 3000 SW [104]. Conventional SEC was found to have many advantages over semimicro SEC, except for the sample size and solvent consumption [103].

6.3.1.4 Process size-exclusion chromatography

When it comes to the production of substances rather than information, the objective is fast processing of solutions for the purpose of purification, and maximum throughput at a predefined level of purity is therefore the key point. The preconceived picture of process, standard, and fast chromatography differing on the basis of the particle size of media is not valid, since even 10-μm media have been utilized for process chromatography [78]. However, this is a very expensive proposition, as the present cost of materials for aqueous SEC is roughly proportional to $1/d_p^3$, while the gain in throughput may be expected to be proportional to $1/d_p^2$ (Section 6.3.4). Thus, the cost/benefit ratio favors larger particle sizes, as long as resolution can be achieved. The major objections to using small-particle-sized media are the high initial cost of the material (at constant cycle lifetime of the media) and the difficulty of packing small-diameter media efficiently [106]. In most cases, the large pressure drops generated over these columns at the high solvent velocities used for high throughput may require special safety precautions.

The general trend in process chromatography over the last few years has been to utilize media with particle sizes in the range of 20 to 70 μm [107,108]. As the general strategy for scaleup of chromatography is to use the same media on a larger scale (to avoid re-validation of the process), there is little demand for the transfer of separation schemes between different media. Still, the option of exploring a separation strategy with one type of medium (e.g., a prepacked column) and then scaleup with in-house packed preparative media of similar separating characteristics may be valuable.

6.3.2 Characteristics of new packing materials

The trend in packing materials for SEC has been towards smaller bead sizes, more rigid materials, and higher selectivities. Whereas the media in the Sixties and Seventies were composed of rather soft and large beads (e.g. Sephadex, Sepharose, BioGel, and Styragel [109]), the media of the Eighties have been rigid or semirigid beads of smaller size. Some characteristics of these media are given in Tables 6.2 and 6.3. These packing materials yield very efficient columns (e.g., optimum platenumbers 10 000-50 000 per meter). The rigidity of these small beads is accompanied by an increased matrix concentration, as compared to the earlier materials. The drawbacks are a reduction in pore-volume and increasing adsorptive properties, due to the large matrix content. Mixed-mode separations have been observed for many of the materials [110-112]. Such solute/matrix interactions may under certain circumstances be utilized to improve the separation. However, since adsorptive properties are not likely to be reproducible from lot to lot, they are generally not desired in a process purification scheme based on SEC. Media for aqueous SEC have been classified with respect to ionic and hydrophobic properties by measurement of the interactions with solutes of different functional properties [113] and the elution of small lipophilic substances [114]. Dubin and Principi [115] were able to attribute a hydrophilicity parameter to different media by measuring the elution volumes of a series of normal alcohols. Solute/matrix interactions may be revealed by varying the ionic strength or by adding small amounts or organic modifiers (e.g., 1% 1-propanol) to the eluent. Drastic changes in elution volume with temperature is also indicative of adsorptive effects. Adams and Bickin [62] determined the enthalpy of interaction of small solutes with Styragel, in toluene and tetrahydrofuran, by SEC at various temperatures (cf. Eqn. 6.1).

6.3.3 Column packing and evaluation

The technical aspects of chromatography have received considerable attention in the past, but chromatography is today a more or less established routine, and attention has shifted to other parts of the isolation process. Thus, chemists now generally purchase ready-made columns instead of preparing them in the laboratory. However, ready-made columns are still rather expensive, and furthermore, they can only be obtained in a limited number of gel/column combinations. Thus, there is still a demand for column packing techniques. It is fairly easy to prepare columns for SEC from modern media, with plate-numbers exceeding 10 000 per m [83,93]. As columns for fast SEC are more difficult to pack due to the small particle size (typically less than 10 μm), this type of columns will probably continue to be mainly factory-made.

From the number of different packing procedures found in the literature it may be concluded that column packing is more of an art than a science [101]. Nevertheless, some basic principles can be identified. The packing solvent should prevent the particles from forming aggregates. If the particle-size distribution of the medium is very broad, the settling time of the gel should be kept short or else the particles must be prevented from gravitational settling. This usually causes no problem when modern media, e.g., those in Table 6.2, are used. The packing procedure should result in a dense, homogeneous bed that is stable at the eluent velocities used. A stabilizing step, in which a high flow velocity is used, is very important, also for rigid materials [116]. This is not critical for axial-pressure columns, as variations in the bed can be compensated for by automatic adjustment of a movable plunger. The exact conditions for the packing procedure depend on many factors [106], the most important of which are the surface properties (e.g., structure and polarity) and the elasticity of the packing material and the column dimensions. A successful strategy for packing semirigid materials is to use a two-step procedure, where a dense, homogeneous bed is created by first packing the bed at constant flow velocity (thus keeping the frictional forces constant) and then stabilizing the bed by applying a high, constant pressure [63]. Large-diameter columns, designed for preparative use, may also be packed conveniently according to this principle, yielding columns of maximum theoretical efficiency [83]. A 60 x 10-cm-ID column, packed with Superdex™ 75 Prep. Grade according to this procedure, yielded a column efficiency of 15 000 plates per m and a symmetry factor of 1.0, determined with acetone at optimum flow velocity [117].

The performance of a column should be tested not only before it is put into regular use but also from time to time to ensure consistent chromatographic properties. The performance of the column may be tested in various respects, e.g., by determination of the maximum efficiency, separation of a mixture of solutes, or a complete characterization of the sorptive properties (Section 6.3.2). The column packing can be judged by observing the separation of a mixture of colored substances or evaluated by monitoring the elution of a sample and calculating the resolution or bandbroadening of the column. The latter test may be performed by chromatographing a solute at optimum flow velocity (Eqn. 6.6). The use of a small solute of high diffusivity will permit the test to be carried out at a high flow velocity. The bandbroadening of the solute may be evaluated from the reduced plateheight, $h = H/d_p$, according to (cf. Fig. 6.1)

$$h = H/d_p = (L/N)/d_p = L/[5.54 \cdot (V_R/W_h)^2 \cdot d_p] \quad (6.13)$$

where W_h is the peakwidth at half peakheight. For an efficiently packed column, the reduced plateheight should be close to 2 [118]. The symmetry of the peak at 10% peakheight should be close to 1 (1.0±0.1). Symmetry below unity, i.e., a leading peak, may indicate the formation of channels in the bed during packing. This can be corrected by repacking the column, using a lower flow velocity for the stabilizing step. A tailing peak may be due to the bed being too loosely packed. In this case, the flow velocity in the stabilizing step should be increased.

6.3.4 Guidelines for optimizing the separation

The most important parameters that influence the separation in SEC are given in Eqns. 6.4 and 6.5. In the most favorable situation the bed is utilized as a "filter", i.e. the solute of interest is eluted at one extreme of the separation range and the ballast material at the other extreme. This is desirable in desalting or buffer exchange and is accomplished by the proper choice of a gel with a suitable exclusion limit. The other parameters will then be of relatively little importance, and a high flow velocity as well as a large sample volume may be applied. This is also the case when the sample of interest can be selectively excluded and the contaminants included. An example of this is plasmid DNA, which can be separated from contaminating RNA and proteins at high sample load and short separation time [119]. In situations where the separation is more difficult (i.e. $R_s < 1.5$), different parameters will be important for the different modes of SEC.

6.3.4.1 Fast size-exclusion chromatography

Separation time is proportional to the flow velocity of the mobile phase and inversely proportional to the column length. At high flow velocities the bandbroadening of macromolecules is predominantely caused by the slow mass transfer, and the resolution factor is approximately inversely proportional to the square root of the flow velocity (Eqn. 6.5). The resolution factor may be restored by increasing the column length, but this will not decrease the separation time. The detrimental effect of a high flow velocity may instead be counteracted by reducing the particle size or by increasing the diffusivity of the solute. The former parameter, being squared, has the largest impact, whereas the latter parameter is of major practical importance when separation schemes are to be transferred between different temperature environments. An increase in temperature from 25°C to 40°C will

References on p. A304

reduce the viscosity of an aqueous solution by 27% and thus allow an equal increase in eluent velocity without any reduction in the resolution factor. The same effect would be obtained from a 14% decrease in particle size. Since the negative effect of a high flow velocity on bandbroadening is more pronounced for the earlier-eluted substances, the bandbroadening over the entire separation range may be minimized by using an increasing flow gradient, starting after the void volume. This may reduce the separation time by a factor of 2 [63].

6.3.4.2 Standard size-exclusion chromatography

Due to the large variety of media available for standard SEC, the selection of an optimum gel/column combination for solving a separation problem is much easier than in fast SEC. The strategy for selecting media may be based on selective exclusion of solutes or working at $K_D = 0.2\text{-}0.4$ for maximum selectivity [63]. The cost depends on the column dimensions and the particle size of materials (Section 6.3.1). The resolution is proportional to the square root of the ratio of the column length to the plateheight, as described by Eqn. 6.4. Thus, a reduction of eluent velocity will permit an approximately equally large reduction in column length, i.e., keeping the separation time constant without impairing resolution. The use of medium-sized particles, e.g., 30 μm, may often provide an economical alternative to smaller-sized-materials. This is especially true when larger sample volumes, e.g., several milliliters, are to be applied (Section 6.2.2.3). The increase in particle size may be compensated for by increasing the column length or decreasing the flow velocity. In practice, a combination of the two parameters is often used. Depending on the reduced flow velocity used, the plateheight may be dominated by either the A or the C term of the Van Deemter equation (Fig. 6.3). The influence of the particle size on resolution is drastically different in these cases (i.e. either proportional to $d_p^{1/2}$ or to d_p). In practice, both terms will contribute. By adjusting the separation time roughly proportional to $(d_{p1}/d_{p2})^{1.5}$, similar resolutions of a protein mixture were obtained with 13-μm, 33-μm, and 110-μm packing materials [74].

6.3.4.3 Process size-exclusion chromatography

In preparative separations the throughput, i.e. the volume processed per hour, is to be optimized. The optimal conditions for this are given by Eqns. 6.11 and 6.12. These equations were used in recent work to predict an optimum injection volume of 2-3% of the column bed volume when a total sample volume corresponding to 0.5% of the bed volume was to be processed per hour [83]. As illustrated in Fig. 6.7, the prediction was in good agreement with experimental results. The simulations also yielded an adequate picture of the separation patterns achieved. The results also indicated that optimum sample volumes are probably to be found in the region 1-4% of the bed volume used. The optimum conditions may yield an excessive resolution factor (i.e. $R_s > 1.5$). This may be utilized for further increasing the throughput by increasing the processing speed (increasing u_{nom}). By combining Eqns. 6.12, 6.5, and 6.4 – and assuming that the C term

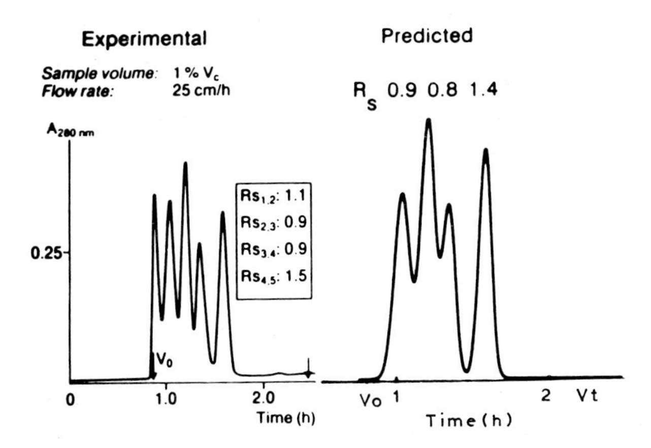

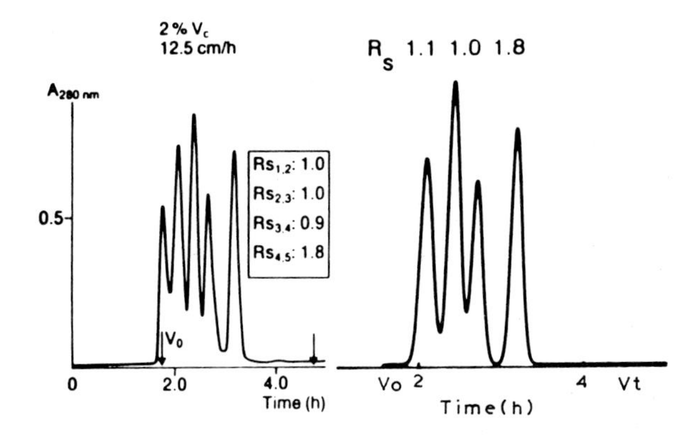

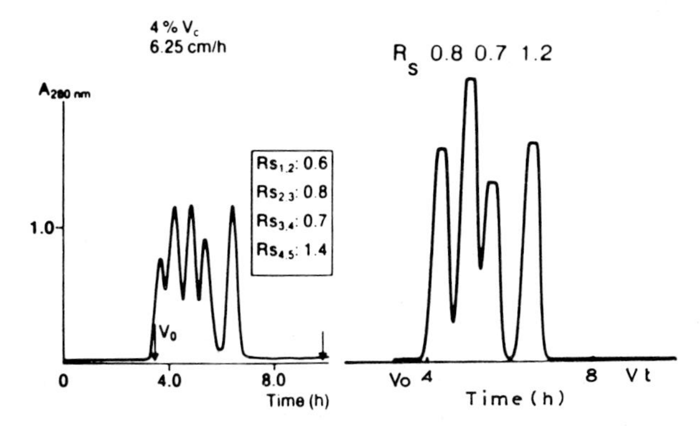

Fig. 6.7. Experimental and predicted separation of a protein mixture at constant processing rate (1.33 ml sample/h) for various sample volumes. Column: 61.3 x 2.6 cm ID of Sephacryl® S-200 HR. Peaks are, from left to right: aggregates; immunoglobulin G; bovine serum albumin; β-lactoglobulin; and myoglobin. The peak of the aggregates was not included in the simulation. (Experimental curves reproduced from Ref. 83, with permission.)

dominates the bandbroadening – it is readily shown that the flow velocity can be increased proportionally to $(R_{s1}/R_{s2})^3$, where R_{s1} is the existing resolution and R_{s2} the desired resolution factor. Under the same circumstances it is possible to trade off the increased resolution that comes from employing a smaller particle size by adjusting the feed (and thus u_{nom}) proportionally to $(d_{p1}/d_{p2})^2$. Once the optimum conditions have been established, the feed may be further increased by proportionally increasing the sample volume and the column cross sectional area.

The maximum sample concentration that can be applied is limited by the overload effects (Section 6.2.2.4) and the hydrodynamic instability of sample zones of relative viscosities exceeding 1.5. It seems to be necessary to keep sample concentrations below 70 mg/ml for proteins and below 10 mg/ml for high-molecular-weight polymers, such as dextran, to avoid column overloading. Intuitively, the overload effects would be reduced by arranging a set of columns in the order of increasing pore size [77]. This arrangement would provide the most rapid initial separation of the species and thus reduce concentration effects. As can be seen in Fig. 6.4, this is indeed the case for the fractionation of glucose oligomers. The best results were achieved when the high-molecular-weight portion was totally excluded (and fractionated in another gel system). This system permits the application of 10-g samples in 50 ml per cycle [117].

The processing time of an optimized procedure may be reduced by 33% by utilizing the deadtime during elution of the void fraction to inject a new sample after only 2/3 of the total column volume is eluted. In situations of selective exclusion, the flow velocity may be maximized to flush the column rapidly after the sample has been eluted, thus minimizing the cycle time.

6.4 APPLICATIONS

Today, 30 years after its commercial introduction, SEC must be regarded as an established separation and characterization technique. It is used routinely in all areas of research, production, and quality control where a separation of molecules according to size and shape is required.

An important application of SEC is the determination of MWDs of polymers. Major developments of this characterization method took place in the early phase of SEC, and the past ten years have seen refinements of the method, as judged from reviews of the subject [27,109]. However, SEC is still the premiere technique when it comes to characterization of MWD, and this application will therefore be briefly covered here.

Four other areas, of particular current interest to biochemists, have been chosen for review. These areas illustrate various extremes in SEC application work. Thus, separations within groups of small molecules, such as peptides, and within groups of very large molecules, such as large fragments of DNA and plasmids, represent problems for both scientists and manufacturers of column packing materials. As will be discussed in detail below, SEC of peptides is of particular concern because as yet there is no ideal medium for this application available on the market. On the other hand, for the separation of very large DNA molecules, wide-pore media have recently been introduced that are useful for separations up to 12 kb. The other set of extremes discussed in this chapter is the application of SEC to very small and very large quantities of sample. Using microcolumns made of fused-silica capillaries, packed with 5-μm-diameter silica-based SEC media, separations can now be accomplished at the subnanogram level. At the other extreme, quantities of proteins of several kilograms per column and per day are now routinely purified in industrial applications of SEC as a final "polishing" step in the processing of valuable chemoterapeutics.

6.4.1 Determination of molecular weight distributions

The need for a method suitable for the rapid determination of MWDs of polymers was identified long before the introduction of SEC, and the report of Porath and Flodin on gel filtration [4] encouraged the exploitation of SEC for this purpose [22]. Granath and Flodin utilized the new tool for the characterization of low-molecular-weight dextrans [120], and Moore employed the technique to prove a 20-year-old statement regarding the MWD of polypropylene glycols [22].

Since polymer properties are closely related to the MWD of the polymer, SEC has established itself as an important technique for the industrial characterization of polymers and polymerization processes [109]. Sometimes, sufficient qualitative information about the MWD is offered by the raw chromatogram. Thus, Hadad [121] succeeded, with the aid of SEC, in identifying the low content of monomer as the cause of a raw material being unsuitable for production of printed circuit boards. The difference between materials delivered by two suppliers was readily disclosed by the raw chromatograms.

The introduction of rigid media of small particle size enabled the use of SEC for fast process analysis. In 1974 MacLean [122] described an automated system for the complete assay of MWD, including calculations and reporting, within 15 min. If high resolution is not important, the speed may be further increased to yield information within seconds [97]. However, it must be realized that, due to the risk of shear degradation of elongated solutes (Section 6.3.1.1), fast SEC is primarily applicable to the characterization of low-molecular-weight compounds and compact solutes.

The importance of SEC for the characterization of aqueous polymers may be illustrated by the assay of clinical dextran, used as a blood plasma expander. Granath and Kvist [123] pioneered by describing the calibration and use of Sephadex for MWD analysis of dextran. The elution curve was constructed from chemical assays of the dextran concentrations of the collected fractions. The separation step was performed by overnight runs. One of the results reported was that the human renal threshold for dextran is ca. 55 000 in M_W. With the introduction of sensitive on-line refractive index detectors [124] the development of a faster separation step became obvious. Hagel [125] showed that the separation could be completed within 6 h with unchanged accuracy by using crosslinked agarose gels. Basedow and co-workers [40] used dextran fractions for studies of the band-broadening of columns packed with controlled-pore glass. With the introduction of rigid and semirigid media of small particle size the time of analysis has been reduced to ca. 1 h, e.g., for Spheron used by Pitz and Le-Kim [126], for TSK PW used by Alsop and Vlachogiannis [127], and for Superose 6 used by Granath [128]. Fast SEC of dextrans with separation times as short as 8 min on a μ-Bondagel E-linear column was described by Hagel [129]. However, column instability, macromolecule shear degradation, and excessive bandbroadening of macromolecules at high flow velocities put a practical limit on the maximum applicable flow velocity. One hour separation time and 10 μm particle size seem to be optimal with respect to preserving the accuracy of uncorrected molecular weight estimates and eliminating the risk of shear degradation (Sections 6.2.1.1 and 6.3.1.1). Alsop et al. [130] concluded in a review that SEC offers a rapid, comprehensive, and reproducible characterization method of dextran, as compared to intrinsic viscosity and light scattering. Furthermore, Nilsson and Söderlund [131] showed that information about the MWD as obtained by SEC was superior to data from traditional methods, i.e. light scattering and viscosimetry, for judging the clinical applicability of different commercial preparations of dextran.

The apparent MWD of a polymer, obtained by SEC, may be affected by various system effects, such as: bandbroadening [46], sample concentration [76], polymer shear degradation [99], temperature fluctuations [28], pump system variability [132], discontinuities in the poresize distribution of the gel medium [133,134], and variation of detector response with solute molecular weight [135,136]. However, with suitable experimental techniques the system effects can be minimized, and the accuracy of the method will then be around 2% [130,137]. A collaborative test in a number of laboratories, using different columns and calibration procedures, showed a fair agreement between results obtained, i.e. a coefficient of variation of ca. 10% [47]. Andersson [138] concluded that, owing to the high precision of SEC, the accuracy of the MWD estimates is predominantly determined by the accuracy of the reference methods used for obtaining data for the calibration samples.

6.4.2 Size-exclusion chromatography of peptides

The ideal SEC medium for any class of molecules should fulfill the following two criteria: Firstly, there should be no interaction between the solutes to be separated and the gel matrix backbone. Secondly, the matrix should be so designed that the separating volume in the gel particles (V_p), is utilized optimally for any desired fractionation range. For peptides in the molecular weight range 200-10 000 D (2-100 amino acid residues), there are still no such SEC media available possessing an adequate poresize distribution or being free of solute/gel interactions. There is probably no other class of solutes exhibiting such a variety of structural features as the peptides, and interactions with other molecules, soluble or insoluble, are the rule rather than the exception. The possibility of various kinds of interactions during SEC of peptides causing anomalous elution behavior must always be kept in mind, notwithstanding the fact that these interactions can sometimes be utilized to advantage.

In discussing the mechanism of adsorption of peptides on different kinds of SEC matrices one must distinguish between pure adsorption, which arises from interactions between the chemical structures of the peptide and the matrix, and superimposed effects, which are due to the composition of the eluent. Thus, the type and concentration of buffer salts, the pH, and the effect of substances added to the eluent, which compete for the adsorption sites of the peptides, must be taken into consideration. The possible interactions are dependent on the following types of sidechains in both the peptides and gel matrix: charged, hydrophobic, and π-electron-rich.

Ionizable groups are often present at low concentrations in SEC media based on silica, certain polymers, and polysaccharides. These are either inherent, as in silica and agarose, or created either during manufacture or use, as in acrylate gels. The effect of charged groups on peptide elution behavior depends on both pH and ionic strength. At low ionic strength, peptides with charges opposite to those on the matrix tend to be retarded, whereas peptides carrying the same charge are often excluded from the gel phase. These effects are completely eliminated by the addition to the eluent of small amounts of an electrolyte, and are of minor importance in most SEC experiments. However, they must be taken into account in desalting experiments on tightly crosslinked gels. In this case, instead of deionized water, a solution of a volatile salt should be used, which can subsequently be removed by lyophilization.

The elution behavior of peptides carrying hydrophobic or π-electron-rich (aromatic or pseudo-aromatic) sidechains is also directly or indirectly dependent on the pH and salt concentration of the mobile phase. Peptides containing arginine with its pseudo-aromatic, π-electron-rich, and strongly charged guanidino group, are expected to be more retarded when the ionic strength is increased because of a reduction in the thickness of the hydration layer, which allows the peptides to come closer to the potential binding sites on the matrix. The same is true for peptides containing aromatic amino acids. This effect is also seen at pH extremes, causing protonization of carboxyl groups and deprotonization of amino groups.

Due to the above-mentioned effects, peptides containing aromatic or hydrophobic amino acids are difficult to separate by true SEC on any tightly crosslinked polymer matrix, unless the gel/solute interactions are quenched in one way or another. Substances known to possess such characteristics are: pyridine, phenol, acetic acid, trifluoroacetic acid, and urea, either individually or in mixtures. Waltz et al. [139] completely separated CNBr fragments of porcine gastrotropin and nondegraded hormone on Sephadex G-50 SF in 1 *M* acetic acid. Carnegie [140] introduced the solvent system phenol/acetic acid/water (1:1:1, w/v/v), which suppresses the adsorption of peptides containing aromatic amino acid residues on Sephadex G-25, and suggested its use for the estimation of their approximate molecular weights. More recently, Vijayalakshmi et al. [141] and Mant et al. [142] used 0.1% trifluoroacetic acid with added organic solvents, such as methanol (35%) and acetonitrile (up to 60%), for the separation of low-molecular-weight peptides on TSK G2000 SW. Stepanov et al. [143] introduced 8 *M* urea for the suppression of nonideal behavior of peptides in SEC. Mant et al. [142] showed that phosphate-buffered 0.5 *M* potassium chloride, containing 8 *M* urea, could be used to obtain a linear log MW versus retention time relationship for horse heart myoglobin, its cyanogen bromide fragments, and five synthetic peptide standards. The same authors also discussed the possibility of using SEC in the presence and absence of denaturating media as a means of demon-

References on p. A304

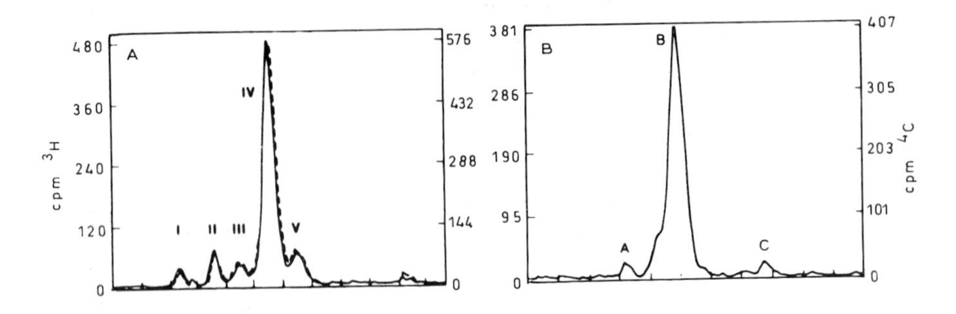

Fig. 6.8. Effect of alkaline borate buffers on the resolution of IgG glycopeptides, obtained after pronase digestion, on Sephadex G-50 SF. Elution profiles in the presence (A) and in the absence (B) of alkaline borate buffer. (Adopted from Ref. 146, with permission.)

strating tertiary or quaternary structures in peptides. That 29- and 36-residue peptides were able to form two-stranded α-helical-coiled coils had previously been shown by Lau et al. [144]. Taneja et al. [145], in a study on the separation of extremely hydrophobic, leucine-rich peptides, containing 13, 22, 26, and 30 residues, used TSK G3000 SW. They observed nonideal behavior, even with 0.1% TFA containing 50% acetonitrile, but the resolution was superior to that obtained on the less porous medium TSK G2000 SW.

Rothman and Warren [146] recently reported that the use of alkaline borate buffers in SEC enhances the resolution of IgG glycopeptides, derived from pronase digestion. A 135 x 1-cm-ID column of Sephadex G-50 SF was equilibrated in a buffer of 45.5 m*M* boric acid/4.5 m*M* sodium tetraborate/2 m*M* EDTA disodium salt/0.02% sodium azide (pH 8.2). The cause of the improved resolution in the presence of borate is not known. It probably has to do with stereospecific interactions of the borate ions with certain oligosaccharide moieties changing their charge and conformation in a favorably way (Fig. 6.8).

Quite a different application of peptide SEC was introduced by Shioya et al. [147]. They experimentally determined a linear relationship between log MW and the optimum plateheight (Section 6.2.2.2) for peptides on TSK G2000 SW. The method required only 1 nmol of substance and was found to be applicable to the estimation of the molecular weight of peptides in the range of 200-10 000 D with a precision of 20%. Interestingly enough, their experimental data are in close agreement with u_{opt} calculated from Eqn. 6.6, with due consideration to the high viscosity of the mobile phase used [50].

In some situations, adsorption phenomena can be used to advantage for the successful separation of molecular species. In an early study, Ruttenberg [148,149] demonstrated how the separation of the cyclic decapeptides tyrocidine A, B, and C (containing 0, 1, and 2 tryptophan residues, respectively) on Sephadex G-25 could be optimized by controlling the acetic acid concentration. Retardation in SEC columns due to hydrophobic interaction has also been observed for low-molecular-weight proteins. Thus, Watson and Kenney [150], using a Protein Pak I-125 column, found that the elution behavior of three recombinant DNA-derived proteins, G-CSF (MW 19 500), IL-2 (ala 125) (MW 15 500) and IFN-gamma (MW 17 000), was strongly influenced by the concentration of 1-propanol, all proteins being eluted at 40%. Obviously, in these cases the columns were not operated in the SEC mode.

6.4.3 Size-exclusion chromatography of DNA

The isolation and purification of double-stranded DNA species, such as plasmids and restriction fragments, is of prime importance in molecular biology. Several methods for the isolation of plasmid DNA have been reported, all with the same objective – their separation from chromosomal DNA. Plasmid DNA is distinguished from its chromosomal counterpart by two structural differences: its closed circular configuration and its much smaller size. These structural differences give rise to physicochemical differences, which have been exploited in the preparative isolation of plasmids by CsCl density gradient centrifugation. However, this technique is very laborious and time-consuming and requires expensive equipment. This is why SEC has recently emerged as an alternative technique, providing, moreover, purer plasmid DNA. Thus, Raymond et al. [151] used Sephacryl® S-1000 to resolve milligram amounts of plasmid DNA in the size range of 4.4-12 kb from residual RNA and chromosomal DNA fragments. These researchers also showed that the columns are reusable, since residual DNA levels were below limits of detection in a transfection assay. McClung and Gonzales [119] separated plasmid DNA (4.1-150 kb) from RNA and protein by selective exclusion, using a less porous gel, Superose 6 Prep. Grade. Plasmid DNA was eluted in the void volume, and this permitted both large sample application and fast separations (0.5 mg plasmid DNA was purified in 20 min). From both studies it was concluded that SEC is a simple, rapid, safe, inexpensive, and reproducible way of obtaining large amounts of highly purified and biologically active plasmid DNA.

Because they are amenable to amplification, the quantities of purified restriction fragments normally required are so small that the demands can be met by extracting submicrogram amounts from agarose gels after electrophoresis. However, the homologous structural feature of DNA fragments should make them ideally suited to fractionation by SEC. There are also certain advantages of using SEC as an alternative to agarose or polyacrylamide gel electrophoresis for the purification of DNA fragments. Thus, there is less risk of contamination by enzyme inhibitors, such as soluble acrylamide or agarose impurities. Also, the base composition of DNA does not influence elution behavior in SEC, while base composition has been demonstrated to have an effect on migration velocity in gel electrophoresis. For fragment length determination, SEC can be accomplished in less than 30 min, and once a column has been calibrated, there is no further need for adding molecular weight markers. In comparison with other chromatographic techniques used for DNA fragment separation, such as ion-exchange chromatography and reversed-phase chromatography, SEC suffers from much lower resolution and should be regarded as a complement to be used in cases when special buffers or separations in the order of the number of base pairs are required. Examples of SEC media that have been successfully applied for separating restriction fragments are TSK G4000 SW, as demonstrated by Kato et al. [152], and Superose 6, as shown by Andersson et al. [153]. It is interesting to note

References on p. A304

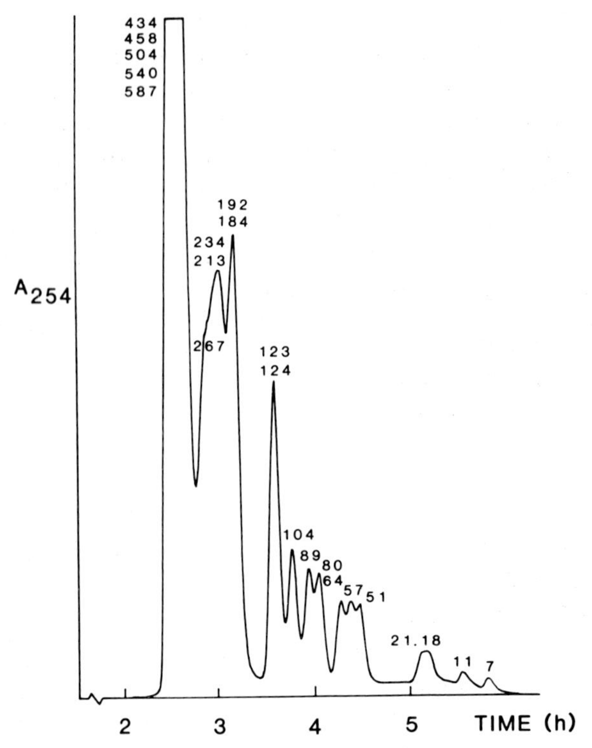

Fig. 6.9. Separation of HaeIII restriction fragments of pBR322 DNA on two Superose® 6 HR 10/30, 30 x 1-cm-ID columns, coupled in series, with 0.05 *M* Tris-HCl (pH 8.0), containing 1 m*M* EDTA at a flowrate of 0.1 ml/min. Peaks are labeled with the number of base pairs of the eluted fragment(s). (Reproduced from Ref. 153, with permission.)

that the separation patterns obtained with the two completely different types of columns are very similar. The latter separation is illustrated in Fig. 6.9. Ellegren and Låås [31] published a report on the use of SEC for accurate chainlength determination of DNA fragments. Addition of 0.15-0.2 *M* NaCl to the eluent buffer suppressed the ionic interactions between the agarose gel matrix and the nucleic acid polymer. The selectivity curve for DNA exhibits different regions, due to the chainlength of DNA, which will cause the molecule to adopt globular, rod-shaped, or flexible-coil configurations (Section 6.2.1.1 and Fig. 6.2).

6.4.4 Microcolumn size-exclusion chromatography

Fused-silica capillaries, 50 cm x 0.25-mm-ID, slurry-packed with 5-μm SynChropac GPC 300, have been applied in the isolation of microquantities of proteins by Flurer et al. [154,155]. Small sample volumes (0.2 μl) were injected, and detection was accomplished with a 150-nl cell at 215 nm. The samples were eluted with a neutral 0.1 *M* phosphate buffer, containing 10% methanol. Nanogram quantities of standard protein mixtures were separated at flowrates of ca. 0.30 μl/min, corresponding to separation times of ca. 1 h. A linear relationship between log MW and elution volume was obtained in the range studied (13.5-165 kD). The extremely low adsorptivity of fused-silica capillary columns was demonstrated in one experiment with a subnanogram sample load of a standard protein mixture. Similar results, but with limited quantitative information regarding the protein load, were reported by Takeuchi et al. [156], who used 80-110 cm x 0.35-mm-ID fused-silica capillary columns, slurry-packed with TSK G3000 SW and TSK G3000 SW_{XL}. Zimina et al. [105] reduced the amount of substances necessary for frontal analysis in the study of protein association by two orders of magnitude by using glass microcolumns (250 x 1-mm-ID) packed with Sephadex G-75 SF or BioGel P-60.

6.4.5 Process size-exclusion chromatography

6.4.5.1 Optimization of productivity

In SEC, productivity is defined as the amount of adequately purified product recovered per unit column cross-sectional area per unit chromatographic cycle time. For any given SEC column packing material the most important parameters governing the resolution are the column length, flowrate, sample volume, and sample concentration, expressed as the relative viscosity of the sample solution. The experimental optimization of these parameters is always carried out in a laboratory-scale column. As discussed in Sections 6.2.2.5 and 6.3.4.3, there is an optimum balance between flowrate and sample volume, which is governed by the requirements for the resolution (i.e. purity) and throughput, and it is possible to use simulations, fine-tuned with data from experiments, for prediction of optimal conditions. This approach was taken by Arve [157] in a study of conditions for the separation of IgG and transferrin with SEC. IgG is often contaminated with transferrin after the ion-exchange chromatography step in the processing scheme, and SEC is an excellent method for the final purification of IgG. The productivity of Superdex 200 Prep. Grade, which is claimed to be designed for these types of separations, and the recovery of IgG at different levels of final purity, were studied as functions of flowrate and sample load volume at varying amounts of contaminating transferrin. Correlations with experimental data were made by computer simulation. The result of one set of conditions, with initial and final IgG purities of 83.3% and 99.9%, respectively, are shown in Fig. 6.10. Of particular interest is the fact that the recovery curves go through a maximum in this operation range. A line drawn through the maxima of these curves should give the combination of load volume and flow velocity at which one should operate to achieve the desired recovery at maximum productivity (Section 6.2.2.5). It is also obvious that in order to increase recovery, the flow velocity and/or the sample volume will have to be decreased at the expense of maximum productivity. When the optimal conditions for maximum productivity have been established, the cycle capacity is increased by proportionally increasing the diameter of the SEC column and the volume of applied sample.

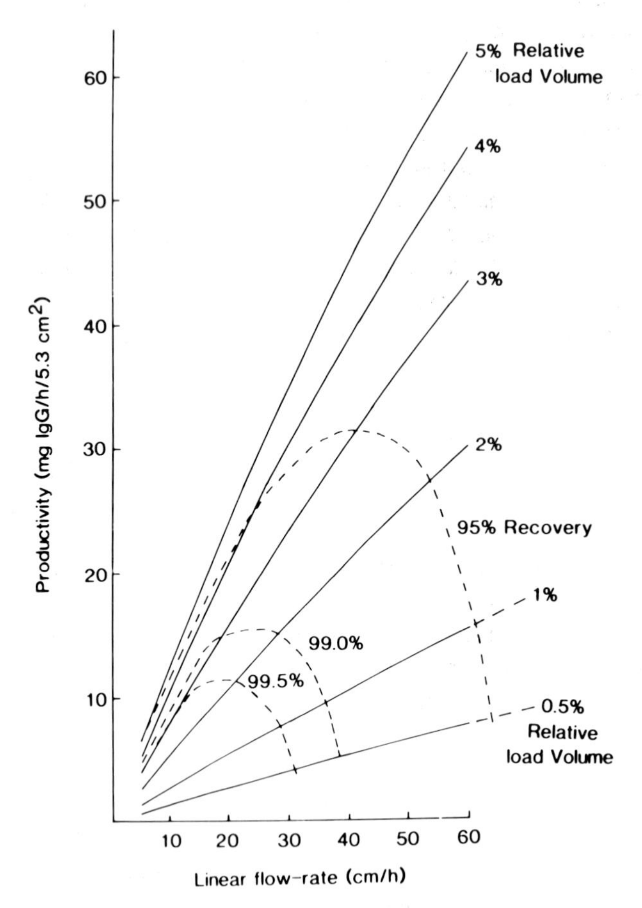

Fig. 6.10. Productivity and recovery of IgG from an artificial mixture, containing transferrin, as a function of eluent velocity and sample volume (in % of total bed volume), purified by SEC on a 60 x 2.6-cm-ID column of Superdex™ 200, Prep. Grade. Computer simulations performed for an initial and final transferrin contamination of 16.7% and 0.01%, respectively. (By courtesy of B. Arve, Pharmacia, Uppsala.)

In order to increase the column productivity in SEC further, deadtimes in the separation scheme should be avoided by leaving no "empty spaces" in the chromatogram. This can only be accomplished in case all sample proteins and contaminants to be separated are eluted within a rather narrow interval. Berglöf et al. [158] described such a favorable situation for the final step in the industrial chromatographic purification of human serum albumin. The peaks to be separated were spaced in such a way that three cycles were performed per eluted column volume, which means that the column productivity could be tripled, as compared to conventional SEC.

6.4.5.2 Selective exclusion (desalting)

It is convenient to distinguish between selective exclusion, or desalting, and molecular fractionation when discussing large-scale SEC. In selective exclusion, gel media are chosen which exclude one of the only two molecular categories to be separated from the gel phase, whereas the other penetrates the gel particles to a larger or lesser extent. The first, and quantitatively the most important, industrial SEC applications of selective exclusion are desalting and deethanolization of protein solutions. Two examples will be given here. In the first example, columns as large as 2500 l (100 x 180 cm ID), packed with Sephadex G-25 Coarse, were used in the middle of the Sixties for the removal of lactose and salts from whey. One such column purified 30 000 kg whey per day, producing a protein solution that generated 28 kg of a 75% protein-rich powder per hour [159]. For economic reasons, the process was soon replaced by membrane-based techniques when the hollow-fiber systems became available. The second example refers to a more long-lived application, the deethanolization of human serum albumin, obtained in the Cohn cold ethanol procedure. Introduced in 1972 by Friedli and Kistler [160] and still in use in many blood fractionation centers around the world, typically 12 l of albumin solution, containing 9% protein and 9% ethanol, is deethanolized in 13.5 min, using a 60 x 40-cm-ID column (75 l bed volume) of Sephadex G-25 Coarse, eluted at a flow velocity of 240 cm/h. The corresponding buffer consumption is 55.5 l and the dilution factor is 1.5-1.9. Periodically and during shutdowns the column is rinsed with 1% NaOH, followed by 1% formaldehyde. This procedure assures a high level of sanitation (bacterial counts usually < 10 per ml).

6.4.5.3 Protein fractionation

The first successful process-scale application of macromolecular separation by SEC was in the industrial purification of insulin. Typically, ca. 50 g of partially purified insulin, dissolved in 2 l of 1 *M* acetic acid, was applied to a stack of six 15 x 37-cm-ID columns, packed with Sephadex G-50 SF Special Grade, and coupled in series (total bed volume, 96 l). With 1 *M* acetic acid as eluent, the operating flowrate was usually 14 l/h, corresponding to a cycle time of approximately 7 h. A second, widespread, large-scale SEC application is the final step in the industrial purification of human serum albumin (HSA). Following two ion-exchange steps and concentration by ultrafiltration, the partially purified HSA solution (6-7% v/w) is fed into a 90-cm column stack, packed with Sephacryl S-200 HR. Column diameters commonly used are 25, 37, 60, and 100 cm, with corresponding total bed volumes ranging from 45 l to 705 l. With a flow velocity of 40 cm/h and a feed sample volume of 4% of the bed volume, applied at optimal intervals, i.e. 3 sample cycles per column elution volume, a productivity of 320 g HSA per 1000 cm^2 column cross-sectional area per h is quite feasible. The corresponding figure for each of the two preceding ion-exchange chromatography steps, run at 120 cm/h, is 250 g per 1000 cm^2 per h. These figures should to some extent dispel the widespread notion that SEC is a low-productivity chromatographic process. Today, there are chromatographic systems for the

References on p. A304

purification of HSA in regular operation, based on one stack of three 30 x 60-cm-ID stainless-steel columns, packed with Sephacryl S-200 HR, which, optimized in the way described above, are producing 5 kg Pharmacopeia-pure protein per day.

REFERENCES

1 *Discuss. Faraday Soc.*, 7 (1949) 114, 135, 321.
2 B. Lindqvist and T. Storgårds, *Nature (London)*, 175 (1955) 511.
3 G.H. Lathe and C.R.J. Ruthven, *Biochem. J.*, 62 (1956) 665.
4 J. Porath and P. Flodin, *Nature (London)*, 183 (1959) 1657.
5 S. Hjertén and R. Mosbach, *Anal. Biochem.*, 3 (1962) 109.
6 S. Hjertén, *Arch. Biochem. Biophys.*, 99 (1962) 466.
7 J.-C. Janson, *Chromatographia*, 23 (1987) 361.
8 A. Polson, *Biochim. Biophys. Acta*, 50 (1961) 565.
9 W. Haller, *Nature (London)*, 206 (1965) 693.
10 J. Uriel, *Bull. Soc. Chim. Biol.*, 48 (1966) 969.
11 J. Porath, J.-C. Janson and T. Låås, *J. Chromatogr.*, 60 (1971) 167.
12 J. Čoupek, M. Křiváková and S. Pokorny, *J. Polym. Sci.*, 42 (1973) 185.
13 R. Epton, C. Holloway and J.V. McLaren, *J. Appl. Polym. Sci.*, 18 (1974) 179.
14 F. Regnier and R. Noel, *J. Chromatogr. Sci.*, 14 (1976) 316.
15 I. Johansson et al., *Sephacryl® S-200 Superfine. For high performance gel filtration*, Pharmacia, Uppsala, 1976.
16 T. Hashimoto, H. Sasaki, M. Aiura and Y. Kato, *J. Polym. Sci., Polym. Phys. Ed.*, 16 (1978) 1789.
17 T. Hashimoto, H. Sasaki, M. Aiura and Y. Kato, *J. Chromatogr.*, 160 (1978) 310.
18 Y. Motozato, *Japanese Kokai Patent*, No. 7807,759 and 7886,749 (1978).
19 E. Brown, R. Joyeau and A. Racois, *Proc. International Symposium Affinity Chromatography, Strasbourg*, Editions INSERM, Paris, Vol. 86, 1979, p. 37.
20 P.-Å. Pernemalm et al., presented at the *Fourth International Symposium on HPLC of Proteins, Peptides and Polynucleotides, Baltimore MD, USA, December 10-12, 1984.*
21 L. Kågedal et al., presented at the *Eight International Symposium on HPLC of Proteins, Peptides and Polynucleotides, Copenhagen, October 31-November 2, 1988.*
22 J.C. Moore, in J. Cazes (Editor), *Liquid Chromatography of Polymers and Related Materials III*, Dekker, New York, 1981, p. 1.
23 J.C. Moore, *Polym. Sci., Part A*, 2 (1964) 835.
24 L. Fisher, in J.Å. Jönsson (Editor), *Chromatographic Theory and Basic Principles*, Dekker, New York, 1987, p. 348.
25 E.F. Casassa and Y. Tagami, *Macromolecules*, 2 (1969) 14.
26 A.M. Basedow, K.H. Ebert, J.H. Ederer and E. Fosshag, *J. Chromatogr.*, 192 (1980) 2590.
27 J.V. Dawkins, in C. Booth and C. Price (Editors), *Comprehensive Polymer Science*, Vol. 1, *Polymer Characterization*, Pergamon Press, Oxford, 1989, p. 231.
28 S.R. Holding, G. Vlachogiannis and P.E. Barker, *J. Chromatogr.*, 261 (1983) 33.
29 J.R. Adams and M.K.L. Bickin, *Anal. Chem.*, 57 (1985) 2844.
30 W.W. Yau and D.D. Bly, in T. Provder (Editor), *Size Exclusion Chromatography (GPC), ACS Symposium Series, American Chemical Society, Washington DC, 1980, p. 197.*
31 H. Ellegren and T. Låås, *J. Chromatogr.*, 467 (1989) 217.
32 P.L. Dubin and J.M. Principi, *Macromolecules*, 22 (1989) 1891.
33 A.C. Ouano and W. Kay, *J. Polym. Sci. Part A-1*, 12 (1974) 1151.
34 C. de Haen, *Anal. Biochem.*, 166 (1987) 235.
35 P. Andrews, in D. Glick (Editor), *Methods of Biochemical Analysis, Vol. 18*, Wiley, New York, 1970, p. 28.

36 H. Determann and J.E. Brewer, in E. Heftmann (Editor), *Chromatography,* Van Nostrand-Reinhold, New York, 3rd Edn., 1975, p. 385.
37 W.W. Fish, J.A. Reynolds and C. Tanford, *J. Biol. Chem.,* 245 (1970) 5166.
38 N. Ui, *Anal. Biochem.,* 97 (1979) 65.
39 W.D. Lansing and E.O. Kraemer, *J. Amer. Chem. Soc.,* 57 (1935) 1369.
40 A.M. Basedow, K.H. Ebert, H. Ederer and H. Hunger, *Makromol. Chem.,* 177 (1976) 1501.
41 T. Ogawa and T. Inaba, *J. Appl. Polym. Sci.,* 20 (1976) 2101.
42 L.H. Tung, *J. Appl. Polym. Sci.,* 10 (1966) 375.
43 T. Ishige, S.-I. Lee and A.E. Hamielec, *J. Appl. Polym. Sci.,* 15 (1971) 1607.
44 W.W. Yau, H.J. Stoklosa and D.D. Bly, *J. Appl. Polym. Sci.,* 21 (1977) 1911.
45 L. Marais, Z. Gallot and H. Benoit, *J. Appl. Polym. Sci.,* 21 (1977) 1955.
46 L.H. Tung, *J. Appl. Polym. Sci.,* 10 (1966) 1271.
47 E.M. Sörvik, *J. Appl. Polym. Sci.,* 21 (1977) 2769.
48 W.W. Yau, J.J. Kirkland, D.D. Bly and H.J. Stoklosa, *J. Chromatogr.,* 125 (1976) 219.
49 W.W. Yau, J.J. Kirkland and D.D. Bly, *Modern Size-Exclusion Liquid Chromatography,* Wiley, New York, 1979, p. 104.
50 L. Hagel, Pharmacia AB, Uppsala, unpublished results.
51 H. Benoit, Z. Grubisic, P. Rempp, D. Decker and J.G. Zilliox, *J. Chim. Phys.,* 63 (1966) 1507.
52 R.P. Frigon, J.K. Leypoldt, S. Uyeil and L.W. Hendersson, *Anal. Chem.,* 55 (1983) 1349.
53 Z. Grubisic, P. Rempp and H. Benoit, *Polym. Lett.,* 5 (1967) 753.
54 D.J. Harmon, in R. Epton (Editor), *Chromatography of Synthetic and Biological Polymers,* Vol. 1, Ellis Horwood, Chichester, 1978, p. 122.
55 D. Berek, I. Novák, Z. Grubisic-Gallot and H. Benoit, *J. Chromatogr.,* 53 (1979) 55.
56 P.G. Squire, *J. Chromatogr.,* 210 (1981) 433.
57 A.E. Hamielec and A.C. Ouano, *J. Liquid Chromatogr.,* 1 (1978) 111.
58 H. Goetz, H. Elgass and L. Huber, *J. Chromatogr.,* 349 (1985) 357.
59 H.G. Barth, *J. Chromatogr. Sci.,* 18 (1980) 409.
60 P.L. Dubin and J.M. Principi, *J. Chromatogr.,* 479 (1989) 159.
61 M.A. Strege and P.L. Dubin, *J. Chromatogr.,* 463 (1989) 165.
62 J.R. Adams and M.K.L. Bickin, *Anal. Chem.,* 57 (1985) 2844.
63 L. Hagel, in J.-C. Janson and L. Rydén (Editors), *Protein Purification, Principles, High Resolution Methods and Applications,* VCH Publishers, New York, 1989, p. 63.
64 J.C. Giddings, *Anal. Chem.,* 39 (1967) 1027.
65 J.H. Knox and H.P. Scott, *J. Chromatogr.,* 316 (1984) 311.
66 J.J. van Deemter, F.J. Zuiderweg and A. Klinkenberg, *Chem. Eng. Sci.,* 5 (1956) 271.
67 J.C. Giddings and K.L. Mallik, *Anal. Chem.,* 38 (1966) 997.
68 J.C. Giddings, in E. Heftmann (Editor), *Chromatography,* Van Nostrand-Reinhold, New York, 3rd Edn., 1975, p. 27.
69 C. Tanford, *Physical Chemistry of Macromolecules,* Wiley, New York, 1961, Chap. 6.
70 A. Rudin and H.K. Johnston, *J. Polym. Sci., Polym. Lett. B,* 9 (1971) 55.
71 K. Granath, *J. Colloid Sci.,* 13 (1958) 308.
72 W.W. Yau, J.J. Kirkland and D.D. Bly, *Modern Size-Exclusion Liquid Chromatography,* Wiley, New York, 1979, p. 336.
73 J.F.K. Huber, *Instrumentation for High-Performance Liquid Chromatography,* Elsevier, Amsterdam, 1978, p. 1.
74 L. Hagel, *J. Chromatogr.,* 324 (1985) 422.
75 S. Mori, *J. Appl. Polym. Sci.,* 20 (1976) 2157.
76 J. Janca, *Anal. Chem.,* 51 (1979) 637.
77 A.R. Cooper, A.J. Hughes and J.F. Johnson, *J. Appl. Polym. Sci.,* 19 (1975) 435.
78 Y. Kato, T. Kametani, K. Furukawa and T. Hashimoto, *J. Polym. Sci., Polym. Phys. Ed.,* 13 (1975) 1695.
79 L. Hagel, *J. Chromatogr.,* 160 (1978) 59.
80 Y. Kato and T. Hashimoto, *J. High Resolut. Chromatogr. Chromatogr. Commun.,* 5 (1982) 577.

81 Y. Kato, T. Hashimoto, T. Murotsu, S. Fukushige and K. Matsubara, *J. High Resolut. Chromatogr. Chromatogr. Commun.*, 6 (1983) 626.
82 O. Chiantore and M. Guaita, *J. Chromatogr.*, 353 (1986) 285.
83 L. Hagel, H. Lundström, T. Andersson and H. Lindblom, *J. Chromatogr.*, 476 (1989) 329.
84 S.H. Chang, K.M. Gooding and F.E. Regnier, *J. Chromatogr.*, 125 (1976) 103.
85 L. Hagel, in P. Dubin (Editor), *Aqueous Size- Exclusion Chromatography,* Elsevier, Amsterdam, 1988, p. 146.
86 F.B. Anspach, A. Johnston, H.-J. Wirth, K.K. Unger and M.T.W. Hearn, *J. Chromatogr.*, 476 (1989) 205.
87 Md.M. Hossain, D.D. Do and J.E. Bailey, *AIChE J.*, 32 (1986) 1088.
88 J.H. Petropoulos, A.I. Liapis, N.P. Kolliopoulos, J.K. Petrou and N.K. Kanellopoulos, *Bioseparation,* 1 (1990) 69.
89 E.P. Kroeff, R.A. Owens, E.L. Campbell, R.J. Johnsson and H.I. Marks, *J. Chromatogr.*, 461 (1989) 45.
90 L.R. Snyder, M.A. Stadalius and M.A. Quarry, *Anal. Chem.*, 55 (1983) 1413.
91 J.W. Dolan and L.R. Snyder, presented at the *12th International Symposium on Column Liquid Chromatography, Washington, DC, June 19-24, 1988.*
92 L. Hagel, presented at the *Symposium on Chromatography of Biopolymers, the 198th ACS Meeting, Miami Beach, September 13, 1989.*
93 L. Hagel and T. Andersson, *J. Chromatogr.*, 285 (1984) 295.
94 P.G. Squire, A. Magnus and M.E. Himmel, *J. Chromatogr.*, 242 (1982) 255.
95 O. Schou and P. Larsen, *J. High Resolut. Chromatogr. Chromatogr. Commun.*, 4 (1981) 515.
96 L.R. Snyder, in E. Heftmann (Editor), *Chromatography,* Elsevier, Amsterdam, 5th Edn., 1991, Chap. 1.
97 C.N. Renn and R.E. Synovec, *Anal. Chem.*, 60 (1988) 200.
98 J.G. Rooney and G. Ver Strate, in J. Cazes (Editor), *Liquid Chromatography of Polymers and Related Materials III,* Dekker, New York, 1981, p. 207.
99 H.G. Barth and F.J. Carlin, Jr., *J. Liq. Chromatogr.*, 7 (1984) 1717.
100 W. Heitz, in J. Cazes (Editor), *Liquid Chromatography of Polymers and Related Materials III,* Dekker, New York, 1981, p. 137.
101 D.C. Shelly and T.J. Edkins, *J. Chromatogr.*, 411 (1987) 185.
102 D.O. Hancock and R.E. Synovec, *Anal. Chem.*, 60 (1988) 1915.
103 S. Mori and M. Suzuki, *J. Chromatogr.*, 320 (1985) 343.
104 A. Hirose and D. Ishii, *J. Chromatogr.*, 411 (1987) 221.
105 T.M. Zimina, V.G. Maltsev and B.G. Belenkii, *J. High Resolut. Chromatogr. Chromatogr. Commun.*, 9 (1986) 111.
106 M. Verzele, C. Dewaele and D. Duquet, *J. Chromatogr.*, 391 (1987) 111.
107 J.H. Knox and H.M. Pyper, *J. Chromatogr.*, 363 (1986) 1.
108 K. Jones, *Chromatographia,* 25 (1988) 437.
109 W.W. Yau, J.J. Kirkland and D.D. Bly, *Modern Size-Exclusion Liquid Chromatography,* Wiley, New York, 1979, p. 2.
110 T. Hashimoto, H. Sasaki, M. Aiura and Y. Kato, *J. Chromatogr.*, 160 (1978) 301.
111 R.A. Jenik and J.W. Porter, *Anal. Biochem.*, 111 (1981) 184.
112 W. Kopaciewicz and F.E. Regnier, *Anal. Biochem.*, 126 (1982) 8.
113 E. Pfannkoch, K.C. Lu, F.E. Regnier and H.G. Barth, *J. Chromatogr. Sci.*, 18 (1980) 430.
114 Å. Haglund, cited in T. Andersson, M. Carlsson, L. Hagel, P.-Å. Pernemalm and J.-C. Janson, *J. Chromatogr.*, 326 (1985) 33.
115 P. Dubin and J.M. Principi, *Anal. Chem.*, 61 (1989) 780.
116 S.A. Karapetyan, L.M. Yakushina, G.G. Vasijarov and V.V. Brazhinkov, *J. High Resolut. Chromatogr. Chromatogr. Commun.*, 8 (1985) 148.
117 T. Andersson, Pharmacia, Uppsala, personal communication, 1989.
118 P.A. Bristow and J.H. Knox, *Chromatographia,* 10 (1977) 279.
119 J.K. McClung and R.A. Gonzales, *Anal. Biochem.*, 117 (1989) 378.
120 K. Granath and P. Flodin, *Makromol. Chem.*, 48 (1961) 160.

121 D.K. Hadad, in J. Cazes (Editor), *Liquid Chromatography of Polymers and Related Materials III,* Dekker, New York, 1981, p. 157.
122 W.W. MacLean, *J. Chromatogr.,* 99 (1974) 425.
123 K.A. Granath and B.E. Kvist, *J. Chromatogr.,* 28 (1967) 69.
124 L. Hagel, *Anal. Chem.,* 50 (1978) 569.
125 L. Hagel, *J. Chromatogr.,* 160 (1978) 59.
126 H. Pitz and D. Le-Kim, *Chromatographia,* 12 (1979) 155.
127 R.M. Alsop and G.J. Vlachogiannis, *J. Chromatogr.,* 246 (1982) 227.
128 K. Granath, Pharmacia, Uppsala, personal communication, 1984, cited in T. Andersson, M. Carlsson, L. Hagel, P.-Å. Pernemalm and J.-C. Janson, *J. Chromatogr.,* 326 (1985) 33.
129 L. Hagel, *Abstracts of Uppsala Dissertation from the Faculty of Science,* 453, Almqvist & Wiksell, Stockholm, 1978, pp. 37-43.
130 R.M. Alsop, G.A. Byrne, J.N. Done, I.E. Earl and R. Gibbs, *Process. Biochem.,* 12 (1977) 15.
131 K. Nilsson and G. Söderlund, *Acta Pharm. Suec.,* 15 (1978) 439.
132 D.D. Bly, H.J. Stoklosa, J.J. Kirkland and W.W. Yau, *Anal. Chem.,* 47 (1975) 1810.
133 M.R. Ambler, L.J. Fetters and Y. Kesten, *J. Appl. Polym. Sci.,* 21 (1977) 2439.
134 P.C. Christopher, *J. Appl. Polym. Sci.,* 20 (1976) 2989.
135 A.M. Basedow, K.H. Ebert and U. Ruland, *Makromol. Chem.,* 179 (1978) 1351.
136 E.M. Barrall II, M.R.J. Cantow and J.F. Johnson, *J. Appl. Polym. Sci.,* 12 (1968) 1373.
137 G. Nilsson and K. Nilsson, *J. Chromatogr.,* 101 (1974) 137.
138 L. Andersson, *J. Chromatogr.,* 325 (1985) 37.
139 D.A. Walz, M.D. Wider, J.W. Snow, C. Dass and D.M. Desiderio, *J. Biol. Chem.,* 263 (1988) 14189.
140 P.R. Carnegie, *Biochem. J.,* 95 (1965) 9P.
141 M.A. Vijayalakshmi, L. Lemieux and J. Amiot, *J. Liq. Chromatogr.,* 9 (1986) 3559.
142 C.T. Mant, J.M.R. Parker and R.S. Hodges, *J. Chromatogr.,* 397 (1987) 99.
143 V. Stepanov, D.H. Handschuh and F.A. Anderer, *Z. Naturforsch.,* 16b (1961) 626.
144 S.Y.M. Lau, A.K. Taneja and R.S. Hodges, *J. Biol. Chem.,* 259 (1984) 13253.
145 A.K. Taneja, S.Y.M. Lau and R.S. Hodges, *J. Chromatogr.,* 317 (1984) 1.
146 R.J. Rothman and L. Warren, *Biochim. Biophys. Acta,* 955 (1988) 143.
147 Y. Shioya, H. Yoshida and T. Nakajima, *J. Chromatogr.,* 240 (1982) 341.
148 M. Ruttenberg, Doctoral Dissertation, Rockefeller University, New York, 1965.
149 J.-C. Janson, *J. Chromatogr.,* 28 (1967) 12.
150 E. Watson and W.C. Kenney, *Biotechnol. Appl. Biochem.,* 10 (1988) 551.
151 G.J. Raymond, P.K. Bryant III, A. Nelson and J.D. Johnson, *Anal. Biochem.,* 173 (1988) 125.
152 Y. Kato, M. Sasaki, T. Hashimoto, T. Morotsu, S. Fukushige and K. Matsubara, *J. Chromatogr.,* 266 (1983) 341.
153 T. Andersson, M. Carlsson, L. Hagel, P.-Å. Pernemalm and J.-C. Janson, *J. Chromatogr.,* 326 (1985) 33.
154 C.L. Flurer, C. Borra, F. Andreolini and M. Novotny, *J. Chromatogr.,* 448 (1988) 73.
155 C.L. Flurer, C. Borra, S. Beale and M. Novotny, *Anal. Chem.,* 60 (1988) 1826.
156 T. Takeuchi, T. Saito and D. Ishii, *J. Chromatogr.,* 351 (1986) 295.
157 B. Arve, Pharmacia, Uppsala, personal communication, 1989.
158 J.H. Berglöf, S. Eriksson and I. Andersson, in S. Karger (Editor), *Develop. Biol. Standard.,* Vol. 67, Basel, 1987, p. 25.
159 L.O. Lindquist and K.W. Williams, *Dairy Ind. Int.,* 38 (1973) 459.
160 H. Friedli and P. Kistler, *Chimica,* 26 (1972) 25.

Chapter 7

Affinity chromatography

T.M. PHILLIPS

CONTENTS

7.1 INTRODUCTION AND HISTORICAL OVERVIEW

Affinity chromatography is an isolation method based on the principle that the molecule to be purified can form a selective but reversible interaction with another molecular species, which has been immobilized on a suitable chromatographic support. The technique is widely used to isolate a variety of biological materials, including enzymes, membrane receptors, antibodies, antigens, viruses, and intact cells by their reactivity with specific immobilized substances.

Bioselectivity as a means of separating biologically active materials was first described in 1910 by Starkenstein [1], who found that α-amylase bound tightly to insoluble starch. This was followed by the work of Holmbergh [2], who reported the separation of the α and β forms of amylase by chromatography on starch gels. Microbial amylases were separated by Thayer [3], who allowed the enzymes to be adsorbed on insolubilized starch and recovered them by bioselective elution, using soluble starch. In a similar fashion, DNA- and RNA-binding proteins have been isolated, by adsorption on nucleic acids immobilized on cellulose [4-6]. In 1951 Campbell et al. [7] introduced immunoaffinity chromatography by using antigens immobilized on diazotized cellulose to purify specific antibodies. Other researchers have immobilized antigens on fine particles of diazotized cellulose [8], ester-derivatized cellulose [9,10] and bromoacetylcellulose [11] to isolate specific antibodies.

The field rapidly expanded with the introduction of (a) support matrices, such as Sephadex [12] and beaded agaroses [13], and (b) the demonstration by Axen et al. [14] that molecules could be immobilized on cyanogen bromide-activated polysaccharide supports via their primary amine groups. The technique was further enhanced by the introduction of "spacer arms", which hold the immobilized ligand away from the support matrix surface to overcome the problem of steric hindrance [15]. The work of Inman and Dintzis [16] demonstrated that synthetic polyacrylamide supports could also be chemically modified to produce suitable supports for ligand immobilization. Since these early experiments, affinity chromatography has rapidly developed and promoted the introduction of new synthetic and inorganic supports, such as Enzacryl [17], Spheron [18], glass [19], silica [19], and methacrylate [20]. The introduction of glass and silica, which are incompressible, allowed affinity chromatography to enter the world of high-performance liquid chromatography.

In addition to biospecific interactions, affinity chromatography has expanded to include a number of associated techniques which are based on other types of interactions as the separation mechanism. These techniques include dye-ligand affinity chromatography [21] and immobilized-metal-ion affinity chromatography (IMAC) [22].

7.2 BASIC CONCEPTS OF BIOSELECTIVE CHROMATOGRAPHY

Bioselective chromatography is usually performed in columns, which are packed with a suitable support. This support is coated with an immobilized molecule or ligand, which is chosen to perform a specific separation because of the specificity and strength of the interaction between itself and the molecule to be isolated (analyte). Once this interaction has taken place, the ligand immobilizes the analyte until the complex is dissociated and the bound material is released. This dissociation usually takes place after the nonreactive molecules in the sample mixture have passed through the column. In this way, the analyte is isolated in a pure, active form. Fig. 7.1 illustrates the general principles of bioselective affinity chromatography.

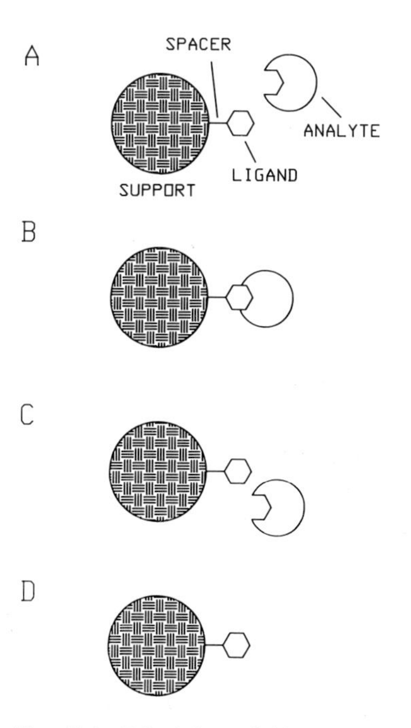

Fig. 7.1. Principles of bioselective sorption. (A) The support presents the immobilized ligand to the analyte to be isolated. (B) The analyte makes contact with the ligand and attaches itself. (C) The analyte is recovered by the introduction of an eluent, which dissociates the complex holding the analyte to the ligand. (D) The support is regenerated, ready for the next isolation.

7.2.1 Bioselective reactions

Although several theoretical models have been formulated to study bioselective reactions [23], no single model can encompass the wide range of interactions which occur in affinity interactions. These interactions arise not only from specific reactions between the substance to be isolated and the immobilized ligand but also as nonspecific interactions between the substance and either the support matrix or the spacer arm.

The basic equilibrium reaction can be simply stated as the formation of a reversible reaction complex (AL) between the analyte (A) and the immobilized ligand (L) [24].

References on p. A335

$$A + L \underset{k_{-1}}{\overset{k_1}{\rightleftharpoons}} AL \tag{7.1}$$

The equilibrium or dissociation constant (K) of this reaction, which is an indication of the strength of the reaction, can be stated as

$$\begin{aligned} K &= k_{-1}/k_1 \\ &= [A][L]/[AL] \end{aligned} \tag{7.2}$$

To use this relationship with a gel matrix which contains immobilized ligand, Eqn. 7.2 must be altered to accommodate the small volume of solution residing within the support matrix. One must also substitute L_O (moles/v) for the concentration of the immobilized ligand and V for the volume of analyte, at concentration A_O, added to the gel. At equilibrium, the concentration of bound analyte in the complex [AL] can be calculated from

$$[AL] = L_O A_O V / K(V + v) + L_O v + A_O V \tag{7.3}$$

This equation can be used in the majority of situations where A_O is less than L_O.

The fraction of analyte bound to the ligand at equilibrium can be estimated from

$$\text{bound analyte / free analyte} = [AL]v / A_O V = L_O v / K(V + v) + L_O v + E_O V \tag{7.4}$$

Once the amount of bound analyte has been established, the next problem is to estimate the amount of bound material that can effectively be recovered during the elution phase. In simple pH change elution, we can assume that (1) v is equilibrated to V and (2) that a volume of elution buffer (V') is added to v and equilibrated at a new, higher K (K').

$$\text{Recovered } A = 1 - \{L_O v\,(Kv + L_O v) \,/\, [K(V + v) + L_O v][K'(V' + v) + L_O v]\} \tag{7.5}$$

The analyte removed in V is calculated from

$$AV / A = [V/(V + v)][1 - L_O v / K(V + v) + L_O v] \tag{7.6}$$

and the difference between Eqns. 7.5 and 7.6 when multiplied by the correct volume fraction gives the amount of analyte recovered. Recovery is strongly influenced by the ratio of V' to v and less by the ratio of V to v.

From Eqns. 7.3 and 7.6 we can show that effective affinity chromatography can be achieved only when we can fulfill the condition $L_O \gg K$. The upper limit value for ligand concentration is in the range 10^{-3} - 10^{-2} M. If we want the separation to be efficient, then the lower limit for K must fall in the range 10^{-3} - 10^{-4} M. It can be further inferred that bioselective complexes with $K < 10^{-4}$ require a minimum ligand concentration.

7.2.2 Immunoaffinity reactions

Immunoaffinity separation depends on the specificity and selectivity of an immobilized immunological reagent (antibody or antigen) to perform the separation. Like other bioselective interactions, immunoaffinity separations are performed in two phases: (a) a primary sorption phase, followed by (b) a desorption or elution phase. In the primary phase, an analyte in solution reacts with the immobilized immunological ligand to form an immobilized antibody/antigen complex. The bound material is recovered during the second or elution phase, by breaking the antibody/antigen complex and thus freeing the analyte. The general reactions governing immunoaffinity chromatography are outlined in the two equations given below

Immobilized antibody + antigen → immobilized complex (7.7)

Immobilized complex + eluting agent → free antigen (7.8)

7.2.2.1 Antibody/antigen interactions

The basis for the formation of an antibody/antigen immune complex is schematically represented in Fig. 7.2. It lies in the potential for the combining sites or antigen receptors on each of the arms of the antibody molecule to form functional cups, which can conform to the shape made by the electron cloud of the antigenic determinant of the antigen. This conformational matching is important for maintenance of "closeness of fit" between the antigen electron cloud shape and the "space-fill" of the area within the antibody combining

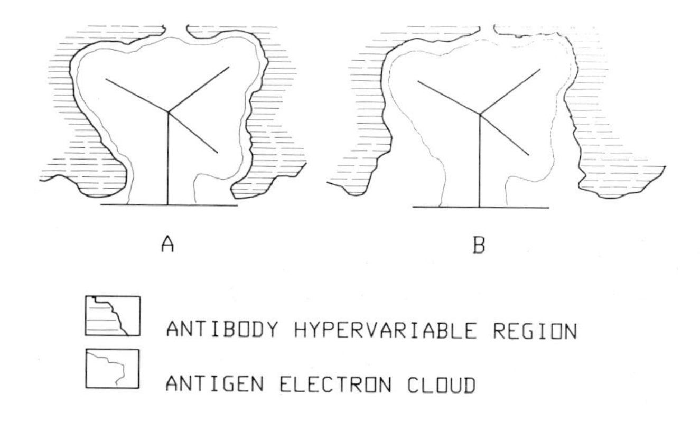

Fig. 7.2. Diagram of the conformational matching which takes place between the antibody receptor and the electron cloud shape of the antigen. (A) Good "closeness of fit", found in strong antibody/antigen reactions. (B) Poor "closeness of fit", resulting in weak reactions. (Reproduced from *Advances in Chromatography,* 29 (1989) 139 with permission.)

site. Once an antigen determinant has come into close contact with the specific antigen receptor, the bonding forces come into play. These forces are ionic attraction, hydrogen bonds, hydrophobic attractive forces, and Van der Waal forces. The latter create attractive forces that are inversely proportional to the 6th power of the distance [25,26]. The "closeness of fit" between the antigenic determinant and the reactive antibody thus dictates the strength of the interaction.

In addition, there are other forces which contribute to the overall strength of the reaction. Forces of steric repulsion arise from the interpenetration of the electron clouds of the unbound parts of the molecules. These forces also affect the reaction binding strength of the antibody/antigen reaction. Therefore, when the shapes of the antibody receptors and the antigenic determinants are complementary, the repulsive forces are weak and the attractive forces are strong. This situation results in a strong affinity between the two reactants [27]. At equilibrium, antibody affinity can be thought of as representing the sum of both the attractive and repulsive forces arising from the interaction between the antigen receptor on the antibody and the antigenic determinant on the antigen.

This equilibrium can be expressed as

$$\mathrm{Ab} + \mathrm{Ag} \xrightarrow[k_d]{k_a} \mathrm{AbAg} \qquad (7.9)$$

where Ab is free antibody, Ag is free antigen, and the antibody/antigen complex is represented by AbAg. The association and dissociation constant are represented as k_a and k_d, respectively.

Applying the Law of Mass Action to the above equation

$$k_a\,[\mathrm{Ab}][\mathrm{Ag}] = k_d\,[\mathrm{AbAg}] \qquad (7.10]$$

and

$$K = [\mathrm{AbAg}] / [\mathrm{Ab}][\mathrm{Ag}] \qquad (7.11)$$

where $K = k_a / k_d$ (the affinity constant of the reaction). Conditions for immunoaffinity are ideal when the K value for the antibody/antigen reaction lies between 10^{-4} and 10^{-8} M in free solution. The general rule for immunoaffinity separations is: When $K > 10^{-4}$ M, the affinity of the reaction is weak and loss of retained antigen will occur at high flowrates. When $K > 10^{-8}$ M, the affinity is extremely strong, and often specialized elution techniques will be required to recover the retained antigen.

7.3 SUPPORT MATRICES

A matrix suitable for affinity chromatography must possess several specialized properties, such as a macroporous structure, presenting a large surface area, which is mechanically stable in water and inert. The latter property is required to ensure that nonspecific adsorption does not take place during the affinity separation. The matrix must be chemically stable and possess suitable sidechains that can easily be modified for ligand attachment. The chemical stability of the matrix must cover a wide pH range (pH 1.5-11) to ensure that the matrix will remain intact during the elution or recovery phase of the separation. Currently, there are many matrices to choose from, ranging from single-composition matrices to dual-composition supports, which are sometimes called "second-generation" matrices. Although particulate and fibrous supports are available, a beaded structure is generally found to be the easiest to work with and the most efficient at maintaining a good flow in column chromatography. In addition, the support material must be easily modifiable to accept suitable sidechains which are stable under elution conditions.

7.3.1 Organic or biopolymer supports

Supports are made of biological materials, such as cellulose, dextran, and the most popular of all, agarose. For affinity chromatographic purposes all of these materials are usually used in a beaded form. Cellulose, the first material to be used for attachment of affinity ligands, was used in either strands or microcrystalline form to isolate either antibodies [7] or a variety of enzymes [28-30]. The primary structure of cellulose, linear units of β-1,4-linked D-glucose with occasional 1,6-bonds, makes the material fibrous and uniform. This does not promote high flowrates when cellulose is packed into a column [31,32]. Cellulose has lost its place as the major affinity matrix since the introduction of beaded polysaccharides (i.e. dextran and agarose), which permit high flowrates. However, when used in suspension as a batch process matrix, cellulose has the advantage of being cheaper than most dextran or agarose beads, being mechanically strong and chemically inert. The preparation of macroporous cellulose in a beaded form was first described by Peske et al. [33]. This type of support has been used for the immobilization of nucleic acids [34] and dyes [35].

The next most popular material is beaded dextran. Dextran is a microbial polysaccharide, composed of α-1,6-linked glucose polymers, the most frequently used of which contains 5% α-1,3-sidebranches. These polymers can be crosslinked in alkaline medium with epichlorohydrin to produce a beaded gel which has been the universal support for size-exclusion chromatography [36]. The beads are highly hydrophilic due to the abundance of OH groups, which can be used for chemical attachment of ligand linkage groups [37]. Although the beads are stable in 0.1 *M* hydrochloric acid and in most alkaline media, they are easily broken down by strong acids and oxidants. The Sephadex series of matrices, produced by Pharmacia, are classical examples of this type of support.

More recently, both cellulose and dextran supports have given way to agarose beads, which possess a reasonable degree of mechanical strength and are easily derivatized. Agarose was first introduced as an affinity support by Cuatrecasas et al. [38] and since then has become one of the most popular materials in affinity chromatography. This popularity has been enhanced by the useful range of commercially available beaded agaroses, offered by such companies as Pharmacia, Bio-Rad, and Pierce. Agarose is a purified, linear, water-soluble polysaccharide, composed of alternating units of 1,3-linked β-D-galactose and 1,4-linked 3,6-anhydro-α-L-galactose [39], which is prepared in a beaded form by the techniques of Hjertén [40] and Bengtsson and Philipson [41]. Beaded agarose is mechanically stable and very hydrophilic. It is easily modified via its abundant OH groups, found throughout the entire matrix, and under most experimental conditions, it does not posses undue nonspecific binding properties. One reported disadvantage is a tendency to bind certain proteins and nucleic acids, such as tRNA, nonspecifically, when used under conditions which involve high salt concentrations, such as high-molarity salt gradients, and chaotropic eluents [42,43].

Several variations are commercially available, such as Sepharose and Superose 6B (Pharmacia), which is a rigid, highly crosslinked gel that can withstand flowrates up to 0.3 ml/min. Large beads of crosslinked agarose (i.e. Sepharose 6MB) are ideal matrices for the affinity isolation of intact cells.

7.3.2 Inorganic supports

Inorganic material such as glass, silica, and alumina have been used as affinity supports for the immobilization of enzymes. These supports are mechanically stable and are resistant to microbial attack. The work of Weetall and Filbert [19,44] has helped to develop this type of support into a useful column material for medium- to high-pressure affinity chromatography. Glass beads have many of the features desired in affinity chromatography; they are mechanically durable and thermally stable. The beads are made by heating mixtures of borosilicate glass to high temperatures for prolonged periods of time. Although controlled-pore glass beads (Pierce) are usually employed, solid glass beads (Sigma Chemical Co.) are becoming popular as affinity supports. Glass beads are usually composed of 96% silica and 4% borate plus traces of other inorganic oxides, with over 30% of the borate at the bead surface. Unfortunately, the presence of these boron groups together with the SiOH groups, which have a slightly negative charge, cause glass to adsorb proteins nonspecifically on its surface [45]. Regnier and Noel [46] described a technique for applying a "glycerol" surface to glass beads, using glyceropropylsilane as the coating agent. In all cases, silanization of the bead surface not only reduces the nonspecific adsorption of proteins on the bead surface but also provides groups for the attachment of reactive sidechains, which can be used for ligand attachment. The use of protein-coated solid glass beads as a suitable support for antibody attachment has been described by Babashak and Phillips [47,48].

Silica is a material which is gaining popularity as a support for high-performance affinity applications, especially in its microparticulate form [49-51]. Silica possesses many of the advantages and disadvantages of glass, the major disadvantage being its solubility in even slightly alkaline buffers. Macroporous silica particles have been chemically modified with a thin coating of hydrophilic polymers, such as 6-acylamidohexanoic acid and *N*-droxysuccinimidylacrylic acid ester, to form both a protective coat and a spacer arm for ligand attachment [52]. Diol-bonded silica, which has superior ligand binding capacities, is commercially available, although the material can easily be made in the laboratory [53,54].

7.3.3 Synthetic supports

Although many different synthetic polymers are available for use as affinity supports, the most popular and readily available supports are made of polyacrylamide. Polyacrylamide beads are composed of a crosslinked hydrocarbon matrix to which carboxyamide groups are atttached. This material is made by the copolymerization of acrylamide with bis-acrylamide. When produced in a beaded form, the matrix contains a hydrocarbon structural framework with abundant carboxamide sidechains. In their classical paper, Inman and Dintzis [16] describe several chemical pathways for the modification of polyacrylamide beads to make matrices suitable for affinity ligand attachment. Scouten also gives an excellent overview of this subject in his book on affinity chromatography [55].

A pre-activated form of polyacrylamide, called Enzacryl, is available for protein immobilization (Koch-Light). The gel is made by copolymerization of *N*-acrylmorpholine and *N,N'*-methylene-bisacrylamide. The support is available in several forms: Enzacryl AH, which contains reactive hydrazide sidechains; Enzacryl AA, which contains aromatic acid residues; and the polyacetal version, which binds proteins via its amino groups.

Methacrylate beads are prepared by copolymerization of hydroxyethylmethacrylate and ethylenedimethacrylates. The beads are rigid and are relatively thermostable up to 250°C, but they are affected by some organic reagents. The rigidity of the beads maintains high flowrates, and this enables them to be used in high-performance liquid chromatography. Ligand coupling is effected via OH groups found on the surface of the beads [56]. Although these beads exhibit a reasonably high degree of nonspecific binding, they can effectively be used for hydrophobic-interaction separations of proteins and peptides [57]. Oxiran or epoxide sidegroups are the most popular attachment groups for the coupling of proteins to plastic affinity matrices, although other forms of the beads contain *p*-nitrophenyl ester groups [20]. Spheron (Koch-Light) is a macroporous bead, made from hydroxyethylmethacrylate. The beads are commercially available with reactive hydroxyl, carbonyl, sulfonyl, and aminoaryl groups, suitable for affinity ligand attachment, derivatized on the bead surface.

Another group of synthetic supports is the Trisacryl (IBF Biotechnics) synthetic gel, made by polymerizing a unique monomer, *N*-acryloyl-2-amino-2-hydroxymethyl-1,3-propanediol. The polymer possesses three available hydroxymethyl groups per repeating

References on p. A335

unit. Trisacryl beads are hydrophilic, not biodegradable, and thermostable between −20°C and 120°C. The beads are easily modified to carry excess amino groups, which can be used to immobilize ligands for the affinity separation of proteins and living cells [58].

Oxirane acrylic beads are produced by the copolymerization of methacrylamide, methylene-bis-methacrylamide and either glycidylmethacrylate or allylglycidyl ether. The beads are electroneutral and moderately hydrophilic and are available in sizes ranging from 30 μm diameter (Sigma Chemical Co.) to 150 μm diameter (Eupergit C, Rohm Pharma). Ligand immobilization is achieved through reaction of amine or hydroxyl groups, on the ligand, with the oxirane (epoxide) groups on the bead matrix over a wide pH range [59], but this is a slow process.

7.3.4 Other supports

The combination of agarose with acrylamide polymers, called Ultrogel, was described as possessing the tensile strength of agarose and the sieving qualities of polyacrylamide. Ultrogels (IBF Biotechnics) are commercially available forms of these gels, containing both the amide groups of the polyacrylamide and the hydroxyl groups of the agarose for ligand attachment. Other combination gels are the Sephacryl series (Pharmacia), which combine the strength of polyacrylamide with the advantages of dextran. The beads are made by copolymerizing *N*,*N*-bisacrylamide with allyl dextran. Although this material is popular in gel chromatography, owing to its extensive porosity, Sephacryl has received little or no attention from affinity chromatographers. Porath [60] described a new type of dense agar-agarose gel, which was made by shrinking the gel in methanol before crosslinking. The particles formed by this technique are mechanically stable and can be activated for ligand attachment by the techniques described above for agarose beads.

Magnetic beads are a ligand support which is becoming popular for batch separation techniques, where the affinity ligand can easily be recovered with a magnet. Magnogel (IBF Biotechnics), a 4% w/v epichlorohydrin-crosslinked agarose bead, which incorporates 7% w/v ferrous tetroxide into its composition, is a typical example of this type of support [61]. Ligand coupling is performed by any of the techniques described for the activation of agarose supports, although the beads are better suited to batch purifications and immunoassays. A magnetic agarose polyacrylamide gel, Magnogel AcA 44 (IBF Biotechnics), which is based on the Ultrogel series, is also available. Other examples are made from iron oxide particles, coated with Enzacryl (Koch-Light). These beads are used in stirred-batch reactors or fluidized beds, where they can be easily recovered after the purification is completed.

7.4 IMMOBILIZATION CHEMISTRY

7.4.1 General considerations

The basis of affinity isolations is the immobilization and presentation of the ligand to the analyte. Ligands can be chemically attached, either directly to activated supports or via a spacer molecule, by a series of different reactions. Most of these reactions involve mixing the ligand with the activated support at a given pH or in the presence of a crosslinking agent. After selecting a suitable support matrix, chemical modification of the reactive sidechains of the support is required in order to attach the specific ligand. Attachment of a ligand can be performed in two ways: (a) chemically activating a matrix in the laboratory or (b) by choosing a commercially available matrix. In the former case, it is necessary to derivatize the support sidechains to form suitable reactive groups before attachment of the ligand can be achieved. When using the commercially available supports, it is a simple matter to activate the pre-derivatized sidechains under suitable conditions and then add the ligand.

When any ligand is attached to an activated support matrix it must always be remembered that not all of the available sidechains may have been coupled to the ligand. This leaves uncoupled sidechains, which can react with and bind both the specific and non-specific components of the materials being separated. To avoid this situation, the investigator must always block any unused sidechains, following ligand attachment. Once the support matrix has been activated and a suitable spacer molecule attached between the support and the ligand, all unreacted groups must be neutralized before using the matrix for affinity separations. In most cases, this is easily accomplished by incubation of the ligand-coated support with a suitable blocking agent, such as 1 *M* 2-aminoethanol (pH 8.5-9.0) for 2-4 h at room temperature. Normal human IgG Fc has been employed as a blocking agent for immunoaffinity supports of antibodies used to isolate human IgE [62].

7.4.2 Activation of biopolymer supports

Biopolymer supports, such as cellulose, dextran, and agarose, may be activated by several different techniques, many of which are applicable to other support types. The most common activating agent is cyanogen bromide. Cyanogen bromide activation of polysaccharide supports, first described by Axen et al. [14], has become one of the most popular methods of introducing attachments in affinity chromatography and can be applied to all biopolymer and some synthetic supports. The activation process involves incubation of the support with a solution of aq. cyanogen bromide (2:1) in potassium phosphate buffer (pH 11.5) for 10 min at 4°C. The support is washed in cold, distilled water before attaching the spacer arm or ligand. Attachment of the ligand is effected via primary aromatic or aliphatic amino groups and involves incubation of the activated support in sodium carbonate buffer (pH 9.0), containing the ligand for 12-24 h at 4°C, with constant mixing. Fig. 7.3 outlines the chemical reactions involved in cyanogen bromide activation of biopolymer supports.

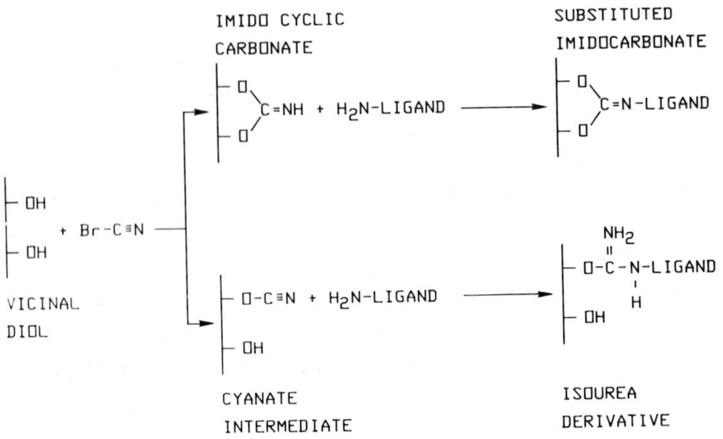

IMIDO CYCLIC CARBONATE

SUBSTITUTED IMIDOCARBONATE

C=NH + H_2N-LIGAND

C=N-LIGAND

OH

+ Br-C≡N

OH

VICINAL DIOL

NH_2

O-C-N-LIGAND

H

O-C≡N + H_2N-LIGAND

OH

OH

CYANATE INTERMEDIATE

ISOUREA DERIVATIVE

Fig. 7.3. Cyanogen bromide activation of biopolymer supports. Reaction of vicinal diol groups, on the support matrix, with cyanogen bromide leads to the formation of imido cyclic carbonate groups, which are predominant only in crosslinked dextrans, and reactive cyanate intermediate groups, which are important in the activation of agarose supports. Interaction between amino groups, on the ligand, and the imido cyclic carbonate leads to ligand immobilization via a substituted imidocarbonate linkage, while reaction with the cyanate intermediate leads to ligand immobilization via an isourea derivative.

Biopolymer supports can also be activated by the triazine method, consisting of treatment of the matrix with 2-amino-4,6-trichloro-*s*-triazine, during which one chlorine atom is replaced by a solubilization group. The ligand is covalently bound via a primary amino group in a secondary step, during which another chlorine atom undergoes nucleophilic substitution. The triazine method is extensively used to bind reactive dyes to polysaccharide surfaces, especially agarose [63].

Oxidation of polysaccharide supports has become a popular activation technique for the immobilization of proteins. Reactive vicinal *cis*-hydroxyl groups within the matrix can be modified by treatment with sodium periodate [64]. This oxidation yields aldehydes, which can easily be converted to secondary amines by reductive amination or to hydrazides by reaction with dihydrazides. Attachment of spacer arms or ligands is effected via primary amine groups. This technique is suitable for activation of only the highly cross-linked dextrans or agaroses and is not applicable to the activation of Sephadex (Pharmacia) beads with high G-numbers.

Reaction of agarose OH groups with a solution containing epichlorohydrin and sodium borohydride produces a highly stable matrix, which can withstand attack by denaturants such as urea, iodide, and isothiocyanate [65]. The reaction will also produce activated supports with oxiran groups. This reactive group has become popular for the attachment of proteins and peptides to both polysaccharide and plastic supports.

7.4.3 Activation of inorganic supports

Before an inorganic matrix, such as glass or silica, can be used as an affinity support it must be treated with silane to reduce the weak ion-exchange properties of the matrix. Silanization or treatment with an amino-functional trialkoxysilane prepares the surface for further derivatization, essential for covalent attachment of the ligand. This process can be performed by several different techniques: (a) by refluxing the matrix overnight in a 10% solution of 3-aminopropyltriethoxysilane; (b) by placing the matrix in a 10% aq. solution of silane, which is boiled for 4 h, followed by drying at 110°C; and (c) treatment with triethoxypropylglycidoxysilane [19,44,47,66]. Silanization can also be easily performed by refluxing glass beads or silica in τ-mercaptopropyltrimethoxysilane, followed by the attachment of oxirane groups. This procedure produces a stable hydrophilic support with reactive epoxide groups for ligand attachment. Fig. 7.4 summarizes the steps involved in the preparation of epoxide-activated mercapto-derivatized glass and silica supports.

Carbodiimide condensation can be used to form a peptide bond between free amino groups on the support surface and free carboxyl groups on the ligand [67]. During the coupling stage, the carbodiimide reacts with carboxyl groups to form *O*-acylisourea at pH 4-5. The activated carboxyl group then condenses with an amino group on the silanized support surface to form a peptide bond plus urea, which needs to be removed from the reaction mixture. Ordinarily, either 1-ethyl-3-(3-dimethylaminopropyl)carbodiimide (EDC) or 1-cyclohexyl-3-(2-morpholinoethyl)carbodiimide metho-*p*-toluenesulfonate (CMC) water-soluble carbodiimides are used for this reaction.

```
A
                        OCH3
                         |
|-SI-OH  +  H3CO-SI-CH2CH2CH2SH
                         |
                        OCH3
          γ-MERCAPTOPROPYLTRIMETHOXYSILANE

B
            OCH3
             |
|-SI-O-SI-CH2CH2CH2SH   +  CH2-CH-CH2O-(CH2)4-O-CH2-CH-CH2
             |               \ /                     \ /
            OCH3              O                       O
                            1,4-BUTADIOLDIGLYCIDOXYETHER

C
            OCH3
             |
|-SI-O-SI-(CH2)3-S-CH2-CHCH2O-(CH2)4-O-CH2-CH-CH2
             |              |                 \ /
            OCH3            OH                 O
```

Fig. 7.4. Silanization of glass and silica with τ-mercaptopropyltrimethoxysilane and attachment of epoxy groups for ligand immobilization. (A) The glass or silica reacts with the silane. (B) The resulting mercapto glass or silica is further derivatized by reaction with 1,4-butadiol diglycidoxy ether to produce (C) epoxy- or oxirane-coated supports ready for ligand attachment.

Many different materials can be attached to silanized inorganic supports by the two-step technique, using glutaraldehyde as the coupling agent [68]. Coupling is achieved by reaction of an aldehyde group with the amino group on the support, leading to the formation of a Schiff base. This is followed by a coupling of the amino group on the ligand with the second aldehyde group.

Carbonylation with 1,1′-carbonyl diimidazole can be performed on both glass [47,48] and silica supports. Following silanization, reactive carbonyl diimidazole groups are attached to amino groups on the support by shaking the support material with diimidazole for 4-6 h at room temperature. The activated support can bind protein ligands following 12-18 h of incubation in 50 m*M* carbonate buffer at pH 9.5. In a similar manner, *N*-succinimide ester activation of inorganic supports can be performed. Attachment of the ester, like carbonyl diimidazole, allows rapid and reliable attachment of protein ligands to the support matrix via amino groups [19,44].

7.4.4 Activation of synthetic supports

Modification of the carboxamide groups on polyacrylamide beads with ethylenediamine or hydrazine hydrate will produce beads with acylazides or acid anhydrides. These modifications are performed at 90°C and 50°C, respectively. The aminoethyl derivatives can then be converted to a series of different reactive groups, suitable for the attachment of many different types of ligand. Usually the attachment is effected via pri-

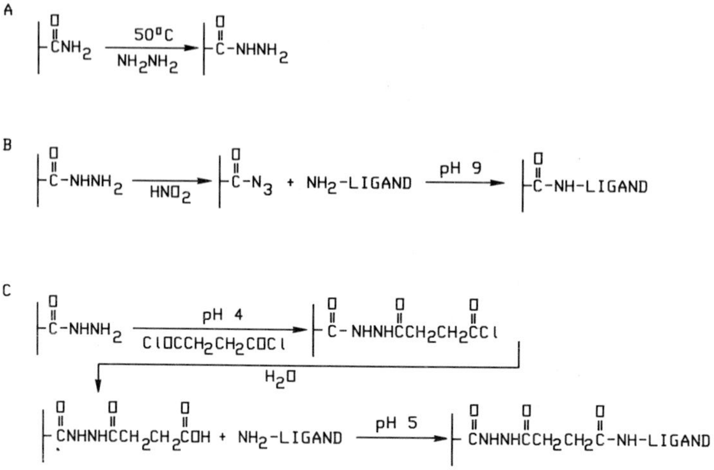

A

$$\vdash\overset{O}{\overset{\|}{C}}NH_2 \xrightarrow[NH_2NH_2]{50^\circ C} \vdash\overset{O}{\overset{\|}{C}}-NHNH_2$$

B

$$\vdash\overset{O}{\overset{\|}{C}}-NHNH_2 \xrightarrow[HNO_2]{} \vdash\overset{O}{\overset{\|}{C}}-N_3 + NH_2\text{-LIGAND} \xrightarrow{pH\ 9} \vdash\overset{O}{\overset{\|}{C}}-NH\text{-LIGAND}$$

C

$$\vdash\overset{O}{\overset{\|}{C}}-NHNH_2 \xrightarrow[ClOCCH_2CH_2COCl]{pH\ 4} \vdash\overset{O}{\overset{\|}{C}}-NHNH\overset{O}{\overset{\|}{C}}CH_2CH_2\overset{O}{\overset{\|}{C}}Cl$$

$$\xrightarrow{H_2O} \vdash\overset{O}{\overset{\|}{C}}NHNH\overset{O}{\overset{\|}{C}}CH_2CH_2\overset{O}{\overset{\|}{C}}OH + NH_2\text{-LIGAND} \xrightarrow{pH\ 5} \vdash\overset{O}{\overset{\|}{C}}NHNH\overset{O}{\overset{\|}{C}}CH_2CH_2\overset{O}{\overset{\|}{C}}-NH\text{-LIGAND}$$

Fig. 7.5. Hydrazine activation of polyacrylamide beads. (A) The beads are incubated with 6 *M* hydrazine hydrate to produce a hydrazine-derivatized support. Once this derivatization has been performed, ligands can easily be immobilized by either of the two reactions shown in the lower half of the figure. (B) The hydrazine derivative reacts with nitrous acid to form an acyl azide intermediate, to which ligands can be directly attached. If a spacer arm is required, Reaction C is used. Succinyl chloride is attached to the hydrazine to form a succinylhydrazine derivative, to which ligands can be attached, via their amino groups, by carbodiimide condensation at pH 5.

mary amino groups on the ligands [16,55]. Activation chemistry of polyacrylamide supports is shown in Fig. 7.5. Methacrylate and trisacryl beads can also be activated by attaching amino groups to the OH sidechains in the bead matrix. Although all synthetic supports can easily be modified to accept ligand attachment, the majority of this type of supports are commercially available with reactive chains already derivatized to the bead or particle surface. In this case, ligand attachment is usually through incubation of the support and ligand in a carbonate or other alkaline buffer at pH $>$ 8.5.

7.4.5 Spacer arms

In many cases, especially when the immobilized ligand is a small molecule, such as a receptor substrate of a chemical antigen, steric hindrance will occur between the immobilization support and the material to be isolated. This phenomenon will cause a reduced or complete lack of specific binding. To overcome this situation, it is often necessary either to select a different support (i.e. one with a spacer arm already attached) or to bind a spacer molecule to the support prior to attaching the ligand. The use of a spacer arm will ensure that the ligand is placed at a suitable distance from the surface of the support. This will allow the analyte ample space for complete attachment to the immobilized ligand. However, if the spacer arm is too long, it will bend under the column flow pressures and fold the ligand into the support surface. These spacer arms can be directly attached to the

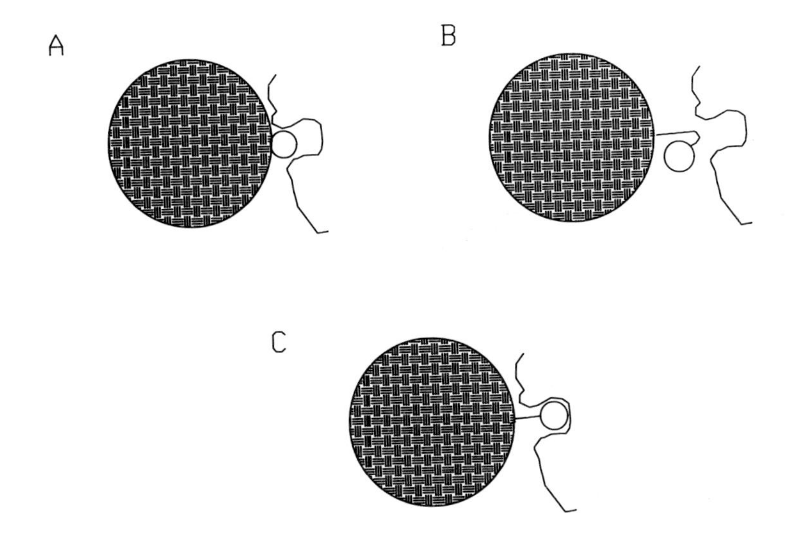

Fig. 7.6. The effects of spacer arms on ligand efficiency. (A) Ligands attached directly to the support surface are not always efficient affinity supports. The analyte to be isolated often encounters steric hindrance from the support surface. (B) Spacer arms which are too long may become ineffective when the units of the spacer interact with each other or bend the ligand away from the mobile phase on the column. (C) A well-designed spacer arm will place the ligand at a suitable distance from the support surface, thus ensuring efficient separation.

TABLE 7.1

COMMERCIALLY AVAILABLE SUPPORT SPACER ARMS AND THEIR CHEMICAL CONFIGURATIONS

Hexamethylenediamine	$\vert -NH-(CH_2)_6-NH_2$
Aminocaproic acid	$\vert -NH-(CH_2)_5-COOH$
N-Hydroxysuccinimide	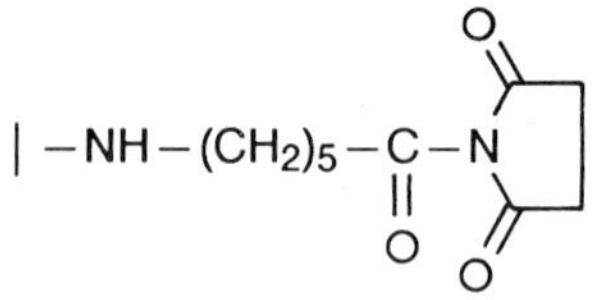 $\vert -NH-(CH_2)_5-C(=O)-N$ (succinimide ring, with two C=O)
Bisoxiran	$\vert -O-CH_2-CH-CH_2$ (epoxide, CH and CH_2 bridged by O)
3,3′-Diaminodipropylamine	$\vert -NH-(CH_2)_3-NH-(CH_2)_3-NH_2$
Hydroxyalkyl	$\vert -C(=O)-O-CH_2-CH_2-OH$

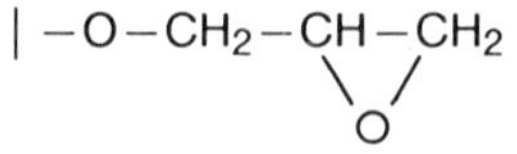

support surface with a suitable ligand attachment sidechain at its farthest end or a spacer can be chemically attached to the reactive sidechain of the support, and the ligand is then attached to the spacer via a secondary reaction. This secondary reaction can be performed by using glutaraldehyde or carbodiimide reactions. Table 7.1 lists some of the commercially available spacer arms and their chemical structures. Fig. 7.6 illustrates the concept of ligand spacers.

7.5 LIGANDS

The selection of any ligand should be influenced by two major concerns: The ligand must specifically and reversibly bind the material to be isolated and must contain groups, within its chemical structure, which can be chemically modified to allow attachment to the support. In addition, the chemical modification must not impair or damage the specific binding activity of the ligand. Although almost any material can be immobilized on the surface of an affinity support, the major ligands are proteins, nucleic acids, lectins, and dyes.

7.5.1 Proteins

Many different classes of proteins have been used as affinity chromatography ligands and are too numerous to list. Protein ligands range from immobilized enzymes and lectins (see below) to antibodies and antigens. The immobilized enzymes have been used to isolate both coenzymes and substrates. An interesting use of immobilized enzymes is used by Pierce, which sells immobilized papain for the affinity isolation and cleavage of IgG antibodies. In addition, proteins with specific binding properties, such as Protein A (Pharmacia; Fermentech), Protein G (Sigma Chemical Co.), avidin (Pierce), and streptavidin (Pierce) have specialized uses in immunochemistry [69-72]. These proteins have been used to isolate monoclonal antibodies from cell cultures as well as biotin-labeled and biotin-containing molecules. Immobilized hormones have been used to isolate specific receptors, and peptide antigens have been used to isolate specific antibodies [73].

7.5.2 Nucleic acids and nucleotides

The bioselective interactions between nucleic acids and their binding proteins has made the use of immobilized nucleic acid ligands a great success. Both DNA and RNA have been used as immobilized ligands for the isolation of proteins that specifically bind to nucleic acids [74,75]. In a similar fashion, RNA can be immobilized for the isolation of specific RNA-binding nucleotides and proteins. Columns containing specific polynucleotides have been used to isolate complementary DNA and RNA [76].

7.5.3 Lectins

Lectins are a special class of proteins, often of plant origin, which exhibit strong affinity binding to certain carbohydrate residues. This specificity, together with the wide range of commercially available lectins, has led to their wide application in glycoprotein separation [77-79]. Since most membrane proteins express some degree of glycosylation, lectins have become popular as selective ligands for affinity isolation of membrane proteins. Concanavalin A (Sigma Chemical Co.), a lectin obtained from the jackbean, reacts with α-D-mannopyranosyl and α-D-glucopyranosyl residues, and this makes it suitable as a general affinity ligand for membrane-associated biological materials [80]. Another lectin, isolated from the red kidney bean, has the ability to bind to glycoproteins containing *N*-acetyl-D-galactosamine residues. The most popular lectins used as affinity ligands for the isolation of membrane proteins and glycoproteins are: Concanavalin A (α-D-mannosyl and α-D-glucosyl terminal residues), wheatgerm agglutinin (*N*-acetyl-β-glucosyl terminal residues and *N*-acetyl-β-glucosamine oligomers), lentil agglutinin (α-D-mannosyl and α-D-glucosyl terminal residues), castor bean agglutinin (β-D-galactosyl residues), peanut agglutinin (β-D-galactosamine(1-3)-galactosylneuraminic acid), soybean (*N*-acetyl-D-galactosamine), and Osage orange lectin (α-D-galactosyl and *N*-acetyl-D-galactosaminyl residues). All of these lectins are commercially available from Sigma Chemical Co.

References on p. A335

7.5.4 Dyes

Synthetic textile dyes, by virtue of their reactivity with a wide variety of biological materials, can be used as affinity ligands. The most popular, Cibacron Blue, has been reported to be suitable for the isolation of serum albumin [81], α-fetoprotein [82], IgG [83], complement components [84], α_2-macroglobulin [85], and human clotting factors [86]. Dye-ligand chromatography can be used to isolate adenine-thymine-specific and guanine-cytosine-specific nucleic acids. The technique is similar to that described above for the isolation of proteins. Other dyes with potential affinity applications, which are available from several commercial sources (Amicon; Sigma Chemical Co.), are: Cibacron Blue F3G-A, Cibacron Brilliant Blue FBR-P, Procion Red HE-3B, Procion Blue MX-3G, Procion Blue MX-R, Procion Red MX-2B, Remazol Brilliant Blue R, Ostazin Brilliant Red S-5B, Orange A, and Green A. Although the majority of the dyes have been used for the isolation of dehydrogenase and kinase enzymes [87-89], Neame and Parikh [63] have used a variety of different immobilized dyes to isolate human lymphoblastoid interferon.

7.5.5 Miscellaneous

In addition to the major ligand groups, outlined and discussed above, other specialized materials, such as heparin, have been used as affinity ligands for the isolation of a variety of different molecules, including enzymes and DNA-dependent RNA polymerases [90].

7.6 SAMPLE RECOVERY TECHNIQUES

Recovery of the ligand-bound material is an important step in affinity separation and can be achieved by several different techniques. Some of these procedures can be generally applied to most affinity separations, but some applications, such as lectin separations, require a more specialized elution technique. This form of elution involves competition elution, i.e. desorption of the ligand-bound material in the presence of a competing agent, such as a specific sugar.

7.6.1 Ligand competition

The use of a free ligand to displace the adsorbed material from the matrix-immobilized ligand is a highly specific technique for affinity elution. However, in many cases the simple addition of free ligand to the running buffer may prove to be very expensive. Often extremely high concentrations of free ligand are required to neutralize the strong interactions between the substance of interest and the immobilized ligand. However, a combination of free ligand in the presence of a chaotropic ion or pH change lowers the amount of ligand required to perform an efficient elution. The use of a co-substrate, such as a sugar in lectin affinity chromatography, has proved to be the most efficient and inexpensive form of this type of elution.

7.6.2 pH manipulation

Increasing or decreasing the pH of the initial buffer is another common technique for eluting bound materials from affinity columns. Changes in pH alter the degree of ionization of charged groups at the binding site. In affinity chromatography, the most popular technique for dissociating the ligand/analyte complexes is manipulation of the pH. Changes in the pH of the running buffer are usually in the acidic range. Elution pH ranges as low as 1-1.5 have been reported. Also, it must be remembered that even with the protein coat, silica is very sensitive to sharp changes in pH, especially at the alkaline end of the pH range, and continuous exposure to conditions of high pH will affect the structure of the silica matrix. When acid elution is used, the pH of the running buffer can be altered by adding a gradient of citric, formic, or acetic acid [91]. A combination of Tris/HCl has also been shown to act as an efficient eluent [92]. Alkali treatment has been used for low-pressure affinity and immunoaffinity techniques but appears to have a greater denaturing effect on the immobilized ligands, especially antibodies [93].

7.6.3 Chaotropic ion elution

This eluent appears to give the most effective results without many of the denaturing effects observed when pH manipulation is used as the elution technique. Chaotropic salts are effective because they can disrupt the structure of water and are highly effective in reducing hydrogen bond and hydrophobic interactions. There are several salts that exhibit chaotropic characteristics. The order of their effectiveness as dissociation agents for immobilized ligand/substance complexes is as follows:

$$CCl_3COO^- > SCN^- > CF_3COO^- > ClO_4^- > I^- > Cl^-$$

These salts, in concentrations between 1.5 and 8 *M*, have been shown to be effective in the dissociation of high-affinity antibody/antigen complexes [93] and recovery of receptors from affinity matrices [94]. The most widely used chaotropic ion is thiocyanate in concentrations up to 3 *M*. Other reagents used in elution techniques are 3 *M* solutions of sodium iodide [95] and 2.5-3 *M* solutions of a polyvinylpyrrolidone/iodide mixture [92]. Sodium chloride is a weak chaotropic ion, which can be used at concentrations of > 2 *M* as an effective eluent. Effective dissociation of antibody/antigen complexes in free solution has been reported with 4-8 *M* solutions of sodium chloride [92,96]. It must be remembered that use of such high-concentration chaotropic ions will also cause some degree of denaturation to both the immobilized antibody and the antigen.

7.6.4 Electrophoretic desorption

This mild, nonionic technique for the recovery of charged materials from immunoadsorbents is well suited to the recovery of antibodies and hormone-binding proteins [97]. However, a special electrophoresis apparatus is required (Bio-Rad), if efficient recoveries are to be performed [98]. Protein desorption can be performed by low-current electro-

References on p. A335

phoresis in a protein elution apparatus, using an acidic buffer, or by isoelectric focusing with the affinity gel placed in a prefocused flat-bed gel. Although this system has not become popular, it is applicable to desorption of reagents from immunoaffinity gels and can dissociate biotin/avidin complexes.

Another electrophoretic desorption system is the Elutrap chamber, made by Schleicher & Schuell [99]. This apparatus is designed for protein and nucleic acid recovery from analytical polyacrylamide gel, but the chamber can also be packed with immunoaffinity gels and the desorption performed under conditions of low-voltage electrophoresis. In this system, the eluted materials are captured and concentrated in membrane "traps".

7.6.5 Zonal elution

Zonal elution [100], like frontal elution, is a technique used more for measuring molecular interactions than as a practical method of recovering isolated substrates from affinity columns. Both techniques are used to study the binding capacity and intermolecular interactions between the ligand and the molecule being separated. In zonal elution, the input signal is a short, simple pulse, which requires small amounts of sample. Kinetic and equilibrium constants can be calculated from the retention volume and the shape of the output signal. Readers interested in the characterization of the affinity process are referred to an excellent practical review of this technique, written by Abercrombie and Chaiken [101].

7.6.6 Other techniques

Elution of bound substances can be performed by a series of physicochemical means, such as increasing the ionic strength of the running buffer by simply raising the buffer salt concentration during the experiment. Temperature changes have also been used in the elution of enzymes from affinity adsorbents, and when combined with other techniques, this can be a sensitive means of eluting weakly bound enzymes from immobilized substrates. In this way, isolated enzymes may be recovered without contamination with elution agents.

7.7 CHROMATOGRAPHIC TECHNIQUES

Affinity separations can be performed by two different techniques, column chromatography and batch methods. The former technique is usually applied to analytical and semipreparative separations, while the latter technique is applied to large-scale preparative separations, following elucidation of the separation parameters by analytical procedures.

7.7.1 Column chromatography

Affinity chromatography is usually performed in small columns at reasonably slow flowrates to allow for slow reactions. The ligand-coated support is packed into the column under gravity to prevent stripping off the ligand coat. Care must be exercised to keep the room temperature below 26°C. If the separation is to be performed in the cold-room, a slower flowrate should be applied. It is always good practice to use a labeled substance to ensure that a test sample of the support is adequately coated with ligand and that the ligand is in an active condition. Excessive pressure will damage the support and will move the analyte through the column at a pace that may not permit completion of the affinity reactions. In the sorption phase the analyte reacts with the ligand and is thus retained in the column. Unreactive material will then pass through the column and form the first or nonspecific chromatographic peak. It is important to let enough time elapse between the initial phase and the start of the second or elution phase. Although pH changes or chaotropic ions will result in the recovery of most bound substances, there are times when a combination of elution techniques is required. Once the ligand/analyte complex is dissociated, the free analyte moves through the column in the mobile phase and is eluted as a sharp, second peak. If the second peak is not sharp, the eluent is dissociating the complex in stages, according to the affinity of the different substance components for the immobilized ligand. In such cases, the elution technique should be changed or a different ligand should be chosen. A typical affinity chromatogram is shown in Fig. 7.7. If the second peak is either a twin peak or very diffuse, then the ligand is possibly also being eluted from the support. In this case, the column needs to be unpacked and the support examined for the presence of the immobilized ligand.

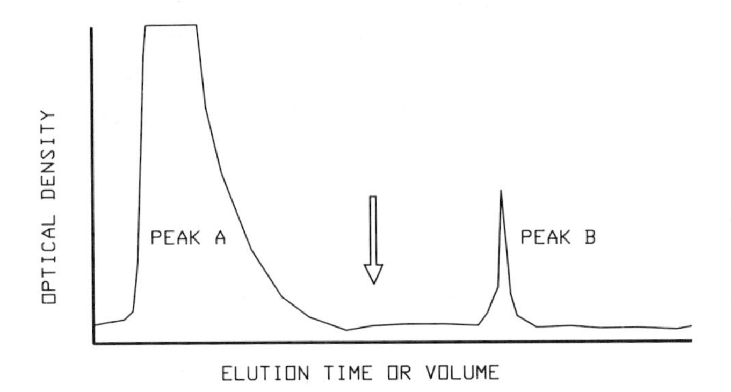

Fig. 7.7. Diagramatic representation of a generic affinity chromatogram. Peak A is formed by the nonaffinity-isolated components of the test sample, while the affinity-isolated analyte is recovered as Peak B during the elution phase. The arrow marks the ideal point at which collection of material was initiated.

7.7.2 Batch techniques

The use of batch separations is becoming more popular as the biotechnology industry becomes familiar with the advantages of affinity and immunoaffinity chromatography. The development of magnetic beads has greatly enhanced this form of purification and made recovery of the ligand-bound analytes easy [102]. Batch techniques are relatively simple to perform. The ligand is attached to the support, as described earlier, and the activated support is mixed with the test solution. Constant stirring is required in this technique, as the supports usually settle, causing a loss of ligand efficiency. Recovery is performed as a separate operation following recovery of the support. Although batch procedures can be used to isolate specific analytes, they are more often used to purify solutions, by affinity removal of specific contaminants, rather than for the isolation of molecules for further analysis. Thus, in large-scale batch purification, recovery, by dissociation of the ligand-bound analyte, is for the purpose of support regeneration rather than recovery of the separated analyte. However, recovery is simple; the ligand-bound material is recovered by incubation of the batch in elution buffer.

7.7.3 Affinity chromatography of living cells

Lectin, antibodies, antigens, and receptor or enzyme substrate can be used as affinity ligands for the separation of intact cells. The chromatographic technique is similar to that described for molecules, but the support matrix needs to be very porous to enable the cells to pass through the column without becoming enmeshed in the matrix. For this reason, most whole-cell affinity isolations are performed by a batch technique in which the ligand-coated matrix is mixed with a suspension of the cells. After reacting with the cells, the adherent matrix and these cells are separated by either centrifugation in density media or by magnetic means if ligand-coated iron oxide beads were used [103].

Alternatively, the cells can be isolated by a column technique. In either case, care must be exercised to maintain the integrity of the cells. Some buffers may cause cell shrinkage or lysis and certain eluents may cause irreversible changes in the cell membranes. Pressures and centrifugal forces must not be so high as to cause harm to the isolated cells.

7.7.4 Specialized affinity techniques

In addition to affinity techniques based on bioselective processes, there are a number of associated affinity techniques. For instance, metal chelate affinity chromatography is based on the ability of transitional metal ions (Cu^{2+} and Zn^{2+}) to form compounds with high- and low-molecular-weight ligands [104]. This form of affinity chromatography has been used to isolate a number of enzymes and plasma proteins [105-108]. Another member of this family of pseudo-affinity techniques is charge-transfer adsorption chromatography, which is based on the interaction of certain pairs of organic compounds that can form a stable complex by electron transfer. The technique has been applied to the separation of nucleotides and nucleic acids [109,110].

Covalent chromatography was introduced in 1973 by Brocklehurst et al. [111]. It involves the formation of disulfide bridges between SH-group-containing biological materials and SH groups on the packing medium. This reaction is then reversed under mild reducing conditions. The matrices available for performing this form of affinity chromatography are activated thiol Sepharose 4B and Thiopropyl Sepharose 6B (Pharmacia). Covalent chromatography has been used successfully to isolate thiol-containing peptides from human erythrocytes [112] and mercurated mRNA [113], and for the immobilization of cysteine-containing antigens in the purification of antibodies [114].

7.8 IMMUNOAFFINITY TECHNIQUES

Immunoaffinity chromatography is a unique separation procedure which depends on immunological reactions to isolate the material of interest. In this form of affinity chromatography an immobilized antibody or antigen is used to separate the specific reactive antigen or antibody. Although the majority of immunoaffinity techniques are based on low- or medium-pressure systems, the development of silica particles and glass beads with reactive sidechains, suitable for protein immobilization, has provided packing material that allows this technique to be applied to high-performance liquid-chromatographic systems [115-118]. Immunoaffinity chromatography can be applied to the isolation of any biological material to which a specific antibody can be raised. In a similar fashion, specific antibodies can be isolated from animal or human body fluids by immunoaffinity chromatography with a specific antigen as the ligand. The introduction of Protein A- and avidin-coated supports, together with the development of monoclonal antibodies has greatly enhanced the use of immobilized antibodies as analytical tools and increased their use in many different fields, including analytical chemistry.

7.8.1 Antibody and antigen ligands

Antibodies, used as the immobilized ligand in immunoaffinity chromatography, can be obtained from many different commercial sources, clinical samples, or animal immunizations. The IgG class of antibody is the most useful for immunoaffinity techniques, although the introduction of the hydrazide biotin-labeled procedure has allowed all classes of glycosylated antibodies to be immobilized on avidin-coated glass beads. Apart from the different classes of antibodies, there are also several different types of antibodies that can be used, each with its own properties.

Polyclonal antibodies are produced when animals are immunized with specific antigens. These antibodies are usually the major immunological component of hyperimmune serum, which also contains antibodies of other specificities. Therefore, the antibody of interest must be isolated before it can be used as an immunoaffinity reagent. Polyclonal antibodies are natural products of an induced immune response and often react with common determinants on the antigen.

Another type of polyclonal antibody, an autoantibody, can be naturally produced by any animal against components of its own body [119]. These antibodies are often made

References on p. A335

against specific molecules, such as cell membrane receptors, subcellular fractions, hormones, and cell secretory byproducts. They can be used to isolate their specific antigens from normal tissue or used to study the immunological mechanisms involved in disease.

The introduction of monoclonal antibodies in 1975 [120] has greatly increased the use of antibodies as specific separating agents. Monoclonal antibodies are the product of cellular fusions, performed at the discretion of the investigator, against both common and discrete antigenic determinants. The advantage of this type of antibody is that all of the antibody molecules are reactive against the same defined region of the antigen. These antibodies often exhibit greater specificity than their polyclonal counterparts and are well suited for isolating subcomponents of proteins and polypeptides.

Idiotypes and their counterparts, anti-idiotypes, are specialized antibodies, which are part of the normal immune system regulatory circuits [121]. They have distinct properties not exhibited by other antibodies, such as being able to act as receptor substrates, and for this reason they are valuable reagents for immunoaffinity isolations [122-124]. Specific anti-idiotypes can be produced by monoclonal antibody technology, and these antibodies exhibit such exquisite specificity that they will react only with a defined area on the antigen and with very little else.

7.8.2 Specialized antibody immobilization techniques

In addition to the techniques outlined above for chemically attaching ligands to their immobilization supports, other specialized supports have been developed for the attachment and immobilization of antibodies. These supports are coated with proteins that are able to bind antibodies. The proteins generally used for this purpose are Protein A, Protein G, avidin, and streptavidin.

Protein A and Protein G are bacterial products which possess the ability to bind antibodies via their Fc or tail portion. Binding via the Fc portion helps to orient the antigen-binding sites of the antibodies into the mobile phase of the column, thus producing maximum antigen-binding efficiency of the ligand. Antibodies naturally bind to the IgG receptors on the Protein A or G molecules [125,126] and can be permanently immobilized by crosslinking the antibody to the bacterial protein with a solution of 50 m*M* carbodiimide.

Avidin is a protein which is derived from egg whites. It possesses high binding affinity for materials labeled with biotin [127,128]. A more refined version, called streptavidin, has been isolated from *Streptomyces avidinii* [129]. Biotin is available in many forms suitable for attachment to proteins or carbohydrates. Protein attachment can easily be achieved by using the *N*-hydroxysuccinimide-derivatized form of biotin (Sigma Chemical Co.), the binding taking place when the protein and biotin are incubated at room temperature for 2 h at pH 9.0 in carbonate buffer. Attachment of biotin molecules to carbohydrates requires modification with 0.1 *M* sodium periodate prior to reaction with hydrazide-derivatized biotin (Pierce) [130,131]. This form of biotinylation has been successfully used to label the carbohydrate portion of IgG antibodies [131-133]. The advantage of this technique is that biotin is attached to the Fc portion of the IgG antibody, and when incubated with the

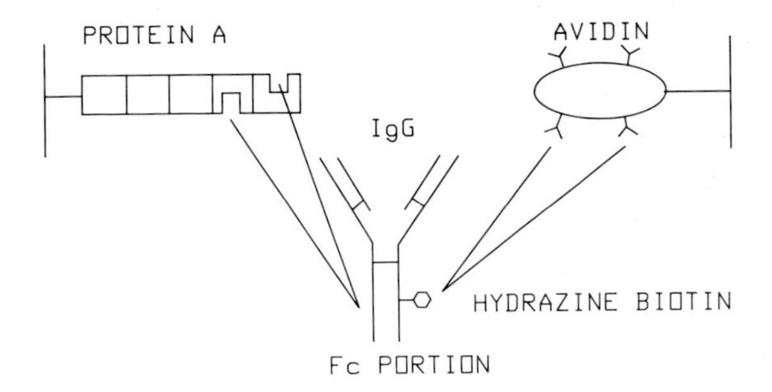

Fig. 7.8. Diagramatic illustration of the ways in which bacterial coat proteins and avidins are used to attach IgG ligands for immunoaffinity chromatography. Unmodified IgG will attach itself to Protein A supports via Fc receptors, while biotinylation is required for attachment of IgG to avidin supports. This latter attachment is via biotin receptors on the avidin molecule.

avidin- or streptavidin-coated supports, the biotinylated antibody becomes attached to the biotin receptors, on the avidin molecule, via its Fc portion. In this way, biotinylated antibodies become oriented in a similar way to those bound to Protein A- or Protein G-coated immunoaffinity supports. Due to the extremely high affinity exhibited by either avidin or streptavidin for biotin-labeled antibodies, no chemical modification is required, such as chemical crosslinking, once the antibody is attached to the support. Fig. 7.8 diagramatically illustrates the binding of IgG antibodies to either Protein A- or avidin-coated supports.

7.8.3 High-performance immunoaffinity chromatography

High-performance immunoaffinity chromatography (HPIAC) can be used to isolate any biological material or chemical to which an antibody can be made. However, since it is a new technique, HPIAC has been used in only a limited number of applications. Immobilized monoclonal antibodies, directed against Substance P, have been used to isolate Substance P receptors from solubilized lymphoblastic cell membranes [134]. Other researchers have used HPIAC to isolate liver membrane proteins, similar to those isolated by concanavalin A lectin chromatography [135]. Immunoaffinity has been used by several research groups as a sample preparation technique prior to conventional HPLC [136] or in tandem with reversed-phase chromatography of the immunoaffinity-isolated materials [137,138].

HPIAC columns can be prepared by slurry-packing activated or protein-coated glass beads or silica particles. The columns used for this type of immunoaffinity chromatography are either short and wide or long, narrow-bore stainless-steel columns. Conventional 4.6-mm-ID columns, ranging in length from 5 to 25 cm, make excellent HPIAC columns [132,139]. The length depends on the amount of available antibody and/or the amount of antigen in the test solution. Long, narrow columns often yield higher amounts of antigen from dilute solutions than do short, wide columns, due to the longer

References on p. A335

TABLE 7.2

ELUENTS USED IN HIGH-PERFORMANCE IMMUNOAFFINITY CHROMATOGRAPHY

Molarity		pH range
Acidic		
0.1-1	Glycine	1.0-1.5
0.1-1	Tris/HCl	1.0-2.0
0.33	Citric acid	1.0-2.5
0.5-1	Acetic acid	1.0-1.5
Alkaline		
0.5-2	NaOH or KOH	10.0-14.0
Chaotropic		
2-3	NaSCN	
2-3	NaI	
2.5-6	NaCl	
2-4	Polyvinylpyrrolidone/iodide complex	

sample residence time. After the column is packed, it can be used with any conventional HPLC apparatus, from a simple isocratic system to a complicated gradient system. Flow-rates of 0.5-1.5 ml/min are commonly used in HPIAC separations; higher flowrates can impede the interactions between the immobilized antibody and the antigen. Temperature control is very important both for maintaining the integrity of the immobilized antibody and for maintaining the activity of the isolated antigen. Most HPIAC experiments must be performed at 4°C, to ensure maximum column life and to maintain the activity of the isolated antigens.

The two most common techniques for recovering isolated materials from HPIAC columns are acid and chaotropic ion elution, although other techniques have been described [132,140]. Table 7.2 summarizes these eluents. Once the antigen has been eluted from the column and collected, the eluent must be removed to protect the antigen from denaturation [141]. Several techniques are available for performing this task. The simplest is dialysis, which is either performed by on-line, continuous-flow dialysis or after the chromatographic separation. Another technique involves desalting the isolated antigen on a small size-exclusion column [142], such as the P10 column from Pharmacia, which is packed with Sephadex G25. This technique is advantageous when the isolated antigen is of large molecular weight and can easily be separated from the low-molecular-weight eluents by size-exclusion chromatography.

REFERENCES

1 E. Starkenstein, *Biochem. Z.,* 24 (1910) 210.
2 O. Holmbergh, *Biochem. Z.,* 258 (1933) 134.
3 P.S. Thayer, *J. Bacteriol.,* 66 (1953) 656.
4 R. Litman, *J. Biol. Chem.,* 243 (1968) 6222.
5 B.M. Alberts and G. Herrick, *Methods Enzymol.,* 21 (1971) 198.
6 G.G. Carmichael, *J. Biol. Chem.,* 250 (1975) 6160.
7 D.H. Campbell, E. Luescher and L.S. Lerman, *Proc. Natl. Acad. Sci. USA,* 37 (1951) 575.
8 A.E. Gurvich, O.B. Kusovlena and A.E. Tumanova, *Biokhimiya,* 26 (1961) 803.
9 E.R. Centeno and A.H. Sehon, *Fed. Proc., Fed. Am. Soc. Exp. Biol.,* 25 (1966) 729.
10 M.M. Behrens, J.K. Inman and W.E. Vannier, *Arch. Biochem. Biophys.,* 119 (1967) 411.
11 J.B. Robbins, J. Haimovich and M. Sela, *Immunochemistry,* 4 (1967) 11.
12 J. Porath and P. Flodin, *Nature (London),* 183 (1959) 1557.
13 S. Hjerten, *Arch. Biochem. Biophys.,* 99 (1962) 466.
14 R. Axen, J. Porath and S. Ernback, *Nature (London),* 214 (1967) 1302.
15 P. Cuatrecasas, M. Wilchek and C.B. Anfinsen, *Proc. Natl. Acad. Sci. USA,* 61 (1968) 636.
16 J.K. Inman and H.M. Dintzis, *Biochemistry,* 8 (1969) 4074.
17 J. Coupek, M. Krivakova and S. Porkorny, *J. Polym. Sci. Polym. Symp.,* 42 (1973) 185.
18 R. Epton, C. Holloway and J.V. McClaren, *J. Appl. Polym. Sci.,* 18 (1974) 179.
19 H.H. Weetall and A.M. Filbert, *Methods Enzymol.,* 34 (1974) 59.
20 J. Turkova, *Methods Enzymol.,* 44 (1976) 66.
21 R.L. Easterday and J.M. Easterday, in R.B. Dunlap (Editor), *Immobilized Biochemicals and Affinity Chromatography,* Plenum Press, New York, 1974, p. 123.
22 J. Porath, *J. Chromatogr.,* 159 (1978) 13.
23 I.M. Chaiken, in I.M. Chaiken (Editor), *Analytical Affinity Chromatography,* CRC Press, Boca Raton, FL, 1987, p. 1.
24 D.J. Graves and Y.-T. Wu, *Methods Enzymol.,* 34 (1974) 140.
25 M.W. Steward, in D.M. Weir (Editor), *Handbook of Experimental Immunology,* Blackwell Scientific Publications, Oxford, 1978, Vol. 1, p. 16.1.
26 J.N. Murrell, S.F.A. Kettle and J.M. Tedder, *The Chemical Bond,* Wiley, New York, 1985, p. 290.
27 F. Karush, *Adv. Immunol.,* 2 (1962) 1.
28 L.S. Lerman, *Proc. Natl. Acad. Sci. USA,* 39 (1953) 232.
29 N. Wellicky and H.H. Weetall, *Immunochemistry,* 2 (1965) 293.
30 C. Arsenis and D.B. McCormick, *J. Biol. Chem.,* 241 (1966) 330.
31 C.S. Knight, *Adv. Chromatogr.,* 4 (1967) 61.
32 J.K. Madden and D. Thom, in T.C.J. Gribnau, J. Visser and R.J.F. Nivard (Editors), *Affinity Chromatography and Related Techniques,* Elsevier, Amsterdam, 1982, p. 113.
33 J. Peske, J. Stamberg, J. Hradil and M. Ilavsky, *J. Chromatogr.,* 125 (1976) 455.
34 J.A. Thompson, S. Garfinkinkel, R.B. Cohen and B. Safer, *Biochromatography,* 2 (1987) 166.
35 D. Mislovicova, P. Gemeiner, L. Kuniak and J. Zemek, *J. Chromatogr.,* 194 (1980) 95.
36 J. Porath, *Biochem. Soc. Trans.,* 7 (1979) 1197.
37 J.F. Kennedy, *Adv. Carbohydr. Chem., Biochem.,* 29 (1974) 305.
38 P. Cuatrecasas, M. Wilchek and C.B. Anfinsen, *Proc. Natl. Acad. Sci. USA,* 61 (1968) 636.
39 J. Porath, R. Axen and S. Ernback, *Nature (London),* 215 (1967) 1491.
40 S. Hjertén, *Biochim. Biophys. Acta,* 79 (1964) 393.
41 S. Bengtsson and L. Philipson, *Biochim. Biophys. Acta,* 79 (1964) 399.

42 W.M. Holmes, R.E. Hurt, B.R. Reid, R.A. Rimerman and G.A. Hatfield, *Proc. Natl. Acad. Sci. USA,* 72 (1975) 1068.
43 M. Zeichner and R. Stern, *Biochemistry,* 16 (1977) 1378.
44 H.H. Weetall, *Sep. Purif. Methods,* 2 (1973) 199.
45 P. Cuatrecasas and C.B. Anfinsen, *Ann. Rev. Biochem.,* 40 (1971) 259.
46 F.E. Regnier and R. Noel, *J. Chromatogr. Sci.,* 14 (1976) 316.
47 J.V. Babashak and T.M. Phillips, *J. Chromatogr.,* 444 (1988) 21.
48 J.V. Babashak and T.M. Phillips, *J. Chromatogr.,* 476 (1989) 187.
49 R.R. Walters, *Anal. Chem.,* 57 (1985) 1099.
50 K. Ernst-Cabrera and M. Wilchek, *J. Chromatogr.,* 397 (1987) 187.
51 B. Anspach, K.K. Unger, J. Davies and M.T.W. Hearn, *J. Chromatogr.,* 457 (1988) 195.
52 J. Schutyser, T. Buser, D. van Olden, H. Tomas, F. van Houdenhoven and G. van Dedem, in T.C.J. Gribnau, J. Visser and R.J.F. Nivard (Editors), *Affinity Chromatography and Related Techniques,* Elsevier, Amsterdam, 1982, p. 143.
53 M.E. Landgrebe, D. Wu and R.R. Walters, *Anal. Chem.,* 58 (1986) 1607.
54 D. Wu and R.R. Walters, *J. Chromatogr.,* 458 (1988) 169.
55 W.H. Scouten, *Affinity Chromatography,* Wiley, New York, 1981, p. 42.
56 J. Turkova and A. Seifertova, *J. Chromatogr.,* 148 (1978) 293.
57 P. Strop, F. Mikes and Z. Chytilova, *J. Chromatogr.,* 156 (1978) 239.
58 E. Boschetti, in P.D.G. Dean, W.S. Johnson and F.A. Middle (Editors), *Affinity Chromatography: A Practical Approach,* IRL Press, Washington, DC, 1985, p. 11.
59 J. Turkova, K. Blaha, M. Malanikova, D. Vancurova, F. Svec and J. Kalal, *Biochim. Biophys. Acta,* 524 (1978) 162.
60 J. Porath, *J. Chromatogr.,* 218 (1981) 241.
61 K. Mosbach and L. Andersson, *Nature (London),* 270 (1977) 259.
62 T.M. Phillips, N.S. More, W.D. Queen and A.M. Thompson, *J. Chromatogr.,* 327 (1985) 205.
63 P.J. Neame and J. Parikh, *Appl. Biochem. Biotechnol.,* 7 (1982) 295.
64 J. Turkova, J. Vajcner, D. Vancurova and J. Stamberg, *Collect. Czech. Chem. Commun.,* 44 (1979) 3411.
65 J. Porath, J.C. Janson and T Laas, *J. Chromatogr.,* 60 (1971) 167.
66 S.-H. Chang, R. Noel and F.E. Regnier, *Anal. Biochem.,* 48 (1976) 1839.
67 H. Hjelm, K. Hjelm and J. Sjoquist, *FEBS Lett.,* 28 (1972) 73.
68 S. Avrameas and T. Ternynck, *Immunochemistry,* 6 (1969) 53.
69 M.D.P. Boyle, *Biotechniques,* November (1984) 334.
70 L. Bjorck and G. Kronvall, *J. Immunol.,* 133 (1984) 969.
71 M. Wilchek and E.A. Bayer, *Immunol. Today,* 5 (1984) 39.
72 B.M. Conti-Tronconi and M. Raftery, *Ann. Rev. Biochem.,* 51 (1982) 491.
73 P.J. Rosenfeld and T.J. Kelly, *J. Biol. Chem.,* 261 (1986) 1398.
74 G.D. Sinclair and G.H. Dixon, *Biochemistry,* 21 (1982) 1869.
75 S. Okamura, F. Crane, H.A. Messner and T.W. Mak, *J. Biol. Chem.,* 253 (1978) 3765.
76 J. Arends, *Methods Enzymol.,* 63 (1981) 116.
77 F.S. Heinemann and J. Ozols, *J. Biol. Chem.,* 258 (1983) 4185.
78 D. Corradini, Z. El Rassi, C. Horvath, G. Guerra and W. Horne, *J. Chromatogr.,* 458 (1988) 1.
79 D. Josic, W. Hofmann, R. Habermann and W. Reutter, *J. Chromatogr.,* 444 (1988) 29.
80 M. Monsigny, in J.M. Egly (Editor), *Affinity Chromatography and Related Techniques,* INSERM, Paris, 1979, p. 207.
81 R.J. Laetherbarrow and P.D.G. Dean, *Biochem. J.,* 189 (1980) 27.
82 G. Berkenmeier, E. Usbeck, L. Saro and G. Kopperschlager, *J. Chromatogr.,* 265 (1983) 27.
83 J. Saint-Blancard, J.M. Kirzin, P. Riberon, F. Petit, J. Foucart, P. Girot and E. Boschetti, in T.C.J. Gribnau, J. Visser and R.J.F. Nivard (Editors), *Affinity Chromatography and Related Techniques,* Elsevier, Amsterdam, 1982, p. 305.

84 A.P. Gee, T. Borsos and M.D.P. Boyle, *J. Immunol. Methods,* 30 (1979) 119.
85 M.A. Bridges, D.A. Applegarth, J. Johannson, A.G.F. Davidson and L.T.K. Wong, *Clin. Chim. Acta,* 118 (1982) 21.
86 E. Gianazza and P. Arnaud, *Biochem. J.,* 201 (1982) 129.
87 D.H. Watson, M.J. Harvey and P.D.G. Dean, *Biochem. J.,* 173 (1978) 591.
88 G. Kopperschlager, H.-J. Bohme and E. Hofmann, *Adv. Biochem. Eng.,* 25 (1982) 101.
89 B. Anspach, K.K. Unger, J. Davies and M.T.W. Hearn, *J. Chromatogr.,* 457 (1988) 195.
90 G. Muszynska, E. Ber and G. Dobrowolska, in P.D.G. Dean, W.S. Johnson and F.A. Middle (Editors), *Affinity Chromatography: A Practical Approach,* IRL Press, Washington, DC, 1985, p. 125.
91 P. Mohr and K. Pommerening, *Affinity Chromatography,* Dekker, New York, 1985.
92 T.M. Phillips, *LC.GC,* 3 (1985) 962.
93 T. Kristiansen, in O. Hoffman-Ostenhof, M. Breitbach, F. Koller, D. Kraft and O. Scheiner (Editors), *Affinity Chromatography,* Pergamon Press, New York, 1978, p. 191.
94 V. Sica, G.A. Puca, M. Molinari, F.M. Buonaguro and F. Bresciani, *Biochemistry,* 19 (1980) 83.
95 S. Avrameas and T. Ternynck, *Biochem. J.,* 102 (1967) 37.
96 S. Nishi and H. Hirai, *Biochim. Biophys. Acta,* 278 (1972) 293.
97 M.R.A. Morgan, P.M. Johnson and P.D.G. Dean, *J. Immunol. Methods,* 23 (1978) 381.
98 M.R.A. Morgan, N.A. Slater and P.D.G. Dean, *Anal. Biochem.,* 92 (1978) 144.
99 E. Jacobs and A. Clad, *Anal. Biochem.,* 154 (1986) 583.
100 H.E. Swaisgood and I.M. Chaiken, in I.M. Chaiken (Editor), *Analytical Affinity Chromatography,* CRC Press, Boca Raton, FL, 1987, p. 67.
101 D.M. Abercrombie and I.M. Chaiken, in P.D.G. Dean, W.S. Johnson and F.A. Middle (Editors), *Affinity Chromatography: A Practical Approach,* IRL Press, Washington, DC, 1985, p. 169.
102 K.E. Howell, J. Gruenberg, A. Ito and G.E. Palade, in D.J. Morris, K.E. Howell and G.M.W. Cook (Editors), *Cell-Free Analysis of Membrane Traffic,* Alan R. Liss, New York, 1987, p. 185.
103 T. Lea, F. Vartdal, K. Nustad, S. Funderud, A. Berg, T. Ellingsen, R. Schmid, P. Stenstad and J. Ugelstad, *J. Mol. Recognition,* 1 (1988) 9.
104 J. Porath and B. Olin, *Biochemistry,* 22 (1983) 1621.
105 E. Sulkowski, *Trends Biotechnol.,* 3 (1985) 1.
106 A.J. Fatiadi, *Crit. Rev. Anal. Chem.,* 18 (1987) 1.
107 T.W. Hutchens and C.M. Li, *J. Mol. Recognition,* 1 (1988) 80.
108 D. Corradini, Z. El Rassi, C. Horvath, G. Guerra and W. Horn, *J. Chromatogr.,* 458 (1988) 1.
109 J.-M. Egly and E. Bochetti, in T.C.J. Gribnau, J. Visser and R.J.F. Nivard (Editors), *Affinity Chromatography and Related Techniques,* Elsevier, Amsterdam, 1982, p. 445.
110 J. Porath, *Pure Appl. Chem.,* 51 (1979) 1549.
111 K. Brocklehurst, J. Carlsson, M.P.J. Kierstan and E.M. Crook, *Biochem. J.,* 133 (1973) 573.
112 A. Kahlenberg and C. Walker, *Anal. Biochem.,* 74 (1976) 337.
113 R.M.K. Dale and D.C. Ward, *Biochemistry,* 14 (1975) 2458.
114 B.S. Parekh, P.W. Schwimmbeck and M.J. Buchmeier, *Peptide Res.,* 2 (1989) 249.
115 R.R. Walters, *J. Chromatogr.,* 249 (1982) 19.
116 P.O. Larsson, *Methods Enzymol.,* 104 (1984) 212.
117 L. Varady, K. Kalghatgi and C. Horvath, *J. Chromatogr.,* 458 (1988) 207.
118 F.L. Zhou, D. Muller, X. Santarelli and J. Jozefonvicz, *J. Chromatogr.,* 476 (1989) 195.
119 M. Blecher, in D. Evered and J. Whelan (Editors), *Receptors, Antibodies and Disease,* Pitman, London, 1982, p. 279.

120 G. Kohler and C. Milstein, *Nature (London),* 256 (1975) 495.
121 C.A. Bona and H. Kohler, in J.C. Venter, C.M. Fraser and J. Lindstrom (Editors), *Monoclonal and Anti-idiotypic Antibodies: Probes for Receptor Structure and Function,* Alan R. Liss, New York, 1984, p. 141.
122 K. Sege and P.A. Patterson, *Proc. Natl. Acad. Sci. USA,* 75 (1978) 2443.
123 A.D. Strosberg, S. Chamat, J.-G. Guillet, A. Schmutz, O. Durieu, C. Delavier and J. Hoebeke, in J.C. Venter, C.M. Fraser and J. Lindstrom (Editors), *Monoclonal and Anti-idiotypic Antibodies: Probes for Receptor Structure and Function,* Alan R. Liss, New York, 1984, p. 151.
124 T.M. Phillips, *Clin. Chem.,* 34 (1988) 1698.
125 K.J. Reis, E.M. Ayoub and M.D.P. Boyle, *J. Immunol.,* 132 (1984) 3091.
126 L. Bjorck and G. Kronvall, *J. Immunol.,* 133 (1984) 969.
127 M. Wilchek and E.A. Bayer, *Immunol. Today,* 5 (1984) 39.
128 J.W. Buckle and G.M.W. Cook, *Anal. Biochem.,* 165 (1986) 463.
129 L. Chaiet and F.J. Wolf, *Arch. Biochem. Biophys.,* 106 (1964) 1.
130 J.D. Rodwell, V.L. Alvarez, C. Lee, A.D. Lopez, J.W.F. Goers, H.D. King, H.J. Powsner and T.J. McKearn, *Proc. Natl. Acad. Sci. USA,* 83 (1986) 2632.
131 D.J. O'Shannessy and R.H. Quarles, *J. Immunol. Methods,* 99 (1987) 153.
132 T.M. Phillips, *Adv. Chromatogr.,* 29 (1989) 133.
133 T.M. Phillips, S.C. Frantz and J.J. Chmielinska, *Biochromatography,* 3 (1988) 149.
134 J.P. McGillis, M.L. Organist and D.G. Payan, *Anal. Biochem.,* 164 (1987) 502.
135 D. Josic, W. Hofmann, R. Habermann, A. Becker and W. Reutter, *J. Chromatogr.,* 397 (1987) 39.
136 Rybacek, M. D'Andrea and J.S. Tarnowski, *J. Chromatogr.,* 397 (1987) 355.
137 L.J. Janis and F.E. Regnier, *J. Chromatogr.,* 444 (1988) 1.
138 L.J. Janis, A. Grott, F.E. Regnier and S.J. Smith-Gill, *J. Chromatogr.,* 476 (1989) 235.
139 R.R. Walters, in P.D.G. Dean, W.S. Johnson and F.A. Middle (Editors), *Affinity Chromatography: A Practical Approach,* IRL Press, Washington, DC, 1985, p. 94.
140 T.M. Phillips, in A. Kervalage (Editor), *The Use of HPLC in Receptor Biochemistry,* Alan R. Liss, New York, 1989, p. 129.
141 A.J. Furth, *Anal. Biochem.,* 109 (1980) 207.
142 J. Porath and P. Flodin, *Nature (London),* 183 (1959) 1657.

Chapter 8

Supercritical-fluid chromatography

P.J. SCHOENMAKERS and L.G.M. UUNK

CONTENTS

8.1 INTRODUCTION

Supercritical-fluid chromatography (SFC) is not a new technique. It was first applied in 1963 by Klesper et al. [1], and subsequently many of the fundamental aspects of the technique were studied by Giddings et al. [2,3], Gouw and Jentoft [4], and Rijnders [5], among others. By the mid Seventies, SFC appeared to be a reasonably well-understood method [6]. However, even though it had been amply demonstrated that SFC could be applied to the separation of nonvolatile solutes, the development of the technique then came to a standstill. It took until the mid Eighties for SFC to gain new attention from researchers and, in contrast to the situation two decades earlier, commercial interest from column and instrument manufacturers soon followed.

The main factor obstructing the advance of SFC after it was first introduced has arguably been the concurrent research in and development of HPLC. While there had not been any rapid and efficient method for the separation of nonvolatile species before, suddenly there were two. Of these, HPLC was instrumentally less difficult. At that time, coping with pressures of 100 bar or more caused considerable problems, which had to be overcome to make HPLC a routine analytical technique. HPLC requires high pressures at the column inlet. SFC requires high pressures throughout the system, and because the viscosity of supercritical fluids is much lower than that of liquids, leak-tight systems are much more difficult to achieve. Moreover, SFC requires temperature control (often at elevated temperatures), while most HPLC experiments were then performed at room temperature. Finally, SFC requires means of controlling the flowrate or pressure at the outlet of the system. Given the greater complexity of SFC instrumentation, it was understandable that HPLC was developed first.

A second reason why HPLC was developed much more quickly and much more thoroughly than SFC is related to the types of solutes that could be analyzed. In liquid chromatography, both nonpolar solvents, such as alkanes, and polar ones, such as water, could be used from the very beginning. In SFC, the use of polar eluents is much more problematic. Although ammonia was used in one of the early SFC studies [7], its routine application is still by no means a practical proposition.

A number of factors have contributed to the recent rise in popularity of SFC. The most important of these are:

(1) the successful application of gas chromatographic detectors, such as the flame-ionization detector, which cannot readily be used in HPLC [8-11];

(2) the application of open-tubular columns, in particular fused-silica columns with immobilized stationary phases and internal diameters of 50 μm or less [12];

(3) the application of very small particles (3 to 10 μm) without serious problems caused by the pressure drop over the column [13];

(4) the availability of commercial SFC instruments;

(5) the successful coupling of SFC to other analytical methods, such as mass spectrometry (MS) and (Fourier-transform) infrared spectroscopy (FTIR) [14-18].

All these factors concern the instrumentation and columns for SFC, and many of the problems in this area have now been solved, or at least greatly ameliorated. The other major problem tempering the progress of SFC – the difficulties associated with the elution and separation of (very) polar solutes – has not yet been fully overcome. This is now probably the most active area of research in SFC, and some of the directions taken and results obtained so far will be described in this chapter.

8.1.1 Definition

Chromatography performed at pressures and temperatures above the critical values of the mobile phase can be called supercritical-fluid chromatography (SFC). A supercritical fluid is neither a liquid nor a typical gas. It can be formed from a conventional liquid by raising the temperature or from a conventional gas by increasing the pressure (Fig. 8.1). Compressed gas or "dense gas" are acceptable alternatives for the term supercritical fluid. The horizontal and vertical lines bordering the domain of supercritical fluids in Fig. 8.1 are

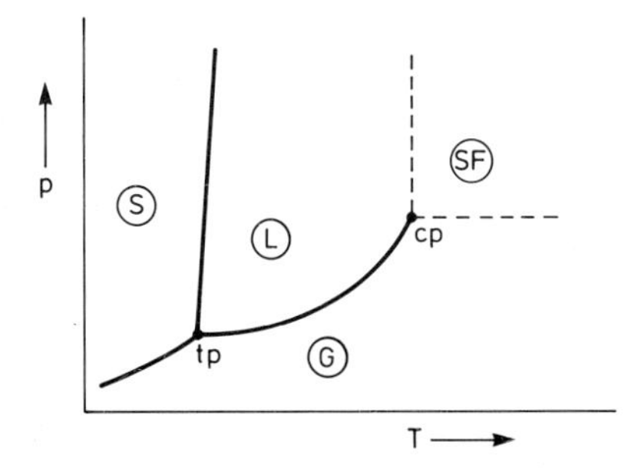

Fig. 8.1. Phase diagram for a pure component, indicating the areas in which solid (S), liquid (L), gas (G), and supercritical fluid (SF) occur. tp = triple point; cp = critical point. (Reprinted from Ref. 19, with permission.)

dashed, indicating that no phase change occurs in going from a liquid (or a gas) to a supercritical fluid. When crossing one of the dashed lines, the physical properties of the fluid change smoothly [20] rather than abruptly, the latter being the case when crossing one of the solid lines.

8.1.2 Advantages relative to other chromatographic methods

Both gases and liquids have their limitations when used as mobile phases for chromatography. Gases cannot be used for the migration of nonvolatile components through a chromatographic column. Also, solutes may not be sufficiently stable at elevated temperatures to allow their elution in GC. Liquids have a high viscosity, a factor in the requirement for high pressures in HPLC. Also, the molecular motion of solute molecules, which is usually characterized by the diffusion coefficient, is much smaller in liquids than it is in gases. Because diffusion coefficients are much smaller in liquids than in gases, very small particles must be used in packed columns for LC. Reducing the particle size also leads to an increase in the required pressure.

Supercritical fluids provide a compromise between gases and liquids. This is illustrated by the data in Table 8.1. In this table, the relevant properties of a typical gas (helium) under GC conditions and a typical liquid (water) under LC conditions are compared with supercritical fluids. Carbon dioxide appears twice in the table at different values of pressure and temperature. However, in both cases the values are above critical (for CO_2 the critical pressure, P_c is ca. 73 bar and the critical temperature, T_c, is ca. 31°C). It appears from the table that the density of supercritical CO_2 is between that of a gas and that of a

TABLE 8.1

RELEVANT PROPERTIES OF A TYPICAL GAS, A TYPICAL LIQUID, AND TWO TYPICAL SUPERCRITICAL FLUIDS THAT CAN BE USED AS MOBILE PHASES FOR CHROMATOGRAPHY

Reprinted from Ref. 19, with permission.

	Solvent			
	Helium	CO_2	CO_2	Water
Pressure, bar	1.5	80	200	–*
Temperature, °C	200	100	35	20
State	G	SF	SF	L
Density, g/cm^3	2 x 10^{-4}	0.15	0.8	1
Diffusion coefficient, cm^2/sec	1	10^{-3}	10^{-4}	10^{-5}
Viscosity, cP	0.02	0.02	0.1	1

*Not relevant.

uid, but that it may vary considerably with variations in pressure and temperature. deed, the density of supercritical fluids may be varied conveniently from gas-like to uid-like values, and the possibility to do so is one of the attractive features of SFC. At v densities, a supercritical fluid behaves like a gas, i.e. the viscosity is low, the diffusivity high, and there is little interaction between the molecules of the fluid either mutually or th other components. Hence, the concentration of such other components will be termined largely by their vapor pressures. On increasing the density of a supercritical id, its behavior will more and more resemble that of a liquid, i.e. the viscosity increases, e diffusivity decreases, and the molecular interactions in the fluid phase become in-easingly pronounced. The concentration of other components in the fluid will be more d more determined by interactions with fluid molecules. They dissolve rather than aporate. As is the case with liquid solvents, the solubility is not restricted by the vapor essure. Hence, highly nonvolatile solutes can be dissolved in supercritical fluids.

By varying the density of a supercritical fluid a compromise can be found between vorable chromatographic conditions and eluotropic strength. Low viscosity (resulting in small pressure drop over the column) and high diffusivity (potentially leading to short alysis times) are desirable from a chromatographic point of view. For the elution of nvolatile components a high density is required. In general, the required density will be termined by the properties of the solutes in the sample. Gases present one extreme oice (highest diffusivity, lowest viscosity, lowest density) and liquids the other (lowest fusivity, highest viscosity, highest density). Supercritical fluids may provide intermediate mbinations (see also Sections 8.5 and 8.3.1.2).

2 PRINCIPLES OF SUPERCRITICAL-FLUID CHROMATOGRAPHY

.1 Retention mechanism

In the Introduction (Section 8.1.2) the mechanism of retention in SFC, i.e. dissolving her than evaporating the solutes, has been discussed in qualitative terms. In this ction we will look more closely and more quantitatively into the effects of various rameters on retention in SFC.

.1.1 Effects of pressure and temperature

Pressure and temperature are the two parameters that can be controlled instrumentally SFC. At constant temperature, pressure has a very strong effect on retention. This is strated in Fig. 8.2, where the variation of the (logarithm of the) capacity factor (k') under C conditions is shown for naphthalene, with CO_2 as the mobile phase, at three different nperatures. At low pressures (and the low temperatures used for these experiments) capacity factors are high. At a certain pressure, which is slightly above the critical essure of CO_2 (ca. 73 bar), the capacity factors start to decrease rapidly, and at high ssures retention is very low. Because the temperatures are above the critical value (ca. C), these parts of the curves refer to SFC. At high pressures, naphthalene is eluted not

erences on p. A389

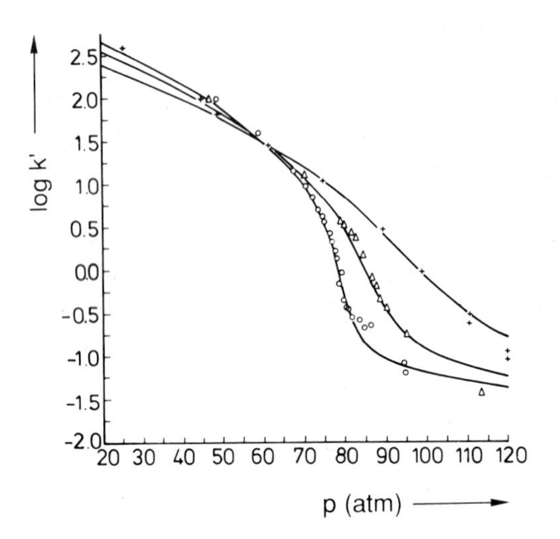

Fig. 8.2. Change of the capacity factor of naphthalene as a function of pressure at three different temperatures (o, 35°C; △, 40°C; +, 50°C). Mobile phase, CO_2; stationary phase, octadecyl-modified silica. (Reprinted from Ref. 21, with permission. Experimental data points are from Ref. 6.)

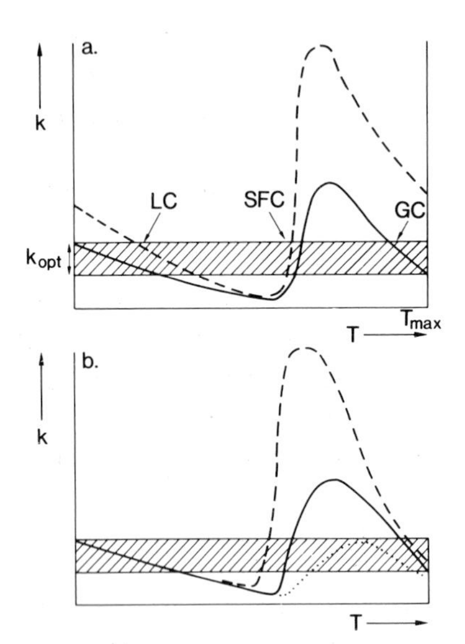

Fig. 8.3. Schematic illustration of the variation of retention with temperature in chromatography, assuming constant pressure throughout the temperature range. (a) "Small" solute (drawn line) vs. "large" solute (dashed line). (b) Intermediate pressure (drawn line) vs. low pressure (dashed line) vs. high pressure (dotted line). (Reprinted from Ref. 22, with permission.)

n the basis of its vapor pressure, but on the basis of interactions with CO_2 molecules of e supercritical fluid.

One noteworthy aspect of Fig. 8.2 is the intersection of the different curves at some termediate pressure. At low pressures, retention decreases with increasing temperature, is commonly observed in GC. At high pressures, retention increases with increasing mperature. This phenomenon is typically observed in SFC. Indeed, one can define SFC a form of interaction chromatography, in which an increase in temperature leads to an crease in retention. Such a definition underlines that the effect of the density of the obile phase forms the distinction between GC and SFC. The effect of temperature on tention in chromatography (at constant pressure!) is schematically illustrated in Fig. 8.3. e will first consider the drawn line in Fig. 8.3a. At low temperatures, we assume our obile phase to be a liquid, so that we are dealing with LC. Because the density of the obile phase is almost independent of the temperature (expansion coefficients are of the der of 10^{-3} K^{-1}), the volume would be constant in a closed system, and we may write for e capacity factor, k'

$$k' = \ln K + \ln \Phi = -\frac{\Delta H}{\mathcal{R}T} + \frac{\Delta S}{\mathcal{R}} + \ln \Phi \tag{8.1}$$

here K is the thermodynamic partition coefficient ($K = c_s/c_m$, i.e. the ratio of solute ncentrations in the stationary and mobile phases), Φ is the phase ratio ($\Phi = V_s/V_m$, i.e. ratio of the volumes of the two phases in the column), ΔH is the partial molar enthalpy the transfer of the solute from the mobile to the stationary phase, ΔS is the corre-onding entropy of transfer, $\mathcal{R}$ is the gas constant, and T is the absolute temperature.

When the temperature is increased further, retention is seen to start increasing. This n be explained by a reduction in the mobile-phase density, resulting in a reduction in extent of interaction between the solvent and the solute. This is the domain of SFC. At gh temperatures, the dominant effect is no longer the density of the eluent, but the latility of the solute. This is the case in GC. A more fundamental treatment of the effects temperature on retention in SFC has been given by Yonker et al. [23].

For certain solvents (and in certain pressure ranges) it is indeed possible to perform , SFC, and GC just by varying the temperature. Ishii et al. [24] have tried to bring all ee techniques together into a single instrument, and gave the technique the name oika chromatography". However, it is difficult to perform all three techniques under timum conditions on a single instrument, mainly because the dimensions required for three different techniques are very different.

The dashed line in Fig. 8.3a indicates that a solute of high molecular weight (but of nilar nature, e.g., a homolog) is likely to give rise to higher capacity factors. For many ch "larger" solutes GC is not feasible, because there is a maximum operating tempera-e (T_{max}). Fig. 8.3b shows the effect of column pressure on solute capacity factors. At v temperatures (LC) this effect is negligible. Under SFC conditions (intermediate tem-ratures) the effect of pressure is significant, higher pressures leading to lower k' values. high temperatures (GC) the effect of pressure diminishes.

8.2.1.2 Effects of density

In the previous section we have related the effects of variations in pressure and temperature on retention in SFC to variations in density. Generally, increasing the pressure will lead to higher densities ("compression") and increasing the temperature results in lower densities ("expansion"). However, the extent to which either process occurs depends on the actual conditions. This is usually expressed in terms of an equation of state. This is a relationship between the molar volume (v) or, equivalently, the density (ρ) and the pressure and temperature, i.e.

$$\rho = MW/v = f(P,T) \quad (8.2)$$

where MW is the molecular weight. Accurate equations of state are usually quite complex, featuring a series of coefficients. Moreover, equations that are applicable to a variety of compounds usually rely on the principle of corresponding states. In our situation this principle implies that temperatures and pressures are expressed as reduced properties (T_r and P_r, respectively), relative to the critical point, i.e.

$$P_r = P/P_c \quad (8.3)$$

and

$$T_r = T/T_c \quad (8.4)$$

According to the principle of corresponding states, the behavior of two fluids is expected to be similar at identical values of P_r and of T_r, even though the actual pressure and temperature may be very different. Frequently, a third parameter, the acentric factor, is introduced to deal with inaccuracies in the principle of corresponding states [25].

Fig. 8.4 illustrates the change of density with pressure at various temperatures for carbon dioxide. In other words, Fig. 8.4 is a graphic illustration of an equation of state. At low (sub-critical) temperatures, CO_2 may exist in two forms, a gas with a low density (bottom left in the figure) and a liquid with a high density (bottom right). Intermediate forms do not exist. Above the critical temperature, a phase separation no longer occurs. At each combination of pressure and temperature the fluid density is uniquely defined. Continuous lines connect low-density (gas-like) conditions to high-density (liquid-like) conditions in the figure. Close to the critical point, there is a very strong increase in density with increasing pressure. Indeed, at the critical point the change of density with pressure is infinitely large.

The change of density with pressure bears some relation to the variation of retention with pressure, which was illustrated in Fig. 8.2. At temperatures just above critical, an increase in pressure just above the critical value simultaneously results in a large decrease in retention and a large increase in density. Fig. 8.5 illustrates that the two observations are closely related. When the data in Fig. 8.2 are combined with those in Fig.

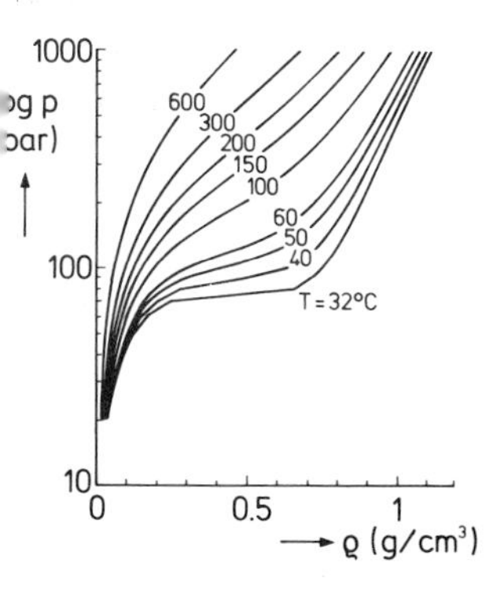

g. 8.4. Relationship between density and pressure for carbon dioxide. (Reprinted from ef. 26, with permission.)

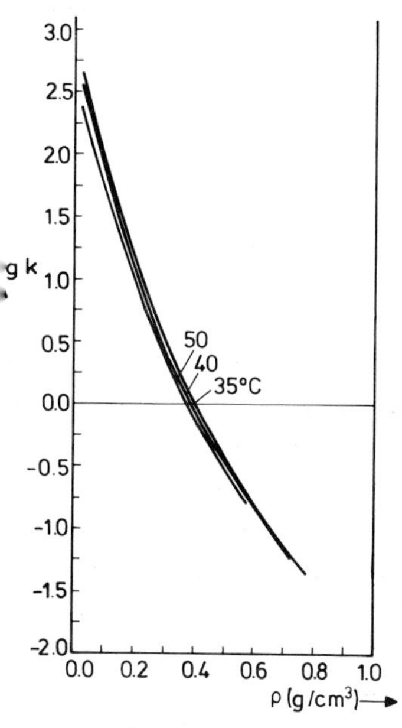

j. 8.5. Variation of the capacity factor of naphthalene as a function of density at three ferent temperatures (same data as in Fig. 8.2). (Reprinted from Ref. 21, with per- ssion.)

4, the result is a set of very smooth, monotonic curves, which relate retention in SFC th mobile-phase density. In Fig. 8.5 the effect of the temperature is seen to be small mpared to that of the mobile-phase density. Fig. 8.5 confirms that our explanation of the C retention mechanism in terms of solvation and density effects is reasonable. The ickest way to develop some understanding of SFC is to memorize the patterns of Figs. 4 and 8.5.

ferences on p. A389

An elegant concept for explaining the retention mechanism in SFC has been described by Martire [27,28]. Within a lattice model of solute/solvent interactions, he pictured a supercritical fluid as a mixture of solvent molecules and empty space ("vacuum"). The number of solvent molecules is related to the density. The empty spaces do not contribute to the interaction with the solute molecules, so that the effect of density on retention can be understood.

8.2.1.3 Influence of the stationary phase

Retention in SFC is mainly determined by the density of the mobile phase, which in turn depends on the pressure and temperature. Nevertheless, the two other components of the chromatographic system, the solute and the stationary phase, are still of great significance. Fortunately, retention does vary considerably with the polarity, shape, functional groups, molecular weight, etc. of the solute molecules. If this were not so, separations would indeed be hard to achieve. Also, the stationary phase may have a considerable effect on retention. However, given the magnitude of the effect of density on retention (see Fig. 8.5), the effects of the stationary phase are usually smaller.

The stationary phase can be an attractive "haven" for the solute molecules, or it can be quite unattractive. The difference will be manifested in different retention times. However, stationary phases that are very unattractive for the solute molecules are likely to show a limited sample capacity (for the solutes in question) and poorly shaped peaks. Therefore, the stationary phase is not the ideal place to look for changing the retention by one order of magnitude or more.

However, the effect of the stationary phase on selectivity in SFC can be quite considerable. In GC, the selectivity is determined in part by the solutes (volatility) and in part by the stationary phase. The mobile phase, which behaves approximately like an ideal gas, has no effect on the selectivity. In LC, the situation is quite different. Many different separations can be performed on a single type of stationary phase, e.g., octadecyl-modified silica. Variations in selectivity are produced by varying the nature and composition of the mobile phase. The position of SFC is intermediate between GC and LC. In SFC, pressure and temperature have dramatic effects on retention (Section 8.2.1.1), but these parameters usually do not influence the selectivity (i.e. differences in retention) very much. A variety of mobile-phase mixtures have been used in SFC (Section 8.2.2.2), the selectivity of which may be modified by varying their composition. Because the solvent strength can also be modified by varying the density, there are various possibilities to optimize separations, even when using a single stationary phase. However, as will be explained in Section 8.2.2, the use of unmodified CO_2 as a solvent is preferred if this is possible, in other words, if a suitable stationary phase can be found. Changes in selectivity are ideally achieved by changing the stationary phase.

Fig. 8.6 illustrates the differences in selectivity that may be induced by the use of different stationary phases in packed-column SFC. In Fig. 8.6a, obtained with a stationary phase of polysiloxane-modified silica and in Fig. 8.6b, where a Carbowax-type stationary phase is used, the elution order of the test compounds is identical, but the ratios of the

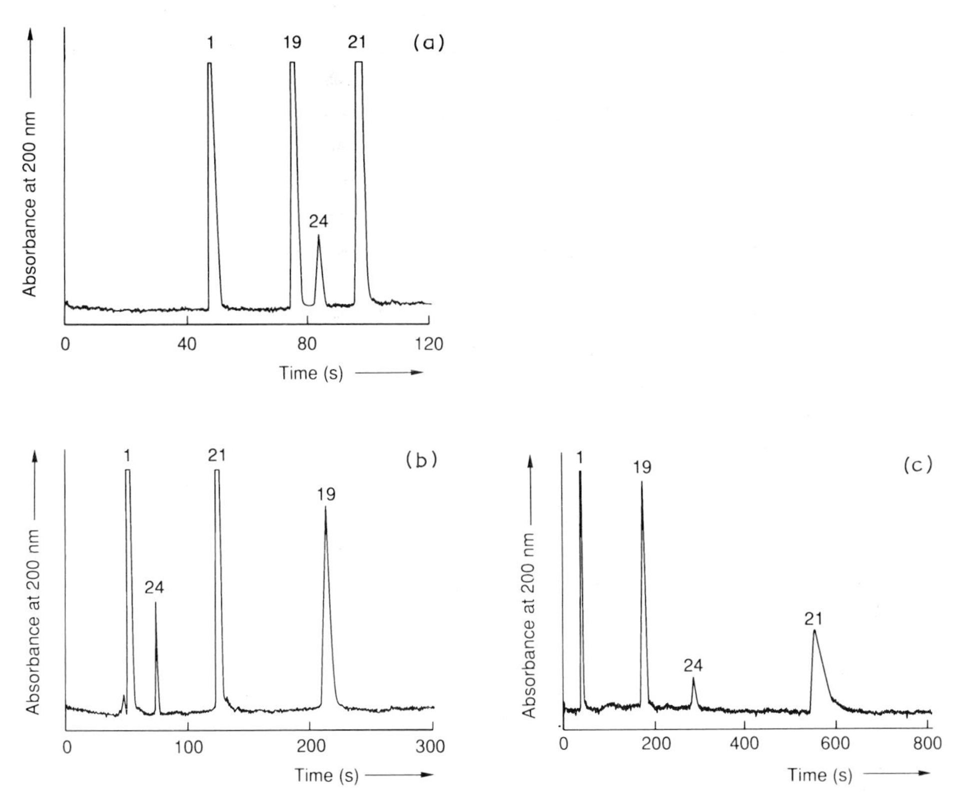

Fig. 8.6. Illustration of the differences in selectivity obtained with various stationary phases in packed-column SFC. Solutes: 1 = dichloromethane (solvent); 19 = nitrobenzene; 21 = naphthalene; 24 = diheptyl ether. Mobile phase, CO_2; stationary phase, (a) Deltabond C-8 (i.e. poly-*n*-octyl-methylsiloxane on silica); (b) "Carbonwax" (i.e. porous carbon, coated with polyethylene glycol); (c) γ-propylamino-modified silica. (Reprinted from Ref. 29, with permission.)

capacity factors (which, by definition, is the selectivity [30]) are quite different for the different pairs of peaks. Fig. 8.6c shows a more striking example. When the stationary phase is γ-propylamino-modified silica, the elution order is completely different. Nitrobenzene is now eluted long after, instead of before, the two other aromatic solutes.

8.2.2 Supercritical fluids

8.2.2.1 Single-component mobile phases

So far, we have spoken about supercritical fluids without discussing their nature. We have used carbon dioxide as a typical example. This choice is not without reason. There are a number of reasons why CO_2 is an attractive mobile phase for chromatography: It is safe (nontoxic, nonflammable), compatible with a variety of detectors, readily available in

TABLE 8.2

CRITICAL PROPERTIES [25] AND SOLUBILITY PARAMETERS [21] OF SOME SINGLE-COMPONENT SOLVENTS THAT HAVE BEEN USED FOR SFC

Solvent	P_c, bar	T_c, °C	δ^*_{SFC}, $(cal/cm^3)^{1/2}$
Ethene	49.7	9.2	5.8
Xenon	57.6	16.6	6.1
Freon 23 (CCl_2F_2)	38.7	28.9	5.4
Carbon dioxide	72.9	31.05	7.5
Ethane	48.2	32.2	5.8
Nitrous oxide	71.5	36.4	7.2
Sulfur hexafluoride	37.1	45.5	5.5
Propane	41.9	96.7	5.5
Ammonia	111.3	132.4	9.3
n-Butene	36.0	134.9	5.2
n-Butane	37.5	152.0	5.3
Diethyl ether	35.9	193.5	5.4
n-Pentane	33.3	196.5	5.1

*The solubility parameter δ is defined as the square root of the cohesive-energy density. It can be used as a measure for the polarity of compounds.

sufficient purity, largely inert, and cheap. As an eluent for SFC, CO_2 has the following additional advantages: It has favorable critical properties ($P_c \cong 73$ bar; $T_c \cong 31$°C), and it can be stored and handled like a liquid. However, CO_2 also has some disadvantages: It is not a good solvent for polar and high-MW solutes, it may react with primary amines, and it is not transparent throughout the mid-IR range.

The safety aspect is one that is often underestimated. While working with compressed gases is an obvious hazard, working with compressed toxic gases is an even greater hazard. Although it is a compressed gas, supercritical CO_2 may be preferable to the toxic liquids, such as acetonitrile and methanol, that are typically used in LC.

The fact that the critical temperature of CO_2 is above ambient means that it can be compressed into a liquid. Cylinders of carbon dioxide contain a liquid and a vapor at a pressure of about 55 bar. By using a rising (or "siphon") tube, liquid CO_2 can be drawn from the cylinder. If the ambient temperature is equal to the cylinder temperature, the solvent is then exactly on the vapor/liquid equilibrium line (solid line between tp and cp in Fig. 8.1). This implies that CO_2 turns into a gas if either the pressure is slightly decreased or the temperature slightly increased. Usually, the liquid state is preferred. When CO_2 is cooled below ambient temperature, it becomes liquid. It is easier to pump liquids at high pressures than it is to compress gases. Many SFC pumps have provisions for cooling the solvent, to facilitate priming the pump and to prevent air bubbles.

TABLE 8.3

SINGLE-COMPONENT SOLVENTS THAT HAVE BEEN USED FOR SFC AND THEIR ADVANTAGES (+) AND DISADVANTAGES (−) (± DENOTES NEUTRAL CHARACTERISTICS)

Solvent	Safety	P_c and T_c	Detection	Cost	Inertness	Solvent strength	
						Polar	High MW
CO_2	+ +	+	+	+	+	−	−
N_2O	±	+	+	+	+	−	−
Alkanes	−	+	−	+	+	−	±
SF_6	±	+	±	±	+	−	−
Xenon	+	±	+	− −	+	−	−
Freons	±	±	±	+	+	−	−
Ammonia	−	±	−	+	− −	+	+

The most serious disadvantage of CO_2 is that it cannot be used to elute all possible solutes. Especially with regard to polar and high-molecular-weight solutes the solvent strength of CO_2 is limited, even at high (liquid-like) densities. For such solutes other supercritical fluids are required.

Table 8.2 gives an overview of the critical properties of CO_2 and other single-component solvents that have been used for SFC. The advantages and disadvantages of the most important of these solvents are summarized in Table 8.3. That table shows that CO_2 is the most advantageous solvent for SFC. Unfortunately, its disadvantage is poor solvent strength toward polar and high-molecular weight solutes. For a more detailed overview of the characteristics and applicability of the various single-component solvents see Ref. 22. Here, we will limit ourselves to the most relevant advantages and disadvantages of the different solvents.

The behavior of nitrous oxide is similar to that of carbon dioxide in many respects. This can be explained on the basis of the similarity between the critical properties of the two solvents (Table 8.2) and the principle of corresponding states. It has been argued that N_2O may be a better solvent for basic solutes [31]. On the negative side, N_2O is more toxic and less compatible with flame-ionization detection, the latter being manifested in a higher noise level [32].

Alkanes may be better solvents for nonpolar solutes than CO_2. However, for the separation of high-molecular-weight samples, such as oligomeric series, certain amounts of polar solvents (e.g., 1,4-dioxane or DMSO [33]) must be added. Alkanes are highly flammable, making their use as supercritical fluids potentially hazardous, and they are not compatible with flame-ionization or infrared detection. Higher *n*-alkane homologs can be handled more easily. For example, *n*-pentane can be stored and – with some precautions to prevent the formation of vapor bubbles – pumped like other liquids. However, the critical temperatures of the *n*-alkanes rapidly increase with increasing carbon number. For *n*-pentane, temperatures of 200°C or above are needed in SFC.

The use of SF_6 has been advocated for group-type separations [34], i.e. the separation of petrochemical fractions into saturated, unsaturated, and aromatic hydrocarbons. It is a nonflammable and relatively low-toxicity solvent. However, toxic products are formed when SF_6 is used in conjunction with flame-ionization detection, which is the method of choice for the group-type separations.

Xenon is an apparently ideal solvent in combination with infrared detection [35]. As a mono-atomic molecule it does not show any absorption bands in the IR region, in contrast with CO_2, which shows considerable absorption, besides useful transparent regions ("windows") [22]. Unfortunately, xenon is horrendously expensive. The critical temperature (see Table 8.2) is below ambient, so that xenon must be stored as a gas and either handled (compressed) like a gas, or cooled considerably before it can be handled like a liquid. Moreover, like SF_6, it is not a good solvent for polar and/or high-molecular-weight solutes.

Freons are in principle nontoxic, nonflammable, nonpolar solvents. However, when used in combination with flame-ionization detection, toxic gases may be formed. Also, some doubt has been raised about the stability of these solvents under supercritical conditions, bearing in mind that the critical temperatures are fairly high [36]. Detection by FID is reasonably successful, and so is UV spectrometry, but Freons are not really compatible with FTIR.

The only solvent that scores high in the last two categories in Table 8.3 is ammonia. Ammonia has long been recognized as a potentially useful solvent for SFC [22]. However, its use in SFC is still very difficult in practice [37]. Ammonia is a good solvent for many polar and high-molecular-weight solutes, but it is also highly corrosive and attacks some of the materials in SFC instruments and most stationary phases. Silica particles or fused-silica columns can withstand ammonia only if water can be scrupulously excluded from the system and samples. Even in the absence of water, column stability and instrument reliability are likely to be problematical when ammonia is used in SFC. More importantly, ammonia may react chemically with a variety of solutes. While the lack of inertness of ammonia towards materials and stationary phases may seem to be a mere technical problem that is open to a solution, the reactivity of ammonia towards a variety of solute classes is a fundamental problem. Finally, ammonia is not compatible with flame-ionization detection and not satisfactory in combination with IR detection. Detectors typically used in LC, such as UV or fluorescence, are best suited for use in combination with ammonia. However, this reduces the potential advantages of ammonia-based SFC over water-based LC.

The use of ammonia in SFC is unlikely to lead to a major breakthrough in analytical chemistry. However, this should not be taken as a plea for directing research efforts towards other goals. If ammonia could be used, it would offer some unique possibilities. For example, if it could be used for the preparative or industrial separation of some very important, highly valuable compounds, the effort may well prove to be worthwhile.

The use of pure, single-component mobile phases offers some general advantages, i.e. inherent reproducibility and instrumental simplicity. However, none of the solvents in Table 8.3 proves to be completely satisfactory. When potentially useful polar solvents are identified [22,37], within the limits of reasonable critical properties ($T_c \leq 200°C$) and reasonable safety risks (thus disregarding, e.g., NO_2), ammonia, despite all its problems, seems to be the only realistic choice.

8.2.2.2 Mixed mobile phases

The alternative to using polar single-component mobile phases is to use mixed solvents. Besides being better solvents for polar solutes, mixed mobile phases may have additional advantages: Polar solvents may enhance solute solubility; polar additives may improve efficiency and reduce peak tailing, and they can be used to "tune" the selectivity. On the other hand, more complex instrumentation is usually required, the reproducibility may be lessened, detector compatibility may be greatly reduced, and column life may be shortened.

The addition of a small amount of a polar organic solvent, such as methanol, to a nonpolar "base" solvent, such as CO_2, often leads to considerably shortened analysis times, narrower peaks, and improved peak shapes. These effects can be observed especially (but not exclusively) when packed columns are being used. The addition of small

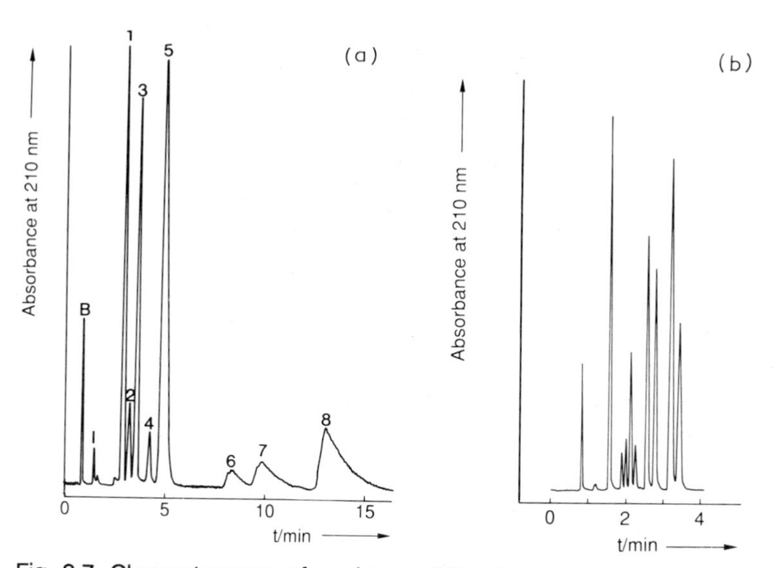

Fig. 8.7. Chromatograms of a mixture of liquid-crystal components with (a) CO_2 and (b) 5% methanol in CO_2 as the mobile phase. Stationary phase, octadecyl-modified silica; temperature, (a) 40°C, (b) 50°C; inlet pressure, (a) 190 bar; (b) 188 bar; average linear velocity, $\bar{u}$, (a) 0.21 cm/sec; outlet pressure, (b) 168 bar; UV detection at 210 nm. [Reprinted from Refs. 38 (a) and 22 (b), with permission.]

TABLE 8.4

MODIFIERS FOR SFC AND THEIR COMPATIBILITY WITH DIFFERENT DETECTORS

+ indicates good compatibility, ± limited compatibility, and – incompatibility.

Modifier	Detection compatibility				Modifier strength
	UV	IR	FID	MS	
Methanol	+	–	–	±	high
Ethanol	+	–	–	+	high
Acetonitrile	+ +	–	–	+	low
Tetrahydrofuran	±	–	–	+	moderate
Water	+	–	+	+	very high
Formic acid	+	–	+	+	high

amounts of polar solvents, usually called modifiers, may have three different effects: (i) increasing the polarity of the mobile phase, (ii) increasing the density, and (iii) dynamically modifying the stationary phase. Of these three effects, only the last one can explain the vast differences in the results between open-tubular (small surface area of the stationary phase) and packed columns (large surface area). For this and a number of other reasons this factor is thought to play an important role. Fig. 8.7 illustrates the effect of the addition of a small amount of modifier to carbon dioxide on a chromatogram obtained by packed-column SFC.

The most significant effect of modifiers in packed-column SFC is their adsorption on the surface of the stationary phase. By assuming that the surface contains a limited number of active groups (e.g., silanols) and a larger number of less active (e.g., octadecyl chains), the poor peak shapes in Fig. 8.7a can be explained as the result of a mixed retention mechanism [39]. This model suggests that peak shapes may be improved, if either the number of active groups or their activity can be decreased. Both aspects may be improved by developing new stationary phases (Section 8.3.4) or by using modifiers.

A large number of different modifiers have been proposed for SFC. An overview is provided by Table 8.4. This table also illustrates the most serious disadvantage of the use of modifiers. namely the reduced compatibility of the mobile phase with different types of detectors. For example, when a few percent of the most popular modifier, methanol, is added to carbon dioxide, a FID can no longer be used. Also, most of the "windows" in the IR spectrum of CO_2 are blocked by absorption bands of methanol. UV detection and MS remain possible, although for the latter ethanol may be preferred to lengthen the filament life. Acetonitrile is a useful modifier for UV detection, because it is transparant down to 190 nm. Conversely, THF is not very useful because of its higher UV cutoff. Moreover, stabilizers interfere with UV detection. However, at low concentrations of modifier, this may not cause any problems.

Most common modifiers cannot be used with the sensitive and universal FID, because they give rise to excessive background and noise levels. The two possible exceptions are water and formic acid, both of which are compatible with the FID but have significant effects on retention and peak shape. However, because of their limited solubility in CO_2, which is the solvent of choice in combination with FID their concentrations may not be high enough to achieve the desired modifier effect. To increase the amount of modifier in the mobile phase, mixtures of water and formic acid have been used [40]. Water may be introduced into a CO_2 stream by using a "saturator column", i.e. a small, packed column containing a wet adsorbent. Water is desorbed from such a column up to its solubility limit [41]. A certain degree of control of the modifier concentration is possible by varying the conditions (P, T) in the saturator column, but the mobile phase must remain homogeneous throughout the system, i.e. conditions must not allow the solubility of water (or formic acid) to decrease after the precolumn.

The modifier strength indicated in Table 8.4 is related to the affinity of a modifier for the active groups on the surface of the stationary phase rather than to its enhancing effect on either the polarity or the density of the mobile phase. The density effect has been shown to be relatively small [42]. The concentration of methanol required to cover 95% of the active sites on a octadecyl-modified silica column (at 45°C and 179/162 bar*) was found to be 2.3 mol %. For ethanol a concentration of 0.66 mol % was needed, while for 2-propanol and THF concentrations of 0.21 and 1.2 mol %, respectively, were sufficient [42]. However, the maximum surface coverage was also different for the different solvents, namely 68 μmol for methanol, 25 μmol for ethanol, 8.6 μmol for 2-propanol, and 16 μmol for THF. These figures show that methanol is the most effective modifier in terms of surface coverage, but that high concentrations are required to achieve the maximum effect, while 2-propanol is more effective at lower concentrations.

Adding a few percent of an organic modifier to a supercritical fluid is not always straightforward. If the "base fluid" is a liquid under ambient conditions, the modifier can be added in the bottle before the solvent is pressurized and heated. For example, mixtures of *n*-pentane and 1,4-dioxane can be prepared in this way. When the "base fluid" is a compressed gas, such as CO_2, the process is more difficult. Premixing the solvents in the cylinder inevitably leads to a large dependence of the modifier concentration on the liquid level in the cylinder [22]. In that case, instrumentation is required for mixing the modifier and the "base fluid" at elevated pressures. Even then, it is difficult to make mixtures of reproducible and stable composition.

It was suggested earlier that the selectivity can be "tuned" to the requirements of the sample by varying the nature and concentration of the organic modifier [43]. This approach has hardly been pursued in SFC, most likely because the exact concentration of modifiers is difficult to control and to reproduce. Obviously, single-component solvents are

* This notation stands for a column-inlet pressure of 179 bar and an outlet pressure of 162 bar.

to be preferred, and the use of mixed mobile phases should be avoided as much as possible.

8.3 COLUMNS

8.3.1 Theory

8.3.1.1 Efficiency

8.3.1.1.1 Open-tubular columns

As in other forms of chromatography, the efficiency of a chromatographic system can be characterized by the number of theoretical plates (N). The height equivalent of a theoretical plate, $H = L/N$ (where L is the column length), can be expressed in terms of the average linear velocity $\bar{u}$ by a plateheight equation, such as the Golay equation for open-tubular (capillary) columns

$$H = \frac{2D_m}{\bar{u}} + f(k')\frac{d_c^{\,2}\,\bar{u}}{D_m} + g(k')\frac{d_f^{\,2}\,\bar{u}}{D_s} \tag{8.5}$$

with H in cm and $\bar{u}$ in cm/sec. In SFC the density may decrease along the column length, causing the linear velocity to increase. The average value can be found by measuring the holdup time, t_o, and using $\bar{u} = L/t_o$. In Eqn. 8.5 D_m (cm^2/sec) is the diffusion coefficient of the solute in the mobile phase and D_s (cm^2/sec) its diffusivity in the stationary phase, k' the capacity factor (dimensionless retention time), d_c the column diameter (cm), and d_f the thickness of the stationary-phase film on the column wall (cm). $f(k')$ and $g(k')$ are functions of the capacity factor, k'. If the mobile-phase density is not too high, the same equations can be used in SFC as in GC, i.e.

$$f(k') = \frac{1 + 6k' + 11k'^2}{96\,(1 + k')^2} \tag{8.6}$$

and

$$g(k') = \frac{2k'}{3\,(1 + k')^2} \tag{8.7}$$

In order to compare columns of different diameters, we introduce the reduced (dimensionless) plateheight

$$h = H/d_c \tag{8.8}$$

and the (average) reduced velocity

$$\bar{v} = \frac{\bar{u}\, d_c}{D_m} \tag{8.9}$$

into Eqn. 8.5. We then find

$$h = \frac{2}{\bar{v}} + f(k')\,\bar{v} + g(k')\left(\frac{d_f}{d_c}\right)^2 \left(\frac{D_m}{D_s}\right)\bar{v} \tag{8.10}$$

Columns with different stationary-phase layer thicknesses may best be compared in terms of the (dimensionless) reduced film thickness [44], which is equal to

$$\delta_f = \frac{d_f}{d_c}\sqrt{\frac{D_m}{D_s}} \tag{8.11}$$

In part, the reduced film thickness is a characteristic of the column (d_c, d_f, and D_s, are parameters of the column and the stationary phase). However, especially in SFC, the mobile-phase diffusion coefficient, D_m, may be strongly affected by the operating conditions. Increasing the density of the mobile phase will tend to decrease the diffusion coefficient. The effect of increasing the temperature (at constant density) is partly canceled, because D_m and D_s may be affected in a similar way. Most of all, it is important to realize that a column does not show any performance (separation efficiency) of its own. Especially in SFC the observed column efficiency is strongly dependent on pressure and temperature. Hence, the optimum conditions (optimum flowrate and maximum efficiency) will also be a function of pressure and temperature.

By substituting the reduced film thickness into Eqn. 8.10 a very simple plateheight equation remains

$$h = \frac{2}{\bar{v}} + f(k')\,\bar{v} + g(k')\,\delta_f^2\,\bar{v} \tag{8.12}$$

Eqn. 8.12 expresses the reduced plateheight in open-tubular chromatography exclusively in terms of dimensionless parameters.

8.3.1.1.2 Packed columns

A different treatment is needed for packed columns. The flow pattern and the path of mobile-phase or solute molecules through the column are less well-defined. As a consequence, there is no analytical $H/\bar{u}$ or $h/\bar{v}$ equation. Also, the increase in plateheight with increasing capacity factor is much less pronounced than it is in open-tubular chromatography. Indeed, this is an advantage of packed columns. To a first approximation, h may be assumed to be independent of k', and the Knox equation may be used

References on p. A389

$$h = A\,\bar{v}^{\,1/3} + \frac{B}{\bar{v}} + C\,\bar{v} \tag{8.13}$$

In Eqn. 8.13 the dimensionless parameters $\bar{v}$ and h are defined as

$$\bar{v} = \frac{u\,d_p}{D_m} \tag{8.14}$$

where d_p is the particle size, and

$$h = H/d_p \tag{8.15}$$

Typical values obtained in LC for well-packed columns (independent of the particle size) are $A = 1$, $B = 2$, and $C = 0.05$ [45].

8.3.1.2 Speed

The time required to perform an analysis, t_{req}, is determined by the retention time of the last-eluted solute (denoted by z). When the width of the last peak is taken into account, we find

$$t_{req} = t_z + 2\sigma_z = t_z + 2\,\frac{t_z}{\sqrt{N}} = t_z\,(1 + \frac{2}{\sqrt{N}}) \tag{8.16}$$

which for $\sqrt{N} >> 2$ reduces to

$$t_{req} \cong t_z = t_o(1 + k'_z) \tag{8.17}$$

With $t_o = L/\bar{u}$ and $L = N_{req}\,H$, where N_{req} is the number of theoretical plates required for accomplishing the separation, this equation becomes

$$t_{req} = \frac{N_{req}\,H}{\bar{u}}\,(1 + k'_z) \tag{8.18}$$

The introduction of the dimensionless parameters $h = H/d_o$ and $\bar{v} = ud_o/D_m$, where d_o is either the inner diameter of an open-tubular column or the particle size in a packed column, now yields

$$t_{req} = \frac{N_{req}\,h\,d_o^2}{\bar{v}\,D_m}\,(1 + k'_z) \tag{8.19}$$

Eqn. 8.19 is very useful for comparing the speed of separation in different forms of chromatography with different columns. Assuming that for a given separation the same mobile phase is used, the same capacity factors are observed and the selectivity is the same, resulting in the same required number of plates, simple and reasonable comparisons can be made. However, it is not easy to satisfy all these assumptions in practice [46]. Therefore, the experimental comparison of different types of columns remains a subject of study.

Eqn. 8.19 can be divided into a sample-independent factor and a sample-dependent factor, as follows

$$t_{req} = \frac{h\, d_o^2}{\overline{v}\, D_m} \times N_{req}\,(1 + k'_z) \qquad (8.20)$$

Because the order of magnitude of D_m is determined by the mobile phase rather than by the solute, it has been included in the sample-independent factor. Obviously, the required analysis time increases with increasing complexity of the separation problem (higher value of N_{req}) and with increasing capacity factors.

It is equally obvious that the required analysis time will increase if the reduced plate-height increases, and it will decrease with increasing velocity. The fastest columns are those for which the ratio $h/\overline{v}$ is as low as possible, i.e. those columns which show very shallow reduced plateheight curves. Open-tubular columns with regular stationary-phase films of constant thickness or packed columns with a good packing of uniformly sized particles will best meet this requirement. In principle, all good columns will have similar values for h and $\overline{v}$. Open-tubular columns are typically operated at higher than optimum velocities to achieve a reasonable speed of analysis. Typical values [46] are $\overline{v} \cong 45$ and $h \cong 4.5$ (cf. Eqn. 8.12 for $k' = 10$ and $\delta_f = 0.3$), i.e. $h/\overline{v} \cong 0.1$. Typical values for packed columns [47] are $\overline{v} \cong 10$ and $h \cong 3$, i.e. $h/\overline{v} \cong 0.3$. Using these typical values and Eqn. 8.20, the required analysis time is seen to be about three times higher for packed columns than for open columns, provided that all other factors, including the characteristic diameter, d_o, are equal. This implies that an open-tubular column with an inner diameter of 20 μm is three times faster than a column packed with 20-μm particles. Conversely, to obtain a comparable speed of analysis, the characteristic (inner) diameter of an open-tubular column may be a factor $\sqrt{3}$ larger than the diameter of the particles in a packed column.

In comparing different chromatographic techniques, such as GC, SFC, and LC, the relevant factor is the ratio d_o^2/D_m in Eqn. 8.20. This implies, for example, that if the difference in diffusion coefficients between gases and liquids is typically of the order of 10^4, the characteristic dimensions in LC should be a factor 10^2 smaller than those in GC. Hence, whereas open-tubular capillary columns in GC have inner diameters of several hundreds of μm, the inner diameter of open-tubular columns for LC may only be a few μm [48].

TABLE 8.5

ANALYSIS TIMES (sec) CALCULATED FROM EQN. 8.19, ASSUMING $N_{req} = 10^4$ AND $k' = 9$, FOR DIFFERENT COLUMNS IN DIFFERENT FORMS OF CHROMATOGRAPHY

(a) Open-tubular columns; (b) packed columns. Diffusion coefficients: $D_m = 5 \times 10^{-2}$ cm^2/sec for GC, $D_m = 2 \times 10^{-4}$ cm^2/sec for low-density SFC [46], $D_m = 3 \times 10^{-5}$ cm^2/sec for high-density SFC [46], and $D_m = 10^{-5}$ cm^2/sec for LC. For open-tubular columns $\bar{v} = 45$ and $h = 4.5$; for packed columns $\bar{v} = 10$ and $h = 3$ [47]. Required analysis times between 10 sec and 1 h are printed in bold digits.

d_c (for *a*) or d_p (for *b*), μm	t_{req} (GC), sec	t_{req} (SFC), sec		t_{req} (LC), sec
		low ρ	high ρ	
a. Open-tubular columns				
500	**500**	1.25×10^5	8.3×10^5	2.5×10^6
250	**130**	3.1×10^4	2.1×10^5	6.25×10^5
100	**20**	5×10^3	3.3×10^4	10^5
50	5	**1250**	8300	2.5×10^4
25	1.3	**310**	**2100**	6250
10	0.2	**50**	**330**	**1000**
5	0.05	**12.5**	**83**	**250**
b. Packed columns				
500	**1500**	3.75×10^5	2.5×10^6	7.5×10^6
250	**375**	9.4×10^4	6.25×10^5	1.9×10^6
100	**60**	1.5×10^4	10^5	3.0×10^5
50	**15**	3750	2.5×10^4	7.5×10^4
25	4	**940**	6250	1.9×10^4
10	0.6	**150**	**1000**	**3000**
5	0.15	**38**	**250**	**750**
3	0.05	**14**	**90**	**270**

Table 8.5 lists the calculated analysis times that can be obtained for a moderately complex separation ($N_{req} = 10^4$ and $k' = 9$) on open and packed columns of different dimensions (d_c and d_p). In this table, values between 10 sec and 1 h are printed in bold digits. We find this range reasonable for a separation of moderate complexity. The table shows that reasonable analysis times are obtained in GC for columns with internal diameters between 100 and 500 μm, which are the columns used in practice. In fact, for the relatively simple separation considered here, even somewhat larger diameters ($d_c \leq 1.3$ mm) can be used, with analysis times within 1 h.

For low-density SFC, a 50-μm-ID column is acceptable, but the analysis time is fairly long. Open columns with smaller diameters lead to shorter analysis times. For high-density SFC, columns with internal diameters of 25 μm or less are required to achieve reasonable analysis times. As is well known, open-tubular columns for LC should have internal diameters of 10 μm or less.

Currently, the most popular internal diameter for open-tubular columns in SFC is 50 μm. Although reducing the diameter would have clear advantages in terms of analysis times, there are some serious obstacles to be overcome:

(1) columns with smaller diameters having a homogeneous stationary-phase film of constant thickness are difficult to prepare;

(2) smaller amounts of sample must be injected, both in terms of mass and in terms of volume;

(3) these smaller amounts need to be detected in a smaller volume.

Therefore, progress in the direction of smaller columns for SFC is understandably slow.

Table 8.5b shows that reasonable particle-size ranges for packed-column SFC are $3 \leq d_p \leq 25\ \mu$m for low-density applications and $3 \leq d_p \leq 10\ \mu$m at high densities. These ranges are similar to those typically used in LC (see last column in Table 8.5b), so that the optimum particle size is readily available. In terms of speed of analysis, packed-column SFC is one or two orders of magnitude better than open-tubular SFC at the moment, i.e. comparing 50-μm-ID open columns with 5-μm particles. This is a practical, rather than a fundamental advantage of packed columns, which will disappear when very narrow open-tubular columns become feasible in practice.

8.3.1.3 Pressure drop

An approximate expression for the pressure drop over the column (ΔP) is the Darcy equation. In this equation the expansion of the mobile phase along the length of the column is neglected, i.e. the mobile-phase density and the linear velocity are assumed to be constant. For LC, this is obviously a valid approximation. For open-tubular SFC and for high-density packed-column SFC, the approximation is quite reasonable [26]. If we assume u to be constant, the Darcy equation reads

$$\Delta P = B^{\circ}\eta L u = B^{\circ}\eta N h \nu D_m \tag{8.21}$$

If ΔP is in Pa units, then viscosity, η, of the mobile phase may be expressed in Pa $\times$ sec, the column length, L, in m, and the velocity, u, in m/sec. B° is the specific permeability coefficient in m^{-2}, which for open-tubular (OT) columns equals

$$B^{\circ}_{OT} = 32 / d_c^2 \tag{8.22}$$

and for packed columns

$$B^{\circ}_{PC} \cong 1000 / d_p^2 \tag{8.23}$$

References on p. A389

TABLE 8.6

REQUIRED PRESSURE DROPS (Pa x 10^6) ACCORDING TO THE DARCY EQUATION (EQN. 8.21) IN HIGH-DENSITY SFC, ASSUMING $\bar{\nu}$ = 45 AND h = 4.5 FOR OPEN-TUBULAR COLUMNS AND $\bar{\nu}$ = 10 AND h = 3 FOR PACKED COLUMNS, D_m = 3 x 10^{-5} cm^2/sec AND η = 7 x 10^{-5} Pa·sec [46]

(a) Open-tubular columns (Eqn. 8.22); (b) packed columns (Eqn. 8.23); (c) μ-packed columns (Eqn. 8.24). Values above 2.5 x 10^7 Pa (≅ 250 bar) are placed in brackets.

(a) Open-tubular columns

d_c, μm	ΔP_{req}, Pa x 10^6		
	$N = 10^3$	$N = 10^4$	$N = 10^5$
10^2	10^{-3}	10^{-2}	10^{-1}
5 x 10^1	5 x 10^{-3}	5 x 10^{-2}	5 x 10^{-1}
2.5 x 10^1	2 x 10^{-2}	2 x 10^{-1}	2.2
10	10^{-1}	1.4	1.4 x 10^1

(b) Packed columns

d_p, μm	ΔP_{req}, Pa x 10^6		
	$N = 10^3$	$N = 10^4$	$N = 10^5$
2.5 x 10^1	10^{-1}	1	10
10	6 x 10^{-1}	6	(6 x 10^1)
5	2.5	2.5 x 10^1	(2.5 x 10^2)
3	7	(7 x 10^1)	(7 x 10^2)

(c) μ-Packed columns

d_p, μm	ΔP_{req}, Pa x 10^6		
	$N = 10^3$	$N = 10^4$	$N = 10^5$
2.5 x 10^1	3 x 10^{-2}	3 x 10^{-1}	3
10	2 x 10^{-1}	2	2 x 10^1
5	8 x 10^{-1}	8	(8 x 10^1)
3	2	2 x 10^1	(2 x 10^2)

Usually, it is assumed that the pressure drop in packed columns is independent of the internal column diameter, provided that the linear velocity is kept constant. This is no longer true if the column diameter approaches the order of the particle size. For packed-capillary columns, in which the internal diameter of the column is typically between 100 and 300 μm, a better estimate for the specific permeability coefficient may be

$$B^{\circ}_{\mu PC} \cong 300 / \ d_p^{\,2} \tag{8.24}$$

Some calculated values for column pressure drops in SFC with open-tubular columns, conventional packed columns, and μ-packed (packed-capillary) columns (μPC) are given in Table 8.6. Although large column pressure drops (of the order of 2.5 x 10^2 bar or 2.5 x 10^7 Pa) have been demonstrated to be feasible [13], at extremely high values the pressure at the column outlet becomes too low. In that case, the solubility of the solute cannot be maintained throughout the systems, and serious losses in efficiency will result [49]. Therefore, values above 2.5 x 10^7 Pa have been placed in brackets in Table 8.6.

Table 8.6a shows that typical pressure drops in open-tubular SFC are of the order of a few bar or less (i.e. $\leq 10^5$ Pa), unless very small internal diameters (e.g., 10 μm) or very large plate numbers (e.g., $\geq 10^5$) are achieved. Table 8.6b illustrates that the situation is different in packed-column SFC. Relatively simple separations ($N = 10^4$) can be accomplished with columns packed with 5-μm particles. Somewhat larger numbers of theoretical plates can be achieved at a given pressure drop by reducing the linear velocity to the value corresponding to the minimum in the plateheight curve. This is the case for both open and packed columns, in both cases at the expense of an increase in analysis time. The number of plates achieved per unit time (N/t) is highest at high linear velocities.

Plate numbers much higher than 10^4 will be hard to achieve with packed columns, because the pressure drop will soon become a limiting factor. Table 8.6c shows that the lower B° value of packed-capillary columns makes them attractive in this respect.

8.3.2 Open-tubular columns

In Section 8.3.1 we have been able to draw some significant conclusions about the use of open-tubular columns in SFC:

(1) In order to obtain reasonable analysis times, the internal diameter of open-tubular columns should be 50 μm or (preferably) less for low-density SFC and 25 μm or less when high densities are required (see Table 8.5a).

(2) The pressure drops are low in open-tubular SFC, so that a large number of theoretical plates (e.g., 10^5) may be obtained in principle (see Table 8.6a).

Open-tubular columns are especially useful when a high number of theoretical plates is needed for complex separations. Such a high number of plates can be obtained (much) more easily for applications that do not require a high mobile-phase density. Open-tubular SFC is at its best for samples that just exceed the possibilities of current high-resolution GC. This may be due to a lack of either volatility or thermal stability of the solutes. If the volatility of the solutes becomes very low, e.g., in the case of polymers, a high mobile-phase density may be required, and open-tubular SFC becomes more problematic.

Open-tubular columns are prepared (almost exclusively) from fused-silica capillary tubing, in which the stationary phase is deposited on the inner wall. In order to obtain a

TABLE 8.7

OPTIMUM FILM THICKNESSES (in μm) FOR A NUMBER OF DIFFERENT OPEN-TUBULAR COLUMNS IN GC, SFC ("LOW" AND "HIGH" DENSITY), AND LC, CALCULATED FROM EQN. 8.11, USING A VALUE OF 0.32 FOR THE REDUCED FILM THICKNESS

The values corresponding to reasonable column diameters (according to Table 8.5a) appear in bold print. Diffusion coefficients: $D_S = 5 \times 10^{-7}$ cm^2/sec for all techniques, mobile-phase diffusion coefficients as in Table 8.5. Data were taken from Ref. 47.

d_c, μm	d_f (GC), μm	d_f (SFC)		d_f (LC), μm
		low ρ	high ρ	
500	**0.5**	8	20	40
250	**0.25**	4	10	20
100	**0.1**	1.5	4	7
50	0.05	**0.8**	2	3.5
25	0.025	**0.4**	**1.0**	2
10	0.01	**0.15**	**0.4**	**0.7**
5	0.005	**0.1**	**0.2**	**0.35**

sufficiently large column capacity and dynamic range, the stationary-phase film should be as thick as possible. On the other hand, according to Eqn. 8.5, the last term in the plateheight equation increases in proportion with the square of the film thickness. Therefore, thick films may lead to a loss of efficiency. It can be shown [22] that the optimum value for the reduced film thickness (Eqn. 8.11) is about 0.3. In Table 8.7 this is translated into actual film thicknesses for a number of different column diameters in GC, SFC, and LC, assuming that in all three forms of chromatography the same stationary phase (e.g., a polysiloxane) can be used. It appears from this table that the stationary-phase films applied in SFC columns can be considerably thicker than in current practice, where films of 0.25 μm are common in 50-μm-ID columns. Again, preparing the columns may be a critical factor. Thicker layers of stationary phases may be more difficult to prepare from polymeric solutions [50]. Also, the optimal layer thickness is reduced if the diffusion coefficient of the solute in the stationary phase (D_S) is lower. However, lower values of D_S are always undesirable in chromatography.

8.3.3 Packed columns

The packed columns that are typically applied in SFC have been developed for HPLC. "Conventional" HPLC columns have inner diameters between 2 and 5 mm, lengths between 50 and 300 mm, and packings with particles of (mean) diameters between 3 and 10 μm. The use of such small particles in SFC, which was pioneered by Gere et al. [13],

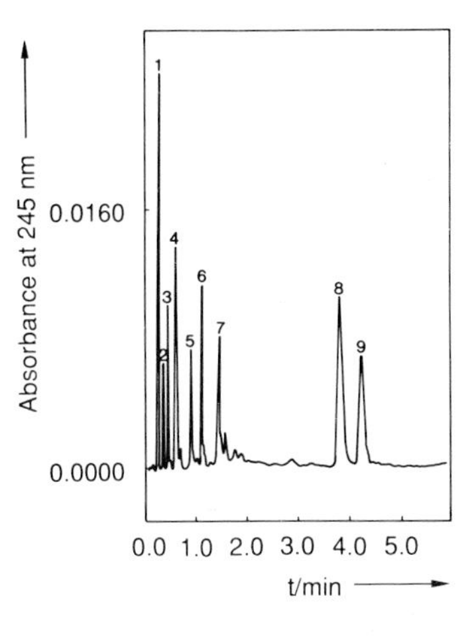

Fig. 8.8. Example of a packed-column supercritical-fluid chromatogram, obtained with a high pressure drop over the column. Stationary phase, octadecyl-modified silica; mobile phase, CO_2; pressure, 400/243 bar; temperature, 33°C; solutes, 1 = toluene, 2 = naphthalene, 3 = fluorene, 4 = anthracene, 5 = fluoranthene, 6 = pyrene, 7 = 1,2-benzanthracene, 8 = perylene, 9 = 1,2,3,4-benzanthracene. (Reprinted from Ref. 13, with permission.)

has meant a major breakthrough for packed-column SFC. We have previously concluded that

(1) Rapid analysis with a reasonable number of theoretical plates can be achieved on columns packed with 3- to 10-μm particles (see Table 8.5b).

(2) The pressure drop over the column becomes a limiting factor when a high number of theoretical plates (e.g., 10^5) is needed (see Table 8.6b).

(3) Packed-capillary columns, with internal diameters of 0.5 mm or less, show a somewhat lesser pressure drop. Therefore, the potential maximum number of plates is higher.

It has been shown by Gere et al. [13] that high pressure drops (up to a few hundred bar) are possible in SFC, provided that the outlet pressure is kept sufficiently high. Fig. 8.8 shows an example of a chromatogram obtained with a pressure drop of 157 bar. Because the temperature is just above the critical value and the outlet pressure is twice the critical pressure, the mobile-phase density remains high throughout the column (cf. Fig. 8.4).

Packed columns are probably most useful for high-density applications, where SFC resembles LC more closely than GC. The choice between open-tubular columns and packed columns is affected most of all by the diffusion coefficient of the solute in the mobile phase, which in turn is closely related to the density [51]. The higher the density, the lower will be the diffusion coefficient. At low densities, the diffusion coefficients are high ($D_m \cong 10^{-2}$ cm^2/sec), characteristic dimensions (d_o; Eqn. 8.19) may be relatively large, and open-tubular columns may be used. At high densities, the diffusion coefficients are low ($D_m \cong 10^{-4}$ cm^2/sec) and the required characteristic dimensions are too small for open-tubular columns to be routinely applicable at present (see also Table 8.5). As a consequence, open-tubular columns are most useful for GC and low-density (gas-like)

SFC, whereas packed columns are most useful for LC and high-density (liquid-like) SFC. Unfortunately, we are not free to choose the mobile-phase density. Its value will almost exclusively be determined by the requirement to elute the solutes of interest with capacity factors in the optimum range.

8.3.4 Stationary phases

In all forms of chromatography, be it GC, LC, or SFC, the essence of retention is the distribution of the solute molecules between the mobile and the stationary phases. This underlines that both phases are equally important for the chromatographic process. In SFC the mobile phase is different from the gases used in GC and the liquids used in LC. Therefore, it is logical that the mobile phase receives most attention in the literature on SFC. Nevertheless, the stationary phase is relevant in many respects:

(1) *Retention* is determined by the distribution of the solute between the mobile and stationary phase. This is the product of two factors, i.e. the thermodynamic distribution coefficient in terms of concentrations, K_c, and the phase ratio, Φ ($k' = (c_s/c_m)(V_s/V_m) = K_c \Phi$, Chapter 1). Both the nature and the volume of the stationary phase in the column will thus affect the retention.

(2) *Selectivity*, α_{ji}, is defined as differences in retention between two different solutes, *i* and *j*, i.e. $\alpha_{ji} = k'_j / k'_i = K_{c,j} / K_{c,i}$. Apparently, the selectivity is affected by the nature, but not by the volume of the stationary phase.

(3) *Efficiency and peak shape*, which may be affected by the nature and volume (film thickness) of the stationary phase (Section 8.3.2). The effect of the stationary phase on the observed peak shapes will be described in more detail in Section 8.3.4.2.

8.3.4.1 Stationary phases for open-tubular columns

8.3.4.1.1 Polysiloxanes

The stationary phases used for open-tubular SFC are almost exclusively immobilized films of polymeric materials. Most commonly, polysiloxanes are used. Their structures are schematically illustrated by the examples in Fig. 8.9. By incorporating various functional groups in the polysiloxane molecules, the polarity of the stationary phase can be varied. The polarity of the phases illustrated in Fig. 8.9 increases from Structure 1 to 4.

The popularity of polysiloxanes in chromatography can largely be explained by two factors, i.e. their high temperature stability and the high diffusion coefficients for solutes. The first of these factors is much more relevant for (high-temperature) GC than it is for SFC, at least when CO_2 is used in the latter technique. The second factor remains valid, so that polysiloxanes are also useful for SFC. The diffusion coefficients in polysiloxanes are higher than in most other polymers.

Diffusion coefficients generally increase with increasing temperatures. For polymeric materials this increase may be quite dramatic at temperatures around the "glass-transition" point. When heated to the glass-transition temperature (T_g), the polymer will change from glass-like to rubber-like. In the latter state the mobilities of molecules, and, thus, the

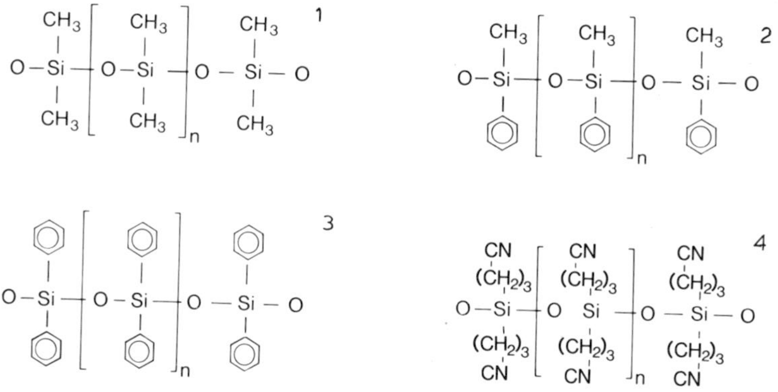

Fig. 8.9. Some examples of the structures of polysiloxane stationary phases. 1 = Polymethylsiloxane (100% methyl), 2 = polymethyl-phenylsiloxane (50% phenyl), 3 = polyphenylsiloxane (100% phenyl), 4 = polycyanopropylsiloxane (100% cyano). (Redrawn from Ref. 22, with permission.)

diffusion coefficients of compounds dissolved in the polymer, are much higher. By operating open-tubular SFC columns at relatively low densities (e.g., $\rho \leq 0.3$) and relatively high temperatures (e.g., $100 \leq T \leq 120°C$) all polysiloxanes used in SFC are in the rubber state. A number of other polymeric materials that are commonly used in GC haver higher glass-transition temperatures and are therefore less useful for SFC.

An important aspect of stationary phases in SFC is that they should be immobilized. This implies that the stationary phases should not be washed from the column by the eluents used in SFC. Because SFC can be used for the elution of nonvolatile solutes, a lack of volatility is not sufficient for a stationary phase to be stable under SFC conditions. The common procedure used to immobilize polymeric films in chromatographic columns is the so-called "crosslinking" [52]. In this process different polysiloxane chains, such as those depicted in Fig. 8.9, are connected into a network, from which the initial polymeric chains cannot be removed without breaking chemical bonds. By conditioning the column under GC conditions prior to crosslinking, all low-molecular-weight components (including residual monomers, small oligomers, and initiator) can be removed, so that a stable, inert, and homogeneous phase can be obtained.

8.3.4.1.2 Other phases

Most of the stationary phases used in open-tubular SFC are polysiloxanes, as described in the previous section. Among the other stationary phases that have been used are polymers, such as polyethylene glycol (Carbowax), which is a polar stationary phase, and liquid-crystalline phases, which offer good possibilities for the separation of molecules that are geometrically different (e.g., isomers).

It is also possible to create a porous layer of a solid adsorbent on the inner wall of a capillary tube (so-called PLOT columns). The resulting stationary phase is a solid with a high surface area, while the mobile-phase flowpath remains an open channel. In principle, the larger surface area may allow a higher retention and a higher sample capacity to be achieved without increasing the film thickness of the stationary phase. However, a large surface area is not always advantageous in SFC, as will be explained in Section 8.3.4.2.

8.3.4.2 Stationary phases for packed columns

8.3.4.2.1 Modified silicas

The most common stationary phases used for packed-column SFC are modified silicas. Silica (or silica gel) is pure silicon dioxide (SiO_2). However, because the silicon atom is surrounded by four oxygen atoms, the structure is necessarily incomplete at the surface. The surface of silica gel contains a significant number of silanol (-SiOH) groups, which can interact strongly with (polar) solute molecules by means of hydrogen bonding. Also, the silanol groups can be used to perform chemical reactions at the surface. Fig. 8.10 illustrates the product of the reaction of the silica support with a functional dimethylsilyl compound. The general structure of such a reagent is X–Si(CH_3)$_2$)–R, where X denotes the functional group involved in the reaction with the surface. For example, X may be an ethoxy group (ethanol being a byproduct of the reaction) or chlorine (byproduct, HCl). The R group remains anchored on the surface, as is illustrated in Fig. 8.10. (For more details on chemically bonded stationary phases for liquid chromatography see Chapter 4.)

In principle, the stationary phases that have been developed for HPLC can also be used for SFC, provided that the phases are (1) stable under SFC conditions and (2) inert with respect to the mobile phases used in SFC. All common modified silicas have been used in SFC, the most popular ones being octadecyl (R = C_{18}), octyl (R = C_8), cyano (R = –(CH_2)$_2$–CN), and amino (R = –(CH_2)$_2$–NH_2) phases. Of these, the amino phases may not be inert toward CO_2, because the solvent may react with the primary amine

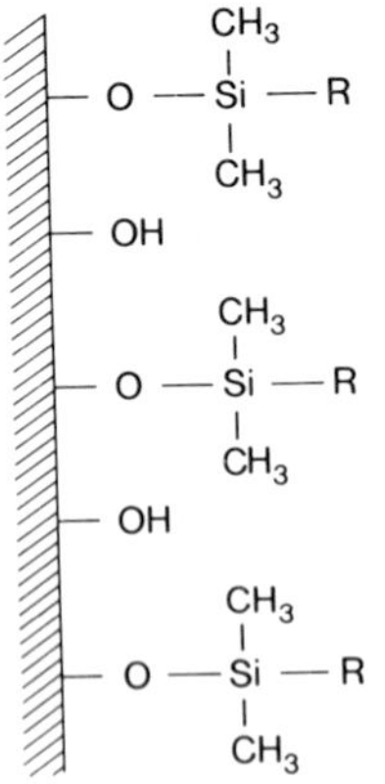

Fig. 8.10. Schematic illustration of the structure of chemically modified silicas used as stationary phases in packed-column SFC. (Redrawn from Ref. 22, with permission.)

group to form a carbamate. It is unclear at the moment whether or not such a reaction does take place, but the fact is that some good separations of (fairly) polar molecules have been demonstrated on amino (or carbamate) columns.

8.3.4.2.1.1 Surface inhomogeneity. Packed-column SFC on modified-silica surfaces may be applied to a variety of samples. However, when CO_2 is used as the mobile phase, poorly shaped peaks are often observed for many polar solutes, especially those which possess functional groups that can participate in hydrogen-bonding interactions. A clear example of this is shown in Fig. 8.7a. The molecular structures of the solutes showing good peak shapes contain hydrocarbon and cyano groups. Solutes containing ether or ester groups show poorly shaped peaks (see also Table 8.8).

When "pure" CO_2 is used as the mobile phase and ODS-modified silica as the stationary phase, poorly shaped peaks are also observed for many other solutes. Some of these solutes show much improved peak shapes when different stationary phases are used in packed-column SFC [29,54]. Moreover, a considerable number of solutes yields much better peak shapes in open-tubular SFC (with polysiloxane films as stationary phases) than in packed-column SFC (with modified silicas). Therefore, it is quite reasonable to assume the stationary phase to be one of the main causes of the poor peak shapes observed in (packed-column) SFC. However, other factors, such as insufficient solubility of the solute in the mobile phase, may also contribute. Different (improved) stationary phases may lead to better peak shapes for a number of (classes of) solutes. However, some other (classes of) solutes may yield poorly shaped peaks on any column, as long as CO_2 is used as the mobile phase.

One way to distinguish between the different causes of broad and asymmetric peaks in SFC is by studying the effects of the amount of sample injected on the peak shape and on the observed retention times (capacity factors). If mobile-phase overloading or solubility

TABLE 8.8

MOLECULAR STRUCTURES OF THE SOLUTES INDICATED IN FIG. 8.12

Solute	Structure	Peak shape*
PCH-5	C_5H_{11}–(H)–(O)–CN	good
ME5NF	C_5H_{11}–(O)–COO–(O)–CN (F)	good
M3	C_3H_7O–(O)–(O)–CN	poor
M6	$C_6H_{13}O$–(O)–(O)–CN	poor

*Conditions as in Fig. 8.7a: mobile phase, CO_2; stationary phase, ODS.

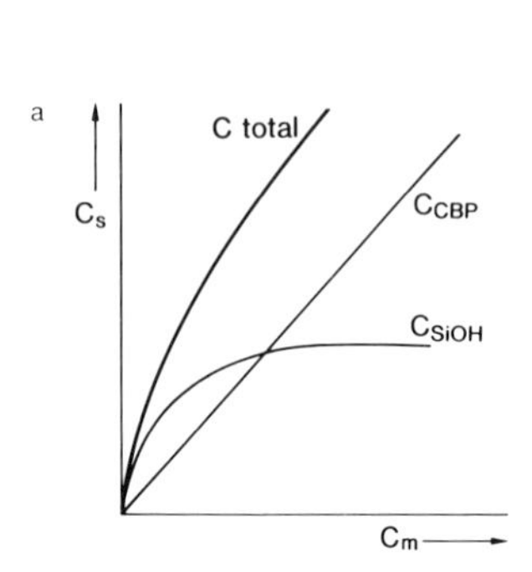

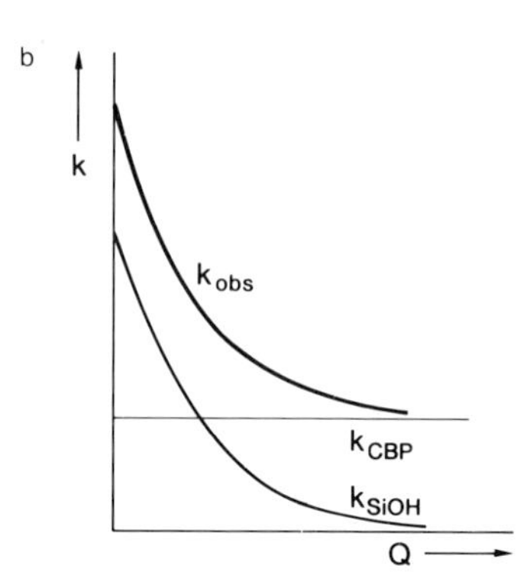

Fig. 8.11. Schematic illustration of the effects of sample size in packed-column SFC, based on the assumption of a mixed-retention mechanism due to residual silanol groups on an octadecyl-modified silica. (a) Adsorption isotherms. (b) Resulting variation of the capacity factor with the amount of sample injected. (Reprinted from Ref. 39, with permission.)

problems are to blame for poorly shaped peaks, then we would expect peak shapes to deteriorate and capacity factors to increase when the amount of injected solute is increased. This is not what is observed in packed-column SFC. On increasing the amount of injected solute, peak shapes tend to improve and capacity factors decrease (Figs. 8.11 and 8.12). These effects have been explained by assuming a mixed retention mechanism, to which both the chemically bonded groups at the surface (e.g., octadecyl chains) and residual (unreacted) silanol groups contribute [39]. It is assumed that the silanol groups exhibit strong interactions with the solute molecules, but that their number is limited. It is assumed further that the number of chemically bonded groups available for interaction is relatively large. This leads to the assumption of adsorption isotherms of the general form shown in Fig. 8.11a. The isotherm in this figure is the sum of a linear isotherm for the interaction of the solute molecules with the chemically bonded groups and a Langmuir-type adsorption isotherm for the silanol-group interaction. The capacity factor is related to the adsorption isotherm. At any concentration, k' is proportional to the slope of a line drawn through the origin and the corresponding point on the adsorption isotherm, the proportionality factor being the phase ratio, Φ. At higher concentrations, the slope of the cord lines decreases, so that the capacity factor decreases with increasing concentrations. In physical terms this can be understood from a saturation of the active silanol groups. Fig. 8.12 illustrates that in packed-column SFC of hydrogen-bonding solutes on octadecyl-modified-silica stationary phases, with "pure" carbon dioxide as the eluent, the capacity factor decreases with increasing amounts of injected solute.

A change of the capacity factor of the solute with the concentration, i.e. a nonlinear adsorption (or distribution) isotherm, generally leads to deformed, asymmetrical peaks. If the capacity factor is lowest at the highest concentrations (as in Figs. 8.11b and 8.12), the top of the peak travels faster through the column than the edges. As a consequence, the

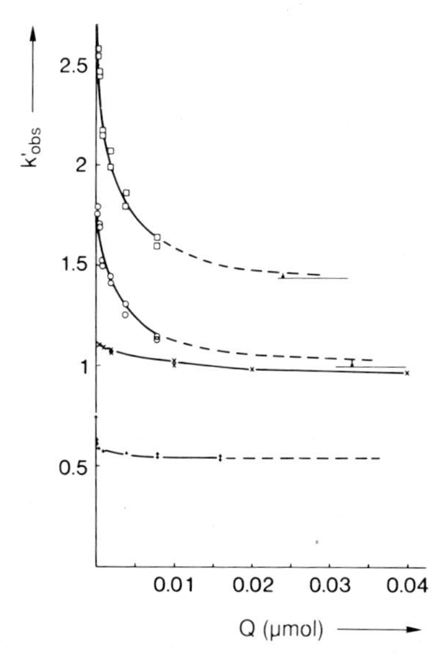

Fig. 8.12. Observed variation of the capacity factor with the amount of solute injected in packed-column SFC. Solutes, liquid-crystal components (see Table 8.8); mobile phase, CO_2; stationary phase, Chromspher C18 (ODS-silica); pressure, 255/240 bar; temperature, 35°C. (Reprinted from Ref. 39, with permission.)

peak maximum tends to catch up with the leading edge of the peak, while the tailing edge drops more and more behind. As a consequence, tailing peaks, such as those of the later-eluted solutes in Fig. 8.7a, can be qualitatively explained by the mixed retention mechanism.

The mixed retention mechanism also aids in understanding the possible remedies for the poor peak shapes in Fig. 8.7a:

(1) Saturate the active (silanol) groups of the stationary phases by adding a polar modifier to the mobile phase. The small modifier molecules have easy access to the active groups, undergo strong interactions, and are present in much larger numbers than the solute molecules, even at low modifier concentrations.

(2) Prepare stationary phases that do not contain different types of interacting groups, so that mixed retention mechanisms do not occur.

Fig. 8.7b shows that Approach 1 does present a good solution. However, the addition of modifiers to the mobile phase does have a number of disadvantages, i.e. an increased complexity of the system, a potential loss in reproducibility, and – most of all – a serious loss in detection options. Therefore, Approach 2 is definitely worth considering. In the remainder of this section and in Section 8.3.4.2.2 some alternative stationary phases for the elution of relatively polar solutes in packed-column SFC will be described.

8.3.4.2.1.2 Polymer-coated silicas. The purpose of chemically modifying silica surfaces is to create a homogeneous stationary phase which interacts uniformly with solute molecules at any place on the surface. If the starting material is silica, then all the silanol groups should be made inaccessible. For steric reasons, this cannot be done by making every single one of them react with a silanizing agent. Only about half the total number of silanols can undergo the type of reaction illustrated in Fig. 8.10 [55]. The other half is not accessible to small silylating agents, such as trimethylchlorosilane (TMCS) ("end-capping"). However, the remaining ("residual") silanols prove to be accessible to a variety of solute molecules.

One way to decrease the number of accessible silanols is by depositing a polymeric layer on top of the silica matrix. In this way, the silanols are not "removed" by a chemical reaction, but "camouflaged" by a polymeric blanket. If the solute molecules interact with the polymeric layer, but do not diffuse through it to interact with the silica surface, a more homogeneous retention mechanism may result.

Polymer-modified silicas may either be prepared by immobilizing a polymeric layer on the silica in a manner similar to that used to prepare stationary phases on the inner wall of fused-silica columns [57], or by a chemical reaction of a polymer containing functional

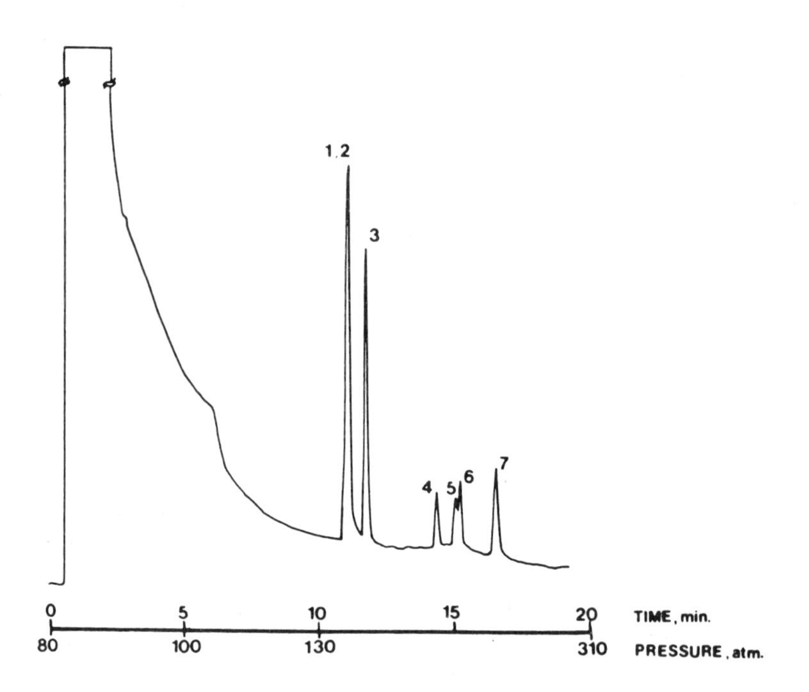

Fig. 8.13. Illustration of the application of polymer-modified silicas in SFC. Sample, nitrated diphenylamine mixture: 1 = *N*-nitrosodiphenylamine; 2 = nitrodiphenylamine; 3 = diphenylamine; 4 = 2,4-dinitrodiphenylamine; 5 = 2,4,6-trinitrodiphenylamine; 6 = 2,4′-dinitrophenylamine; 7 = 4-nitrodiphenylamine. Mobile phase, CO_2; stationary phase, Deltabond-CN (cyanopropylsiloxane-coated silica); pressure programed as indicated in the figure; temperature, 70°C. (Reprinted from Ref. 56, with permission.)

groups with the silica surface, i.e. in a manner similar to the reaction illustrated in Fig. 8.10 [58].

An example of the application of polymer-modified silicas in packed-column SFC is shown in Fig. 8.13. Previous studies on this nitrated biphenylamine mixture had indicated that in could not be eluted from conventional, chemically modified silicas by 100% CO_2 [56]. In our experience, the surface homogeneity is greater with polymer-coated silicas, such as the one used for Fig. 8.13, than it is with conventional, chemically bonded phases. Better peak shapes are obtained for certain classes of molecules. However, camouflaging the silica surface does not appear to suffice for abating the effects of residual surface activity completely.

8.3.4.2.2 Other materials

8.3.4.2.2.1 Polymer-based phases. Instead of silica, other solid adsorbents can be used as a basis for stationary phases in packed-column chromatography. One obvious alternative is alumina. Both modified and unmodified alumina have been used in SFC [59]. However, alumina itself interacts more strongly with many classes of solutes than does silica (basic solutes being exceptions). Thus, the problem of residual surface activity is likely to be more serious with modified aluminas than it is with modified silicas.

Another approach is to use adsorbents based on organic polymers. The most commonly used phases are based on copolymers of styrene and divinylbenzene (PS/DVB polymers). Because divinylbenzene contains two polymerizable groups, it acts as a crosslinking agent. Thus, polymeric networks are formed that are insoluble in organic solvents and in supercritical fluids, such as CO_2. PS/DVB phases exhibit very high retention times. When such phases are being used in LC, stronger eluents (e.g., higher volume fractions of organic modifier in RPLC) are used to compensate for the higher adsorptivity of the stationary phase. In SFC, PS/DVB requires much higher mobile-phase densities than do octadecyl-modified silicas to elute the same compounds with similar retention times. Because there is a practical limit to the density that can be achieved (see Fig. 8.4), very large, nonvolatile solutes cannot be eluted from PS/DVB phases. For example, using pure CO_2 as the eluent, it was barely possible to elute naphthalene with a high mobile-phase density. Since solutes that can thus be eluted are usually compatible with GC (unless their thermal stability is a limiting factor), PS/DVB itself is of limited interest for SFC.

A possible remedy is to modify the PS/DVB surface chemically, e.g., by introducing alkyl groups. However, this material is not stable when used with supercritical CO_2. Even with the most careful handling of the phase, large voids form at the top of the column. A plausible explanation for this is that despite the crosslinking with divinylbenzene, these phases may swell considerably in organic solvents and in supercritical fluids, and that, moreover, the swelling may vary significantly under different conditions. Modified PS/DVB columns cannot be regenerated by washing with conventional (liquid) solvents. Since (modified) PS/DVB phases are quite expensive, their use in SFC is not advisable, unless the applicability and stability of a particular phase have been amply demonstrated by the manufacturer.

8.3.4.2.2.2 Porous-carbon phases. Porous (graphitic) carbon is a second alternative to silica as a solid adsorbent. Again, as with PC/DVB, the unmodified adsorbent exhibits much greater retention than do octadecyl-modified silicas and, again, high mobile-phase densities are required as a compensation. Unlike PS/DVB phases, porous carbon does not show swelling behavior, let alone differences in swelling behavior between different solvents. Columns packed with porous carbon are very stable under SFC conditions.

It does not seem easy to modify the surface of porous carbon by performing a reaction at the surface. Reactive groups do not appear to be available. However, because of the very strong retention of the adsorbents, it has been possible to create stable stationary phases by physically coating a layer of polyethylenglycol (Carbowax 20M, nominal MW 2×10^4) on the surface. Results with this so-called "Carbonwax" phase have been encouraging [29]. Symmetrical peaks were obtained for a variety of (classes of) compounds. However, retention of "poly"aromatic solutes, such as naphthalene and anthracene, remains high, probably due to residual activity on the surface.

8.4 INSTRUMENTATION

8.4.1 SFC systems

SFC systems consist of one or more high-pressure pumps, an injection device, a column, a detector, and a restrictor. The latter is installed either before or after the detector, for low- or high-pressure detection, respectively. The pump obtains the solvent from a cylinder or flask. A second pump may be used for adding a modifier to the mobile phase. Between the pump and the injector, some kind of heat-equilibration facility (e.g., a thermostated coil) and, in case of two solvents, some kind of mixing device are usually incorporated. Together with the injector and the column, these parts are usually maintained at a constant temperature by an air-circulation oven or a water bath. After high-pressure detection, the pressure on the system is reduced by some kind of pressure-control or flow-control device. Because there is no danger of peak broadening after detection, this device can be relatively large and rugged, and variable restrictors can be used. This allows the inlet and outlet pressures to be varied independently.

In low-pressure detection, decompression of the solvent to a gas takes place before it enters the detector. In this case, it is critical to let decompression occur rapidly but smoothly, to ensure that (1) solutes do not precipitate from the gas phase, and (2) the peakwidth is not increased significantly. For this purpose, a number of different decompression devices have been developed (Fig. 8.14), none of which is adjustable. Hence, the word "restrictor" is commonly used. The flow of the supercritical fluid is restricted to such an extent that a very large pressure drop occurs almost instantaneously. In this way, the two demands formulated above can best be met.

Until recently, a length of (fused-silica) tubing with a very narrow internal diameter (e.g., 5 or 10 μm; Fig. 8.14A) was often used. However, in this type of restrictor decompression takes place relatively slowly, leading to significant precipitation problems. These are either manifested in a loss of solute or in the occurrence of spikes (bursts of precipi-

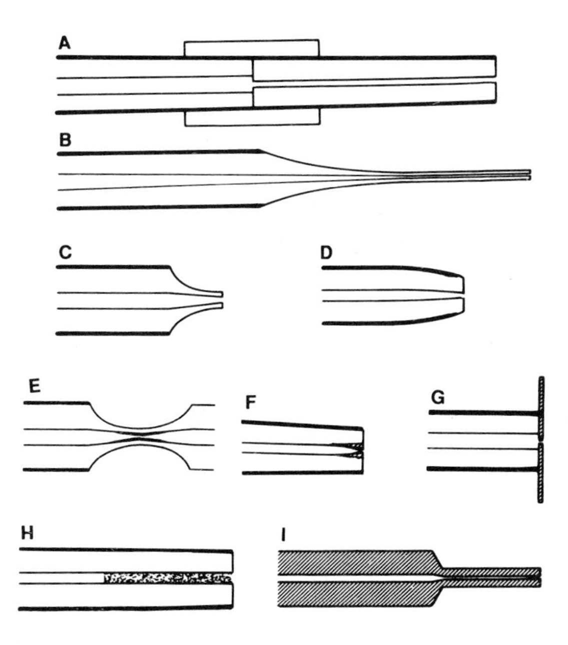

Fig. 8.14. Schematic illustration of a number of fixed restrictors used in combination with low-pressure detection in SFC. A = linear restrictor; B = tapered restrictor; C = drawn capillary; D = polished (or integral) restrictor; E = converging/diverging restrictor; F = deposition restrictor; G = pinhole orifice; H = frit restrictor; I = pinched restrictor. (Reprinted from Ref. 60, with permission.)

tated molecules) in the detector signal. Small orifices (Fig. 8.14G) are seemingly ideal restrictors, but they are very vulnerable and difficult to prepare reproducibly. Extremely vulnerable is the tapered restrictor (Fig. 8.14B), which is made by drawing the end of a fused-silica column into a very thin capillary. The integral restrictor (Fig. 8.14D) is more rugged. It is made by closing a fused-silica tube in a flame and then polishing the end until a very small hole appears and the desired flowrate is obtained. This is now a very popular restrictor, but because the hole is so small, it is liable to (partial) blockages. This problem occurs less frequently with the frit restrictor (Fig. 8.14H), but here again reproducibility is poor.

The main disadvantage of fixed restrictors is that the column inlet pressure (or more importantly, the density) and the outlet pressure (more importantly, the flowrate) cannot be controlled independently. Schwartz et al. [61] have pointed out that the higher the mobile-phase density, the lower is the optimum mobile-phase velocity. Understandably, when a fixed restrictor is being used, higher densities (i.e. higher column inlet pressures) lead to higher linear velocities, opposite to what is desired. As a result, the linear velocity increases during pressure or density programing, causing the efficiency to decrease gradually from the beginning to the end of the chromatogram. A means of regulating the flowrate, independent of density, in SFC with low-pressure detection is high on the wish list of developments in SFC.

References on p. A389

8.4.1.1 Solvent delivery

8.4.1.1.1 Types of solvents

Three types of supercritical solvents may be distinguished: (1) those that are liquids at ambient conditions (e.g., *n*-pentane), (2) those that are vapors at ambient conditions (e.g., CO_2), and (3) those that are gases at ambient conditions (e.g., Xe). Each category is handled differently.

Liquid solvents are easily stored, and pumps designed for HPLC can be used without modification. Vapors can be compressed to liquids, so that they can be stored as liquids in high-pressure (e.g., $\cong$ 55 bar for CO_2) cylinders. At the cylinder temperature, the solvent is then exactly at the vapor/liquid equilibrium line (see Fig. 8.1). As a consequence, if the same pressure is maintained in the supply cylinder and in the system, a slight increase in temperature will cause the fluid to be a gas, while a slight decrease in temperature will change it into a liquid. The latter condition is commonly used. Compressed vapors are pumped as liquids at temperatures below ambient.

For permanent gases other measures need to be taken. Since the critical temperature for xenon is 16°C, it cannot be stored as a liquid at ambient temperatures. High pressures and low temperatures are needed to convert "permanent" gases to liquids. Usually, permanent gases cannot be compressed by a HPLC pump, and gas compressors are needed. Apart from difficulties in delivering stable flowrates, gas compressors tend to be expensive and noisy. Therefore, permanent gases are less desirable than liquids or vapors as mobile phases for SFC.

8.4.1.1.2 Pumps

HPLC pumps are either reciprocating-piston pumps or syringe pumps. The first type of pump is equipped with one, two, or three pistons, which go through refill/delivery cycles. For single-piston pumps (i.e. reciprocating-piston pumps with small pistons) the refill stroke must be very fast in order to approximate the delivery of a constant flow. However, in principle, a constant flow cannot be obtained by this type of pump. Dual-piston pumps offer this feature by delivering with one piston while the other one is refilling. The total flow delivered by the two pump heads may be designed to be constant. Corresponding designs with three-piston pumps are the same in principle.

However, although two-piston and three-piston pumps may be designed to provide a constant flow, this is not usually achieved in practice. Because the compressibility of different solvents is different, a variable part of the piston stroke is "wasted" on compression. This requires some kind of correction (usually electronic) to the operation of the pump. Moreover, the valves that must be opened and closed when the direction of piston movement is reversed never work instantaneously [62]. Mainly on account of these two reasons, a certain variation in pressure ("pump noise") can always be observed when the different pistons take over duties.

The increasingly popular solution to the problem of "pump noise" is the use of large pistons, which are big enough to ensure that (at least) a complete analysis can be performed without refilling the piston. Because pistons for capillary SFC may already be considered large if their volume is of the order of a few ml and because early pumps were equipped with syringes instead of metal pistons, large-piston pumps are often referred to as syringe pumps. Syringe pumps may provide more constant flowrates with less pressure fluctuations than reciprocating-piston pumps. The main disadvantages are the time needed to refill the piston and the resulting lack of flexibility (see also Section 8.4.1.1.3). To reduce the problem of long refill times, the piston may be cooled. Alternatively, dual-syringe pumps may be used.

8.4.1.1.3 Modifiers

Three ways of adding modifiers can be distinguished: (1) premixing the solvents in the reservoir flask or cylinder; (2) mixing the solvents in or just before the compression chamber ("low-pressure mixing"); (3) mixing the effluents of two separate pumps ("high-pressure mixing"). All three methods are commonly used in HPLC. In SFC, there are some additional problems that the user should be aware of. Premixing the solvents is likely to result in considerable variations in the mobile-phase composition, if the base solvent is a compressed vapor, such as CO_2 [22]. For example, if a full cylinder at 20°C contains 5% (by volume) of methanol in CO_2, then the methanol concentration will have increased to almost 6% by the time the cylinder is half empty, to 7% when the cylinder is 80% empty, and to much higher concentrations upon further usage. Given the dramatic effects of the modifier concentration on retention and peak shape in SFC (Section 8.2.2.2), accurate control of the mobile-phase composition is highly desirable.

Modifiers can also be mixed with the "base" solvent in the compression chamber or pump head. In this case, a measured amount of, e.g., methanol is introduced into the pump, followed by a measured amount of, e.g., CO_2. In LC this method of mixing is known as "low-pressure mixing" (ahead of the pump), but if CO_2 is used in SFC, the

TABLE 8.9

ILLUSTRATION OF THE VOLUMETRIC FLOWRATE OF MODIFIERS IN DIFFERENT SFC SYSTEMS

Modifier concentrations are in % v/v; F_{CO_2} is the volumetric flow of liquid CO_2 ($\rho \cong 1$ g/cm^3).

Column		F_{CO_2}, min^{-1}	Modifier flow, min^{-1}		
Type	Diameter		10%	1%	0.1%
Packed	4.6 mm	1 ml	0.1 ml	10 μl	1 μl
Packed	1 mm	50 μl	5 μl	0.5 μl	50 nl
Packed Capillary	0.3 mm	5 μl	0.5 μl	50 nl	5 nl
Open	50 μm	1 μl	0.1 μl	10 nl	1 nl

pressure at the pump inlet is usually around 55 bar. If the dimensions of the pump relative to that of the column are such that one piston stroke takes a long time (e.g., when a syringe pump is used in capillary SFC), the "low-pressure mixing" method offers limited flexibility, and composition gradients cannot be used.

The third method – the equivalent of what is known as "high-pressure mixing" in HPLC – offers the greatest flexibility. However, flowrates in SFC may be small when small columns are used, and small concentrations of modifiers then pose very severe demands on the capacity of the modifier pump for reproducibly delivering very small volumes (see Table 8.9). For example, at modifier concentrations of 1% the modifier pump should be able to deliver a constant and reproducible flowrate of the order of 10 μl/min. In open-tubular SFC the corresponding modifier flowrate is of the order of 10 nl/min.

8.4.1.2 Sample introduction

8.4.1.2.1 Packed columns

Sample introduction in SFC systems depends greatly on the type of column used and on the column dimensions. In packed-column SFC, sample introduction methods are very similar to those practiced in LC. The sample can be introduced into the (open) loop of an injection valve, provided that this loop is not very large. Sample loops having volumes of several ml are common in some systems. Such valves are less suitable for SFC, because the entire volume of the loop must be compressed to the operating pressure once the valve is turned. For packed columns of "conventional" diameters (i.e. 2 to 5 mm ID), sample loops of 10 and 20 μl are satisfactory. For smaller packed columns, HPLC injection valves with smaller loops can readily be used. Injection valves with (internal) loops of 100 or even 50 nl are commercially available for use in combination with packed-capillary columns.

The most critical aspect of sample introduction is the injection time. Because very fast analyses are possible, the sample must be injected in the shortest time possible. One way to achieve this is to open the injection valve for only a short time, e.g., by electrical activation. However, the compression of the volume of the sample loop may now be an advantage. When the valve is switched to the "load" position, the mobile phase present in it (usually CO_2) expands to atmospheric conditions. If the time before injection is long, the mobile phase in the loop will be replaced by ambient air through diffusion. When, for example, 1 μl of a liquid sample is introduced into the loop, which is subsequently turned to the "inject" position, the air behind (or around) the sample will be compressed, and this may result in a sharp injection band. This may explain why HPLC injection valves appear to work better in SFC than in LC. An additional advantage of mobile phases such as CO_2 is the virtual absence of particulate matter, so that both columns and injection valves can be operated more reliably.

8.4.1.2.2 Open-tubular columns

In open-tubular SFC, sample introduction is much more difficult. The amounts of sample that can be introduced into the small-diameter columns used for SFC

($d_c \leq 50\,\mu m$) are extremely small. The volume of one theoretical plate may serve as a guideline. For $d_c = 50\,\mu m$ and $h = 4.5$ the plate volume is approximately 0.5 nl. When the diameter decreases by a factor of 2, the plate volume decreases by a factor of 8, i.e. for a 25-μm column the plate volume becomes about 50 pl. This illustrates both the difficulty of sample introduction into open-tubular SFC columns, as well as the problems that need to be overcome if the column diameter is to be reduced. Because of the relatively thin films of stationary phases currently used, the absolute amounts of solutes that can be injected (and need to be detected!) are also quite small. This can be ameliorated, if good columns with thicker films, closer to the optimum values (Section 8.3.2) can be prepared.

In comparison with capillary (open-tubular) GC, injection in open-tubular SFC is not only complicated by the smaller column diameters, but also by the much higher column-inlet pressures. Split injection appears to be attractive because of the constraints on sample size. However, this is not straightforward, because a reproducible flow through the split line must be maintained. This requires an efficient restrictor at the end of it. Also, because supercritical solvents tend to cool upon expansion, split lines usually should be heated. To increase the reliability and reproducibility of injection in capillary SFC, other injection techniques have been advocated [11,63]. The most important of these appears to be the time-split injection, which is based on switching a sample valve to the "inject" position for a short and controlled period of time. There is still some discussion on the reliability and reproducibility of injection in capillary SFC, but progress is being made in this area.

8.4.1.3 Flow and pressure control

Because the mobile-phase density greatly depends on the pressure in the region just above the critical point (Fig. 8.4), pressure control is extremely important in SFC. While an error of 1% in the flowrate may lead to a 1% change in retention at constant pressure, a change in pressure of 1% at constant flowrate may lead to much larger effects. Therefore, SFC pumps are typically operated in the constant-pressure rather than the constant-flow mode. In principle, this may imply that retention-time reproducibility is fundamentally less in SFC than it is in HPLC. However, there is no indication that retention data in SFC are not sufficiently reproducible in practice.

Once the pressure at the column inlet is carefully controlled, there is one other parameter that may be controlled, i.e. either the pressure or the flowrate at the column outlet. Fixed restrictors are typically used in combination with low-pressure detection (Section 8.4.1). The flowrate through the column is then a function of the mobile-phase pressure and temperature, and actual control of the outlet pressure or the flowrate is impossible.

When high-pressure detection is used, backpressure valves, mass-flow controllers, or needle valves may be used to control either pressure or flowrate. However, such devices are not very rugged under SFC conditions. While variable controllers do exist, some progress in this area would be welcomed by SFC users.

References on p. A389

8.4.1.4 *Programed elution*

Programed elution can be defined as chromatography in which the elution conditions are varied during the experiment. Since decreasing migration speeds are of little use, programed elution usually implies that the elution of components is speeded up during the experiment. There are a number of ways to do this, i.e. by programing pressure, temperature, density, solvent composition, or flowrate. Detailed reviews of programed elution in SFC can be found elsewhere [64-66]. Flow programing will not be discussed further. It is of limited use for several reasons:

(1) The flowrate has only a linear effect on retention, whereas the effect of all the other programable parameters is (in one way or the other) exponential.

(2) An increase in the flowrate leads to an increase in the pressure drop over the column.

(3) For a given column under given conditions (P, T, composition) there exists an optimum linear velocity (Section 8.3.1.1), above which the efficiency will decrease.

(4) Unlike other programing techniques, flow programing does not increase the solute concentration at the peak maximum. Consequently, no increase in sensitivity may be expected when using flow programing in combination with concentration-sensitive detectors (UV, IR), but for mass-flow-sensitive detectors (FID, MS) an increase in the flowrate will lead to an increase in sensitivity.

Although flow programing as such will not be discussed, it should be mentioned that variations in the flowrate may well be the unintended result of variations in other parameters. For example, when a fixed restrictor is used, increasing the pressure will increase the flowrate and linear velocity [46]. This is unfavorable according to (3) above, and even more so because the optimum velocity decreases upon increasing the pressure (density). Thus, fixed restrictors are quite inadequate in combination with pressure or density programing, and useful alternatives would be of great value for SFC.

8.4.1.4.1 *Pressure (density) programing*

Retention decreases with increasing pressure in SFC, and the effects are quite considerable (Fig. 8.2), so that pressure programing can obviously be of use. However, Fig. 8.2 also shows that under conditions close to the critical point the change of retention with pressure is very drastic. Because the effect of density on retention is more regular (Fig. 8.5), density programing may be considered to be more elegant. However, there are two arguments to be made in favor of pressure programing rather than density programing:

(1) retention (and density) vary rather regularly with pressure in regions well away from the critical point, and

(2) pressure can be controlled directly, but density is an indirect parameter, which is controlled by varying pressure and/or temperature.

The first of these arguments implies that linear pressure programs can be quite adequate as long as either $T \gg T_c$ or $P \gg P_c$. The second argument implies that, in order to use density programing, an equation of state is needed to calculate (at a given temperature) the required pressure for obtaining a certain density. During a programed elution, such calculations need to be performed at various times, either before or during the analysis. The instrument controls the pressure to yield the desired density. Only pressure programing can be achieved in SFC. Density programing is a special form of pressure programing, such that the desired density program is defined and the pressure program is derived from it.

Density programing is possible only if a suitable equation of state is available. This is the case for CO_2 and for other pure solvents, but for mixtures the calculation of density as a function of pressure and temperatures is extremely difficult. Therefore, for mixed mobile phases it appears to be much more logical to use a well defined and, if necessary, nonlinear pressure program than to attempt density programing. In this way, the problem of calculating the density is not a serious factor, and data obtained by different laboratories can be meaningfully compared.

8.4.1.4.2 Temperature programing

For all but the very volatile solutes, an increase in the temperature in SFC typically leads to an increase in retention times (see Fig. 8.3). Therefore, in order for temperature programs to be useful in SFC, the temperature should decrease during the analysis. This technique has indeed been applied in SFC and is usually referred to as "negative temperature programing". Negative temperature programing is sometimes preferable to ("positive") pressure programing for practical reasons, e.g., if varying the temperature is easier in practice than varying the pressure. This will not usually be the case if high-pressure pumps are used for pressurizing the solvent. Changes in pressure can then be produced much more rapidly than changes in temperature. There may also be a fundamental advantage in negative temperature programing in cases where the selectivity increases with decreasing temperature. In such a case, the resolution towards the end of the chromatogram may be better with negative temperature programing than with positive pressure programing.

8.4.1.4.3 Solvent programing

Solvent programing for SFC has been reviewed extensively by Klesper and Schmitz [64]. It is perhaps the most powerful, but certainly the most complex way to perform programed elution in SFC. All the aspects of mixed mobile phases and the addition of modifiers to the mobile phase (Sections 8.2.2.2 and 8.4.1.1) are equally applicable to isocratic elution (constant mobile-phase composition) and to solvent-programed elution. To all this, the time factor should then be added.

A gradual increase in the concentration of a (polar) organic modifier in a (nonpolar) SFC solvent may have the following effects: (1) a gradual increase in the mobile-phase polarity, (2) an increase in the mobile-phase density, (3) a decrease in the stationary-phase activity, (4) a change in the linear velocity.

The first effect is directly due to the altered composition. The second effect is due to an increase in the critical temperature of the mixture, so that the reduced temperature decreases and the density increases. This increase may be gradual, but also quite drastic in the region close to the critical point of the mixture. The third effect has been discussed

References on p. A389

extensively for isocratic separations in Section 8.2.2.2. The last effect is due to a combination of variations in the mobile-phase density and in the viscosity.

In order to adapt the various effects of a composition gradient to the requirements of a separation, other parameters may be programed simultaneously. For example, an increase in the concentration of organic modifier may be accompanied by an increase in the temperature. Thus, both "negative" and "positive" temperature programing may have their place in SFC.

8.4.2 Detection

Detection in SFC can either take place at high pressures (i.e. approximately the column outlet pressure) or at low pressures (atmospheric or less). In low-pressure detection, pressures higher than the atmospheric conditions may occur just after an "open" restrictor, as the velocity of the expanding fluid will not exceed the speed of sound [67]. These two categories of detectors lead to different instrumental requirements. High-pressure detection in SFC corresponds to a high density of the mobile phase in the detector, as is the case when detection takes place in the liquid phase. Thus, many LC detectors can be used in SFC if the density of the mobile phase is kept sufficiently high by maintaining a high pressure in the detector. Low-pressure detection in SFC corresponds to a low density of the mobile phase in the detector, similar to gas-phase detection. Most GC detectors can be used in SFC after the density of the mobile phase has been reduced by reducing the pressure.

UV-absorbance, fluorescence, and FTIR spectrometers can be used as high-pressure detectors in SFC. The FID, FPD, mass spectrometers, and FTIR spectrometers can be used as low-pressure detectors. In high-pressure FTIR detection a flowcell is used, whereas in the low-pressure mode the solvent is first removed (so-called solvent-elimination or matrix-isolation techniques; Section 8.4.3.3). A detailed discussion on the use of low-pressure and high-pressure detectors in SFC can be found in Refs. 68 and 69, respectively.

8.4.2.1 Low-pressure detectors

8.4.2.1.1 Flame-ionization detector

When the effluent from a chromatographic column is passed into a (hydrogen) flame, the solute molecules may be ionized. By collecting the ions at an electrode, an electric current may be measured, which may be displayed as a chromatogram. A FID for SFC has been described in Ref. 70. Critical aspects of FID detection are a stable flame, a smooth flow of the (decompressed) mobile phase into the flame, and good positioning of the collector electrode. A FID can be a very sensitive detector with a very wide dynamic range. However, the detection principle applies almost universally. While this implies that almost all solutes can be detected by a FID, it also implies that almost all possible mobile phases are excluded. Ionizable solvents produce a very large number of ions, and this results in very large currents, exceeding the linear range of the detector. Also, very high noise levels are obtained with ionizable mobile phases.

For these reasons, flame-ionization detection is not feasible with most liquid solvents, and the FID is hardly ever used in combination with LC. Many gases (He, N_2, H_2, and others) do not produce ions in the flame, so that the FID is the most popular detector in GC. If sufficiently pure, CO_2 is another mobile phase that is compatible with flame-ionization detection. Thus, the FID has been used in SFC after decompression of the CO_2 mobile phase. Other SFC solvents that are compatible with the FID include N_2O and SF_6, although in the latter case, HF is formed in the flame, which may cause severe corrosion, unless the detector (and the collector electrode in particular) are made of or plated with corrosion-resistant metals, such as gold [34]. The use of modifiers (except water or formic acid [40]) usually prohibits the use of FID detection.

Early applications of the FID in SFC were hampered by the frequent occurrence of spikes on the baseline [1,71]. This is now believed to be due to the formation of clusters (or tiny droplets) of solute molecules and CO_2. The key to avoiding such clusters from being formed is in the design of the restrictor. A restrictor that allows a rapid, but smooth decompression to take place will not give rise to spikes in a SFC/FID system.

8.4.2.1.2 Other GC detectors

Besides the FID, several other flame-based detectors have been used for SFC. The most significant of these are the flame-photometric detector (FPD) and the thermionic detector. In a FPD, the light emitted by a flame is measured. This is an especially sensitive means for detecting compounds containing sulfur or phosphorus. For sulfur the response of the detector is fundamentally nonlinear.

The thermionic detector is similar to the FID, except that alkali ions are introduced into the flame, typically by means of a heated rubidium bead. Because the thermionic detector is especially sensitive for compounds containing nitrogen or phosphorus atoms, it is also known as the nitrogen/phosphorus detector (NPD). Other common names are rubidium-bead detector and alkali-flame detector. Nitrogen/phosphorus detectors are difficult to operate in GC, and it appears to be even more difficult to maintain optimum operating conditions in a "jet" of the rapidly expanding mobile phase at the exit of a SFC column.

Two other popular GC detectors are of limited interest for use in SFC, at least when CO_2 is used as the mobile phase. Both the electron-capture detector and the thermal conductivity detector are not easily compatible with SFC.

8.4.2.2 High-pressure detectors

8.4.2.2.1 Ultraviolet detectors

UV spectrophotometers are easily the most popular detectors in liquid chromatography. The principle of detection is the absorption of UV light (typically in the range between 190 and 390 nm) by the solution of the solute in the mobile phase. Many liquid and supercritical solvents show little absorption in the UV range. Most of the currently used supercritical fluids, including carbon dioxide, exhibit no absorption throughout the

References on p. A389

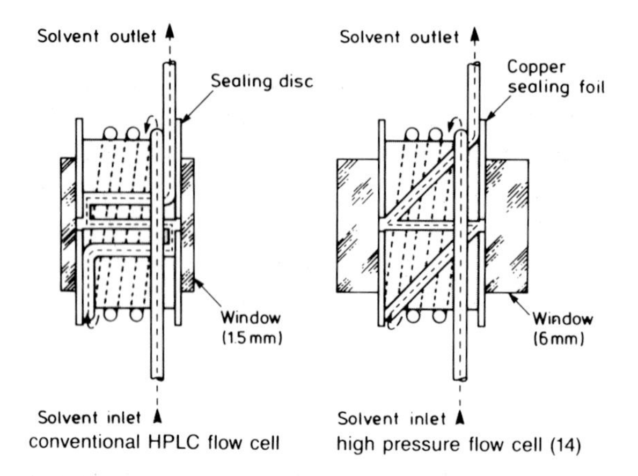

Fig. 8.15. Schematic illustration of a high-pressure flowcell for UV detection in packed-column SFC (right), in comparison with a conventional cell, as used in HPLC (left). (Reprinted from Ref. 19, with permission.)

entire range. A straightforward extension of UV spectrophotometers is the addition of a second lamp to facilitate measurements in the visible range of the spectrum.

The application of UV detectors to SFC requires a high-pressure flowcell. Fig. 8.15 schematically illustrates a flowcell designed for packed-column SFC. When capillary or packed capillary columns are used, UV detection may be achieved "on-column", by removing the (polyimide) coating from the end of the (fused-silica) column, usually just after the end of the packed section of the column. Alternatively, another piece of fused-silica tubing, equipped with such a "window", may be connected to the column. The disadvantages of this method are a less-than-optimal optical configuration, which may give rise to substantial scattering of light, and the very short optical pathlength. Especially in open-tubular SFC the sensitivity of UV detection is therefore limited. A possible solution is to direct the light axially rather than radially through a part of the column [72].

8.4.2.2.2 Other LC detectors

Fluorescence is usually measured in cells similar to those used for UV detection, including the on-column type. However, the emitted light is usually measured at a 90° angle of the excitation light. On-column fluorescence detection may suffer from fluorescent impurities present in the fused-silica tubing.

Two other detectors that are frequently used in LC cannot easily be used in SFC. A refractive-index detector requires very precise flow and temperature control to yield sufficient sensitivity. This is much more difficult to achieve in SFC (and with a high-pressure flowcell) than it is in LC (with a low-pressure flowcell). Electrochemical detection requires much more polar mobile phases to be used than are common (and feasible) in SFC.

8.4.3 Identification methods

SFC can be coupled on-line or off-line to a variety of solute identification methods. The off-line methods, which we will not discuss in this chapter, essentially consist of the collection of a certain fraction of the eluate and evaporating the solvent, usually by decompression. Such a coupling implies that SFC is used in a (semi-)preparative way. There are advantages to using SFC rather than LC in such situations, mainly related to (a) the ease of solvent removal and (b) the potential high purity of the solvent. Below, we will discuss three of the most important on-line combinations of SFC and a solute identification method.

8.4.3.1 Ultraviolet absorption

The most straightforward way to obtain additional information on the nature and chemical structure of the solute is by using the complete UV spectrum, rather than the absorption at a single wavelength. The same (or a similar) UV flowcell may be used as for conventional UV detection. Spectra can be obtained either by rapid-scanning spectrometers or by using a diode-array detector. The former technique is likely to give the highest spectral resolution, whereas the second enables more spectra per second to be recorded. Moreover, when scanning takes place at the edge of a peak, the shape of the spectrum may be distorted, because the concentration of the sample is different at different wavelengths.

SFC/UV was first demonstrated by Sugiyama et al. [73]. The main disadvantage of SFC/UV for solute identification is the limited information content of UV spectra. Therefore, UV spectra alone are usually not enough for a positive identification. Even within a small class of solutes a combination of spectra and retention data will be needed for qualitative analysis [38].

8.4.3.2 Mass spectrometry

A number of reports in the literature suggest that SFC may be more compatible with MS than is LC. Much research has been devoted to the LC/MS combination in the last ten years, without the technique approaching the perfection of GC/MS. Unfortunately, GC/MS is only applicable to volatile solutes, and an alternative is needed for nonvolatiles. Again, the advantages of SFC are the ease of solvent removal (evaporation in the vacuum chamber) and the purity of the solvent. Several ways have been described for coupling SFC with MS. These have been reviewed in Refs. 15-17. Although SFC/MS may be easier than LC/MS, a number of problems do remain:

(1) SFC is not the best possible technique for (very) polar solutes. In the foreseeable future, LC/MS with aqueous mobile phases will be needed for such solutes.

(2) Often, chemical-ionization spectra are obtained, which have a low information content and offer little information on the solute other than its molecular weight.

(3) Information-rich spectra, resembling those obtained with electron-impact (EI) ionization may be obtained on certain instruments, but the reproducibility is much less than with conventional EI spectra.

(4) Due to Problems 2 and 3, it is difficult to use in SFC/MS the library-search programs, which have contributed greatly to the success of GC/MS.

8.4.3.3 SFC/IR

SFC and (FT)IR can be coupled "on-line" in two different ways. Detection can take place both at high pressures (prior to decompression) or at low pressures (after decom-

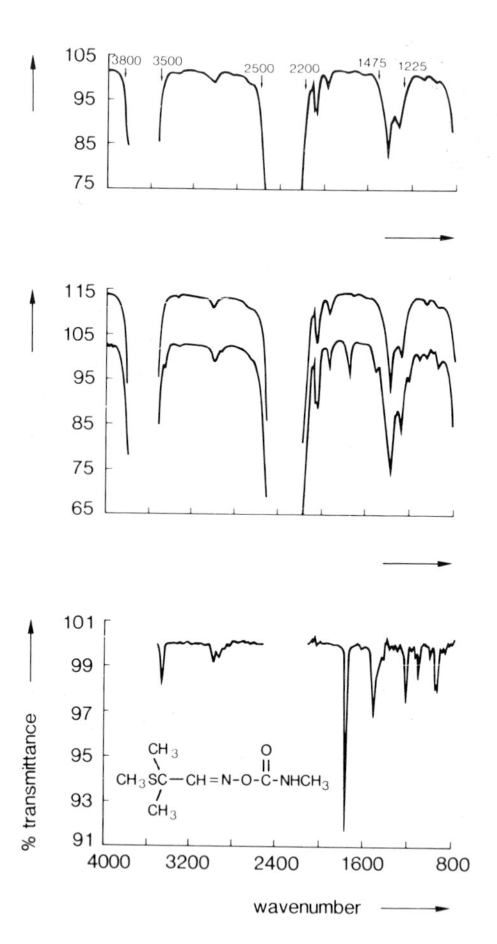

Fig. 8.16. (a) IR spectrum of supercritical CO_2; (b) spectrum obtained of a solute peak (aldicarb) by high-pressure FTIR detection after SFC with CO_2 as the eluent; (c) solute spectrum, obtained by subtracting Spectrum a from b. Mobile-phase density, 0.36 g/ml. Nicolet SXC IR spectrometer interfaced with Lee Scientific SFC system. A 100-μm-ID open-tubular column was used with a 0.5-μm-thick film of SB-methyl-100 (crosslinked polymethylsiloxane). (Reprinted from Ref. 74, with permission.)

pression). High-pressure detection is conceptually simpler, but suffers from the absorptivity of the eluent. Fig. 8.16a shows the IR absorbance spectrum of carbon dioxide under supercritical conditions ($\rho \cong 0.36$ g/ml). The figure shows that CO_2 has two strong absorption bands, i.e. from ≅ 3800 cm^{-1} to ≅ 3500 cm^{-1} and from ≅ 2500 cm^{-1} to ≅ 2200 cm^{-1}. The Fermi-resonance bands between ≅ 1470 cm^{-1} and ≅ 1200 cm^{-1} are less intense, so that these band can usually be compensated for by subtracting the spectrum of CO_2 (Fig. 8.16a) from that obtained at (or around) the peak maximum (Fig. 8.16b). Because the intensity of the absorption bands in the spectrum of carbon dioxide increases with increasing pressure, it is especially critical in pressure-programed (or density-programed) SFC/IR to obtain a reference spectrum for CO_2 at a pressure (density) as close as possible to the elution pressure (density). The resulting spectrum (Fig. 8.16c) resembles the IR spectrum of the solute, with the exception of the blank areas, which correspond to regions in which all IR radiation is absorbed by the CO_2. Although most of the relevant solute information in the mid-IR range can be obtained from spectra such as that in Fig. 8.16c, complete spectra are still desirable. This may be achieved by using completely transparent solvents, such as Xe, or by removing the mobile phase prior to detection.

In the latter case, FTIR is used as a low-pressure detector. In contrast to flowcell SFC/IR, all solvents (including modifiers) may, in principle, be used in these solvent-elimination or matrix-isolation techniques. Because the solutes are precipitated on some surface or powder, more time can be spent on measuring the spectra, thus greatly increasing the sensitivity. Preferably, the solutes should be deposited on an IR-transparent substrate (e.g., KBr, ZnSe), because otherwise reflection spectra rather than transmission spectra (such as those in Fig. 8.16) are obtained. Reflection spectra tend to be somewhat more difficult to interpret, and library-search techniques cannot as easily be applied as to transmission spectra. A disadvantage of solvent-elimination techniques may be their increased complexity and cost.

8.5 APPLICATION AREAS

In this chapter we do not intend do discuss extensively all the possible applications of SFC. Very useful compilations have arisen from two workshops on SFC in Utah [75,76]. The various advantages of SFC in comparison with GC and LC lead to different application areas of SFC. In the following sections the application areas of SFC are grouped according to these advantages.

8.5.1 Advantages relative to gas chromatography

(1) Lower elution temperatures

Because the retention mechanism in SFC is based on interactions between the solute and the mobile phase, rather than on solute volatility, the operating temperatures are typically much lower in SFC than in GC. Two types of applications of SFC arise from the fact that much lower temperatures can be used: (1) Thermo-labile solutes, which de-

References on p. A389

compose at the temperatures required for GC, e.g., explosives. (2) Another consequence of lower operating temperatures may be higher selectivities. In a number of situations in which the selectivity is largely determined by specific interactions, lower temperatures may be beneficial, e.g., in the separation of isomers and enantiomers.

(2) Elution of nonvolatiles

Samples that are not sufficiently volatile for GC (even at the high temperatures currently feasible) can often be analyzed by SFC. Although the upper temperature at which GC can be performed has been increased considerably in recent years, separation temperatures of 400°C no longer being exceptional, there will always be an upper limit. Consequently, there will also be a minimum volatility required for samples to be compatible with GC. Nonvolatile samples can be analyzed (at much lower temperatures) by SFC. This has resulted in the use of SFC for simulated-distillation studies and for the characterization of very heavy petrochemical fractions.

8.5.2 Advantages relative to liquid chromatography

(1) Compatibility with low-pressure detection

Upon decompression, supercritical fluids turn into gases. Depending on the nature of the supercritical solvent, typical GC detectors, such as the FID, may be used. SFC/FID can be used in situations in which a universal detector is needed. For example, alkanes and alkenes cannot easily be detected by UV detectors in LC. In group-type separations (paraffins, olefins, aromatics, etc.) use can be made of SFC with flame-ionization detection. For solutes requiring high elution densities in SFC the use of the FID is not straightforward. Thus, the characterization of high-molecular-weight polyethylene is not yet routinely performed by SFC/FID. In comparison with LC, SFC is thought to be more readily compatible with MS. However, relatively large amounts of modifier are often added to the mobile phase to facilitate the elution of (moderately) polar solutes. SFC should also be more useful than LC in combination with solvent-elimination FTIR interfaces. In this case, large amounts of modifier could cause more problems than in SFC/MS.

(2) Very fast separations

SFC analyses are typically two or three times faster than LC analyses on the same column. Some pharmaceutical industries appear to use SFC instead of LC for simple, routine applications. An added advantage of SFC appears to be the virtual absence of particulate contamination from the mobile phase. This may increase the reliability of columns packed with very small (i.e. $d_p \leq 3\,\mu$m) particles.

(3) High-efficiency separations

Because open-tubular SFC is more within practical reach than open-tubular LC, SFC may be the preferred method for high-resolution separation of nonvolatiles. In this area, SFC is still promising rather than useful. On the one hand, this is because the theoretical platenumbers in open-tubular SFC have not yet been approached. On the other hand, progress in LC (e.g., the use of packed-capillary columns) and other techniques (e.g., electromigration methods) have also been targeted at high-efficiency separations. Improvements in column technology may open more applications for SFC in this area.

REFERENCES

1 E. Klesper, A.H. Corwin and D.A. Turner, *J. Org. Chem.,* 27 (1962) 700.
2 J.C. Giddings, M.N. Myers, L. McLaren and R.A. Keller, *Science,* 162 (1968) 67.
3 J.C. Giddings, M.N. Myers and J.W. King, *J. Chromatogr. Sci.,* 7 (1969) 276.
4 T.H. Gouw and R.E. Jentoft, *J. Chromatogr.,* 68 (1972) 303.
5 G.W.A. Rijnders, *Chem. Ing. Tech.,* 42 (1970) 890.
6 U. van Wasen, I. Swaid and G.M. Schneider, *Angew. Chem.,* 92 (1980) 585.
7 L. McLaren, M.N. Myers and J.C. Giddings, *Science,* 159 (1968) 197.
8 J.C. Fjeldsted, R.C. Kong and M.L. Lee *J. Chromatogr.,* 279 (1983) 449.
9 T.L. Chester, *J. Chromatogr.,* 299 (1984) 424.
10 B.E. Richter, *J. High Resolut. Chromatogr. Chromatogr. Commun.,* 8 (1985) 297.
11 E.J. Guthry and H.E. Schwartz, *J. Chromatogr. Sci.,* 24 (1986) 236.
12 P.A. Peaden, J.C. Fjeldsted, M.L. Lee, S.R. Springston and M. Novotny, *Anal. Chem.,* 54 (1982) 1090.
13 D.R. Gere, R. Board and D. McManigill, *Anal. Chem.,* 54 (1982) 736.
14 L.G. Randall and A.L. Wahrhaftig, *Anal. Chem.,* 50 (1978) 1705.
15 B.W. Wright, H.T. Kalinowski, H.R. Udseth and R.D. Smith, *J. High Resolut. Chromatogr. Chromatogr. Commun.,* 9 (1986) 145.
16 R.D. Smith, H.T. Kalinowski and H.R. Udseth, *Mass Spectrom. Rev.,* 6 (1987) 445.
17 K.J. Voorhees, S.D. Zaugg and S.J. DeLuca, in C.M. White (Editor), *Modern Supercritical Fluid Chromatography,* Hüthig, Heidelberg, 1988, pp. 59-79.
18 K.H. Shafer and P.R. Griffiths, *Anal. Chem.,* 55 (1983) 1939.
19 P.J. Schoenmakers and F.C.C.J.G. Verhoeven, *Trends Anal. Chem.,* 6 (1987) 10.
20 H.H. Lauer, D. McManigill and R.D. Board, *Anal. Chem.,* 55 (1983) 1370.
21 P.J. Schoenmakers, *J. Chromatogr.,* 315 (1984) 1.
22 P.J. Schoenmakers and L.G.M. Uunk, in J.C. Giddings, E. Grushka and P.R. Brown (Editors), *Advances in Chromatography,* Vol. 30, Dekker, New York, 1989, pp. 1-80.
23 C.R. Yonker, B.W. Wright, R.C. Petersen and R.D. Smith, *J. Phys. Chem.,* 89 (1985) 5526.
24 D. Ishii, T. Niwa, K. Ohta and T. Takeuchi, *J. High Resolut. Chromatogr. Chromatogr. Commun.,* 11 (1988) 801.
25 R.C. Reid, J.M. Prausnitz and T.K. Sherwood, *The Properties of Liquids and Gases,* McGraw-Hill, New York, 3rd Edn., 1977.
26 P.J. Schoenmakers, P.E. Rothfusz and F.C.C.J.G. Verhoeven, *J. Chromatogr.,* 395 (1987) 91.
27 D.E. Martire and R.E. Boehm, *J. Phys. Chem.,* 87 (1983) 1045.
28 D.E. Martire, *J. Liq. Chromatogr.,* 10 (1987) 1569.
29 P.J. Schoenmakers, L.G.M. Uunk and J.G.M. Janssen, *J. Chromatogr.,* 506 (1989) 563.
30 P.J. Schoenmakers, *The Optimization of Chromatographic Selectivity. A Guide to Method Development,* Elsevier, Amsterdam, 1986.
31 B.W. Wright, H.T. Kalinowski and R.D. Smith, *Anal. Chem.,* 57 (1985) 2823.
32 E. Lundanes, B. Iversen and T. Greibrokk, *J. Chromatogr.,* 366 (1986) 391.
33 F.P. Schmitz and E. Klesper, *Polym. Commun.,* 24 (1983) 142.
34 H.E. Schwartz and R.G. Brownlee, *J. Chromatogr.,* 353 (1986) 77.
35 S.B. French and M. Novotny, *Anal. Chem.,* 58 (1986) 164.
36 W. Asche, *Chromatographia,* 11 (1978) 411.
37 J.C. Keui, K.E. Markides and M.L. Lee, *J. High Resolut. Chromatogr. Chromatogr. Commun.,* 10 (1987) 257.

38 P.J. Schoenmakers, F.C.C.J.G. Verhoeven and H. M. van den Bogaert, *J. Chromatogr.,* 371 (1986) 121.
39 P.J. Schoenmakers, L.G.M. Uunk and P.K. de Bokx, *J. Chromatogr.,* 459 (1987) 201.
40 H.E. Schwartz, P.J. Barthel, S.E. Moring, T.L. Yates and H.H. Lauer, *Fresenius Z. Anal. Chem.,* 330 (1988) 204.
41 H. Engelhardt, A. Gross, R. Mertens and M. Petersen, *J. Chromatogr.,* 477 (1989) 169.
42 J.G.M. Janssen, P.J. Schoenmakers and C.A. Cramers, *J. High Resolut. Chromatogr. Chromatogr. Commun.,* 12 (1989) 645.
43 P.J. Schoenmakers, *J. Liq. Chromatogr.,* 10 (1987) 1865.
44 P.J. Schoenmakers, *J. High Resolut. Chromatogr. Chromatogr. Commun.,* 11 (1988) 278.
45 P.A. Bristow, *Liquid Chromatography in Practice,* HETP, Wilmslow, 1976, p. 25.
46 H.E. Schwartz, P.J. Barthel, S.E. Moring and H.H. Lauer, *LC-GC Mag.,* 5 (1987) 490.
47 P.J. Schoenmakers, in R.M. Smith (Editor), *Supercritical-fluid Chromatography,* Royal Society of Chemistry, London, 1988, pp. 102-136.
48 J.H. Knox and M. Gilbert, *J. Chromatogr.,* 186 (1979) 405.
49 P.J. Schoenmakers and L.G.M. Uunk, *Chromatographia,* 24 (1987) 51.
50 R.C. Kong, S.M. Fields, W.P. Jackson and M.L. Lee, *J. Chromatogr.,* 289 (1984) 105.
51 P.J. Schoenmakers and L.G.M. Uunk, *Eur. Chromatogr. News,* 1 (1987) 14.
52 B.W. Wright, P.AS. Peaden, M.L. Lee and T.J. Stark, *J. Chromatogr.,* 248 (1982) 17.
53 B.E. Richter, J.C. Kuei, N.J. Parks, S.J. Crowley, J.S. Bradshaw and M.L. Lee, *J. High Resolut. Chromatogr. Chromatogr. Commun.,* 6 (1983) 371.
54 M. Ashraf-Khorassani and L.T. Taylor, in C.M. White (Editor), *Modern Supercritical Fluid Chromatography,* Hüthig, Heidelberg, 1988, pp. 115-134.
55 G.E. Berendsen, *Preparation and Characterization of Well-defined Chemically Bonded Stationary Phases for HPLC,* Thesis, Delft Technical University, 1980.
56 M. Ashraf-Khorassani and L.T. Taylor, *J. High Resolut. Chromatogr. Chromatogr. Commun.,* 12 (1989) 40.
57 G. Schomburg, A. Deege, G. Breitenbruch and W. Roeder, in P. Sandra and G. Redant (Editors), *Proceedings 10th International Symposium on Capillary Chromatography, Riva del Garda,* Hüthig, Heidelberg, 1989, pp. 1194-1204.
58 M. Ashraf-Khorassani, L.T. Taylor and R.D. Smith, *Chromatographia,* 28 (1989) 569.
59 R.P. Khosah, in C.M. White (Editors), *Modern Supercritical Fluid Chromatography,* Hüthig, Heidelberg, 1988, pp. 155-187.
60 B.W. Wright and R.D. Smith, in C.M. White (Editor), *Modern Supercritical Fluid Chromatography,* Hüthig, Heidelberg, 1988, pp. 189-210.
61 H.E. Schwartz, P.J. Barthel, S.E. Moring and H.H. Lauer, *LC-GC,* 5 (1987) 490.
62 T. Greibrokk, J. Doehl, A. Farbrot and B. Iversen, *J. Chromatogr.,* 371 (1986) 145.
63 B.E. Richter, D.E. Knowless, M.R. Andersen, N.L. Porter, E.R. Campbell and D.W. Later, *J. High Resolut. Chromatogr. Chromatogr. Commun.,* 11 (1988) 29.
64 E. Klesper and F.P. Schmitz, *J. Chromatogr.,* 402 (1987) 1.
65 A. Wilsch and G.M. Schneider, *J. Chromatogr.,* 357 (1986) 239.
66 K.H. Linnemann, A. Wilsch and G.M. Schneider, *J. Chromatogr.,* 369 (1986) 39.
67 R.W. Bally, *Instrumental Aspects of Capillary Supercritical-Fluid Chromatography,* Thesis, Technical University of Eindhoven, 1987.
68 B.E. Richter, D.J. Bornhop, J.T. Swanson, J.G. Wangsgaard and M.R. Andersen, *J. Chromatogr. Sci.,* 27 (1989) 303.
69 D.J. Bornhop and J.G. Wangsgaard, *J. Chromatogr. Sci.,* 27 (1989) 293.
70 M.G. Rawdon, *Anal. Chem.,* 56 (1984) 831.
71 S.T. Sie, W. van Beersum and G.W.A. Rijnders, *Separ. Sci.,* 1 (1966) 459.
72 J.P. Chervet, M. Ursem, J.P. Salzmann and R.W. Vannoort, *J. High Resolut. Chromatogr. Chromatogr. Commun.,* 12 (1989) 278.
73 K. Sugiyama, M. Saito, T. Hondo and M. Senda, *J. Chromatogr.,* 332 (1985) 1107.
74 R.C. Wieboldt and R.J. Rosenthal, *Microchim. Acta,* II (1988) 203.

75 K.E. Markides and M.L. Lee (Editors), *SFC Applications, Workshop on SFC, Park City, Utah, January 12-14, 1988.*
76 K.E. Markides and M.L. Lee (Editors), *SFC Applications, Workshop on SFC, Snow Bird, Utah, June 13-15, 1989.*

Chapter 9

Gas chromatography

COLIN F. POOLE and SALWA K. POOLE

CONTENTS

9.1 INTRODUCTION

Gas chromatography (GC) is the principal method of analysis for thermally stable, volatile, organic compounds present in mixtures that can span a wide range of complexity [1-9]. No other analytical technique can provide equivalent resolving power and sensitivity to sample components in low concentrations. The primary limitation of gas chromatography is that the sample, or some convenient derivative of it, must be thermally stable at the temperature required for its volatilization. The thermal stability of the sample and column materials then represents the fundamental limit to expanding the technique to other samples. In contemporary practice an upper temperature limit of about 400°C and a molecular weight less than 1000 is indicated, although higher temperatures have been used and higher-molecular-weight samples have been separated in a few instances.

In gas chromatography samples are separated by distribution between a stationary phase and a mobile phase, by adsorption, partition, or a combination of the two. When a solid adsorbent serves as the stationary phase, the technique is called gas/solid chromatography (GSC). When a liquid phase is spread on an inert support or coated as a thin film onto the wall of a capillary column, the technique is designated gas/liquid chromatography (GLC). The separation medium may be a coarse powder, coated with a liquid phase through which the carrier gas flows. This is an example of packed-column gas chromatography. If the adsorbent, liquid phase, or both, are coated onto the wall of a narrow-bore column of capillary dimensions, the technique is called wall-coated open-tubular (WCOT), support-coated open-tubular (SCOT), or porous-layer open-tubular (PLOT) column gas chromatography. A distinction between these column types will be made later.

In this chapter we will adhere closely to the recent developments and trends in the practice of gas chromatography. Theoretical considerations are treated in Chapter 1. Packed-column gas chromatography has reached maturity, and only a few new developments have taken place in this area in recent years. However, during that time spectacular developments have occurred in column fabrication and instrument design for open-tubular columns, which will be our main focus. However, we caution against the current trend of demeaning packed columns as being obsolete. Packed columns have been used for many of the theoretical and practical developments in gas chromatography. They are less expensive, require little training in their use, are better suited to isolating preparative-scale quantities, and can better tolerate samples containing involatile or thermally labile components. For stable samples the quantitative accuracy and precision of results with packed columns for simple mixtures are generally superior to those with open-tubular columns, as there are fewer injection problems. Only a limited number of stationary phases have been immobilized successfully in open-tubular columns; thus, the range of selectivity values is more limited than in the case of packed columns, for which a large number of phases are available. Packed columns may not be used as often as open-tubular columns in contemporary practice, but they have not fallen into disuse either and are always likely to find a significant number of applications in spite of the trend towards open-tubular columns as the first choice for analytical separations. Also, emerging technology for combining liquid and supercritical-fluid chromatography with a separate, high-resolution gas-chromatographic step to maximize the information obtainable from the sample has achieved some general acceptance and is worthy of detailed consideration.

9.2 COLUMN TYPES IN GAS CHROMATOGRAPHY

Five types of columns are routinely used in gas chromatography:

(1) classical, packed columns with internal diameters greater than 2 mm, containing particles 100-250 μm in diameter;

(2) micropacked columns, having diameters < 1 mm with a packing density similar to classical packed columns (d_p/d_c < 0.3, where d_p is the particle diameter and d_c the column diameter);

(3) packed capillary columns, having a column diameter < 0.5 mm and a packing density less than classical packed columns (d_p/d_c = 0.2-0.3);

(4) SCOT columns, where the liquid phase is coated on a surface covered with a layer of solid support material, leaving an open passageway through the center of the column; and

(5) WCOT columns, in which the liquid phase is coated directly on the smooth or chemically etched column wall.

Some characteristic properties of the various column types are given in Table 9.1 [9-11]. The most significant difference between the various column types is their permeability. The open-tubular columns offer much lower flow resistance and can therefore be much longer to give very high total platecounts. The minimum plateheight of the best packed column in gas chromatography is about 2-3 particle diameters, whereas that of an open-tubular column will be similar to the column diameter. Thus, using a column packed with 10-μm particles and an open-tubular column of about 30 μm ID should give a similar number of theoretical plates per unit length. The intrinsic efficiency of open-tubular columns is not necessarily greater than that of packed columns, but because of their greater permeability at a fixed column pressure drop, a greater number of total theoretical plates may be obtained, since longer columns can be used.

Any optimization strategy that considered only efficiency is inadequate to describe accurately resolution, which is a strong function of the capacity factor at low capacity factor values. The phase ratio (ratio of the volumes of gas to liquid phase in a column) for

TABLE 9.1

REPRESENTATIVE PROPERTIES OF DIFFERENT COLUMN TYPES IN GAS CHROMATOGRAPHY

Column type	Phase ratio	H_{min} (mm)	u_{opt} (cm/sec)	Permeability (10^7 cm^2)
Classical packed	4-200	0.5-2	5-15	1-50
Micropacked	50-200	0.02-1	5-10	1-100
Packed capillary	10-300	0.05-2	5-25	5-50
SCOT	20-300	0.5-1	10-100	200-1000
WCOT	15-500	0.03-0.8	10-100	300-20 000

References on p. A442

TABLE 9.2

CHROMATOGRAPHIC PROPERTIES OF COMMERCIALLY AVAILABLE COLUMNS

Column Type	Length (m)	ID (mm)	Film thickness (μm)	Phase ratio	Capacity factor*	HETP (mm)	Total platecount	Plates per meter
Classical	2	2.16	10%(w/w)	12	10.4	0.55	3640	1820
Packed	2	2.16	5%(w/w)	26	4.8	0.50	4000	2000
SCOT	15	0.50		20	6.2	0.95	15 790	1050
SCOT	15	0.50		65	1.9	0.55	27 270	1820
WCOT	30	0.10	0.10	249	0.5	0.06	480 000	16 000
WCOT	30	0.10	0.25	99	1.3	0.08	368 550	12 285
WCOT	30	0.25	0.25	249	0.5	0.16	192 000	6400
WCOT	30	0.32	0.32	249	0.5	0.20	150 000	5000
WCOT	30	0.32	0.50	159	0.8	0.23	131 330	4380
WCOT	30	0.32	1.00	79	1.6	0.29	102 080	3400
WCOT	30	0.32	5.00	15	8.3	0.44	68 970	2300
WCOT	30	0.53	1.00	132	0.9	0.43	70 420	2340
WCOT	30	0.53	5.00	26	4.8	0.68	43 940	1470

*Undecane at 130°C.

TABLE 9.3

FACTORS AFFECTING RESOLUTION IN GAS CHROMATOGRAPHY

R_S = resolution, α = selectivity factor, n = the number of theoretical plates, k = capacity factor.

Value of n needed for $R_S = 1$ at $k = 3$ for different values of α		Value of n needed for $R_S = 1$ at different k values for $\alpha = 1.05$ and 1.10		
α	n	k	$\alpha = 1.05$	$\alpha = 1.10$
1.005	1 150 000	0.1	853 780	234 260
1.01	290 000	0.2	254 020	69 700
1.015	130 000	0.5	63 500	17 420
1.02	74 000	1.0	28 220	7740
1.05	12 500	2.0	15 880	4360
1.10	3 400	5.0	10 160	2790
1.20	1 020	10.0	8540	2340
1.50	260	20.0	7780	2130
2.00	110			

a number of typical columns is given in Table 9.2 [10]. At a constant temperature the partition coefficient will be the same for all columns prepared from the same stationary phase. Consequently, for a column with a large phase ratio the capacity factor will be small and vice versa. The number of plates required for a separation becomes very large at small capacity factor values (Table 9.3) [9]. Columns with low phase ratios, i.e. thick-film columns, have a lower intrinsic efficiency than thin-film columns but give better resolution of low-boiling compounds, because they provide a more favorable capacity factor value. The opposite arguments apply to high-boiling compounds that have long analysis times, because their capacity factor values are too large. Increasing the phase ratio lowers the capacity factor to a value within the optimum range so that there is little deterioration in resolution but a substantial saving in analysis time. Phase ratios are usually lower (and analysis times longer at constant temperature) for packed columns than for WCOT columns, SCOT columns occupying an intermediate position. Since several combinations of film thickness and column radius can be used to generate the same phase ratio, there are other factors that need to be considered in selecting these variables for a particular separation.

In general, increasing the column radius and film thickness for a WCOT column will lead to an increase in the column plateheight and a decrease in efficiency. However, the relationship between the variables is very complex and depends on the capacity factor, since this term appears explicitly in the contribution of the resistance to mass transfer to the plateheight. For thin-film columns ($d_f < 0.25\ \mu m$) resistance to mass transfer in the stationary phase is small (frequently negligible) so that decreasing the radius of thin-film columns by minimizing the mobile-phase mass transfer term leads to increased efficiency

TABLE 9.4

COMPARATIVE COLUMN PERFORMANCE DATA FOR SOME MICROPACKED AND PACKED CAPILLARY COLUMNS

Particle diameter (mm)	Ratio of particle diameter to column diameter	H_{min} (mm)	Pressure drop per meter (atm)	Length for n = 10 000 (m)
0.175	0.70	1.00	1	10
0.150	0.17	0.26	3	2.6
0.140	0.18	0.50	1	5
0.125	0.39	0.60	0.5	6
0.113	0.23	0.16	0.8	1.6
0.030	0.03	0.10	7	1
0.030	0.03	0.10	8	1
0.025	0.067	0.09	25	0.9
0.010	0.008	0.02	240	0.2

[12,13]. Narrow-bore, thin-film WCOT columns are the most intrinsically efficient, provided that the inlet pressure is not limited. (This does not mean that these columns will always provide the highest resolution, since they may not provide optimum values of the capacity factor.) If the column radius is held constant and the film thickness is increased, the column efficiency will decline, because the resistance to mass transfer in the stationary phase will eventually dominate the plateheight equation (Table 9.2).

Representative column performance data for some micropacked and packed capillary columns are given in Table 9.4 [11,14]. As anticipated, the column efficiency increases as the particle diameter is decreased, but only with a simultaneous increase in the column pressure drop. Assuming a fixed available pressure, high total column efficiencies can be obtained by reducing the particle diameter ($d_p/d_c < 0.03$) or by using coarser particles and longer columns. In either case, it is not easy to exceed a limit of ca. 60 000 theoretical plates without resorting to unusually high operating pressures. Increasing the column length also increases the analysis time, which can become long when a very large number of theoretical plates is needed for the separation. For high-speed separations columns packed with small particles are preferred, but because of pressure limitations the total number of plates available for the separation will be small, perhaps 10 000. Comparing thick-film WCOT and micropacked columns of similar phase ratio, similar performance is obtained, but in general, the column pressure drop is at least one order of magnitude higher for the micropacked columns. Thus, for difficult separations requiring a large number of theoretical plates with an optimum value for the capacity factor thick-film WCOT columns would be clearly preferred. Micropacked columns should find applications for those samples requiring a greater number of theoretical plates than is available for classical packed columns and/or larger sample capacity or the use of selective stationary phases that are difficult to coat on open-tubular columns.

9.3 CARRIER GAS SELECTION

The carrier gas primarily provides the column transport mechanism by which the sample is moved through the column. The common carrier gases used in gas chromatography do not influence the selectivity of the chromatographic system under normal operating conditions, except for a generally small contribution arising from the nonideal behavior of the gas. The choice of carrier gas influences resolution through its effect on column efficiency, which arises from differences in the solute diffusion rates. It can also affect analysis time, because the optimum carrier gas velocity decreases as the solute diffusivity decreases and also plays a role in pressure-limiting situations due to differences in gas viscosities.

The viscosity of the carrier gas determines the column pressure drop for a given velocity. For hydrogen, helium, and nitrogen it can be approximated by Eqns. 9.1-9.3 in the temperature range of interest for gas chromatography [15,16]

Hydrogen $$\eta_t = 0.1827\,t + 83.9899 \tag{9.1}$$

Helium $$\eta_t = 0.3993\,t + 186.6169 \tag{9.2}$$

Nitrogen $$\eta_t = 0.3838\,t + 167.3534 \tag{9.3}$$

where η_t is the carrier gas viscosity (micropoise) at temperature t (°C). For a given inlet pressure and temperature the average linear gas velocity for an identical column will be highest for hydrogen and the lowest for helium. Viscosity increases with temperature for gases, causing a decrease in the carrier gas velocity at constant pressure. The rate of change of viscosity with temperature is approximately the same for the three gases.

Solute diffusivity influences both the plateheight and the position of the optimum linear velocity, described by the Van Deemter equation [16-19]. Comparing the Van Deemter curves in Fig. 9.1, nitrogen provides the lowest plateheight, but it occurs at an optimum gas velocity, which is rather low and leads to long analysis times. In the optimum plate-height region the ascending portions of the curve are much shallower for hydrogen and helium. Thus, for operation at carrier gas flowrates above the optimum in the Van Deemter curve (the situation which applies in practice) hydrogen – and to a lesser extent helium – show only a modest sacrifice in column efficiency compared to nitrogen. It is clear that in terms of the rate of production of effective theoretical plates hydrogen provides the highest column efficiency per unit time. In addition, because equivalent column efficiencies are obtained at higher average linear gas velocities, compounds exhibit lower elution temperatures during temperature-programed runs and, as the peaks are taller and narrower, sample detectability is improved.

The above observations are sound for thin-film (film thickness, d_f, $< 0.5\ \mu m$) and low-loaded packed columns. For these columns hydrogen is clearly the preferred carrier gas when efficiency and speed of analysis are considered together. Different conclusions are reached for thick-film columns, since diffusion in the stationary phase contributes

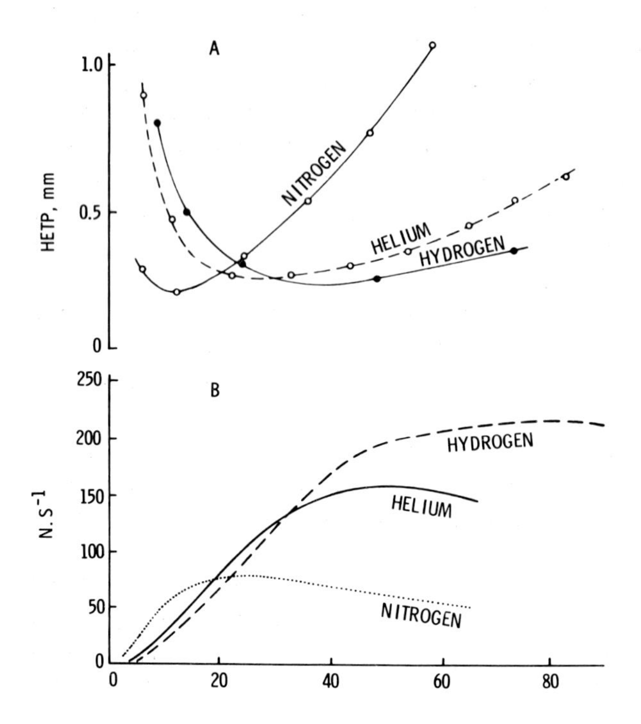

Fig. 9.1. (A) Van Deemter curves for three different carrier gases. (B) Same data replotted as the rate of production of effective plates as a function of the average carrier gas velocity. (Reproduced with permission from Ref. 1.)

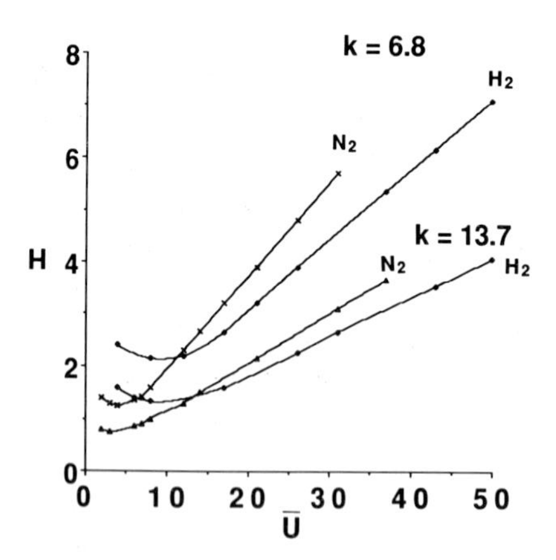

Fig. 9.2. Van Deemter curves for a thick-film open-tubular column for two solutes with different capacity factors and nitrogen and hydrogen as carrier gases.

significantly to the bandbroadening mechanism. For thick-film columns the capacity factor, temperature, and solute diffusion coefficient in the stationary phase need to be taken into consideration when selecting a carrier gas for a particular analysis [20]. In going from a thin-film to a thick-film column (which corresponds to reducing the phase ratio) the efficiency of the columns, as measured by the minimum plateheight, H_{min}, decreases substantially and is dependent on the identity of the carrier gas. The optimum linear carrier gas velocity is reduced by perhaps a factor of 2 and also depends on the identity of the carrier gas, and the slopes of the ascending portion of the Van Deemter curves at high velocity are much steeper for the thick-film columns and depend primarily on the liquid-phase diffusion coefficient (Fig. 9.2). In contrast to thin-film columns, for maximum efficiency, thick-film columns should be operated at their optimum linear velocity with nitrogen as the carrier gas. These conditions will maximize resolution while compromising analysis time.

For packed columns with normal loadings the liquid-phase mass-transfer term is always significant. In situations where the pressure drop is the limiting condition hydrogen is preferred, because its viscosity is only about half that of helium and nitrogen. In terms of efficiency nitrogen is slightly superior at low temperature and flowrates, while hydrogen seems to be superior at higher temperatures and at above-optimum flowrates [18,19]. The only disadvantage to the use of hydrogen as a carrier gas is the real or perceived explosion hazard from leaks within the column oven. Gas sensors are available to switch off the column oven and carrier gas flow at air/hydrogen mixtures well below the explosion threshold limit.

9.4 SELECTION OF THE STATIONARY PHASE

Separations occur in GLC primarily because of differences in solute vapor pressures and because of selective interactions between the solutes and the stationary liquid phase. At the molecular level the principal intermolecular forces acting between a solute and a solvent are dispersion, induction, orientation, and donor/acceptor interactions [21-23]. Dispersion (or London) forces arise from the electric field generated by rapidly varying dipoles, formed between nuclei and electrons at zero-point motion of the molecules, acting upon the polarizability of other molecules to produce induced dipoles in phase with the instantaneous dipoles forming them. Dispersion forces are universal and independent of temperature. Induction (or Debye) forces arise from the interaction of a permanent dipole with a polarizable molecule. Orientation (or Keesom) forces arise from the net attraction between molecules or portions of molecules possessing a permanent dipole moment. Induction and orientation forces decrease with increasing temperature. At a sufficiently high temperature these disappear entirely, as all orientations of the dipoles become equally probable. Complementing the above physical interactions are donor/acceptor interactions of a chemical nature. Donor/acceptor complexes involve special chemical bonding interactions that arise from the partial transfer of electrons from a filled orbital on the donor to a vacant orbital on the acceptor molecule. Important examples in

References on p. A442

GLC are hydrogen-bonding interactions and coordination forces between pi-electron-rich systems and metal ions. Although the above forces provide a suitable framework for the qualitative understanding of the separation process, the concert of interacting forces between polyatomic molecules is far too complex for a quantitative description.

For nonpolar solutes (e.g., hydrocarbons), dispersive forces are the sole forces of interaction in the pure liquid state. Dispersive forces are nonselective and increase approximately with solute molecular weight. Therefore, the elution of nonpolar samples from any chromatographic column occurs, to a first approximation, in the order of increasing boiling points. Dispersive forces are not considered important for the separation of polar solutes of similar boiling point. Here, the important interactions are orientation and induction forces, which depend on the polarizabilities, ionization potentials, and dipole moments of the solute and liquid phase, and on specific electron donor/acceptor interactions of a chemical nature. Although not specifically discussed above, when fine-tuning a separation, it may be necessary to take molecular size and shape into account as well.

The characteristics of a liquid phase of general interest to the chromatographer are its operating temperature range, film-forming properties, mass-transfer characteristics, and solvent strength and solvent selectivity. Film-forming and mass-transfer properties primarily influence peakshapes and column efficiency. Only those liquid phases that at least meet average standards are useful in gas chromatography. The minimum operating temperature for a stationary phase is usually established by its melting point or the temperature at which a plateau region in the column efficiency is obtained, if higher. Some phases have been used below their melting point as selective adsorbents for the separation of positional isomers which were not resolved at temperatures above the melt, but this is the exception rather than the rule. Some polymers and liquid crystals exhibit multiple phase transitions, each one associated with a change in retention corresponding to a change in surface area, volume, or viscosity of the phase above the transition temperature [24].

The upper temperature limit for a liquid phase is established as the highest temperature at which the phase can be maintained without decomposition or significant bleed from the column. Many phases are polydisperse substances containing low-molecular-weight oligomers. These oligomers may be selectively evaporated during column conditioning, leaving a much lower amount of phase on the column than expected. This is less of a problem for chemically defined phases and for phases of low polydispersity. In this case, the maximum allowable operating temperature of the phase is defined as the highest temperature at which the phase can be maintained for 24 h without changing the retention and performance characteristics of the components in a test chromatogram, obtained at a lower temperature before and after the conditioning period. Alternatively, the maximum allowable operating temperature can be assumed to be that temperature at which the liquid phase exhibits a vapor pressure of 0.5 torr.

9.4.1 Solvent strength and selectivity

Stationary-phase solvent properties depend on the solvent strength (polarity), defined as the capacity of a solvent to enter into various intermolecular interactions, and solvent selectivity, defined as the relative capacity of the solvent for a particular intermolecular interaction [21,22,25]. The polarity criterion of a liquid phase is the least satisfactory measure of the properties of a phase, as it is difficult to define rigorously. Since polarity is not a unique property of a molecule but a composite expression for several different interactions, there is no single substance that can be defined as polar. Several ways around this problem have been suggested. Among the most widely used is the reluctance of a stationary phase to dissolve a hydrocarbon and the sum of the retention index differences for the solutes benzene, butanol, 2-pentanone, nitropropane, and pyridine on the polar phase and squalane (P_M value of McReynolds) [22,26,27]. The P_M value has little theoretical support, unlike the reluctance of a phase to dissolve a *n*-alkane. That, at least qualitatively, is not at variance with the common notion of polarity and is usually expressed as the partial molar Gibbs free energy or excess free energy of solution for a methylene group or, by the solvent strength parameter, defined as the partial molar Gibbs free energy of solution for a methylene group per unit solvent volume [22,27,28]. Some typical solvent strength values for common liquid phases are given in Table 9.5; a more comprehensive collection of values is available in Ref. 27. The two solvent-strength scales

TABLE 9.5

SOLVENT STRENGTH PROPERTIES FOR SOME COMMON STATIONARY PHASES AT 121°C

Stationary phase	$\Delta G^\circ(CH_2)$ (cal/mol)	Solvent strength parameter (cal.cm^3/g.mol)
Squalane	- 521	- 728
Apolane-87	- 518	- 669
SE-30	- 463	- 578
Di-*n*-decylphthalate	- 511	
OV-7	- 467	- 504
OV-17	- 470	- 463
Poly(phenyl ether)	- 487	- 436
OV-225	- 418	- 410
OV-25	- 431	- 396
Carbowax 20M	- 400	- 387
Tetra-*n*-butylammonium-4-toluenesulfonate	- 377	- 377
QF-1	- 393	- 337
DEGS	- 324	- 275
1,2,3-Tris(cyanoethoxypropane)	- 280	- 273
OV-275	- 265	- 243

TABLE 9.6

PARTIAL MOLAL GIBBS FREE ENERGY OF SOLUTION FOR A SERIES OF TEST SOLUTES ON DIFFERENT STATIONARY PHASES AT 121°C

Stationary phase	Partial molal Gibbs free energy of solution (kcal/mol)			
	n-Butylbenzene	*n*-Octanol	Nitrobenzene	Benzodioxan
Squalane	– 5.14	– 4.62	– 4.42	– 5.44
Apolane-87	– 5.00	– 4.41	– 4.34	– 5.37
SE-30	– 4.46	– 4.48	– 4.52	– 4.87
Di-*n*-decylphthalate	– 5.17	– 5.01	– 5.77	– 6.04
OV-7	– 4.62	– 4.64	– 4.98	– 5.33
OV-17	– 4.60	– 4.60	– 5.18	– 5.56
Poly(phenyl ether)	– 4.84	– 4.93	– 5.71	– 6.06
OV-225	– 4.30	– 4.77	– 5.49	– 5.60
OV-25	– 4.50	– 4.38	– 5.17	– 5.63
Carbowax 20M	– 4.39	– 5.22	– 5.91	– 6.18
Tetra-*n*-butylammonium-4-toluenesulfonate	– 4.29	– 6.31	– 6.23	– 6.20
QF-1	– 3.77	– 3.99	– 4.83	– 4.57
DEGS	– 3.71	– 4.54	– 5.55	– 5.84
1,2,3-Tris(cyanoethoxypropane)	– 3.76	– 4.61	– 5.81	– 5.94
OV-275	– 3.11	– 3.87	– 5.25	– 5.30

produce a similar, although not identical, order in the ranking of phases. The solvent-strength parameter is less disturbed by molecular-weight differences and, at least intuitively, provides a more logical ranking of phases [28].

Solvent selectivity is a measure of the capacity of a solvent to enter into specific solute/solvent interactions, characterized as dispersion, induction, orientation, and complexation interactions. Unfortunately, fundamental approaches have not advanced to the point where an exact model can be put forward to describe the principal intermolecular forces between complex molecules. Chromatographers, therefore, have come to rely on empirical models to estimate the solvent selectivity of stationary phases. The Rohrschneider/McReynolds system of phase constants, solubility parameters, the solvent selectivity triangle of Snyder, Hawkes polarity indices, and various thermodynamic approaches have been the most widely used [21-23,25-31].

Fundamental problems in the use of retention indices for determining solvent selectivity scales have been exposed which impact strongly on the reliability of the McReynolds phase constants, the most widely used method of stationary-phase characterization, as well as some of the other methods, such as the solvent selectivity triangle approach and Hawkes polarity indices [22,27-30,32]. The interaction determined by the difference in retention indices of a test solute on two compared phases is a composite expression of the properties of the retention index standards and test solute; in most cases, it is dominated by the solubility of the *n*-alkanes (the substances generally selected as retention index standards) in the stationary phase and does not explicitly express the selective interactions of the test solute [22,33]. The reliability of the experimental data used to determine the retention indices is also compromised by the poor retention of the test solutes on many phases and by neglect of the importance of interfacial adsorption as a significant retention mechanism, particularly for *n*-alkanes on polar phases.

Thermodynamic approaches to the measurement of selectivity are based on the determination of the partial molar or molal Gibbs free energy of solution of either functional groups or specific test solutes, such as the first five test solutes suggested by McReynolds, or the series *n*-butylbenzene, *n*-octanol, nitrobenzene, and benzodioxan suggested by Poole and co-workers [22,27,28]. The free energies are calculated directly from the gas/liquid partition coefficient and are independent of retention contributions from interfacial adsorption. Since they are absolute measurements, great care is needed in establishing the experimental conditions, but the results obtained are not compromised by the use of the retention index system, as are those of McReynolds [27]. A comparison between the McReynolds scale and the free energy scale indicates that there is no correlation between the two scales in the general ordering of stationary phases for a particular interaction [29]. Typical values for the thermodynamically derived selectivity ranking of some common stationary phase are summarized in Table 9.6 [27,28]. The greater the negative value, the stronger or more selective the interaction expressed by the test solute.

n-Butylbenzene is retained largely by dispersive interactions, and the free energy scale based on *n*-butylbenzene correlates very well with that of the partial molar Gibbs free energy of solution for a methylene group [27]. It was included originally as a test solute to

References on p. A442

allow for deviations in solubility behavior with respect to the *n*-alkanes as a result of the additional polarizability of aromatic and unsaturated systems. In this case, it is either a poor probe for such interactions or there is no need in the selectivity scheme for such a probe. *n*-Octanol is used as a probe for solvent proton-acceptor capacity and shows a very wide range of interactions. The strongest interactions are found for the liquid organic salts and those phases that contain proton-acceptor functional groups, in keeping with chemical intuition. Nitrobenzene is used as a probe of orientation interactions and is most strongly retained by the liquid organic salts and those phases, such as Carbowax 20M, poly(cyanopropylsiloxanes), poly(esters), and 1,2,3-tris(cyanoethoxypropane), that have large dipole moments. Benzodioxan was selected as the test solute for solvent proton-donor capacity. However, it has been shown that the selective interactions of benzodioxan are governed primarily by orientation interactions [27]. Dibutylformamide has been suggested as a replacement test solute but, as yet, it has not been thoroughly evaluated.

9.4.2 Strategies for optimizing stationary-phase selectivity

The parameters described in Tables 9.5 and 9.6 are meant to be used as an aid in selecting stationary phases for a particular separation, based on a knowledge of the complementary properties of the sample. They are the first step in ensuring sample and phase compatibility and in the identification of phases that have significantly different separation characteristics to avoid needless repetition of experiments with phases of similar properties. As was indicated in the preceding section, these studies are presently incomplete, particularly the scale for solvent proton-donor capacity and also with respect to the influence of temperature on selectivity. However, selectivity optimization is very important, since the selectivity factor has a dramatic impact on the ease with which a certain separation can be achieved, as illustrated in Table 9.3. Practically all chromatographic separations have to be carried out in the efficiency range of 10^3-10^6 theoretical plates and, therefore, it is obvious from Table 9.3 just how important the optimization of the selectivity factor is for controlling the ease of obtaining a certain resolution. For example, with a resolution of 1 and a capacity factor of 3, over 1 million theoretical plates are required for a selectivity factor of 1.005, but only about 12 000 and 250 theoretical plates are required for selectivity factor values of 1.05 and 1.5 respectively.

Three approaches are in common use to optimize liquid-phase selectivity further:

(1) the use of phases with unique selectivity for certain problems, such as the separation of positional and optical isomers;

(2) the window-diagram approach to select binary stationary-phase mixtures for achieving a given separation; and

(3) selectivity tuning with two columns coupled in series, containing phases of complementary solvent properties. These approaches have emerged in the last few years as effective optimization strategies in gas/liquid chromatography for many difficult problems.

.4.2.1 Shape- and structure-selective liquid phases

The separation of enantiomers on a chiral stationary phase is the result of interactions etween the enantiomers and the stationary phase, which form a transient diastereomeric ssociation complex having different sorption enthalpies and hence different retention haracteristics. Compared to indirect methods employing derivative formation, the direct pproach has the advantage that it can be applied to enantiomers lacking reactive funconal groups and is not limited by the need for reagents of high enantiomeric purity. The rst chiral phases introduced for gas chromatography were either amino acid ester, ipeptide, diamide, or carbonyl-bis(amino acid ester) phases [34-36]. In general, these hases exhibit poor thermal stability and are not often used today. Real interest and rogress in enantiomer separations resulted from the preparation of diamide phases, rafted onto a polysiloxane backbone. The L-valine-*t*-butylamide-containing phase, comercially available as Chirasil-Val, has been used to resolve a wide range of racemic mino acid derivatives, amino alcohols, amines, hydroxyketones, 2-hydroxy acids and eir esters, 3-hydroxy acids, lactones, and sulfoxides [35,37,38]. It permitted for the first me the separation of all the enantiomers of the common protein amino acids as their -pentafluoropropionylamide isopropyl ester derivatives, in a single separation (Fig. 9.3) 39], and led to the development of the enantiomer-labeling technique (the unnatural -enantiomers being used as internal standards) for quantitative amino acid analysis [40]. further phase, a L-valine-*S*-α-phenylethylamide-containing polysiloxane, provided a facile ethod for the configurational analysis of sugars as well as for resolving enantiomers of nino acids, amino alcohols, hydroxy acids, and some pharmaceutical compounds 35,41,42]. Shurig et al. [43,44] have developed a number of dicarbonyl metal 3-(trifluorocetyl)-*R*-camphorates of Rh, Mn, Cu, Ni, etc., as additives in a noncoordinating solvent, ıch as a poly(dimethylsiloxane), for the separation of a wide range of stereoisomers of ydrocarbons and oxygen-, nitrogen-, and sulfur-containing electron-donor solutes, such s cyclic ethers, 1-chloroaziridines, thiranes, thiethanes, ketones, alcohols, and insect neromones by gas chromatography. Selective retention results from the fast and reverse chemical interaction of the solute and the metal coordination compound, which gives se to increased retention that is linearly related to the chemical equilibrium constant for olecular association and the concentration of the metal complex dissolved in the stationy phase. Thermal instability of the coordination complex (70-110°C) restricts the method low-molecular-weight stereoisomers.

Liquid-crystal stationary phases are widely used in gas chromatography for the sepation of close-boiling positional and geometric isomers of rigid molecules, such as ıbstituted aromatic hydrocarbons, polychlorinated biphenyls, dibenzodioxans, and steids [45]. In the liquid-crystalline state, the phase exhibits the mechanical properties of a uid, while maintaining some of the anisotropic properties of the solid, due to the presertion of a higher degree of order than that associated with isotropic liquids. Well over 200 uid-crystal phases have been used in gas chromatography. They are of a variety of emical types but all have in common a markedly elongated, rigid, rod-like structure and nerally polar terminal groups. In most cases liquid-crystal phases exhibit a lower effi-

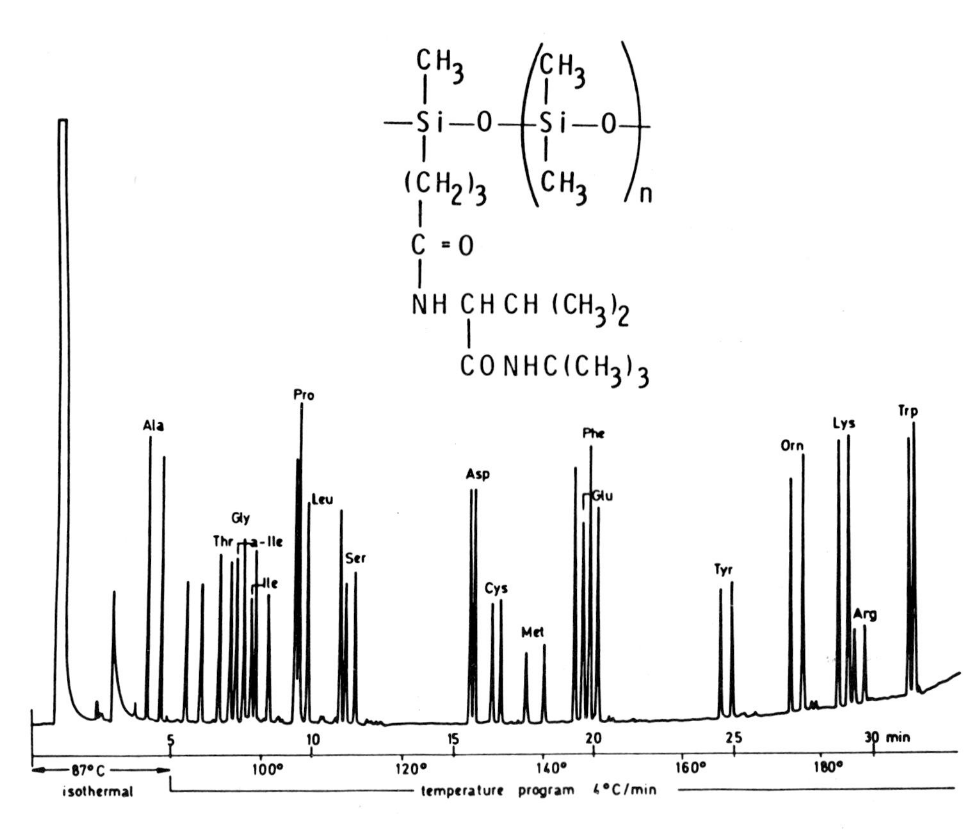

Fig. 9.3. Separation of the enantiomers of the common protein amino acids (*N*-perfluoropropionylamide isopropyl esters) on a 20 m x 0.25-mm-ID open-tubular column, coated with Chirasil-Val. (Reproduced with permission from Ref. 39.)

ciency than common isotropic liquid phases, due to their high viscosity and poor mass transfer characteristics and irreproducible retention data, unless packings of high phase loadings on low-surface-area supports are used [46]. New phases with a polysiloxane backbone, containing liquid-crystal functional groups have been shown to provide higher efficiency and a wider working temperature range. Moreover, they are compatible with modern technology for preparing immobilized phases on fused-silica capillary columns [47,48]. Columns coated with one of these phases, a smectic biphenyl carboxylate ester liquid-crystal polysiloxane stationary phase, are commercially available and provide reasonably favorable selectivity for the separation of polycyclic aromatic compounds and dibenzodioxin isomers, for example [49]. The additional shape selectivity of the liquid-crystal phase results from the preferential solubility for rigid, linear molecules and steric discrimination against bulky molecules. Long and planar molecules fit better into the ordered structure of the liquid-crystal phases, whereas nonlinear and nonplanar mole-

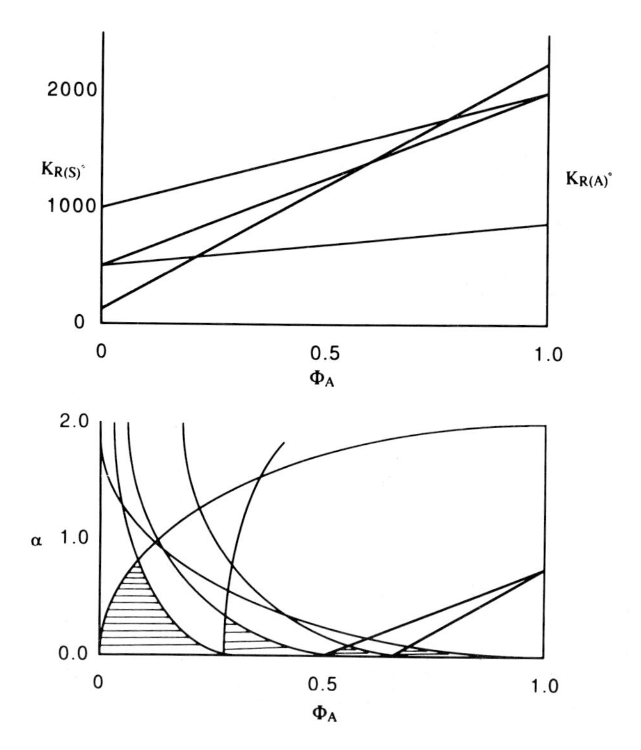

Fig. 9.4. Plot of $K_R°$ against Φ_A for four hypothetical solutes and of α against Φ_A, based on Eqn. 9.4. The best separation is predicted to occur at a volume fraction of 0.12.

cules do not intercalate so easily between the liquid-crystal molecules of the stationary phase and, therefore, are more easily eluted from the column.

9.4.2.2 Window diagram method

A method for predicting binary phase mixtures capable of completely separating samples that cannot be adequately resolved on either phase alone has been developed by Purnell and Laub [50,51]. It is based on the theory of diachoric solutions and is commonly known as the window diagram method.

The basis of the window diagram approach is that the relative retention of a solute on a mixed phase depends only on the volume fractions of the individual phases and the partition coefficients of the solute in the pure liquids. For a binary mixture of two phases, A and S, this can be expressed as

$$K_R° = K_{R(S)}° + \Phi_A \, (K_{R(A)}° - K_{R(S)}°) \tag{9.4}$$

where $K_R°$ is the infinite dilution gas/liquid partition coefficient for a solute in the liquid mixture (A+S), $K_{R(A)}°$ the infinite dilution gas/liquid partition coefficient in pure liquid A, $K_{R(S)}°$ the infinite dilution gas/liquid partition coefficient in pure liquid S, and Φ_A the volume fraction of A. A plot of $K_R°$ against Φ_A will be linear (Fig. 9.4), and furthermore, a plot of α, the separation factor, against Φ_A for all solute pairs provides a window diagram. For n solutes there are $n!/2(n-2)$ solute pairs. To simplify the calculation, all solute pairs with α greater than 1.2 are ignored, as their separation represents a relatively trivial problem. Also, for the purpose of visualizing the data, when plots of $K_R°$ against Φ_A cross for any solute, the calculated value of α is inverted so that $\alpha \geq 1$ at all times. The resultant window diagram is a series of approximate triangles, rising from the volume fraction axis, that constitute the windows within which complete separation for all solutes can be achieved. The optimum solvent composition for the separation corresponds to Φ_A with the largest value of α, provided that the most-difficult-to-separate pairs are reasonably well retained [52]. Further, the number of plates required for complete separation can be calculated from Eqn. 9.5

$$n_{req} = 36[\alpha/(\alpha - 1)]^2[(k + 1)/k]^2 \quad (9.5)$$

where n_{req} is the number of theoretical plates required to give baseline resolution. For a difficult-to-separate mixture with k greater than 10 the capacity factor term in Eqn. 9.5 is ca. 1 and can be neglected. Thus, knowing the values of n attainable with the pure liquid-phase columns, the required length of the mixed-phase column can be calculated with reasonable accuracy, given the predicted value of α. Finally, reference to the straight-line diagram of $K_R°$ against Φ_A corresponding to the optimum phase mixture yields the elution order of the solutes.

It is not necessary to blend liquid phases prior to coating, as it has been shown that mechanically mixed column packings of the pure phases provide the same results. The window diagram approach can be used to optimize the temperature for a separation, but the method itself is intended primarily for use in isothermal gas chromatography. Extension to temperature-programed separations is not straightforward. The window diagram approach has not been widely used in open-tubular column gas chromatography, since the greater intrinsic efficiency of such columns allows the separation of mixtures with low α values and, in addition, the number of calculations required for an average capillary column run is prohibitive [53]. The method has been applied to the design of specific stationary phases for particular common applications, such as the prediction of the optimum composition of poly(cyanopropylphenylmethylsiloxane) gum phases for the resolution of purgeable halocarbon compounds found in contaminated water [54,55]. Mechanically mixing the desired volume fraction of liquid phases for preparing open-tubular columns is not always successful due to the possibility that the phases may be immiscible and that finding a deactivation procedure compatible with the requirements of both phases simultaneously may be difficult. These combined effects result in poor film stability and a tendency for droplet formation as the temperature is varied. A more generally successful approach to selectivity optimization in open-tubular column gas chromatography has been the use of serially coupled columns.

9.4.2.3 Serially coupled columns

Serially coupled columns of different selectivity can result in dramatic changes in resolution compared with the relatively modest improvements obtained by coupling two columns of the same selectivity [56]. Changing the order of the columns or changing the flowrate through the column system results in changes in resolution and possibly elution order due to changes in selectivity of the complete system. Kaiser et al. [57] explained this phenomenon by demonstrating that the critical parameter was not the retention time for the coupled column system but rather the residence time of each solute in the different column segments. The residence time, in turn, will depend on the interplay between the capacity factors characterizing each column unit and the compressibility of the carrier gas on local velocities and will, therefore, depend on the system operating variables, in particular, on the ratio of the column inlet to outlet pressure. In theory, the selectivity of the chromatographic system can be varied continuously over the selectivity range, represented by the selectivity of the individual columns as limiting values. In practice, experimental constraints due to large flowrate differences make the extreme single-col-

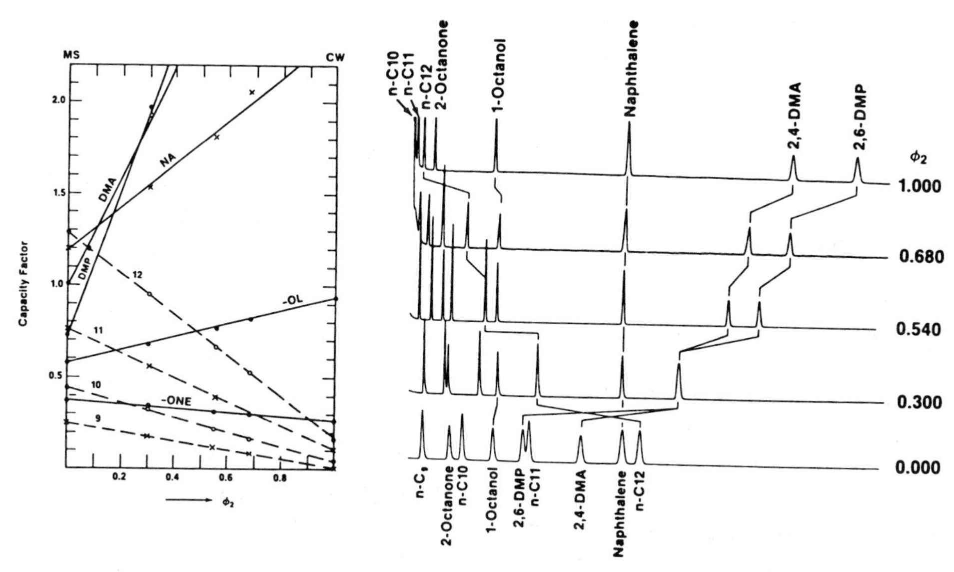

Fig. 9.5. Separation of a polarity test mixture on two 25 m x 0.25-mm-ID (d_f = 0.25 *mm*) serially coupled open-tubular columns, coated with a bonded methylsilicone (Column 1) and stabilized Carbowax 20M (Column 2). The plot of k_S against Φ_2 is shown on the left and the experimental data on the right. Those probes identified by a number are *n*-alkanes, 2,6-DMP = 2,6-dimethylphenol, and 2,4-DMA = 2,4-dimethylaniline. (Reproduced with permission from Ref. 58.)

umn selectivity values difficult to attain without loss of column efficiency. However, most of the selectivity range is available for use, and of course, the columns can be used independently to generate the single-column selectivity values.

The residence time of the sample in the two columns can be adjusted by changing the relative column lengths, by adjusting the average linear velocity of the carrier gas in the two columns to preselected values, or by operating the columns at different temperatures in a dual-oven gas chromatograph [54,56]. For practical convenience, the junction between the two columns is both pressure- and flow-controlled with – also for convenience of accurately setting the experimental conditions for the first column – the possibility of diverting a fraction of the sample from the first column to a detector [58,59]. The residence time of the sample in the first column of the tandem will depend on the pressure drop between the column inlet and the column connection point. For the second column the residence time depends on the outlet pressure of the first column (pressure at the point of the column connection) and on the pressure at the column outlet of the second column, which will normally be atmospheric when the flame-ionization detector is used. It is also possible to have a higher or lower flow in the second column than in the first. If a higher flow is required, carrier gas is added at the column connection point, while, if a lower flow is desired, a part of the gas flow (and consequently, the sample) is vented.

An example of the relative change in peak position for a polarity test mixture with two identical, serially coupled capillary columns, coated with poly(methylsiloxane) and Carbowax 20M stationary phases as a function of their relative retentivity on the second column, Φ_2 is shown in Fig. 9.5 [58]. A nearly optimal separation of the peak-to-peak position occurs at a value of Φ_2 of about 0.40. The peak positions are readily predictable from Eqn. 9.6.

$$k_S = \Phi_2(k_2 - k_1) + k_1 \tag{9.6}$$

where k is the capacity factor and the subscripts S, 1, and 2 refer to the system, Column 1, and Column 2, respectively [60].

9.5 ADVANCES IN OPEN-TUBULAR COLUMN TECHNOLOGY

Today, virtually all open-tubular columns are prepared from either soda-lime, borosilicate, or fused-silica glasses [61]. Fused silica, prepared by oxidizing pure silicon tetrachloride in an oxygen electric plasma oven, is essentially pure silica, containing less than 1.00 ppm of metal impurities [62]. Since it has a relatively high melting point, both special equipment and precautions are required in fabricating columns from this material. Soda-lime and borosilicate glasses contain large quantities of metal oxides (e.g., Na_2O, CaO, B_2O_3) and have lower melting points and working temperatures than fused silica. The surface of fused silica is relatively inert, containing primarily silanol and siloxane groups. The presence of silanol groups is responsible for the residual acidic character of the glass [63]. Soda-lime glasses are slightly alkaline, due to the high content of Na_2O, while the borosilicate glasses are somewhat acidic as a result of the presence of B_2O_3. The metal impurities at the surface of the above glasses act as Lewis acid sites and, unless these glasses are exhaustively leached to remove these impurities, they exhibit much higher activity toward lone-pair donor solutes, such as ketones and amines, than do fused-silica glasses.

9.5.1 Fused-silica columns

The flexible fused-silica columns of the type introduced by Dandeneau and Zerenner in 1979 [62,64] cannot be drawn on standard glass drawing machines. These thin-walled polymer-coated columns are drawn by techniques similar to those used for manufacturing optical fibers on equipment that is not normally found in analytical laboratories. The tubing, as drawn, is inherently straight but sufficiently flexible to be coiled on a spool for collection purposes. The thin-walled tube must immediately be protected from moist air or dust particles that promote the growth of fissures or cracks, making the capillary weak and friable, by coating it with a protective film of polyimide or aluminum.

Strength, flexibility, relative inertness, and the inherent straightness of column ends recommend fused silica as the column material of general choice for column preparation. However, as will be explained later, not all polar stationary phases form stable films on fused-silica surfaces, and columns with internal diameters $>$ ca. 0.6 mm are not available. For these phases and wide-bore columns, borosilicate glass is normally the choice. Greater availability of the preform in the desired size, greater uniformity of composition, and increased mechanical strength tend to dictate the use of borosilicate glass over soda-lime glass in contemporary practice, except perhaps for the separation of basic solutes. Metal columns are rarely used, except for hydrocarbon-type separations where their activity is not a problem, and especially when hydrocarbon phases that do not coat fused silica successfully are required for the separation.

During the drawing process, the fused silica is subject to various temperature changes that cause hydration and dehydration, yielding a material with an uneven silanol concentration. Residual silicon tetrachloride, trapped in the fused-silica preform, and oxides of nitrogen, formed at the high temperatures required to melt the fused silica, can result in the production of acidic impurities (HCl and HNO_3 on hydrolysis). Conditioning and hydrothermal treatment are generally required to remove these impurities and to provide a uniform and defined silanol concentration for column preparation [4,7,65-67]. These procedures are usually performed easily in the laboratory and are essential in minimizing batch-to-batch variation in the properties of the fused-silica columns, which can lead to failure of optimized column-deactivation and coating procedures.

9.5.2 Stationary-phase film-forming properties

To achieve a high separation efficiency in any type of open-tubular column it is essential that the stationary phase be deposited as a smooth, thin, and homogeneous film. This film must also maintain its integrity without forming droplets when the column temperature is varied. The ability of a liquid phase to wet a glass surface depends on whether the critical surface tension of the glass, a measure of the free energy of the surface, is greater than the surface tension of the liquid [68,69]. The stability of the film depends on the viscosity of the liquid and its resistance to disruption by forces tending to minimize the gas/liquid interface (Rayleigh instability). More viscous films provide greater stability whenever the time needed for the film to rearrange becomes long compared to the duration of the disturbance causing it to undergo rearrangement [70]. This is because the film is allowed insufficient time to break up or form ripples under unstable conditions, and because the viscosity of the film acts as a dampening mechanism in opposition to the disturbance. To maintain temperature stability, phases showing little change in viscosity with temperature are preferred. Many of the common oil-like phases developed for packed-column gas chromatography have too low a viscosity, particularly at high temperatures, to enjoy equal popularity in contemporary open-tubular column technology. Specially synthesized analogs of high viscosity, many of them gums, are widely used instead. The surface tension of typical stationary phases are in the range 30-50 dyne/cm. Clean fused silica and glass are high-energy surfaces and should be wet by most stationary phases. However, adsorption of moisture and other impurities generally lowers the critical surface tension to about 25-50 dyne/cm. Also, columns coated without deactivation of the surface often show undesirable activity towards sensitive analytes. Deactivation, then, is a common column preparation procedure, performed to maintain or enhance the wettability of the surface.

The energy of the smooth glass surface can be increased by roughening or chemical modification, or the surface tension of the stationary phase can be lowered by the addition of a surfactant. Roughening and/or chemical modification are the most widely used techniques for column preparation; the addition of a surfactant, although effective, modifies the separation properties of the stationary phase and may also limit the thermal stability of columns prepared with high-temperature stable phases.

Roughening, achieved by chemical treatment of the glass surface, enhances the wettability of the glass surface by increasing the surface area over which interfacial forces can act and dissipate the cohesive energy of the drop. Acid leaching or hydrothermal treatment restores the high energy of the glass surface (40-50 dyne/cm) and provides a stable and uniform surface to achieve maximum deactivation. Numerous procedures have been investigated for surface modification reactions, but only a few are in common use [4,7,65,66,71,72]. Of these, the most important reactions are etching by gaseous hydrogen chloride, leaching with aqueous hydrochloric acid, formation of whiskers, and solution deposition of a layer of solid particles. Because of the high purity and thinness of the column wall, leaching is the only treatment generally used with fused-silica columns. In this case, leaching is useful in maximizing the concentration of surface silanol groups in order to enhance deactivation and possible bonding in subsequent column preparation procedures [65,67].

9.5.3 Column deactivation

Surface modification reactions used to improve the wettability of glasses by the stationary phase and to deionize some glasses result in an increase in the activity of the glass surface. Without subsequent deactivation, columns coated with nonpolar and moderately polar stationary phases exhibit undesirable chromatographic properties, such as tailing, incomplete elution of polar solutes, and in extreme cases, sample decomposition. Although polar stationary phases may act as their own deactivating agents, column pretreatment is still advisable to ensure complete deactivation and to improve the thermal stability of the stationary-phase film.

The effectiveness of a particular deactivating reagent depends on the properties of the glass, the identity of the stationary phase and the sample to be analyzed, as well as on the method of surface roughening. No universal method of deactivation exists, but some techniques have emerged as more useful than others. The most widely used methods include coating with the thermal degradation products of stationary phases [73-75], high-

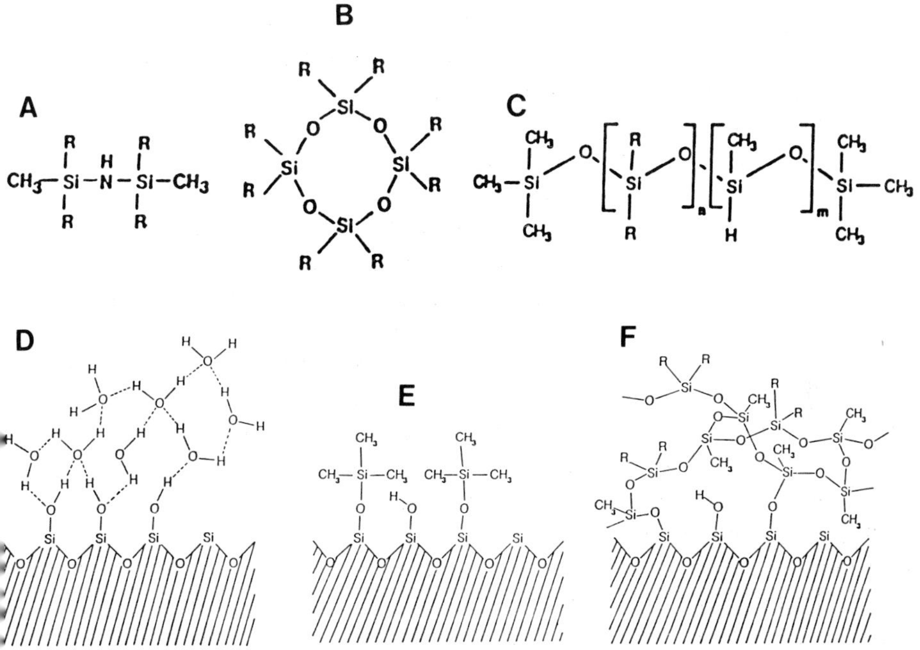

Fig. 9.6. Reagents used for the deactivation of silanol groups on glass surfaces. A = Disilazanes, B = cyclic siloxanes, and C = silicon hydride poly(siloxanes) in which *R* is usually methyl, phenyl, 3,3,3-trifluoropropyl, 3-cyanopropyl, or some combination of these groups. The lower portion of the figure provides a view of the surface of fused silica with adsorbed water (D), fused-silica surface after deactivation with a trimethylsilylating reagent (E), and fused-silica surface after treatment with a silicon hydride poly(siloxane) (F).

temperature silylation [76], and reaction with silicon hydride poly(siloxanes) [77]. Exhaustive silylation with a trialkylsilyl reagent forms surfaces that are highly deactivated but are of low energy (ca. 21-24 dyne/cm), which can only be coated with nonpolar liquid phases. For more polar phases silylation reagents containing functional groups similar or identical to those in the stationary phase are used to maintain the desired degree of wettability. Formation of a nonextractable film of polar polymer, such as Carbowax 20M, by thermal treatment results in a high-energy surface (ca. 45 dyne/cm), which can be coated with many polar phases.

High-temperature silylation with disilazanes, disiloxanes, or cyclosiloxanes, thermal degradation of poly(siloxanes), and the dehydrocondensation of silicon hydride poly(siloxanes) are now established as the most effective procedures for masking silanol groups on deionized glass and fused-silica surfaces for subsequent coating with nonpolar and moderately polar poly(siloxane) stationary phases [65,66,76-78]. Some typical reagents are shown in Fig. 9.6 in which the *R* group is varied to change the wettability of the glass surface and is generally the same as the substituent attached to the poly(siloxane) backbone of the stationary phase. Not all aspects of the high-temperature silylation mechanism for glass surfaces are fully understood. There is probably a large commonality in the way the various reagents act, although they were introduced and optimized separately before a more general picture emerged. Careful hydroxylation and dehydration of the glass surface is an essential prerequisite for maximizing the number of silanol groups and for establishing a fixed concentration of water which will be consumed during the silylation reaction. At low and moderate temperatures the steric bulk of the trimethylsilyl group prevents complete reaction of all silanol groups, resulting in a surface which retains some residual activity that is not useful for preparing inert, nonpolar columns. Above about 400°C several alternative reactions to simple silylation can also occur, the most significant of which are:

(1) condensation of neighboring surface silanol groups to form siloxane groups;

(2) nucleophilic displacement of the organic group from the bonded reagent by a neighboring silanol group to form short, cyclic siloxanes, bonded to the surface by two or three bonds; and

(3) reactions with water, forming short-chain polymers bonded to the surface [76,79].

These reactions result in a decrease in the number of unreactive silanol groups on the deactivated surface and the formation of a protective umbrella-type film, diminishing access of the analyte to the remaining unreacted silanols. In the case of the disilazane reagents the byproduct of the reaction with silanol groups, ammonia, can react with siloxane groups on the glass surface to generate additional silanol groups, which are subsequently silylated, yielding a high density of bonded organic groups. If the experimental conditions are less than optimal an active surface may result because of the production of silanol groups that are not subsequently reacted or shielded at the completion of the deactivation reaction. The thermal decomposition of poly(siloxane) phases occurs with the liberation of cyclic siloxanes, and this explains the equivalence of results observed for deactivation by thermal degradation of poly(siloxanes) and the direct application of cyclic siloxane reagents. Deactivation with silicon hydride poly(siloxanes) occurs primarily by dehydrocondensation with silanol groups on the fused-silica surface, establishing surface-to-polymer bonding [77]. In addition, physically adsorbed water hydrolyzes silane bonds to silanols, which further dehydrocondense with silane bonds or condense with other silanol groups to form a highly crosslinked resin film. The methods of high-temperature silylation yield inert and neutral surfaces that were unobtainable by other surface preparation techniques and are compatible with the coating requirements of the widely used poly(siloxane) stationary phases.

9.5.4 Coating procedures

The static coating method yields efficient columns of predictable film thickness with gum or solid phases [65,80-82]. However, phases of low viscosity cannot be coated successfully by the static method due to their propensity to flow and accumulate in the lowest portions of the capillary column during the relatively long time required for solvent evaporation. For static coating the column is filled entirely with a dilute solution, 0.02-4.0% v/v, of the stationary phase in a volatile solvent which is then evaporated by sealing one end of the column and attaching the other to a vacuum source, elevating the temperature of the column and performing the evaporation step at about atmospheric pressure, or by screwing the open end of the filled capillary into a high-temperature oven to force solvent evaporation. The vacuum technique remains the most widely used variation. In all cases, the solvent is evaporated under quiescent conditions, leaving behind a thin film of liquid phase of predictable thickness. Certainly, one of the least desirable features of the static coating method is the long time required for column preparation, particularly for long and/or narrow-bore columns. Only volatile solvents, such as pentane, dichloromethane, and Freons provide reasonable column preparation times at ambient temperatures. Further, the Rayleigh instability is magnified for narrow-bore columns, reducing the success rate for coating these columns.

In the dynamic coating procedure a solution of the stationary phase is passed through the column at a constant velocity, followed by gas to evaporate residual solvent. The thickness and uniformity of the stationary-phase film produced by dynamic coating is dependent on the concentration and viscosity of the coating solution, the surface tension of the coating solution, the velocity and the constancy of the velocity of the coating solution, temperature, and the rate of solvent evaporation. According to Marshall and Parker [83], the principal cause of low column efficiency for dynamically coated columns, provided that the stationary phase wets the column wall, is the formation of "lenses" (small plugs of liquid left behind the main plug during coating). Small temperature fluctuations may cause condensation of solvent vapors, forming a lens which is able to bridge the bore of the capillary column. Since the lens consists of pure solvent, it will dissolve the film already laid down. To improve the success of the dynamic coating method Schomburg et al. [84,85] suggested a modification of the above method, known as the mercury plug dynamic coating procedure. In this method a mercury plug is interposed between the plug of coating solution and the driving gas. Mercury, which has a high surface tension, wipes most of the coating solution off the surface of the column wall as it moves through the

References on p. A442

column, leaving a more even and thinner film behind. More concentrated coating solutions (10-50% v/v) are used in this procedure, producing films which are more resistant to drainage during the drying process. Certainly, one of the major inconveniences of the dynamic coating method is the difficulty of predicting the film thickness. Generally speaking, it must be calculated after the column is prepared and may not correspond to the thickness desired [1].

9.5.5 Gum and immobilized phases

The gradual evolution of column technology revealed that the low- to medium-viscosity oils used as stationary phases for packed-column gas chromatography were inadequate for preparing thermally stable and efficient open-tubular columns. New phases of much higher viscosity were required to resist film disruption at elevated temperatures. For poly(dimethylsiloxanes) this was relatively easy to achieve by increasing the molecular weight. For poly(methylsiloxanes), containing bulky or polar functional groups, the regular helical conformation of the polymers is distorted, resulting in a greater change in viscosity with temperature. For poly(dimethylsiloxanes) increasing temperature causes the mean intermolecular distance to increase, but at the same time expansion of the helices occurs, tending to diminish this distance. As a result, the viscosity appears to be only slightly affected by temperature. A second breakthrough was achieved when it was realized that crosslinking of the gum phases to form a rubber provided a means of further stabilizing the polysiloxane films without destroying their favorable diffusion characteristics. The unmatched flexibility of the Si-O bond imparts great mobility to the polymer chains, providing openings that permit diffusion, even for crosslinked phases. Stationary phases for open-tubular column preparation in contemporary practice are characterized by high viscosity, good diffusivity, low glass transition temperatures, and generally are suitable for crosslinking. For the present, the above properties are only apparent for the polysiloxane and poly(ethylene glycol) phases, and this limits the achievable range of selectivities.

The polysiloxane phases with various substituent groups are generally prepared by either the catalytic hydrolysis of dichlorosilane or dimethoxysilane monomers or by the platinum-catalyzed addition of an alkene to a poly(alkylhydrosiloxane) polymer (the addition of an alkene to a Si-H bond) [86,87]. The dimethoxysilanes are preferred for the preparation of polar polysiloxanes containing cyanopropyl substituents, e.g., to avoid conversion of functional groups and trapping of hydrogen chloride, generated by hydrolysis of the chlorosilanes. Copolymers containing vinyl, tolyl, or *n*-octyl groups to assist in the crosslinking of the more polar polysiloxanes are prepared by hydrolyzing and polymerizing the appropriate amounts of the relevant monomers. Eventually, a high-molecular-weight polysiloxane gum with terminal silanol groups results, which is suitable for use after fractionation for some column preparation procedures. Alternatively, the polymer can be endcapped by treatment with 1,3-divinyltetramethyldisilazane and a chlorosilane catalyst. Endcapping introduces an additional functionality into the polymer to aid crosslinking, and by terminating the polymerization process it ensures a narrow molecular-weight distribution for the polymer. Thermal curing of the silanol-terminated polymer is a common secondary route to increase the average molecular weight of the polymers [88].

Two different procedures have been used to immobilize polysiloxane phases during column preparation. Thermal condensation of nonpolar and medium-polar silanol-terminated polysiloxanes at high temperature with silanol groups on the glass surface results in bonding of the phase to the surface, yielding inert and thermally stable columns. Columns prepared with poly(dimethylsiloxanes) and poly(methylphenylsiloxanes) are easily made by static coating of the silanol-terminated phase, which is then heated by temperature programing up to 300-370°C and maintained at that temperature for about 5 to 15 h with a slow flow of carrier gas. Bonded, nonpolar polysiloxane phases are suitable for high-temperature gas chromatography with isothermal operation at 400-425°C [89]. Medium-polar phases are more difficult to bond and require careful surface preparation to avoid film disruption during the bonding process and generally yield columns of lower thermal stability.

An alternative approach to stationary-phase immobilization, and the most popular method in contemporary practice, is the free-radical crosslinking of the polymer chains, with peroxides [90,91], azo-compounds [90,92], ozone [93], or gamma radiation [94] as free-radical generators. In this case, crosslinking occurs through the formation of Si-C-C-Si bonds, as shown below. Very little crosslinking (0.1-1.0%) is required to render polysiloxanes with long polymer chains insoluble. Of the reagents indicated above dicumyl peroxide and azo-*t*-butane have emerged as the most widely used. Radiation-induced crosslinking has the advantage that no extraneous groups are introduced into the phase during crosslinking. On the debit side, gamma ray sources, usually ^{60}Co, are not generally available in analytical laboratories. Radiation-induced crosslinking is not successful for polar phases.

$$\left(\begin{array}{c} CH_3 \\ | \\ 2\ \ -Si-O- \\ | \\ CH_3 \end{array}\right)_n \xrightarrow{2R} \begin{array}{c} CH_3 \\ | \\ -Si-O- \\ | \\ \cdot CH_2 \\ \\ \cdot CH_2 \\ | \\ -Si-O- \\ | \\ CH_3 \end{array} \longrightarrow \begin{array}{c} CH_3 \\ | \\ -Si-O- \\ | \\ CH_2 \\ | \\ CH_2 \\ | \\ -Si-O- \\ | \\ CH_3 \end{array} + 2RH$$

Poly(dimethylsiloxane) phases are relatively easy to crosslink, but with increasing substitution of methyl by bulky or polar functional groups the difficulty of obtaining complete immobilization increases. For this reason, moderately polar polysiloxane phases are prepared with various amounts of vinyl, tolyl, or octyl groups that increase the success of the crosslinking reaction. Even so, it has proven impossible to immobilize poly(cyanopropylsiloxane) phases with a high percentage of cyanopropyl groups completely. Commercially available poly(cyanopropylsiloxane) columns are either completely immobilized and contain, generally, less than 33% cyanopropyl groups as substituents, or they are "stabilized", indicating incomplete immobilization with a higher incorporation of cyanopropyl groups. Stabilized columns usually have greater thermal stability than physically coated columns but are not resistant to solvent rinsing or as durable as immobilized columns. Poly(ethylene glycol) phases are also available as bonded or stabilized phases, since, as in the case of the poly(cyanopropylsiloxane) phases, it has proven difficult to immobilize them successfully by using free-radical crosslinking.

Another general advantage of immobilization is that it permits the preparation of open-tubular columns having very thick films. The normal film thickness in open-tubular-column gas chromatography is 0.1-0.3 μm. It is very difficult to prepare conventionally coated columns with stable films thicker than ca. 0.5 μm. However, thick-film columns of 1.0-8.0 μm can easily be prepared by immobilization. Thick-film columns permit the analysis of volatile substances with reasonable retention times without the need for subambient temperatures. They also permit the injection of much larger sample volumes without loss of resolution. This factor is important for identifying trace components by combined techniques, such as GC/FTIR and GC/MS, where the small sample sizes tolerated by conventional columns may preclude positive identification. Immobilization also improves the resistance of the stationary phase to phase stripping by large-volume injection of solvent and allows solvent rinsing to be used to free the column from nonvolatile sample byproducts or from active breakdown products of the liquid phase. Bonding and crosslinking procedures also provide a useful increase in the limit of column operating temperature. Thus, bonding and immobilization techniques have been useful not only in advancing column technology but they have also contributed to advances in the practice of gas chromatography by facilitating injection, detection, and other instrumental developments that would have been difficult without them.

9.6 INSTRUMENTATION FOR GAS CHROMATOGRAPHY

The principal function of the gas chromatograph is to provide those conditions required by the column for achieving a separation without lowering the performance of the column in any way. This means providing a regulated flow of carrier gas to the column, an inlet system for vaporizing the sample, a thermostated oven for optimizing the temperature, an on-line detector for monitoring the separation in real time, and associated electronic components for controlling and monitoring instrument conditions and for recording the chromatogram (Fig. 9.7) [1,3-5,8,95-98]. Although the basic components remain the same, open-tubular columns place greater demands on instrument performance than packed columns due to their lower sample capacity, lower carrier gas flowrates, and the need for detectors of low deadvolume with fast response times. In many cases, it is possible to convert an older, packed-column gas chromatograph to an instrument for capillary-column chromatography by changing pneumatic components to provide a stable, pressure-regulated supply of carrier gas, split gas (if a splitter is in-

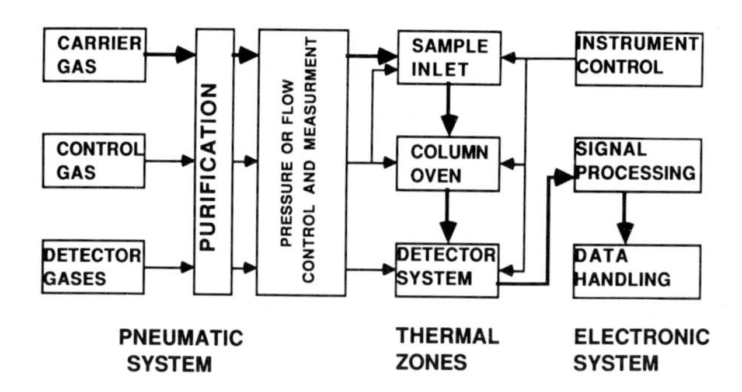

Fig. 9.7. Schematic diagram of the principal components of a gas chromatograph. The solid line shows the path taken by sample and carrier gas, resulting in the production of a chromatogram. The dotted lines represent support and control functions.

stalled), and detector makeup gas by adding a new injector for split, splitless, or cold on-column injection, and by additional minor changes in plumbing, fittings, etc., to accommodate the added components [7,99,100]. Before attempting such a conversion, the response time of the detector electronics should be determined to ascertain the adequacy of time constants. The temperature stability of the oven in both time and space should also be determined to ensure the absence of temperature gradients and cycling or overshooting of preset or programed temperatures, which would result in poor column performance, even if the conversion was carried out successfully. Several companies offer kits for converting packed-column instruments to capillary-column use.

9.6.1 Column ovens

The column ovens is generally a forced-circulation air thermostat of sufficient size to allow comfortable installation of the longest columns normally used. In the design of a column oven it is important to ensure a uniform temperature throughout the column coil region. A temperature stability of ± 0.1°C in time and ± 1.0°C in space are reasonable requirements, particularly for use with capillary columns. Poor column temperature stability has been identified as a distinct source of peak distortion or splitting with open-tubular columns [101-103]. In severe cases, christmas-tree-like peaks are observed. When a column experiences oscillating temperatures, the distribution constants of solutes between the mobile and stationary phases will fluctuate, and this is believed to be responsible for the observed peak distortion. Splitting is most likely to occur when fused-silica capillary columns are used, since their thin walls rapidly transmit temperature fluctuations from the oven to the phase boundary, allowing little time for cross-wall thermal averaging to dampen the oscillations. By keeping temperature fluctuations around 0.05% of the actual oven temperature, peak splitting is practically eliminated. Most modern gas chromatographs utilize microprocessors, sensors, and proportional-heating networks not only for maintaining constant temperature but also for controlling the initial temperature lag, the linearity of the program rate, and the final temperature overshoot in temperature-programed gas chromatography.

9.6.2 Sample inlets

The most common method of introducing samples into a gas-chromatographic inlet is by means of a microsyringe. Typically, this consists of a calibrated glass barrel with a close-fitting metal plunger, which is used to dispense a chosen volume of sample by displacement through the syringe needle. Gastight syringes are available for injecting gases and vapors with Teflon-tipped plungers for improved sealing of the plunger with the barrel in the face of the backpressure created by the inlet pressure of the injector. However, microsyringes may cause problems that have frequently gone unnoticed. The accuracy of quantitative injection is known to depend on the rate of sample introduction, syringe deadvolume, heating of the syringe needle by the injector, and sample handling techniques. At the time of injection, the sample volume delivered to the inlet is equivalent to the calibrated amount, determined by the graduations on the syringe barrel plus some fraction of the sample volume retained in the needle. The correction for the needle volume is difficult to determine for vaporizing injectors, since it will depend on the column inlet temperature and pressure, and in a more complex manner, on the sample concentration [104]. After injection, estimating the noninjected sample volume by drawing the remaining sample up into the barrel is subject to error, due to loss of vapors through the syringe needle while equilibration to ambient conditions occurs. Using the solvent flush technique, a film of sample is formed in the air segment between the sample and flush solvent [105]. This film is taken up by the solvent plug and dispensed backwards and diluted by the flush solvent. During rapid injection, the sample plug breaks up due to excessive shear forces and spreads as small droplets, which are mixed with the solvent plug. For fast injections no improvement in accuracy can be expected by using the solvent flush method. The accuracy of the sample volume injected can only be improved by reducing the deadvolume of the syringe.

A common problem encountered in sample introduction with a syringe into a hot vaporizing injector used with open-tubular columns is sample discrimination [106,107]. The sample leaves the syringe and enters the vaporizer as a stream of droplets, formed by the movement of the plunger and by evaporation of the remaining sample from the syringe needle. It is at this evaporation stage that discrimination is most likely to occur, the solvent and more volatile sample components distilling from the syringe needle at a greater rate than the less volatile components. As a consequence, the sample reaching the column contains more of the more volatile and less of the less volatile components of the sample. The influence of losses inside the syringe needle decreases as the sample volume injected increases, and thus it may be hardly noticed with packed columns, but it may be critical with open-tubular columns where the sample volume can be on the same order as that of the syringe needle.

The limited sample capacity and low carrier gas flowrates characteristic of capillary-column GC give rise to certain difficulties in sample introduction. Direct sample introduction of large volumes or concentrated samples can cause column overloading, leading to a decrease in column efficiency and/or the production of distorted peaks. The most difficult problems arise when mixtures spanning a wide concentration range or volatility range must be analyzed. No single injection method will solve all of the problems likely to be encountered. A number of different techniques are in routine use, the most important of which are split, splitless, on-column, and programed-temperature vaporization [4,107-112].

9.6.2.1 Split injection

For many applications split injection is the most convenient sampling method, since it allows injection of mixtures virtually independent of the choice of solvent for the sample, at any column temperature, with little risk of disturbing solvent effects or bandbroadening in time due to slow sample transfer from the injector to the column [107,111-114]. The classical or hot split injector is really and isothermal vaporizing injector, in which the evaporated sample is mixed with carrier gas and divided between two streams of different flow, one entering the column (carrier gas flow) and the other vented to the outside (split flow). In the case of mixtures containing sample components of unequal volatility split injection discriminates against the less volatile sample components due to selective vaporization from the syringe needle, incomplete sample vaporization, and inhomogenous mixing of sample vapors with the carrier gas in the vaporization chamber. When the sample is vaporized, it generates an instantaneous pressure pulse and rapid change in the viscosity of the carrier gas/sample mixture, altering the flow of gas between the column and split line in a manner that is hardly reproducible. In general, hot split injection has been used as the preferred sampling technique for qualitative analysis under high-resolution conditions and less frequently for quantitative analysis.

9.6.2.2 Splitless injection

The hot splitless injection technique was devised to overcome some of the deficiencies of split sampling for the quantitation of trace components that is achieved through the introduction of relatively large sample volumes into the column [107,111,112,115-117]. Classical splitless injection is a hot-vaporization technique subject to some of the same problems as hot split injection, namely, discrimination due to selective vaporization from the syringe needle and thermal decomposition of labile substances due to prolonged heating at relatively high temperature and contact with catalytic surfaces, including the metal syringe needle. The velocity of gas flow through a splitless injector is relatively low, and the sample is introduced into the column over a comparatively long time, relying on coldtrapping and/or solvent effects to refocus the sample at the column inlet. Consequently, solutes eluted before the solvent are not refocused, and peaks are generally very distorted.

For splitless sampling, the sample is injected through a septum by microliter syringe into a thermostated vaporization chamber, swept by the flow of carrier gas. The split flow is turned off at the start of the injection and is restarted only at the end of the sampling period. The transport of sample vapors to the column is relatively slow, since the linear velocity of the carrier gas through the vaporization chamber is low and dictated by the separation conditions for the column. During sample vaporization, transfer of vapors into the column is negligible. Thus, the vaporization chamber must be large enough to hold the whole volume of vapor produced by the evaporated sample. If the volume of the vaporization chamber is too small, sample vapors will be lost by backflushing through the septum purge exit or by deposition in the carrier gas lines. The sample should be introduced at the bottom of the vaporization chamber, close to the column, so that it fills with vapor from the bottom up, displacing the carrier gas backwards. Sample transfer to the column should be virtually complete but requires a comparatively long transfer, e.g., from several seconds to 2 min. Complete sample transfer is difficult to achieve, since the sample vapors are continually diluted with carrier gas, and some sample vapors accumulate in areas poorly swept by the carrier gas. At the end of the sample transfer period, the split flow is re-established to purge the inlet of remaining solvent vapors. If the split flow is started too soon, sample will be lost; if too late, the solvent peak will trail into the chromatogram.

The slow introduction of sample into the column causes bandbroadening in time, in which all solute bands are broadened equally in terms of gas-chromatographic retention time in the absence of a refocusing mechanism [107,118-120]. Bandbroadening in time can be counteracted or minimized by temporarily increasing the retention power of the column inlet during sample introduction. For this to be effective, the migration velocity of the sample entering the column at the start of the injection period must be sufficiently retarded to allow portions of the sample entering at the end of the injection period to catch up with it. This may be achieved by lowering the column temperature (coldtrapping) or by a temporary increase of the film thickness of the stationary phase by using the solvent as a liquid phase (solvent effects). Coldtrapping is usually performed in the absence of solvent effects by maintaining the column inlet at a temperature not less than 15°C below the solvent boiling point to minimize solvent recondensation in the column inlet. A minimum temperature difference of ca. 80°C between the inlet temperature and the elution temperature of any of the solutes of interest is normally required to ensure efficient coldtrapping. Coldtrapping is frequently used when temperature-programed analysis is required.

A prerequisite for achieving refocusing of bands broadened in time by solvent effects is recondensation of solvent in the column inlet to form a temporary liquid-phase film. The column temperature during the splitless period must be low enough to ensure that the concentration of solvent vapors in the carrier gas entering the column exceeds the saturation vapor pressure of the solvent and that sufficient solvent is recondensed so that some liquid remains in the column inlet up to the end of the sample introduction period. In most cases, these conditions will be met if the column temperature is at least 20-30°C below the solvent boiling point. The solvent should also provide strong interactions with the sample to avoid partial solvent trapping, which can lead to distorted peaks. The solvent should also wet the stationary phase, otherwise a stable film will not be formed. Distortion due to bandbroadening in space may be significant for solute bands eluted at temperatures at least 50°C above the injection temperature.

The attractive feature of splitless injection techniques is that they allow the analysis of dilute samples without preconcentration (trace analysis) and the analysis of dirty samples, since the injector is easily dismounted for cleaning. However, success with individual samples depends on the selection of experimental variables, the most important of which are: sample size, sample solvent, syringe position, sampling time, initial column temperature, injection temperature, and carrier gas flowrate. These must often be optimized by trial and error. The absolute accuracy of retention times is also generally not as high as that for split injection. The reproducibility of retention times for splitless injection depends not only on chromatographic interactions but also on the reproducibility of the sampling period and the evaporation time of the solvent in the column inlet, if solvent effects are employed. For quantitative analysis with adequate injection control, the precision of repeated sample injections is normally acceptable but the method is subject to numerous systematic errors, which may affect accuracy. Internal standards are usually preferred to external standards for improving accuracy.

9.6.2.3 Programed-temperature vaporization

Programed-temperature vaporization (PTV) injectors have become available recently. They provide technical solutions to some of the problems in the classical hot split and splitless injectors discussed above [109-111,116,121-124]. They were originally developed for large-volume injections with solvent elimination and for obviating discrimination arising from selective sample volatilization from the syringe needle occurring in hot-vaporizing injectors. The latter problem can be solved by using special syringes with forced cooling of the syringe needle [112]. In the PTV injector (Fig. 9.8) the liquid sample is introduced into the vaporizing chamber, which is maintained at a temperature below the boiling point of the solvent. Compared to classical, hot-vaporizing injectors, the vaporization chamber is much smaller, about one-tenth the size, and is of lower mass to allow rapid heating and cooling by circulated air. Typically, the injector can be raised from ambient temperature to 300°C in about 20 to 30 sec. The PTV injector can be operated in three modes:

(1) split injection with a hot- or cold-vaporization chamber;

(2) as a cold-solvent split injector for solvent elimination; and

(3) as a cold splitless injector for total sample introduction.

Heating of the injector is usually started a few seconds after withdrawal of the syringe needle for split injection and occurs rapidly to a temperature sufficiently high to ensure rapid volatilization of the highest-boiling sample components. Compared to the classical hot-vaporizing injectors, there is far less discrimination, and the sample split ratio and split flow ratio are usually similar. In the solvent elimination mode, the sample is introduced into the cold injector at a temperature close to the boiling point of the solvent with the split vent open. The solvent is concurrently evaporated in the glass insert and swept through the split vent by the carrier

References on p. A442

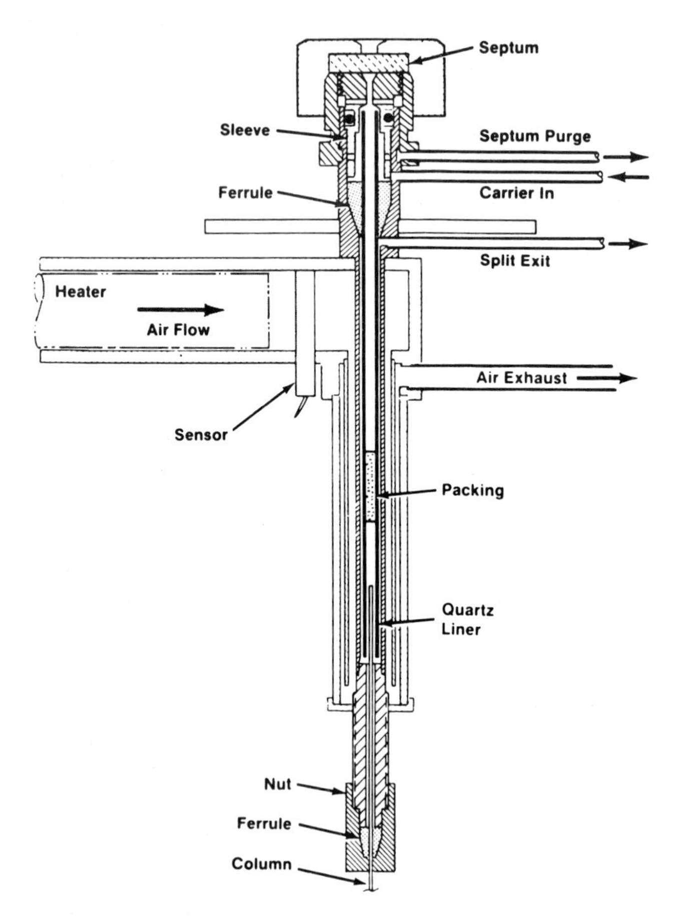

Fig. 9.8. Schematic diagram of the Perkin-Elmer PTV injector. (Reproduced with permission from Ref. 142.)

gas. After completion of the solvent evaporation, the split vent is closed and the injector is ballistically heated to the temperature required to vaporize the sample and transfer it to the column. This injection technique permits the injection of large volumes of up to about 20 μl rapidly and 300 μl volumes slowly. (The maximum sample sizes indicated are not definitive and depend on many factors that are not well understood.) This injection technique is limited to high-boiling solutes. For cold splitless injection the sample is introduced into the vaporization chamber at a temperature close to the boiling point of the solvent with the split vent closed. A few seconds after injection, the injector is rapidly heated to the temperature required to complete the transfer of vapors to the column (30-90 sec). The column temperature program is started, and the vaporizer is purged by opening the split vent to exhaust any solvent residues. As for hot splitless injection, coldtrapping and solvent effects are employed as refocusing mechanisms, since sample transfer takes place in two steps: transfer of solvent and volatiles, followed by transfer of less volatile components at a later time. These refocusing mechanisms are generally most important for the separation of the volatile sample components.

Certainly, the precision and accuracy of the PTV techniques are generally superior to those of the classical hot-split and splitless techniques and approach those obtained by cold on-column injection [111,121,125,126]. However, less is known concerning optimization of PTV injection, and probably more parameters have to be considered than for cold on-column injection. The latter is, consequently, the generally preferred injection technique for most samples, except those contaminated by large amounts of involatile impurities and for headspace vapors.

9.6.2.4 Cold on-column injection

Cold on-column injection differs from the vaporization techniques discussed above in that the sample is introduced directly as a liquid into the column inlet, where it is subsequently vaporized. In this way, discrimination is virtually eliminated and quantitation of mixtures of different volatilities is facilitated. Sample decomposition due to thermal and catalytic effects is minimized. An attractive feature of cold on-column injection is that the technique is easy to implement and capable of being optimized by controlling a few experimental variables. Its popularity has increased with the wider use of immobilized stationary phases that eliminate the problem of phase stripping. The technique can be automated by using wide-bore precolumns, connected to the analytical column. It also can be adapted to the introduction of large sample volumes by using the retention-gap technique. The injection device must provide a mechanism for guiding the fragile needle into the column, for positioning it at the correct height within the column, and for sealing the needle or at least restricting the outflow of carrier gas through the syringe entry port. A number of different methods based on the use of valves, septa, or spring-loaded O-ring seals are used for this purpose (Fig. 9.9) [4,5,108,112,127]. Secondary cooling by circulating air is used in some designs to ensure that the temperature of the needle passageway can be controlled to avoid solvent evaporation. With efficient secondary cooling the oven temperature can be maintained well above the boiling point of the solvent, while maintaining the column inlet at a much lower temperature. This is important for using on-column injection in high-temperature GC [108,111,128,129]. Alternatively, the column inlet can be housed in a separate injection oven, which is temperature-programable and can be thermostated independent of the column oven [4].

For injection volumes of 0.2-2.0 μl with column inlet temperatures at or below the boiling point of the solvent at the column inlet pressure, very few unexpected problems are generally encountered for cold on-column injection. For larger volumes, solvents that are incompatible with the properties of the sample or stationary phase, or for solutes that are eluted more than about 50°C above the injection temperature, peak distortion resulting from bandbroadening in space may be observed [108,130,131]. The plug of liquid leaving the syringe flows into the column inlet, closing the column bore. This plug is then pushed

References on p. A442

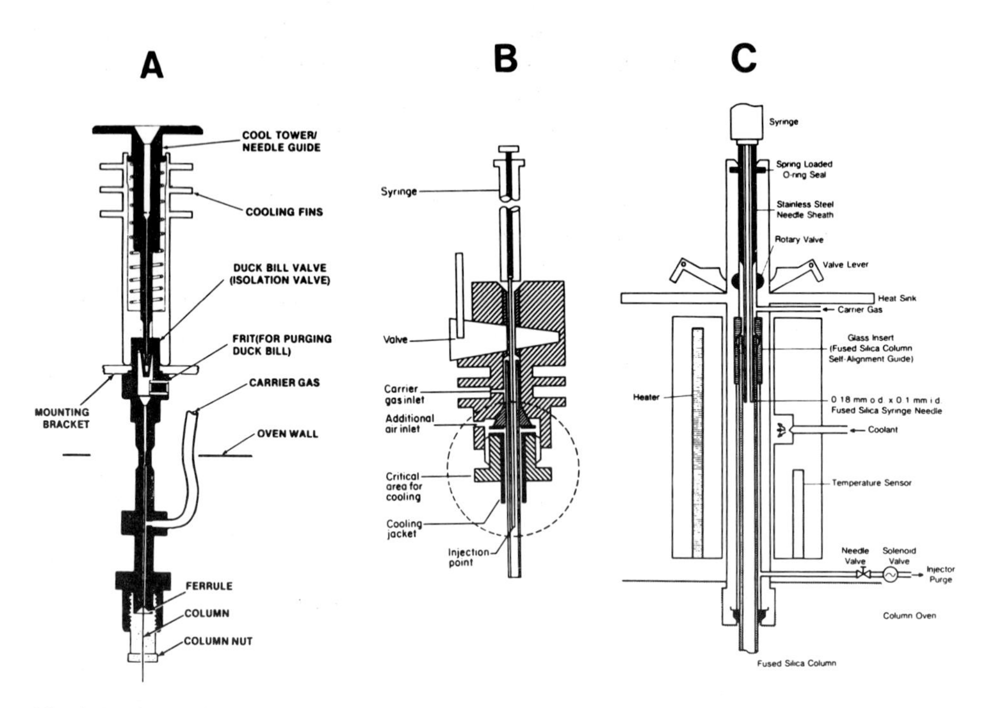

Fig. 9.9. Alternative designs for cold on-column injectors. (A) Injector with a duck-bill valve (Hewlett-Packard), (B) an injector with provision for secondary cooling of the column inlet (Carlo Erba), and (C) a temperature-programable on-column injector with its own oven, isolated from the column oven (Varian Associates).

further into the column by the carrier gas until the liquid is spread over the surface of a considerable length of the column inlet. During the primary flow of liquid, a layer of sample is left behind the plug, reducing the length of the plug until, eventually, the whole sample is spread as a stable film on the column wall. This layer of sample is called the flooded zone. Its length depends primarily on the volume of sample injected, whether the solvent wets the stationary phase or not, and the column temperature. High-boiling or strongly retained solutes are distributed throughout the flooded zone with an initial bandwidth corresponding to the zone length. At the injection temperature, volatile solutes do not remain where they are deposited by the evaporating solvent but migrate with the unsaturated carrier gas. If the volatile solutes move as rapidly as or more rapidly than the rear of the sample layer during solvent evaporation, they end up being reconcentrated at the place where the last portion of solvent evaporates from and form a sharp band. Such volatile solutes are not subjected to bandbroadening in space in the flooded zone. Solutes that migrate more slowly than the rate at which the rear of the sample layer withdraws, are only partially reconcentrated, since materials deposited at the rear of the flooded zone have migrated only part of the way to the front when the last portion of the solvent evaporates.

A general solution to the problem of peak distortion due to bandbroadening in space is to make the injection into a retention gap [108,132-135]. A retention gap is a length of uncoated (but, often deactivated) column, used as an inlet which has a reduced retention power compared to the column used for the separation. Some advantages of retention gaps are: they function as a guard column to protect the analytical column from contamination by involatile residues; wide-bore retention gaps permit the use of autoinjectors with regular syringe needles; and long retention gaps facilitate the injection of large sample volumes. As a general guide, typical retention gaps should be about 25-30 cm for each microliter of solvent injected and should be sufficiently inert to minimize adsorption and decomposition of the sample. It is also important that the solvent wet the retention gap surface in order to produce a film of sample liquid on the capillary wall. When a retention gap is used, refocusing of zones occurs because the retention power of the retention gap is 100 to 1000 times less than that of the column. Solutes migrate through the retention gap at temperatures well below those required to cause elution from the analytical column. Thus, they arrive at the entrance to the coated column at a temperature too low for significant migration and remain and accumulate there until a temperature is reached at which migration starts. Retention gaps are normally used when sample volumes $>$ ca. 5 μl and up to several 100 μl are injected.

9.6.3 Coupled-column and multimodal gas chromatography

Coupled-column chromatography involves the separation of a sample by means of two or more columns in series where the individual columns differ in their capacity or selectivity. Multimodal separations employ two or more chromatographic methods in series, such as the on-line combinations of liquid chromatography/gas chromatography or supercritical-fluid chromatography/gas chromatography. Both methods employ the transfer of one part of the effluent from the first column to another via some suitable interface as a means of affecting trace enrichment of selected analytes, with the purpose of improving the resolution of parts of a complex mixture or of decreasing the time needed for analysis by employing such techniques as heartcutting, backflushing, foreflushing, coldtrapping and recycle chromatography.

9.6.3.1 Coupled-column gas chromatography

The interface in coupled-column gas chromatography must provide for the quantitative transfer of the effluent from one column to the next without altering the composition of the transferred sample or decreasing the resolving power of the second column [9,136,137]. Two types of interfaces are commonly used for this purpose, those employing microvolume multiport switching valves and those using pneumatic switches. Pneumatic switches, also referred to as Dean's switches, are based on the balance of flow at different positions in the chromatographic system and employ in-line restrictors and precision pressure regulators [138]. The direction of flow between columns can be changed by opening and closing the valves located outside to the column oven. Thus, no moving parts are located inside the oven and deadvolume effects are not important, since there are no unswept volumes. Problems can arise from backdiffusion of sample into the switching lines, causing memory effects, and from instability of the pressure controllers, which must remain reproducible under all conditions. Since switching is controlled by pressure, widely varying flows, as might exist in interfacing a packed column with an open-tubular capillary column, are easily handled [139,140].

Intuitively, the simplest approach to switching the effluent from one column to another is by means of a mechanical valve [141,142]. Since the valve forms part of the sample passageway, the valve chosen must be chemically inert, free from outgassing products, gastight at all temperatures, must have a low heat capacity and small internal dimensions, and operate without lubricants. Modern miniature multifunctional valves meet most of these requirements, with some exceptions. They are generally more straightforward to operate than a pneumatic switch and can be coupled to pneumatic or electric actuators for automated operation.

An intermediate cryogenic trap is essential whenever a band-refocusing mechanism is required as part of the switching process [137]. Examples include sample transfer from packed to capillary columns, preconcentration by multiple injections, and accurate determinations of retention index values on the second column. The trap is usually a short length of platinum/iridium tubing, some fused-silica tubing, or a short portion of the second column. In some cases the trap may be packed or coated with liquid phase to enhance the recovery of volatile analytes. It is necessary to provide a temperature gradient within the trap, otherwise breakthrough is observed, due to aerosol formation. A trap should not only effectively retain the substances that are directed into it but should also be of low mass so that it can be rapidly heated so that the trapped fraction is instantaneously introduced into the second column. The trap is usually cooled by circulating the vapors from dry ice or liquid nitrogen through a shroud surrounding the trap. The trap is heated either by forced convection with preheated nitrogen, by ballistic temperature-programing

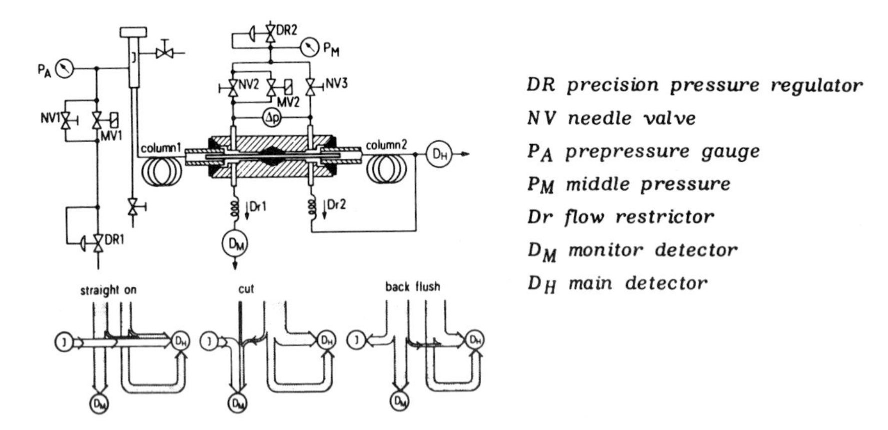

Fig. 9.10. Schematic diagram of a dual-oven coupled-column gas chromatograph with live switching between two capillary columns. (Reproduced with permission from Siemens AG.)

of the column oven, containing the trap, or in the case of metal traps, by resistance heating with a high-amperage circuit.

One design of a commercially available two-oven coupled-column gas chromatograph, which employs live switching for sample transfer, is shown in Fig. 9.10 [137,143-145]. The direction of flow through the double T-piece is controlled by slight pressure differences applied between its ends (Δp in Fig. 9.10). The flow through the makeup gas lines is adjusted with the two needle valves, NV2 and NV3, and the pressure difference, ΔP, is indicated by the manometer, P_M. The T-piece is constructed in such a way that the gas must pass through the connecting capillary, which is inserted loosely into both the first and second column. Effluent gas flow from the first column can be directed to a detector via a restrictor, Dr1, or through the second column and then to a detector. Moreover, the first column can be backflushed by activating solenoid valves MV1 and MV2. Quantitative sample transfer occurs when the switch time is equal to ca. 6 times the peak standard deviation, measured symmetrically around the peak center [146]. For analytical peaks that are overlapped by matrix components, a compromise must be made between the size and position of the fraction switched. If overlap on the second column is to be avoided, quantitative sample transfer may not be possible.

9.6.3.2 Liquid chromatography/gas chromatography

The separation mechanisms in liquid chromatography and gas chromatography are complementary, providing a powerful combination for the separation of complex mixtures. The combined technique is well suited to the analysis of a single component or of components of similar characteristics, that form a single peak or narrow fraction in the liquid chromatogram. This fraction can then be transferred directly to the gas chromatograph, avoiding manual manipulations and reducing the possibility of contamination or losses during solvent removal. Since the final determination of the analyte is by well-established gas-chromatographic detectors, greater sensitivity and selectivity as well as more straightforward interfacing with a mass spectrometer is possible than is the case for liquid chromatography. A more detailed discussion of the advantages of LC/GC has been given by Grob [147,148]. Because of certain incompatibilities of the direct coupling of LC to GC there has been very little interest in this technique until recently [147-150]. Separations by liquid chromatography occur with concurrent dilution of the sample and thus require a fairly large volume of solvent to be transferred to the gas chromatograph. That volume may easily exceed the conventional injection volume by several hundred fold. This problem was only solved by the development of miniaturized liquid-chromatographic systems and immobilized stationary phases and retention gaps in gas chromatography. One approach to quantitative eluent transfer is to miniaturize the LC system by using a packed capillary column and injecting a larger volume of sample into the gas chromatograph than is conventionally applied [149,151]. In this case, a multifunctional valve and a short retention gap was used as the interface, and the LC eluent flowrates were generally $< 50\,\mu$l/min. Using concurrent evaporation of the solvent, a fairly large section of the chromatogram can be transferred to the retention gap while maintaining a well-defined

References on p. A442

solute band for reinjection. The major limitations of this approach are associated with the specialized technology of the LC column and instrument, the need to maintain low dead-volumes in all regions of the chromatograph and interface, and the small sample size tolerated by the packed-capillary column.

To extend the LC/GC coupling to wider-bore liquid-chromatographic columns a short retention gap with complete or partial concurrent solvent evaporation is the most convenient approach [148,152,153]. During concurrent solvent evaporation, the eluent is introduced into the retention gap at a speed allowing immediate evaporation and discharge of the solvent through the column. The length of the flooded zone is generally < 1 m, and short retention gaps of 1-2 m are adequate, independent of the volume of eluent introduced. Optimum conditions for concurrent solvent evaporation are high carrier gas flowrates and an injection (oven) temperature close to the boiling point of the solvent at the prevailing inlet pressure. Under these conditions, solvent trapping and phase soaking are inefficient band reconcentration mechanisms, and coldtrapping must be relied upon for refocusing of the initial solute bands. Consequently, sharp peaks will be obtained only for solutes with a boiling point at least 50-100°C higher than the boiling point of the eluent. The use of concurrent solvent evaporation, therefore, is limited to the analysis of solutes of moderate and low volatility.

Partial concurrent solvent evaporation is a hybrid of the two methods discussed so far. In this case, the eluent is introduced at a rate exceeding the amount of simultaneously evaporating solvent and, hence, there is a steadily expanding flooded zone during eluent introduction. However, the rates of eluent introduction and solvent evaporation are adjusted to each other to cause substantial concurrent solvent evaporation and a corresponding shortening of the flooded zone. Concurrent solvent evaporation is simplified if the carrier gas is introduced behind the plug of liquid such that the carrier gas pushes the liquid into the thermostated retention gap [152]. This is easily achieved by using a sample loop and switching valve to isolate the column fraction of interest from the liquid chromatograph and then transferring it as a plug to the retention gap. The column temperature must be high enough to cause the solvent vapor pressure to exceed the carrier gas inlet pressure, with the effect that the solvent vapor pressure at the front of the plug prevents the liquid from flowing far into the thermostated column. Without loop-type introduction, carrier gas and eluent are mixed in the retention gap. This requires a careful balance between carrier gas flowrate, sample introduction rate, and temperature to establish a steady-state situation.

9.6.3.3 Supercritical-fluid extraction/gas chromatography

The revived interest in supercritical-fluid chromatography has stimulated interest in integrated systems where supercritical-fluid extraction is combined with GC and LC for on-line analysis [150,154,155]. Here, the principal attraction of supercritical-fluid extraction is the possibility of performing extractions more rapidly than with more conventional methods, such as Soxhlet extraction, the ability to control the extraction efficiency for various analytes by changing the fluid density (selective isolation is also possible in this case), and the simplicity with which the fluid phase can be removed by depressurization. However, in common with all extraction procedures, careful optimization is required for quantitative recovery of the analytes and for minimizing matrix effects. As far as the criteria for the design of a suitable interface for coupling to GC is concerned, the rate of analyte extraction and the sample size to be extracted are important considerations. For small extraction cells (volume ca. 0.2 ml), which can be conveniently fabricated from a modified column endfitting, and an extraction time of about 10 min, a small-bore capillary is an adequate interface. The capillary can be inserted directly into the open-tubular column of the gas chromatograph through a standard on-column injection port. The GC oven is cooled to – 30°C during the extraction to coldtrap the analytes at the head of the separation column. Wright et al. [156] used a similar approach with a 15-μl extraction cell and a retention gap for sample focusing, thus eliminating the need for cryogenic cooling. A switching valve enabled the sample extract to be either collected or sent to the GC column for analysis. For both systems there was no additional bandbroadening attributable to the interface.

9.6.4 Detectors

Numerous methods have been described for detecting organic vapors in the effluent from a gas chromatograph [1,4,5,157-160]. The principal methods of detection are based on ionization, bulk physical property, optical, or electrochemical detection. In fact, most of

TABLE 9.7

CHARACTERISTIC PROPERTIES OF COMMON GAS-CHROMATOGRAPHIC DETECTORS

Detector	Minimum detection limit	Linear response range	Selectivity
Thermal conductivity	3×10^{-9} g/ml	10^4	
Flame ionization	10^{-12} g/sec	10^6	
Thermionic ionization	10^{-13} g/sec (N)	10^4	4×10^4 g C/g N
	10^{-14} g/sec (P)		7×10^4 g C/g P 0.5 g N/g P
Electron capture	10^{-13} g/ml	10^4	
Flame photometric	10^{-11} g/sec (S)		10^3-10^6 g C/g S
	10^{-12} g/sec (P)	10^5 (P)	5×10^5 g C/g P
Electrolytic conductivity	10^{-12} g/sec (N)	10^3-10^5	10^4-10^9 g C/g (N, S, or Cl).
	10^{-13} g/sec (Cl)		
Photoionization	0.3 coulombs/g	10^7	

References on p. A442

the widely used detectors for gas chromatography have been in use for many years and have undergone only evolutionary changes in recent years. Emerging detectors include chemiluminescence and plasma atomic emission detectors, which have been commercialized in recent years in a form convenient for routine use at an affordable price. On-line gas chromatography with mass spectrometry is now used routinely and well supported by commercial vendors. Gas chromatography with Fourier transform infrared spectroscopy is moving closer to becoming a routine method. The last two detectors are rich in qualitative information from which the structure of unknowns can be deduced.

Individual detectors for gas chromatography are usually compared in terms of their operational characteristics, defined by the minimum detectable quantity of standards, the selectivity-response ratio between standards of different composition or structure, and the range of the linear portion of the detector-response calibration curve. Typical figures of merit for the most common detectors are summarized in Table 9.7.

9.6.4.1 Ionization-based detectors

At the temperatures and pressures generally used in gas chromatography the common carrier gases employed behave as perfect insulators. In the absence of conduction by the gas molecules themselves, the increased conductivity due to the presence of very few charged species is easily measured, providing the high sensitivity characteristic of ionization-based gas-chromatographic detectors [161]. Examples of ionization detectors in current use include the flame-ionization detector (FID), thermionic ionization detector (TID), photoionization detector (PID), the electron-capture detector (ECD), and the helium-ionization detector (HID). Each detector employs a different method of generating an ion current, but in all cases the signal corresponds to the fluctuation of this ion current in the presence of the volatilized sample in the carrier gas stream.

The response of the FID results from the combustion of organic compounds in a small hydrogen/air diffusion flame. The mechanism of ion production is only poorly understood. The ionization efficiency of the flame is not particularly high, perhaps a few ions per million molecules, consistent with the fact that the thermal energy of the flame is too low to explain the production of ions. It is generally believed that organic compounds are ionized by a chemical ionization mechanism. Energy released in strongly exothermic reactions is retained by one of the fragments as internal vibrational energy and leads to ionization before thermal randomization of the energy occurs. The ionization process would thus be a first-order reaction, and this explains the linear response of the FID. The ionization mechanism would be placed in a low-probability reaction pathway, explaining the low ion yields. Two steps are thought to be important in the above process: radical formation, requiring the absence of oxygen, and chemical ionization of radicals formed by excited atomic or molecular oxygen states. At the end of a chain of reactions, in which methane or the methyl radical is thought to be a key intermediate, the dominant ion-producing reaction is

$$\dot{C}H + O^{*} \rightarrow CHO^{+} + e^{-} \tag{9.7}$$

As a consequence of the FID mechanism, each carbon atom capable of hydrogenation yields the same signal, and the overall FID response to the analyzed substance is proportional to the sum of the "effective" carbon atoms. The FID response is highest for hydrocarbons, being proportional to the number of carbon atoms, while substances containing nitrogen, sulfur, or halogens yield smaller responses, depending upon the heteroatom/carbon character and the electron affinity of the combustion products. The lower response can be explained by a complex series of recombination reactions and electron-capture processes, resulting in a lower ionization current.

A nearly universal response to organic compounds, high sensitivity, long-term stability, simplicity of operation and construction, low deadvolume, fast signal response, and exceptional linear response range have contributed to the FID being the most popular detector in current use. Only the fixed gases (e.g., He, Xe, H_2, N_2), certain nitrogen oxides (e.g., N_2O, NO), compounds containing a single carbon atom, bonded to oxygen or sulfur (e.g. CO_2, CS_2, COS), inorganic gases (e.g., NH_3, SO_2), water, and formic acid do not provide a significant detector response.

The current generation of thermionic ionization detectors (TID) grew out of earlier studies on the properties of alkali-metal-doped flame-ionization detectors [159,162]. Adding an alkali-metal salt to a flame (or plasma in modern detectors) enhances the response of the detector to compounds containing certain elements, such as nitrogen, phosphorus, sulfur, boron, and the halogens. The detector is most commonly used for the selective detection of nitrogen- and phosphorus-containing compounds in complex environmental and biological samples, owing to its favorable response to these compounds and large discrimination against other organic compounds.

All modern TIDs employ a solid surface, composed of a ceramic or glass matrix, doped with an alkali-metal salt, molded onto an electrical heater wire [163,164]. The detector, originally proposed by Kolb et al. [163], contains an electrically heated rubidium silicate bead, situated a few millimeters above the detector jet tip and below the collector electrode. The temperature of the source is independently controlled and based on a negative voltage that dampens the response of the detector in the FID mode. The TID design proposed by Patterson [164] consists of an alkali-metal-doped ceramic cylinder, containing an embedded heater surrounded by a cylindrical collector electrode. The ceramic thermionic emitter is biased at a negative potential with respect to the collector electrode, and it is heated to a surface temperature of 400-800°C, depending on the mode of detection. The response of the detector to different elements depends on the electronic work function of the thermionic surface, the chemical composition of the gas environment immediately surrounding the thermionic surface, and the operating temperature of the thermionic surface.

Several models have been proposed to account for the selectivity of the TID response to nitrogen- and phosphorus-containing compounds [159,161,164-167]. They differ principally in whether the interaction between the alkali-metal atoms and organic fragments occurs as a homogeneous reaction in the gas phase or is purely a surface phenomenon.

References on p. A442

Recent studies have tended to emphasize the complexity of the process rather than to indicate a definitive model.

Early photoionization detectors were operated at low pressures to maximize the photon intensity but this was not very convenient. The PIDs only became popular after the introduction of detectors that physically separated the source and ionization chamber from each other, allowing independent optimization of ion production and ion collection [159,168,169]. The detector is selective by virtue of the availability of several sources of different energy. Ionization will occur only if the source energy is close to or exceeds that of the ionization potential of the sample. The UV source is a discharge lamp, containing an inert gas or gas mixture at low pressure, that emits monochromatic light of a specific energy, depending on the choice of fill gases and the window material. The discharge compartment is mechanically separated from the ionization chamber by an optically transparent window made of metal fluoride. The effluent from the column passes through the thermostated ionization chamber and between two electrodes, positioned at opposite ends of the chamber. Detectors with ionization chamber volumes of 40 and 175 μl are available for use with capillary columns and of 175 and 225 μl for packed columns.

The PID is nondestructive, relatively inexpensive, of rugged construction, and easy to operate. With the 10.2-eV photon source most molecules are ionized; the exceptions are the permanent gases, C_1-C_4 hydrocarbons, methanol, acetonitrile, and chloromethanes. The PID is 5 to 10 times more sensitive than the FID for alkanes and about 50 times so for aromatic compounds [168,169]. Freedman [170] has shown that the ionization potential of the molecule is the most important single factor determining the PID response, the relative number of pi-electrons having little significance. This argument is well supported by the available experimental data, although other authors claim a much greater role for pi-electrons in the ionization process [171]. The most comprehensive collection of response data, relative to benzene, for more than a hundred compounds has been compiled by Langhorst [172]. Since the PID requires no auxiliary or support gases, it can be used in environments where combustion gases may be considered hazardous or in portable instruments, where the additional weight of several gas bottles is undesirable.

The electron-capture detector is one of the easiest-to-operate but least-understood of the ionization detectors. It is very sensitive to certain compounds, but its response can be rendered unreliable by sidereactions involving contaminants in the carrier gas, bleed from the column, or saturation by sample components. Operational characteristics, such as detector temperature, method and mode of collecting electrons, and detector design also affect the response of the detector. The detector consists of an ionization chamber with two electrodes and an isotopic source of beta-electrons (^{63}Ni, Ti^3H_2, etc.) [173-175]. The detector baseline current results from the bombardment of the carrier gas by beta-electrons forming a plasma of positive ions, radicals, and thermal electrons. The application of a potential difference to the ionization chamber, either continuously or pulsed, allows the thermal electrons to be collected. When an electron-capturing compound enters the ionization chamber, it captures a thermal electron to produce a negative ion. The diminution in background current due to the formation of negative ions, subsequently neutralized by positive ions, constitutes the quantitative basis by which detector response is related to solute concentration.

Miniaturization of the ionization chamber is important for use with open-tubular columns [175]. The effective lower limit of current designs is approximately 100-400 μl – still too large to eliminate completely extracolumn bandbroadening [176]. The effective detector deadvolume can be reduced by adding makeup gas at the end of the column to preserve column efficiency at the expense of some loss in detector sensitivity due to sample dilution. Since some ECDs designed for packed columns have cell volumes of 2.0 to 4.0 ml, they may not be suitable for use with open-tubular columns.

The majority of the commercially available ECDs are designed for use with a modified version of the pulsed-sampling technique, termed the variable-frequency constant-current mode [177]. Instead of measuring the cell current at a constant pulse frequency, the cell current is fixed with respect to a reference value, and the frequency of the pulse is changed so that the difference between the cell current and the reference current is zero throughout the chromatographic separation. Since pulse frequency is the variable quantity in this mode of operation, the detector signal is a voltage proportional to the frequency. The two principal advantages of this method are that the linear response range is approximately 10^4 to 10^5, much greater than for the constant-frequency-pulse sampling mode, and secondly, the detector operation is less disturbed by traces of interferents entering the detector cell. A disadvantage of this mode of operation is that for compounds with ultrafast electron-attachment-rate constants (e.g., between 2.8×10^{-7} and 4.6×10^{-7} ml/molecule x sec) the detector response is nonlinear [178-180]. Examples of compounds with ultrafast rate constants are: CCl_4, SF_6, $CFCl_3$, and CH_3I. The ECD is most responsive to these types of compounds and would normally be selected for their determination. For compounds with rate constants less than the above, the detector response is normally linear over the full operating range. This group includes most of the moderately strong and weak electrophores that make up the bulk of compounds determined with the ECD. It has been shown that operating the constant-current ECD at moderately fast baseline frequencies will linearize the low-concentration end of the response range for compounds with ultrafast rate constants at the expense of a reduction of the dynamic response range [180].

The response of the ECD to organic compounds covers approximately seven orders of magnitude [173-175]. The detector responds most strongly to compounds containing halogens or nitro groups, to organometallic compounds, and to conjugated electrophores. This latter group is structurally the least well-defined, and is comprised of compounds containing two or more weakly electron-capturing groups, connected by some specific bridge that promotes a synergistic interaction between the two groups [181]. Examples of conjugated systems with a high detector response include conjugated carbonyl compounds (benzophenones, quinones, phthalate esters, coumarins), some polycyclic aromatic hydrocarbons, some sulfonamides, and certain steroids. The response of the ECD to halocarbon compounds decreases in the order $I > Br > Cl >> F$ and increases synergistically with multiple substitution on the same carbon atom. The identity and number of halogen substituents is more important than subtle variations in the

References on p. A442

geometric framework of the alkyl portion of the molecule, although even these small changes will have a measurable influence on the detector response in many cases [174]. The response of the ECD to haloaromatic and nitroaromatic compounds shows similar trends to that for the alkyl compounds. Much of what is known about the structure response of the ECD is based on empirical observations. Clearly, the ability to correlate the response of the detector to fundamental molecular parameters would be useful. Chen and Wentworth [174,182] have shown that the information required for this purpose is the electron affinity of the molecule, the rate constant for the electron attachment reaction and its activation energy, and the rate constant for the ionic recombination reaction. In general, the direct calculation of detector response factors has rarely been carried out, since the electron affinities and rate constants for most compounds of interest are unknown.

The helium ionization detector is probably the least used of the ionization detectors. It is a universal and ultrasensitive detector, primarily used for the trace analysis of permanent gases and some volatile organic compounds that have a poor response to the FID and are present in too low a concentration for detection with the thermal conductivity detector [183]. The principal mechanism for analyte response is ionization due to collision with metastable He atoms, created by bombarding He atoms with beta-electrons in an ionization chamber. Since the ionization potential of He (19.8 eV) is higher than that of all other species except Ne, all species entering the ionization chamber will be ionized. Some similarities with the design of the ECD are obvious, except that the specific activity of the radioisotopic source is usually much higher and the field potential within the cell is increased to the multiplication region. The polarity, linearity, and sensitivity of the detector response are profoundly influenced by detector contamination levels. The HID has a reputation as being among the more difficult of the gas-chromatographic detectors to operate reliably for long periods of time or when continual changes of the chromatographic conditions are required. To avoid problems caused by column bleed the HID has frequently been used in GSC.

9.6.4.2 Bulk physical property detectors

The bulk physical property detectors respond to some difference in the properties of the mobile phase due to the presence of the analyte. Usually, a large signal for some response characteristic of the mobile phase is required to provide a reasonable working range, but the sensitivity of such devices is invariably noise-limited, as small analyte concentrations are reached. The most important of the bulk physical property detectors are the thermal conductivity detector (TCD), the gas density balance, and the ultrasonic detector. Of these, only the TCD is widely used, primarily for those compounds which have an inadequate response for detection by a FID. In recent years, there have been few changes in the basic design of the TCD [1,5,184-186], which consist of a heated, thermostated cavity that contains a sensing element, either a heated metal wire or thermistor, through which carrier gas flows. With pure carrier gas flowing through the cavity the heat loss from the sensor is a function of the temperature difference between the sensor and cavity and depends on the thermal conductivity of the carrier gas. When an organic solute enters the cavity, there is a change in thermal conductivity of the carrier gas and a resultant change in the temperature of the sensor. This temperature change in the sensor results in an out-of-balance signal in a Wheatstone bridge circuit, which constitutes the detector signal. A novel detector design by Hewlett-Packard employs a single-filament TCD and uses flow modulation to switch the carrier gas flow between two channels, one of which contains a filament. Every 100 msec a switching valve fills the filament channel alternately with carrier gas and column effluent. No reference column is required and, with an effective detector volume of 3.5 μl, it can be used with even narrow-bore capillary columns. Digital data processing is included, and more reliable temperature compensation provides greater sensitivity and stability with either packed or open-tubular columns than that obtained with conventional detector designs.

Theoretical models have been advanced to explain the response characteristics of the TCD under different operating conditions [186]. These models take into account the effects of conduction, convection, and radiation on the loss of heat from the sensor but do not lead to any simple mathematical expression to describe the operation of the detector. Several compilations of relative response data for the TCD are available [181,187]. These values are usually expressed on a weight or molar response basis relative to benzene. They depend on the nature of the carrier gas used for their determination but are generally sufficiently accurate for approximating sample concentrations. For precise quantitative analysis it is necessary to calibrate the detector for each substance determined.

9.6.4.3 Photometric detectors

The use of flames as atom reservoirs for the spectroscopic determination of elements is a well-established technique and is particularly valuable for metal analysis. Most nonmetallic compounds, which account for the majority of samples analyzed by GC, have their principal emission or absorption lines in the ultraviolet region, where flame background contributions are troublesome. In addition, the diffusion flames used in GC lack sufficient stability and thermal energy to be useful atom reservoirs. For direct optical emission detection, microwave-induced and inductively coupled plasmas provide more appropriate atom sources for organic compounds [188-190]. However, the determination of phosphorus and sulfur by a flame-photometric detector (FPD) is widely used in GC. For these elements, a hydrogen diffusion flame provides optimum excitation conditions. A number of chemiluminescent reaction detectors are also used in GC. They are based on pyrolytic or catalytic reactions, which release species that can be subsequently detected in a chemiluminescent reaction [191]. The thermal-energy analyzer is used as a selective detector for nitrosamines after low-temperature pyrolysis to release nitric oxide, which is determined by its chemiluminescent reaction with ozone. In the redox chemiluminescence detector the effluent from a gas-chromatographic column is dynamically mixed with a metered flow of dilute nitrogen dioxide and passed through a thermostated catalyst chamber. Compounds capable of reducing nitrogen dioxide to nitric oxide are subsequently detected by the chemiluminescent reaction between nitric oxide and ozone.

References on p. A442

The response mechanism of the FPD is known superficially, even if the finer details remain obscure [192]. In the low-temperature hydrogen-rich flames favored for use with the FPD, sulfur-containing compounds are combusted and interconverted by a large set of bimolecular reactions to species such as H_2S, HS, S, S_2, SO, and SO_2 in relative proportions that depend on the instantaneous and fluctuating flame chemistry. In the presence of carbon radicals in the flame various carbon/sulfur-containing species might also be formed. The presence of hydrocarbons eluted with the sulfur compounds or variations in the hydrogen/oxygen ratio can change the relative concentration of either H or OH radicals in the flame and will also alter the distribution of S species in the flame. The detector signal results from the formation of excited state S_2^* species, created by a number of possible three-body collisions. Only a small fraction of the total S entering the flame will be converted to S2 species, and this fraction depends on the experimental conditions. The detector response in the sulfur mode can be described by Eqn. 9.8

$$I(S_2) = A[S]^n \tag{9.8}$$

where $I(S_2)$ is the signal intensity, A an experimental constant, $[S]$ the mass flowrate of sulfur atoms, and n the exponential factor. The theoretical value for n is 2, but in practice values between 1.6 and 2.2 are frequently observed. Non-optimized flame conditions, incomplete combustion, hydrocarbon quenching, competing flame reactions that lead to de-excitation, and effects of sample structure all contribute to this deviation [193,194]. In the case of phosphorus, incoming P-containing compounds are first decomposed to PO molecules, which are then converted to HPO^* by three-body collisions, involving flame radicals. HPO^* is the emission species, and in this case, a linear dependence between detector response for P and the amount of sample entering the detector is expected and generally found.

Chemiluminescence reactions important for GC occur in the gas phase and are often combined with a reaction detector to liberate thermally or catalytically a species that is easy to detect [191]. By far the most important reaction is the release of nitric oxide, which is subsequently combined with ozone to generate electronically excited nitrogen dioxide. The instrumentation is simple, consisting of a reaction chamber and a light sensor (photomultiplier tube) with associated amplifiers, a system to generate the reagent gas, and possibly a vacuum pump to operate the reaction chamber at a reduced pressure. Reduced pressure operation has two advantages: it improves sensitivity by diminishing collisional deactivation of the excited reaction product and reduces the effective deadvolume of the detector, maintaining compatibility with the use of open-tubular columns. The redox chemiluminescence detector used to detect a wide range of volatile compounds in environmental samples employs a gold-coated, glass-bead catalyst. Many of the species that can be detected are oxygen-containing compounds, such as alcohols, ethers, ketones, and phenols, or sulfur-containing compounds, such as sulfides, mercaptans, hydrogen sulfide, and sulfur dioxide, that are difficult to detect reliably by other detectors in a complex hydrocarbon matrix [191,195].

Plasma sources are capable of exciting intense emission from the elements C, H, D, O, N, S, P, and the halogens, and are thus uniquely suited for the analysis of organic compounds [188-190]. The plasma consists of a mass of predominantly ionized gas at a temperature of 4000-10 000 K. This state can be maintained directly by an electrical discharge through the gas (DC plasma) or indirectly via inductive heating of the gas. The latter is established by an electromagnetic field, using power generated at radio frequencies (inductively coupled plasma) or microwave frequencies (microwave-induced plasma). Inductively coupled argon plasmas effectively decompose large amounts of organic materials but provide poor excitation efficiency for the nonmetallic elements that are of primary chromatographic interest. Helium is the preferred plasma gas, as it provides a simpler background spectrum, higher excitation energy, and improved linear response range. The development of the Beenakker plasma source has significantly improved the reliability and stability of both the low-pressure and atmospheric-pressure He plasma. Emission from the source is viewed by a rapid-scanning monochromator or polychromator. Certainly, one limitation of early equipment was the need for high-resolution monochromators and multiple detectors, leading to high equipment cost and fairly complex operating requirements. A recently introduced moderate-cost instrument from Hewlett-Packard makes use of a flat focal plane, medium-resolution polychromator, and a mechanically movable photodiode-array detector for monitoring up to four element channels simultaneously [196]. The photodiode-array technology used in this instrument, combined with real-time multipoint background correction, yields high selectivity for most elements. Detection limits vary from 2-120 pg/sec for individual nonmetallic elements with a dynamic range of 10^3 to 10^4. Research instruments may be capable of improved sensitivity and selectivity but are substantially more complex.

At plasma temperatures, molecular breakdown is complete, and the measured response for each element is thus proportional to the number of atoms in the plasma and independent of the structure of the parent compound. Ratioing of the element response channels during the passage of a chromatographic peak will, after response standardization with compounds of know elemental composition, enable the empirical formula of unknown compounds to be directly determined [190-197]. The precision of such measurements is limited by plasma instability, resulting, mainly, from fluctuations in the power input and plasma gas flowrates. In addition, accuracy is limited by the linearity of elemental responses and the possibility of incomplete compound destruction. Under favorable circumstances, empirical formulae can be determined with acceptable accuracy, on sample sizes much smaller than those needed for conventional combustion analysis.

9.6.4.4 Electrochemical detectors

There are several general problems to be solved before electrochemical detection can be applied to GC. First of all, few electrochemical detectors are gas-phase sensing devices, and the sample must, therefore, be transferred into solution for detection. Secondly, the majority of organic compounds separated by GC are neither electrochemically active nor highly conducting. The Hall electrolytic conductivity detector (HECD) [162,198] and

References on p. A442

the microcoulometric detector [199] solve both of these problems by decomposing the gas-phase sample into low-molecular-weight electrochemically active fragments that are readily soluble in a support solvent.

Sample decomposition is carried out either by pyrolysis or by catalytic oxidation or reduction in a low-volume flow-through tube furnace. For oxidation or reduction, O_2, air, or H_2 is mixed with the carrier gas leaving the column and passed over a catalyst, usually a Ni or a quartz tube with a Ni wire inside, maintained at a temperature of 500 to 1000°C. Organic compounds entering the furnace are decomposed into low-molecular-weight fragments, such as CO_2, H_2O, HX (where X = halogen), SO_2, CH_4, NH_3, and H_2S. A chemical scrubber is sometimes added to the flow system at the furnace exit to improve the selectivity of the detection process. Examples of chemical scrubbers include $Sr(OH)_2$ or K_2CO_3 to remove HX or SO_2 selectively, silver wire to remove HX or H_2S, alumina to remove PH_3, and aluminum silicate to remove SO_x. With careful selection of pyrolysis conditions, mode of operation, and chemical scrubber, electrochemical detection permits highly specific and sensitive element detection.

The most recent version of the HECD employs a series conductivity cell in which the conductivity of the solvent is monitored in the first portion of the cell, while the conductivity of the solvent plus that of the reaction products is measured in the second part of the cell. The output of the two conductivity cells are differentially summed. Changes in the concentration of conductivity solvent and temperature fluctuations, which represent the principal source of daily response variations, are thus minimized [200]. The selectivity and sensitivity of the detector response to individual elements can be improved by using an element-selective potentiometric electrode in place of the standard conductivity cell [201].

The HECD is capable of high sensitivity and selectivity, although optimizing detector conditions and maintaining constant sensitivity at low sample levels can be troublesome. Detector selectivity values are variable, depending on the heteroelement and the detector operating conditions, but values in the range of 10^4 to 10^9 for nitrogen-, chlorine-, or sulfur-containing compounds in the presence of hydrocarbons have been obtained. Compared to the FPD in the sulfur mode, the HECD has about the same or greater sensitivity and selectivity [202,203]. In comparison with the ECD, the HECD has a greater selectivity for chlorine-containing compounds and a more predictable response on a per-gram-of-chlorine basis [204].

REFERENCES

1 C.F. Poole and S.A. Schuette, *Contemporary Practice of Chromatography*, Elsevier, Amsterdam, 1984.
2 R.J. Laub and R.L. Pecsok, *Physiochemical Applications of Gas Chromatography,* Wiley, New York, 1978.
3 J.R. Condor and C.L. Young, *Physicochemical Measurements by Gas Chromatography,* Wiley, New York, 1979.
4 M.L. Lee, F.J. Yang and K.D. Bartle, *Open Tubular Column Gas Chromatography: Theory and Practice,* Wiley, New York, 1984.
5 R.L. Grob (Editor), *Modern Practice of Gas Chromatography,* Wiley, New York, 2nd Edn., 1985.

6 J.A. Jonsson (Editor), *Chromatographic Theory and Basic Principles*, Dekker, New York, 1987.
7 W.A. Jennings, *Analytical Gas Chromatography,* Academic Press, New York, 1987.
8 G. Guiochon and C.L. Guillemin, *Quantitative Gas Chromatography for Laboratory Analyses and On-Line Process Control,* Elsevier, Amsterdam, 1988.
9 C.F. Poole and S.K. Poole, *Anal. Chim. Acta,* 216 (1989) 109.
10 L.S. Ettre, *Chromatographia,* 18 (1984) 477.
11 T. Herraiz, G. Reglero, M. Herraix, R. Alonso and M.D. Cabezudo, *J. Chromatogr.,* 388 (1987) 325.
12 L.S. Ettre, *J. High Resolut. Chromatogr. Chromatogr. Commun.,* 8 (1985) 497.
13 W. Seferovic, J.V. Hinshaw and L.S. Ettre, *J. Chromatogr. Sci.,* 24 (1986) 374.
14 G. Reglero, M. Herraiz, M.D. Cabezudo, E. Fernandez-Sanchez and J.A. Garcia-Dominguez, *J. Chromatogr.,* 348 (1985) 338.
15 L.S., Ettre, *Chromatographia,* 18 (1984) 243.
16 W. Kimpenhaus, F. Richter and L. Rohrschneider, *Chromatographia,* 15 (1982) 577.
17 L. Rohrschneider and E. Pelster, *J. Chromatogr.,* 186 (1979) 249.
18 L.S. Ettre, *Chromatographia,* 12 (1979) 509.
19 M.Y.B. Othman, J.H. Purnell, P. Wainwright and P.S. Williams, *J. Chromatogr.,* 289 (1984) 1.
20 F. David, M. Proot and P. Sandra, *J. High Resolut. Chromatogr. Chromatogr. Commun.,* 8 (1985) 551.
21 R.V. Golovnya and T.A. Misharina, *J. High Resolut. Chromatogr. Chromatogr. Commun.,* 3 (1980) 4 and 51.
22 C.F. Poole and S.K. Poole, *Chem. Rev.,* 89 (1989) 377.
23 G.E. Bailescu and V.A. Ilie, *Stationary Phases in Gas Chromatography,* Pergamon Press, Oxford, 1975.
24 P.R. McCrea, in J.H. Purnell (Editor), *New Developments in Gas Chromatography,* Wiley, New York, 1973, p. 87.
25 J.A. Yancey, *J. Chromatogr. Sci.,* 23 (1985) 161.
26 S.R. Lowry, H.B. Woodruff and T.L. Isenhour, *J. Chromatogr. Sci.,* 14 (1976) 129.
27 S.K. Poole and C.F. Poole, *J. Chromatogr.,* 500 (1990) 329.
28 B.R. Kersten, S.K. Poole and C.F. Poole, *J. Chromatogr., 468 (1989) 235.*
29 S.K. Poole, B.R. Kersten and C.F. Poole, *J. Chromatogr.,* 471 (1989) 91.
30 B.R. Kersten and C.F. Poole, *J. Chromatogr.,* 452 (1988) 191.
31 E. Chang, B. de Bricero, G. Miller and S.J. Hawkes, *Chromatographia,* 20 (1985) 293.
32 B.R. Kersten, C.F. Poole and K.G. Furton, *J. Chromatogr.,* 411 (1987) 43.
33 C.F. Poole, S.K. Poole, R.M. Pomaville and B.R. Kersten, *J. High Resolut. Chromatogr. Chromatogr. Commun.,* 10 (1987) 670.
34 R.W. Souter, *Chromatographic Separations of Stereoisomers,* CRC Press, Boca Raton, FL, 1985.
35 W.A. Konig, *The Practice of Enantiomer Separation by Capillary Gas Chromatography,* Huethig, Heidelberg, 1987.
36 D.W. Armstrong and S.M. Han, *CRC Revs. Anal. Chem.,* 19 (1988) 175.
37 B. Koppenhoefer and E. Bayer, *Chromatographia,* 19 (1985) 123.
38 H. Frank, *J. High Resolut. Chromatogr. Chromatogr. Commun.,* 11 (1988) 787.
39 H. Frank, G. Nicholson and E. Bayer, *J. Chromatogr.,* 367 (1987) 187.
40 J. Gerhardt, G. Nicholson, H. Frank and E. Bayer, *Chromatographia,* 19 (1985) 251.
41 W.A. Konig, *J. High Resolut. Chromatogr. Chromatogr. Commun.,* 5 (1982) 588.
42 W.A. Konig and U. Sturm, *J. Chromatogr.,* 328 (1987) 357.
43 V. Shurig and R. Weber, *J. Chromatogr.,* 298 (1984) 321.
44 V. Schurig, U. Leyrer and R. Weber, *J. High Resolut. Chromatogr. Chromatogr. Commun.,* 8 (1985) 459.
45 Z. Witkiewicz, *J. Chromatogr.,* 466 (1989) 37.
46 W. Marciniak and Z. Witkiewicz, *J. Chromatogr.,* 324 (1985) 299 and 309.
47 K.E. Markides, H.-C. Chang, C.M. Schregenberger, B.J.Tarbet, J.S. Bradshaw and M.L. Lee, *J. High Resolut. Chromatogr. Chromatogr. Commun.,* 8 (1985) 516.

48 J.S. Bradshaw, C.M. Schregenberger, H.-C. Chang, K.E. Markides and M.L. Lee, *J. Chromatogr.,* 358 (1986) 95.
49 K.E. Markides, M. Nisheoka, B.J. Tarbet, J.S. Bradshaw and M.L. Lee, *Anal. Chem.,* 57 (1985) 1296.
50 G.J. Price, *Adv. Chromatogr.,* 28 (1989) 113.
51 R.J. Laub, in T. Kuwana (Editor), *Physical Methods of Modern Chemical Analysis,* Academic Press, New York, Vol. 3, 1983, p. 249.
52 M.Y.B. Othman, J.H. Purnell, P. Wainwright and P.S. Williams, *J. Chromatogr.,* 289 (1984) 1.
53 C.-F. Chien, M.M. Kopecni and R.J. Laub, *J. Chromatogr. Sci.,* 22 (1984) 1.
54 P. Sandra, F. David, M. Proot, G. Diricks, M. Verstaape and M. Verzele, *J. High Resolut. Chromatogr. Chromatogr. Commun.,* 8 (1985) 782.
55 R.R. Freeman and D. Kukla, *J. Chromatogr. Sci.,* 24 (1986) 392.
56 J.V. Hinshaw and L.S. Ettre, *Chromatographia,* 21 (1986) 561.
57 R.E. Kaiser, R.I. Reider, L. Leming, L. Blomberg and P. Kusz, *J. High Resolut. Chromatogr. Chromatogr. Commun.,* 8 (1985) 580.
58 J.V. Hinshaw and L.S. Ettre, *Chromatographia,* 21 (1986) 669.
59 H.T. Mayfield and S.N. Chesler, *J. High Resolut. Chromatogr. Chromatogr. Commun.,* 8 (1985) 595.
60 J.H. Purnell, J.R. Jones and M.-H. Wattan, *J. Chromatogr.,* 399 (1987) 99.
61 W.G. Jennings, *Comparison of Fused Silica and Other Glass Columns in Gas Chromatography,* Huethig, Heidelberg, 1981.
62 D.H. Desty, *J. Chromatogr. Sci.,* 25 (1987) 552.
63 H. Saito, *J. Chromatogr.,* 243 (1982) 189.
64 R.D. Dandeneau and E.H. Zerenner, *J. High Resolut. Chromatogr. Chromatogr. Commun.,* 2 (1979) 351.
65 K. Grob, *Making and Manipulating Capillary Columns for Gas Chromatography,* Huethig, Heidelberg, 1986.
66 B. Xu and N.P.E. Vermeulen, *J. Chromatogr.,* 445 (1988) 1.
67 M.W. Ogden and H.M. McNair, *J. High Resolut. Chromatogr. Chromatogr. Commun.,* 8 (1985) 326.
68 K.D. Bartle, B.W. Wright and M.L. Lee, *Chromatographia,* 14 (1981) 387.
69 M.W. Ogden and H.M. McNair, *J. Chromatogr.,* 354 (1986) 7.
70 K.D. Bartle, C.L. Woolley, K.E. Markides, M.L. Lee and R.S. Hansen, *J. High Resolut. Chromatogr. Chromatogr. Commun.,* 10 (1987) 128.
71 M.L. Lee and B.W. Wright, *J. Chromatogr.,* 184 (1980) 235.
72 S.C. Dhanesar, M.E. Coddens and C.F. Poole, *J. Chromatogr.,* 349 (1985) 249.
73 G. Schomburg, H. Husmann and H. Borwitzky, *Chromatographia,* 13 (1980) 321.
74 L. Blomberg and T. Wannman, *J. Chromatogr.,* 148 (1978) 379.
75 R.C.M. de Nijs, J.J. Franken, R.P.M. Dooper, J.A. Rijks, H.J.J.M. de Ruwe and F.L. Schulting, *J. Chromatogr.,* 167 (1978) 231.
76 T. Welsch, *J. High Resolut. Chromatogr. Chromatogr. Commun.,* 11 (1988) 471.
77 C.L. Woolley, K.E. Markides and M.L. Lee, *J. Chromatogr.,* 367 (1986) 9.
78 G. Schomburg, H. Husmann, S. Ruthe and T. Herraiz, *Chromatographia,* 15 (1982) 599.
79 M. Hetem, G. Rutten, L. van de Ven, J. de Haan and C. Cramers, *J. High Resolut. Chromatogr. Chromatogr. Commun.,* 11 (1988) 510.
80 K. Janak, V. Kahle, K. Tesarik and M. Horka, *J. High Resolut. Chromatogr. Chromatogr. Commun.,* 8 (1985) 843.
81 R.C. Kong and M.L. Lee, *J. High Resolut. Chromatogr. Chromatogr. Commun.,* 6 (1983) 319.
82 B. Xu, N.P.E. Vermeulen and J.A.M. Smit, *Chromatographia,* 22 (1986) 213.
83 J.L. Marshall and D.A. Parker, *J. Chromatogr.,* 122 (1976) 425.
84 G. Redant, P. Sandra and M. Verzele, *Chromatographia,* 15 (1982) 13.
85 G. Schomburg and H. Husmann, *Chromatographia,* 8 (1975) 517.
86 L.G. Bolmberg and K.E. Markides, *J. High Resolut. Chromatogr. Chromatogr. Commun.,* 8 (1985) 632.

87 J.S. Bradshaw, M.W. Adams, R.S. Johnson, B.J. Tarbet, C.M. Schregenberger, M.A. Pulsipher, M.B. Andrus, K.E. Markides and M.L. Lee, *J. High Resolut. Chromatogr. Chromatogr. Commun.,* 8 (1985) 678.
88 F. David, P. Sandra and G. Diricks, *J. High Resolut. Chromatogr. Chromatogr. Commun.,* 11 (1988) 256.
89 S.R. Lipsky and M.L. Duffy, *J. High Resolut. Chromatogr. Chromatogr. Commun.,* 9 (1986) 725.
90 B.W. Wright, P.A. Peaden, M.L. Lee and T.J. Stark, *J. Chromatogr.,* 248 (1982) 17.
91 S.R. Lipsky and W.J. Murray, *J. Chromatogr.,* 239 (1982)61.
92 B.E. Richter, J.C. Kuel, J.I. Shelton, L.W. Castle, J.S. Bradshaw and M.L. Lee, *J. Chromatogr.,* 279 (1983) 21.
93 J. Buijten, L. Blomberg, S. Hoffmann, K. Markides and T. Wannman, *J. Chromatogr.,* 289 (1984) 143.
94 G. Schomburg, H. Husmann, S. Ruthe and M. Herraiz, *Chromatographia,* 15 (1982) 599.
95 F.L. Bayer, *J. Chromatogr. Sci.,* 24 (1987) 549.
96 J.Q. Walker, S.F. Spencer and S.M. Sonchik, *J. Chromatogr. Sci.,* 23 (1985) 555.
97 C.J. Cowper and A.J. DeRose, *The Analysis of Gases by Chromatography,* Pergamon Press, Oxford, 1985.
98 J.N. Driscoll, *Crit. Rev. Anal. Chem.,* 17 (1987) 193.
99 R.F. Severson, R.F. Arrendale and O.T. Chortyk, *J. High Resolut. Chromatogr. Chromatogr. Commun.*, 3 (1980) 11.
100 H.J. Spencer, *J. Chromatogr.,* 260 (1983) 164.
101 G. Schomburg, *J. Chromatogr. Sci.,* 21 (1983) 97.
102 F. Munari and S. Trestianu, *J. Chromatogr.,* 279 (1983) 457.
103 G.D. Reed and R.J. Hunt, *J. High Resolut. Chromatogr. Chromatogr. Commun.,* 9 (1986) 341.
104 O.K. Guha, *J. Chromatogr.,* 292 (1984) 57.
105 J. Roeraade, G. Flodberg and S. Blomberg, *J. Chromatogr.*, 322 (1985) 55.
106 K. Grob and G. Grob, *J. High Chromatogr. Chromatogr. Commun.,* 2 (1979) 109.
107 K. Grob, *Classical Split and Splitless Injection in Capillary Gas Chromatography,* Huethig, Heidelberg, 1986.
108 K. Grob, *On-Column Injection in Capillary Gas Chromatography,* Huethig, Heidelberg, 1987.
109 J.V. Hinshaw, *J. Chromatogr. Sci.,* 25 (1987) 49.
110 J.V. Hinshaw, *J. Chromatogr. Sci.,* 26 (1988) 142.
111 P. Sandra (Editor), *Sample Introduction in Capillary Gas Chromatography,* Huethig, Heidelberg, Vol. 1, 1985.
112 G. Schomburg, U. Hausing and H. Husmann, *J. High Resolut. Chromatogr. Chromatogr. Commun.,* 8 (1985) 566.
113 A.E. Kaufman and C.E. Polymeropoulos, *J. Chromatogr.,* 454 (1988) 23.
114 J. Bowermaster, *J. High Resolut. Chromatogr. Chromatogr. Commun.,* 11 (1988) 802.
115 K. Grob, M. Biedermann and Z. Li, *J. Chromatogr.,* 448 (1988) 387.
116 K. Grob, Th. Laubli and B. Brechbuhler, *J. High Resolut. Chromatogr. Chromatogr. Commun.,* 11 (1988) 462.
117 K. Grob and M. Biedermann, *J. High Resolut. Chromatogr. Chromatogr. Commun.,* 12 (1989) 89.
118 C.A. Saravalle, F. Munari and S. Trestianu, *J. Chromatogr.,* 279 (1983) 241.
119 K. Grob, *J. Chromatogr.,* 279 (1983) 225.
120 K. Grob, *J. Chromatogr.,* 324 (1985) 251.
121 J.V. Hinshaw and W. Seferovic, *J. High Resolut. Chromatogr. Chromatogr. Commun.,* 9 (1986) 69.
122 C.A. Saravalle, F. Munari and S. Trestianu, *J. High Resolut. Chromatogr. Chromatogr. Commun.,* 10 (1987) 288.
123 K. Grob and Z. Li, *J. High Resolut. Chromatogr. Chromatogr. Commun.,* 11 (1988) 626.

124 M. Termania, B. Lacomblez and F. Munari, *J. High Resolut. Chromatogr. Chromatogr. Commun.*, 11 (1988) 890.
125 H.-J. Stan and H.M. Muller, *J. High Resolut. Chromatogr. Chromatogr. Commun.*, 11 (1988) 140.
126 F.I. Onuska, R.J. Kominar and K. Terry, *J. Chromatogr. Sci.*, 21 (1983) 512.
127 F. Pacholec and C.F. Poole, *Chromatographia,* 18 (1984) 234.
128 M. Termonia, F. Munari and P. Sandra, *J. High Resolut. Chromatogr. Chromatogr. Commun.*, 10 (1987) 263.
129 K. Grob and T. Laubi, *J. Chromatogr.*, 357 (1986) 345 and 357.
130 L. Ghaoui, F.-S. Wang, H. Shanfield and A. Zlatkis, *J. High Resolut. Chromatogr. Chromatogr. Commun.*, 6 (1983) 497.
131 K. Grob, *J. Chromatogr.*, 251 (1982) 235.
132 K. Grob and B. Schilling, *J. Chromatogr.*, 391 (1987) 3.
133 K. Grob, G. Karrer and M.-L. Riekkola, *J. Chromatogr.*, 334 (1985) 129.
134 K. Grob, *J. High Resolut. Chromatogr. Chromatogr. Commun.*, 7 (1984) 461.
135 K. Grob, *J. Chromatogr.*, 328 (1985) 55.
136 D.E. Willis, *Adv. Chromatogr.*, 28 (1989) 65.
137 G. Schomburg, F. Weeke, F. Muller and M. Oreans, *Chromatographia,* 16 (1982) 87.
138 D.R. Deans, *J. Chromatogr.*, 203 (1981) 19.
139 B.M. Gordon, C.E. Rix and M.F. Borgerding, *J. Chromatogr. Sci.*, 23 (1985) 1.
140 R.J. Philips, K.A. Knauss and R.R. Freeman, *J. High Resolut. Chromatogr. Chromatogr. Commun.*, 5 (1982) 546.
141 S.M. Sonchik, *J. Chromatogr. Sci.*, 22 (1986) 22.
142 S.T. Adam, *J. High Resolut. Chromatogr. Chromatogr. Commun.*, 11 (1988) 85.
143 M. Oreans, F. Muller and D. Leonhardt, *J. Chromatogr.*, 279 (1983) 357.
144 G. Schomburg, H. Husmann and E. Hubinger, *J. High Resolut. Chromatogr. Chromatogr. Commun.*, 8 (1985) 395.
145 M. Ahnoff, M. Ervik and L. Johansson, *J. Chromatogr.*, 394 (1987) 419.
146 J.F.K. Huber, E. Kenndler, W. Nyiry and M. Oreans, *J. Chromatogr.*, 247 (1982) 211.
147 K. Grob and B. Schilling, *J. High Resolut. Chromatogr. Chromatogr. Commun.*, 8 (1985) 726.
148 K. Grob, *Trends Anal. Chem.*, 8 (1989) 162.
149 H.J. Cortes (Editor), *Multidimensional Chromatography: Techniques and Applications,* Dekker, New York, 1989.
150 I.L. Davies, M.W. Raynor, J.P. Kithinji, K.D. Bartle, P.T. Williams and G.E. Andrews, *Anal. Chem.*, 60 (1988) 683A.
151 H.J. Cortes, C.D. Pfeiffer and B.E. Richter, *J. High Resolut. Chromatogr. Chromatogr. Commun.*, 8 (1985) 469.
152 F. Munari and K. Grob, *J. High Chromatogr. Chromatogr. Commun.*, 11 (1988) 172.
153 D. Duquet, C. Dewaele, M. Verzele and S. Mckinley, *J. High Resolut. Chromatogr. Chromatogr. Commun.*, 11 (1988) 824.
154 S.B. Hawthorne, M.S. Krieger and D.J. Miller, *Anal. Chem.*, 60 (1988) 472.
155 S.B. Hawthorne and D.J. Miller, *J. Chromatogr.*, 403 (1987) 63.
156 B.W. Wright, S.R. Frye, D.G. McMinn and R.D. Smith, *Anal. Chem.*, 59 (1987) 640.
157 E.R. Adlard, *CRC Crit. Revs. Anal. Chem.*, 5 (1975) 1 and 13.
158 L.S. Ettre, *J. Chromatogr. Sci.*, 16 (1978) 396.
159 M. Dressler, *Selective Gas Chromatographic Detectors,* Elsevier, Amsterdam, 1986.
160 E. Katz (Editor), *Quantitative Analysis Using Chromatographic Techniques,* Wiley, New York, 1987.
161 P.L. Patterson, *J. Chromatogr. Sci.*, 24 (1986) 466.
162 R.C. Hall, *CRC Crit. Revs. Anal. Chem.*, 8 (1978) 323.
163 B. Kolb, M. Auer and P. Pospisil, *J. Chromatogr. Sci.*, 15 (1977) 53.
164 P.L. Patterson, *J. Chromatogr. Sci.*, 24 (1986) 41.
165 K. Olah, A. Szoke and Zs. Vajta, *J. Chromatogr. Sci.*, 17 (1979) 497.
166 P. van de Weijer, B.H. Zwerver and R.J. Lynch, *Anal. Chem.*, 60 (1988) 1380.
167 C.S. Jones and E.P. Grimsrud, *J. Chromatogr.*, 409 (1987) 139.
168 J.N. Davenport and E.R. Adlard, *J. Chromatogr.*, 290 (1984) 13.

169 J.N. Driscoll, *J. Chromatogr. Sci.*, 23 (1985) 488.
170 A.N. Freedman, *J. Chromatogr.*, 236 (1982) 11.
171 M.K. Casida and K.C. Casida, *J. Chromatogr.*, 200 (1980) 35.
172 M.L. Langhorst, *J. Chromatogr. Sci.*, 19 (1981) 98.
173 E.D. Pellizzari, *J. Chromatogr.*, 98 (1974) 323.
174 A. Zlatkis and C.F. Poole (Editors), *Electron Capture. Theory and Practice in Chromatography,* Elsevier, Amsterdam, 1981.
175 C.F. Poole, *J. High Resolut. Chromatogr. Chromatogr. Commun.*, 5 (1982) 454.
176 G. Wells, *J. High Resolut. Chromatogr. Chromatogr. Commun.*, 6 (1983) 651.
177 R.J. Maggs, P.L. Joynes, A.J. Davies and J.E. Lovelock, *Anal. Chem.*, 43 (1971) 1966.
178 P. Rotocki and B. Drozdowicz, *Chromatographia,* 27 (1989) 71.
179 J. Connor, *J. Chromatogr.*, 200 (1980) 15.
180 W.B. Knighton and E.P. Grimsrud, *J. Chromatogr.*, 288 (1984) 237.
181 J. Vessman, *J. Chromatogr.*, 184 (1980) 313.
182 E.C.M. Chen, W.E. Wentworth, E. Desai and C.F. Batten, *J. Chromatogr.*, 399 (1987) 121.
183 F. Andrawes and R. Ramsey, *J. Chromatogr. Sci.*, 24 (1986) 513.
184 D.M. Rosie and E.F. Barry, *J. Chromatogr. Sci.*, 11 (1973) 237.
185 C.H. Lochmuller, B.M. Gordon, A.E. Lawson and R.J. Mathieu, *J. Chromatogr. Sci.*, 16 (1978) 523.
186 G. Wells and R. Simon, *J. Chromatogr.*, 256 (1983) 1.
187 J.W. Carson, G. Lege and R. Gilbertson, *J. Chromatogr. Sci.*, 16 (1978) 507.
188 P.C. Uden, *Trends Anal. Chem.*, 6 (1987) 238.
189 L. Ebdon, S. Hill and R.W. Ward, *Analyst (London),* 111 (1986) 1113.
190 P.C. Uden, Y. Yoo, T. Wang and Z. Cheng, *J. Chromatogr.*, 468 (1989) 319.
191 R.S. Hutte, R.E. Sievers and J.W. Birks, *J. Chromatogr. Sci.*, 24 (1986) 499.
192 S.O. Farwell and C.J. Barinaga, *J. Chromatogr. Sci.*, 24 (1986) 483.
193 E.C. Quincoces and M.G. Gonzales, *Chromatographia,* 20 (1985) 371.
194 G.H. Liu and P.R. Fu, *Chromatographia,* 27 (1989) 159.
195 S.A. Nyarady, R.M. Barkley and R.E. Sievers, *Anal. Chem.*, 57 (1985) 2074.
196 J.J. Sullivan and B.D. Qumiby, *J. High Resolut. Chromatogr. Chromatogr. Commun.*, 12 (1989) 282.
197 P.C. Uden, K.J. Slatkavitz, R.M. Barnes and R.L. Deming, *Anal. Chim. Acta,* 180 (1986) 401.
198 R.C. Hall, *J. Chromatogr. Sci.*, 12 (1974) 152.
199 J. Sevcik, *Chromatographia,* 4 (1971) 102.
200 R.K.S. Good, H. Kanai, V. Inouye and H. Wakatsuki, *Anal. Chem.*, 52 (1980) 1003.
201 J.N. Driscoll, D.W. Conron and P. Ferioli, *J. Chromatogr.*, 302 (1984) 269.
202 B.J. Ehrlich, R.C. Hall, R.J. Anderson and H.G. Cox, *J. Chromatogr. Sci.*, 19 (1981) 245.
203 S. Gluck, *J. Chromatogr. Sci.*, 20 (1982) 103.
204 T.L. Ramus and L.C. Thomas, *J. Chromatogr.*, 328 (1985) 342.

Chapter 10

Field-flow fractionation

JOSEF JANČA

CONTENTS

10.1 INTRODUCTION

Field-flow fractionation (FFF) is a separation method for the fractionation of macromolecular or particulate samples, usually in a long and narrow channel between two parallel plates. A carrier liquid flows through this channel, while a force acts at right angle to the flow. Separation results from a combination of the flow velocity profile in the carrier liquid and the lateral field force. The sample is injected at the beginning of the channel, and the fractionated sample in the eluent is detected at its end. The dimensions of the channel are typically of the following order of magnitude: length, 10-100 cm; width, 1-2 cm; height, 10-100 μm. The actual sizes vary, depending on the FFF technique used.

The effective field forces act through the height (thickness) of the channel. If, e.g., natural gravitational or the more powerful centrifugal forces interact with suspended particles of various sizes inside the fractionation channel, the particles will sediment and tend to concentrate at the accumulation wall of the channel. The concentration gradient thus formed induces a diffusion flux in the reverse direction. After a certain time, the sedimentation flux of the particles will just balance the diffusion flux. A steady state will be reached and a stable, exponential concentration distribution of each sample component across the channel will be established. Since the sedimentation velocity as well as the rate of diffusion are dependent on the size of the particles, particles of different sizes will form different exponential concentration distributions, their centers of gravity being at different distances from the accumulation wall.

The linear longitudinal velocity of the carrier liquid varies across the channel thickness, due to the viscosity effects that accompany the flow processes. At isoviscous laminar flow conditions, a parabolic flow velocity profile across the channel will be established. As the centers of gravity of concentration distributions across the channel of different particulate components of the sample will be located at different distances from the accumulation wall, they will move with the flow in the longitudinal direction at different velocities, and thus differential migration, elution, and separation will result.

The techniques of normal or classical FFF depend on the physical nature of the field forces applied to interact with the macromolecular or particulate components of the sample [1]. In focusing FFF techniques, the driving field forces are used to concentrate sample components at positions where the intensity of these forces is zero. This focusing principle was originally described for a special case of focusing in a density gradient, formed inside the channel during rotation in a centrifuge [2].

In a density-gradient-forming carrier liquid, sample components of different densities will be focused at different isopycnic layers, located at different distances from the channel wall. The longitudinal separation of the focused zones is again due to the flow velocity profile established in the flowing carrier liquid. This method has been called sedimentation/flotation focusing FFF (SFFFFF). The SFFFFF method was modified further by means of an asymmetrical shape of the flow velocity profile, formed inside a channel of non-rectangular cross section [3]. Later, a method equivalent to focusing FFF, called hyperlayer FFF, was proposed [4].

Numerous review articles, chapters, and monographs have chronicled the progress in this rapidly developing methodology [5-46].

10.2 PRINCIPLES, TECHNIQUES, AND INSTRUMENTATION

10.2.1 Principles of the zone formation

As mentioned in Section 10.1, according to the basic separation principle, two groups of methods can be distinguished: *classical FFF* and *focusing FFF*. In focusing FFF the intensity of effective field forces interacting with sample components depends on the coordinate in the direction of the field action, whereas in classical FFF it does not. With focusing methods this force must acquire zero value at the equilibrium point.

10.2.1.1 Classical FFF

Principles of both classical FFF and focusing FFF can be best explained with the aid of Fig. 10.1. The separation of components in a sample is effected by the interplay of two forces, acting at right angle: The flow in a narrow channel, usually of a rectangular cross section with two main parallel walls, and field forces which concentrate the sample com-

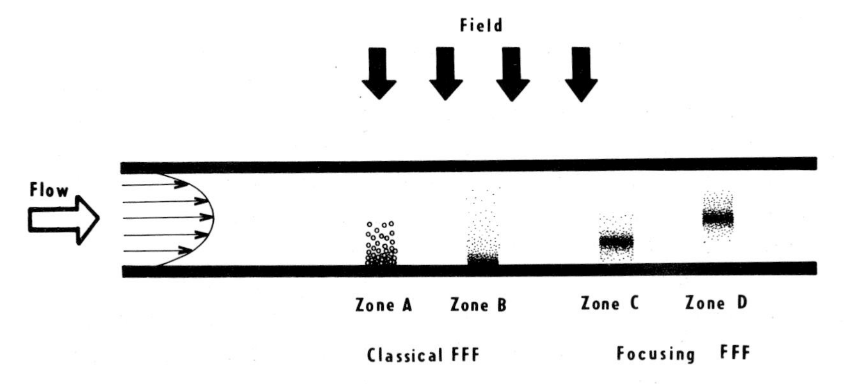

Fig. 10.1. Schematic representation of separation mechanisms of classical and focusing field-flow fractionation.

ponents against one of the channel walls. The concentration gradient established due to the field forces generates a diffusion flux in the opposite direction. After a certain time, a dynamic equilibrium is reached and the resulting exponential distribution of the concentration of sample components across the channel can be characterized by the mean layer thickness, l. The exponential distribution of the concentration of each sample component across the channel in the direction of the field action (x axis) can be described quantitatively by

$$c(x) = c_0 \exp(-x/l) \tag{10.1}$$

where $c(x)$ is the sample component concentration in the x coordinate and c_0 is the maximum concentration at the channel accumulation wall, i.e. at $x = 0$.

With laminar isothermal flow of a Newtonian fluid along the channel in the direction of its longitudinal axis a parabolic flow velocity profile is usually formed between the channel walls. Macromolecules or particles of the fractionated sample are carried in the direction of the longitudinal axis of the channel at various velocities, depending on the distance from the channel accumulation wall, in which the center of gravity of the sample component exponential concentration distribution across the channel occurs. For instance, as shown in Fig. 10.1, if there are two sample components in the channel, A and B, being distributed across the channel due to the field action to different extents, which can be expressed quantitatively by the values l_A and l_B, then the component that is compressed against the accumulation wall to a lesser extent, Component B in our case, will be retained less and move along the channel faster. Zones of Components A and B are thus eluted at different times, and due to this mechanism, separation is accomplished.

10.2.1.2 Focusing FFF

Focusing FFF makes use of the sample component migration, caused by effective field forces, to a position where the intensity of these forces is zero. Diffusion opposes the focusing forces, which concentrate a sample component in the equilibrium point. After reaching dynamic equilibrium between focusing and dispersive processes, a narrow zone of concentrated sample component is created. The longitudinal separation of different sample components is due to different positions of their focused zones in the flow velocity profile. The zone usually shows a Gaussian distribution, describing the shape of the concentration distribution of the sample component focused across the channel [46,47]. The use of a Gaussian distribution function to describe the shape of the focused zone in the direction of focusing forces action is a rough approximation. It assumes that the effective focusing forces change linearly around the focusing point and that the final focused zone is not very broad. These assumptions represent a reasonable approximation to real conditions in focusing FFF and simplify the mathematical treatment of the separation processes. The normalized Gaussian sample component distribution can then be described quantitatively by the error function, e.g.,

$$c(x) = c_0 \exp\left\{ -\frac{1}{2\mathcal{R}T} \left[\frac{dF(x)}{dx}\right]_{x=x_0} \left[x - x_0\right]^2 \right\} \tag{10.2}$$

where $\left[\frac{dF(x)}{dx}\right]_{x=x_0}$ is the gradient of focusing forces in the position of the focused zone, x_0 is the coordinate for which $F(x) = 0$ and $c = c_0$, $\mathcal{R}$ is the Rydberg constant, and T is the absolute temperature. If two or more focused zones have different positions in the direction across the channel, they are carried along the channel at different velocities and thus separated.

The separation of focused zones C and D inside the channel with an axially symmetrical flow velocity profile is illustrated schematically in Fig. 10.1.

10.2.2 Classification of various techniques

Various FFF techniques differ from each other by the nature of the physical field. Any field which interacts with macromolecules or particles can be used and every physical field gives rise to a different FFF technique. The basic techniques of classical FFF are classified by the character and intensity of the applied field.

Based on the character of the field used, the following techniques of classical FFF have been proposed theoretically: thermal FFF (TFFF) [48], sedimentation FFF (SFFF) [49], electrical FFF (EFFF) [50], flow FFF (FFFF) [51], magnetic FFF (MFFF) [52], concentration FFF (CFFF) [53], pressure FFF [54], and shear FFF [55]. Some of them have been implemented experimentally.

Based on the intensity of the applied field, the techniques of classical FFF can be classified into three groups:

(a) $F = 0$ applies to hydrodynamic chromatography

(b) $F > 0$ applies to normal FFF

(c) $F \approx kT/w$, where k is the Boltzmann constant, applies to steric FFF.

In the first group the external physical field is not operating and, consequently, there is no transverse sample component migration. However, formation of a concentration distribution across the channel may occur under the conditions of laminar flow, due to the shear forces, the phenomenon being called tubular pinch effect. Separation may also occur if the dimensions of the particles are comparable with the channel thickness. In this case, the particle centers cannot approach the channel wall more closely than the distance equal to the particle radii. Consequently, larger particles spend more time in the region of higher linear velocities in the flow velocity profile formed inside the channel and are eluted faster than smaller ones.

The second group of classical or normal FFF will be treated in more detail in subsequent paragraphs.

The third group comprises steric FFF [56]. It has an exclusive position among FFF techniques and features the upper limit of applied field intensities at which the retention mechanism changes. If the strength of the field used is increased, particles continue to be compressed more closely against the channel wall. At the moment when the mean distance of the Brownian migration is less than the particle radius, a, steric FFF occurs. The word "steric" expresses the fact that the mean thickness of the sample component layer is controlled by steric exclusion of particles from the space defined by the wall and the particle radius. Larger particles thus reach a region of higher velocities of the carrier liquid than do the smaller ones and are therefore eluted faster.

In steric FFF any effective physical field can be applied in theory (electrical, sedimentation, etc.), but the gravitational field represents the most practical instance of the application of steric FFF to the fractionation of 1- to 100-μm particles. Steric FFF extends the application of FFF to the larger-diameter particles (up to hundreds of μm) by at least one order of magnitude.

Focusing FFF methods have so far been investigated neither experimentally nor theoretically as classical FFF. Nevertheless, it has been possible to classify focusing FFF techniques into two categories. The first category includes such focusing FFF techniques in which focusing forces result from anisotropy of the carrier liquid or, in other words, from a gradient in an effective property of the carrier liquid (in which macromolecules or particles are dissolved or dispersed) in the direction of the field action and from the action of an external or internally generated field of effective forces. Such an anisotropic property can be a density gradient or pH gradient.

Elutriation focusing field-flow fractionation (EFFFF) techniques belong to the second group. In this case, focusing is obtained by combining field forces with a gradient of flow in the opposite direction. Sedimentation flotation focusing FFF [2], isoelectric focusing FFF (IEFFFF) [57-59], elutriation focusing FFF (or flow-focusing FFF [60,61]), and thermal focusing FFF [62] are focusing FFF techniques that have so far been proposed only theoretically and verified experimentally in only a few instances.

10.2.3 Instrumentation

The basic instrumentation for FFF is the same as for liquid chromatography. It consists of a carrier liquid reservoir, a pump, an injection system, a FFF channel, a detector, and a registration device. The FFF channel can be supplemented by a centrifuge for SFFF, by a power supply and other electronic devices for TFFF, EFFF, or MFFF, etc., by a flow meter, a fraction collector, etc. Equipment for special FFF techniques can be assembled from parts of a liquid chromatograph. Since much of the conventional liquid chromatography equipment is described in Chapter 4, it is unnecessary to repeat this description in this chapter. Only some important details concerning the conventional liquid chromatography equipment that should not be neglected when it is applied to FFF will be mentioned, and special FFF equipment (i.e. channels and accessories) will be described in the following paragraphs.

10.2.3.1 Carrier liquid reservoirs

Carrier liquid reservoirs are required, if they are not already accessories of the pump. The technical parameters of the reservoirs are similar to those in liquid chromatography. They must be manufactured of an inert material. Filling and emptying the reservoirs must be easy. In some cases, they must permit keeping carrier liquids under an inert gas. In most cases, simple glass, metal, or plastic vessels will satisfy the needs of routine laboratory operation.

10.2.3.2 Pumps

The channels for FFF generally have a low hydrodynamic resistance. Consequently, the carrier liquid delivery systems should not operate at high pressures, but it is very important that the flow be pulseless. Stability of the flowrate can be ensured by the use of constant-pressure pumps or constant-flowrate pumps.

Hydrodynamic resistance increases during injection of viscous polymer solutions or concentrated particle suspensions, and this may have a considerable influence on the flowrate when constant-pressure pumps are used. Impurities or aggregated particles can completely stop the flow.

Reciprocating pumps, piston pumps, linear-displacement pumps, and peristaltic pumps are all constant-flowrate pumps. High flowrates and high pressures can be achieved with most pumps of this type. The flow pulsation can be partially suppressed by using a bi- or trifunctional pump, together with a damper. Positive-displacement pumps (syringe type) are very suitable. Slow movement of the piston inside the reservoir produces a pulseless flow and makes for easy regulation of the carrier liquid flowrate by electronic control of motor speed. Peristaltic pumps operate well over a wide range of flowrates. They are unable to produce high pressures, but this is not an important factor in FFF.

10.2.3.3 Carrier liquid gradient systems

Use of carrier liquid gradients in FFF is justifiable if the carrier liquid properties influence the extent of the interaction between the physical field and the fractionated sample. This is the case, for example, when the carrier liquid density changes in sedimentation FFF. Both, discontinuous gradients (step gradients) and exponential gradients are usually applied to decrease the range of retention of polydisperse samples [63].

10.2.3.4 Injectors

Samples can be injected with a syringe through a septum, or with four-, six-, or eight-port valves, equipped with a loop of constant or varying volume in bypass. The latter variant, having a valve, is more suitable than injection with a syringe through a septum because of easier operation. Special injectors or sampling devices have been described for some FFF techniques. When applying the stop-flow technique, which serves to eliminate zone broadening due to relaxation processes, a known and stabilized speed of sampling may be required.

10.2.3.5 Elution volume measurement devices

A majority of commercially available pumps provide a high constancy of flowrate and reproducibility. However, an independent flowrate or retention volume measurement at the end of the separation system is necessary in some cases. These special cases are SFFF, SFFFFF, and FFFF. Small carrier liquid leakage through the rotary passage seal at the head of the separation system may occur in the first two cases. If the retention volume measurement is derived from the flowrate set in the pump, serious errors may occur. If the retention volume is measured independently, the retention data are more reliable and their reproducibility is higher.

In FFFF, the carrier liquid flows partly across the channel through semipermeable membranes. The flowrate at the channel outlet in the longitudinal direction must be controlled so that the retention volume can be measured accurately. Control of the cross-flow at the outlet through the wall membranes is also desirable. Several commercially available devices can be employed. A convenient system of retention volume measurement uses the air bubbles injected into the carrier liquid flowing through a capillary. A photo-optical sensor situated at a certain distance from the air injection point registers the passing bubble and causes injection of the next bubble. The time between the passage of two subsequent bubbles is measured, and both the flowrate and the retention volume can be calculated. A modified device utilizes a short thermal pulse at a given point of the capillary, registered by a sensitive thermocouple placed at a determined distance along the capillary. As the temperature maximum in the heated carrier liquid passes along the thermocouple, the next thermal pulse is generated. The system thus resembles the device based on the injection of gas bubbles.

10.2.3.6 Detectors

Some of the liquid chromatography detectors are also applicable to FFF, but not all of them are applicable to macromolecular and particulate samples. The use of some detectors is complicated by response nonlinearity. This is associated with the dependence of the detector response not only on the concentration of the measured samples, but also on their molecular or particle size.

10.2.3.6.1 Refractive-index detectors

Refractive-index detectors of various types are most popular for soluble macromolecules. They measure differences in refractive indices of eluate relative to pure eluent. This difference is proportional to the sample concentration. Problems can be caused by the refractive index dependence on temperature and pressure. The refractive index increment is independent of the molecular mass of macromolecules. This assumption holds true in the range of higher molecular masses, but for lower molecular masses corrections must be introduced. The use of the refractive-index detector for larger colloid particles is complicated by light scattering. Its intensity is a function of the ratio of particle diameter to the wavelength of the light used.

10.2.3.6.2 Photometric detectors

Photometric detectors are based on light interaction with the macromolecular or particulate sample. This interaction gives rise to absorption of light, fluorescence, optical rotation, or light scattering. A combination of these phenomena may sometimes occur.

The sensitivity of the photometric detector can be increased in some instances by fractionated sample derivatization. As most polymers do not fluoresce, application of fluorescence detectors is limited. However, fluorescence in the eluate can be measured indirectly by adding a suitable fluorescence marker to the fractionated sample.

Finally, a group of photometric detectors for macromolecules and particles are instruments that measure the intensity of scattered light either in nonhomogeneous eluates by dispersion (turbidimetric detectors) or in molecularly homogeneous systems (scattered-light photometers). Turbidimetric detectors can alternatively be used for the measurement of polymers with the addition of a precipitating agent to the carrier liquid after fractionation.

10.2.3.6.3 Automatic viscometers

A viscometric detector, together with a concentration detector can provide information on molecular masses of macromolecules exiting the FFF separation system. The viscometric detectors are suitable for FFF separations of high-molecular-mass polymers that are present in solution in the form of statistical coils in good solvents. The relative viscosities of these solutions are so high that they can be measured with sufficient accuracy and precision. Viscometric detectors of this type are not suitable for particles (e.g., latexes), as they are not sufficiently viscous in suspension.

10.2.3.6.4 Other detectors

Detectors capable of continuously measuring the density of flowing liquids have been designed on diverse principles. These detectors are promising and applicable to the detection of macromolecules in solution and apparently also to suspended particles.

It is practically impossible to enumerate all possible detectors and detection principles that can be employed in FFF. In the preceding sections the most widely used detectors were mentioned. All of them are intended for on-line use with the separation channel for FFF. Another possibility that has been suggested is the measurement of the fractionated sample concentration directly inside the channel. Optical detectors with various lasers as light sources will probably be the most useful ones. However, other principles can be envisioned, including those in which acoustic waves are used to determine the fractionated sample concentration distribution [64].

References on p. A477

10.2.3.7 Channels

In this section the channel design for various FFF methods and techniques will be described. The review will deal with typical examples in order to explain the main principles of the construction of the separation channels for FFF. Details can easily be found in the reviews cited.

10.2.3.7.1 Thermal FFF

The channel for TFFF, illustrated in Fig. 10.2, is composed of two metallic bars of thermally highly conductive material, preferably copper [48]. The upper bar is heated with electric cartridges, while the lower bar is cooled with water that flows through longitudinal holes. The surface forming the walls of the channel is perfectly even and highly polished. Electrolytic nickel and chromium plating increases the mechanical and corrosion resistance. Thermistors serve for control and regulation of the temperature gradient. One or both blocks must also be equipped with holes to which capillaries are soldered for the carrier liquid inlet, including sample injection, and eluate outlet. Fig. 10.2 illustrates the case in which inlet and outlet capillaries are situated in the lower cooled block. The lower wall is also the accumulation wall of the channel. The sample, after being injected, is transferred to the vicinity of this wall, and at the channel end, the sample is concentrated at the lower, cooled accumulation wall, and this facilitates its exit.

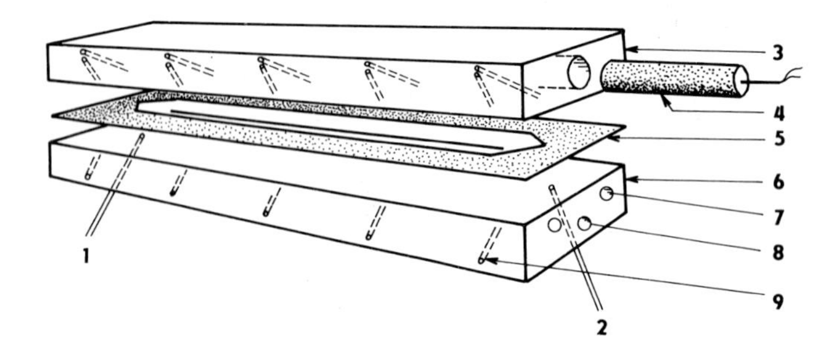

Fig. 10.2. Channel for thermal field-flow fractionation; 1, 2 = inlet and outlet capillaries; 3 = heated copper bar; 4 = electric cartridge; 5 = spacer; 6 = cooled copper bar; 7, 8 = holes for cooling liquid; 9 = holes for thermocouples.

A foil of low-thermal-conductivity material, into which the shape of the channel has been cut, is inserted between the blocks. The thickness of this foil determines the thickness of the channel. The foil material must be resistant to the carrier liquids used within the range of temperatures used. The overall system is clamped with screws, which should be tightened evenly. This provides a constant channel thickness in both longitudinal and transverse directions. Experimental systems for TFFF, equipped with common power supplies and regulators, can easily be controlled.

The experimental procedures for TFFF, from setting up to operation, through the stabilization of required experimental conditions, to analytical fractionation, starting with sample injection and ending with fractogram evaluation, do not represent extraordinarily difficult aspects to which attention would have to be directed.

10.2.3.7.2 Sedimentation FFF

The channel in which separation proceeds, is coiled into a ring and situated in the rotor basket of a centrifuge. Details of the design of the rotor with the channel for SFFF are illustrated in Fig. 10.3 [65]. The rotor consists of a main disk (1) and a supporting ring (2). These two parts are kept together by axial screws (3). Two rings, an outer (4) and an inner (5) one, are placed in peripheral grooves of the rotor. The channel of rectangular cross section (6) is formed between them. The entire system is sealed by two Teflon rings (7). The rotary passage (8), a part of the rotor, provides for flow of the carrier liquid through the channel during rotation. The rotary passage is connected with the inlet and outlet of the peripheral channel by capillaries (9). The rotor is placed on the axle of the centrifuge by means of a holder (10).

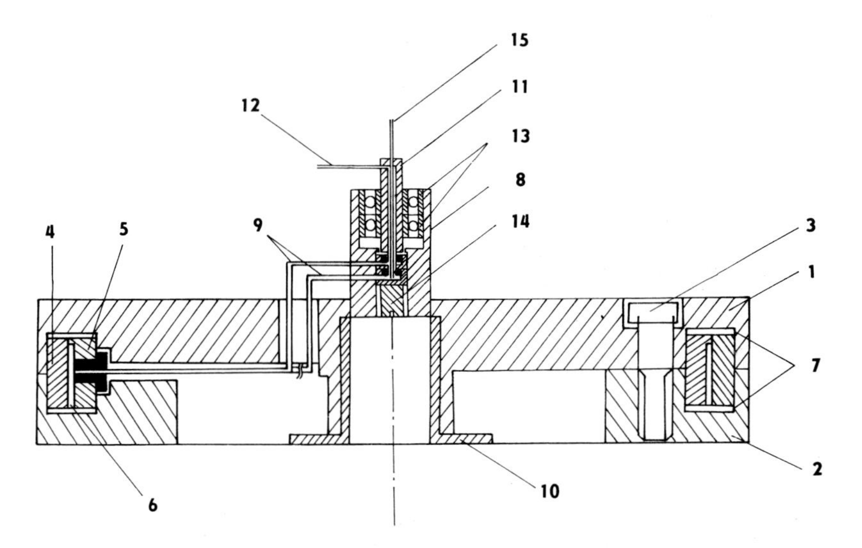

Fig. 10.3. Channel for sedimentation field-flow fractionation; 1 = main rotor body; 2 = supporting ring; 3 = screws; 4 = outer channel wall; 5 = inner channel wall; 6 = channel; 7 = sealing rings; 8 = rotating passage body; 9 = inlet and outlet capillaries; 10 = holder; 11 = stationary part of the passage; 12 = inlet capillary; 13 = ball bearings; 14 = screws; 15 = outlet capillary.

10.2.3.7.3 Electrical FFF

The channel for EFFF with flexible membrane walls is illustrated in Fig. 10.4 [14]. The main body consists of two Plexiglas blocks, into which chambers are machined that

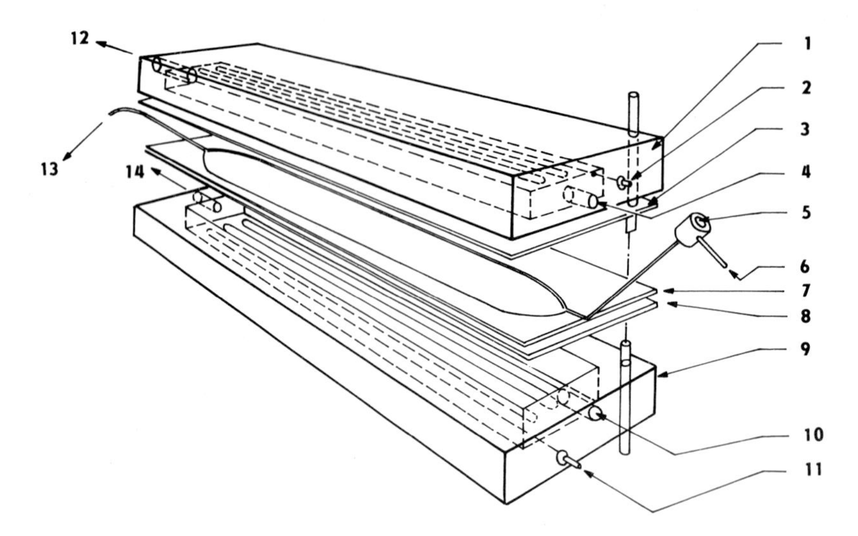

Fig. 10.4. Channel for electrical field-flow fractionation; 1 = upper electrode compartment; 2 = electrode; 3 = membrane; 4 = electrolyte inlet; 5 = injection port; 6 = carrier liquid inlet; 7 = spacer; 8 = membrane; 9 = lower electrode compartment; 10 = electrolyte inlet; 11 = electrode; 12 = electrolyte outlet; 13 = carrier liquid outlet; 14 = electrolyte outlet.

enable buffer solution to flow through. The channel walls are semipermeable, flexible membranes, made of wet-reformed cellulose. The spacer into which the channel is cut is inserted between the membranes. Inlet and outlet capillaries are led into the spacer by Teflon tubing. The entire channel is clamped together by a system of screws. Alternatively, the channel walls are composed of membranes of cellulose acetate, cast on the surface of plastic frits. It is a mechanically resistant configuration, which is an advantage, but its increased electric resistance is a drawback. A source of stabilized DC voltage, generating an electric field across the channel, is also needed.

10.2.3.7.4 Flow FFF

The channel for FFFF is similar to that for EFFF. The membrane material must permit crossflow of the electrolyte, but it must not be permeable to the separated sample components. A channel of suitable shape is cut into a spacer made of Teflon, stainless steel, or other suitable material. The entire channel is clamped together between two Plexiglas or aluminum blocks [14].

Both transverse and longitudinal flows are currently provided by two pumps. To provide the required ratio of cross and longitudinal flowrates, the exits of cross and longitudinal flows should be equipped with metering valves. However, a considerable change in the ratio of flowrates can occur with a change in the resistance at one of the outlets. The use of an accessory pump at the longitudinal flow exit is advantageous. The overall experimental procedure, starting with sample injection through relaxation with the longitudinal flow stopped, is identical to that for other FFF techniques.

10.2.3.7.5 Pressure FFF

The channels of rectangular cross section used for FFFF and described in the preceding section can also be used for pressure FFF. Pressure FFF in a cylindrical capillary can also be used [54]. The exit of the capillary is equipped with a metering valve so that the required portion of the carrier liquid passes through the capillary wall. Various semipermeable cylindrical capillaries are available commercially.

10.2.3.7.6 Steric FFF

The simplest case of steric FFF is the separation of large particles that sediment in a natural gravitational field. The channel for this technique can be formed by cutting the required shape into a spacer, which is placed between two glass plates. This spacer is equipped with inlet and outlet capillaries, and together with glass plates forming the walls, it can be clamped between two plates of Plexiglas with the aid of screws. The channels described in previous sections are usable in the steric FFF mode, taking into account suitable modifications for high-intensity-field application.

10.2.3.7.7 Focusing FFF

A very simple channel with modulated cross-sectional permeability can be constructed for SFFFFF in a natural gravitational field [3]. In a modulated cross-sectional permeability channel the thickness or the distance between the channel walls changes in the direction of focusing field forces, resulting in a shape required by the flow velocity profile. A trapezoidal cross-section channel can serve as an example of a modulated cross-sectional permeability channel. Two strips of foil of different thicknesses are placed longitudinally between two glass plates so as to form a channel of trapezoidal cross section. The ends of the channel are sealed and provided with stainless-steel capillary tubes for the inlet and outlet of the carrier liquid. The entire system is clamped between Plexiglas plates with several screws. Conventional, rectangular cross-section channels are also applicable to this technique.

The channels for various FFF techniques described in the preceding sections demonstrate examples of possible designs. Of course, these examples can be modified, and entirely different designs can be developed.

10.3 BASIC THEORY OF RETENTION AND ZONE SPREADING

In preceding paragraphs the principles, techniques, and instrumentation were discussed in order to present the FFF method as a real analytical tool. Before going into a more detailed explanation of possible applications of FFF methodology, it is necessary to know something about the basic theory of retention and dispersion in FFF. The theory of FFF has been developed extensively, but it is impossible to discuss all the fine details of

References on p. A477

different approaches in this chapter. The theoretical treatment used here is simplified to the level where it still gives a real picture of the FFF separation processes, but it was necessary to neglect some details that are unimportant for the understanding of FFF fundamentals.

Retention ratio and height equivalent to a theoretical plate (HETP) are basic parameters describing quantitatively any separation method. Whereas the retention ratio, R, describes directly processes leading to the separation, the height equivalent to a theoretical plate, H, expresses the efficiency of the system, i.e., processes of dispersive entropic character, which strive to suppress the degree of separation achieved, i.e. to spread, by spontaneous processes, the concentration gradient established as a result of separative transport.

10.3.1 Retention in classical FFF

The retention theory for FFF separations was elaborated by Giddings [5] on the principles corresponding to those known from chromatography. Sample components in a moving liquid migrate due to the combined action of flow and applied physical field. The retention ratio, R, i.e. the ratio of the velocity of the retained component movement to average fluid velocity, is defined by

$$R = \int_0^w c(x)v(x)\mathrm{d}x \Big/ \left(\int_0^w c(x)\mathrm{d}x \int_0^w v(x)\mathrm{d}x \right) \tag{10.3}$$

where $v(x)$ is the streamline velocity in the x coordinate. It holds true for isothermal isoviscous flow of Newtonian liquids between two parallel infinite planes not influenced by any external physical field that

$$v(x) = \Delta P \, x \, (w - x) / 2L\eta \tag{10.4}$$

where ΔP is the pressure drop along a channel of length L, w is the distance between the planes forming the main channel walls, and η is the liquid viscosity. Solving and substituting into Eqn. 10.3, we obtain the final relationship

$$R = 6\lambda \, [\coth(1/2\lambda) - 2\lambda] \tag{10.5}$$

where $\lambda = l/w$. This equation is the basic theoretical relationship describing retention in classical FFF quantitatively. It holds within the limit when λ approaches zero for highly retained components

$$\lim_{\lambda \to 0} R = 6\lambda \tag{10.6}$$

In a number of practical applications the approximation given by Eqn. 10.6 is fully justified. The relationship between the values R and λ is thus often very simple. The retention ratio in steric FFF is controlled by the radius, a, of the fractionated sample component. For the value of R a relationship was derived under the limiting conditions when λ_a approaches zero, which is a real condition, in view of the described character of steric FFF [56].

$$R = 6\lambda_a(1 - \lambda_a) \tag{10.7}$$

where $\lambda_a = a/w$.

10.3.2 Retention in focusing FFF

A generalized retention theory of focusing FFF was formulated by Janca and Chmelík [47]. Theoretical analysis was performed for the system with axially symmetrical and asymmetrical shapes of the flow velocity profile. An asymmetrical flow velocity profile can be obtained, either if the channel cross section is trapezoidal, or if its longer walls are in the cross section parabolically curved. For the retention ratio in a rectangular cross-section channel, i.e. for an axially symmetrical parabolic flow velocity profile, the following relationship holds true

$$R = \frac{3}{2}\,(1 - \Phi_0{}^2 - \sigma^2) \tag{10.8}$$

The dimensionless coordinate Φ in the direction of focusing forces is defined so that it lies in the plane of the channel cross section, its origin, $\Phi = 0$, lies in the channel center, and the value $\Phi = 1$ is reached at one channel wall and the value $\Phi = -1$ at the opposite wall; σ is the standard deviation of the Gaussian focused zone, expressed in terms of the fraction of the total channel width in the direction of the Φ coordinate. The following relationship was derived for the retention ratio in the trapezoidal cross-section channel [47]

$$R = \frac{3}{3 + tg^2\,\alpha}\,[(1 + \Phi_0\,\mathrm{tg}\,\alpha)^2 + 2\,\sigma^2\,\mathrm{tg}^2\,\alpha] \tag{10.9}$$

where α is the angle included by the trapezoidal channel walls and Φ_0 is the coordinate of the Gaussian focused zone maximum. It holds true for the retention ratio in channels having cross sections in the shape of a parabola or of its segment [47] that

$$R = 1 + \frac{c\,\Phi_0}{c + d} \tag{10.10}$$

where c and d are geometric parameters of the parabola.

10.3.3 Zone spreading in classical FFF

Spreading of the zone is characterized quantitatively by the height equivalent to a theoretical plate, *H*.

$$H = 2D / R\,\bar{v}(x) + \chi\, w^2\, \bar{v}(x) / D + \Sigma H_i \qquad (10.11)$$

where χ is a dimensionless parameter [5]. The first term in Eqn. 10.11 describes longitudinal diffusion in the direction of the longitudinal channel axis, the second one describes nonequilibrium effects, and the third one describes the sum of various contributions resulting from relaxation processes, sample injection, etc. Eqn. 10.11 is analogous to that describing chromatographic processes. However, in view of the character of FFF, it does not include the term corresponding to eddy diffusion in a packed chromatographic column. The term describing nonequilibrium processes expresses the most significant contribution to the total value of *H* [5] in Eqn. 10.11. An argument for it being so is the fact that the diffusion coefficients of macromolecules in solution – and even more so the diffusion coefficients of large suspended particles – are so small that, with common linear velocities of fluids in the channel, the contribution of longitudinal diffusion is practically negligible. It follows from nonequilibrium theory [5] that the dimensionless parameter χ is determined by the limit

$$\lim_{\lambda \to 0} \chi = 24\lambda^3 \qquad (10.12)$$

Eqn. 10.12 expresses a fact of great significance. The dimensionless parameter χ increases with the third power of the retention parameter, λ. In view of Eqns. 10.5 or 10.6 and 10.11 this means that, e.g., a two-fold increase in retention, i.e. a decrease in retention ratio, *R*, or in the value of λ to one-half, will cause a two-fold increase in the total separation time. However, in view of Eqn. 10.12, it will also at the same time cause an eight-fold decrease in *H*. Since efficiency or *H* values are linearly dependent on the average linear velocity, $\bar{v}(x)$, the same time required for the separation at a two-fold increase in retention can be reached by increasing $\bar{v}(x)$ twice, and still the efficiency of the system, characterized by the value of *H*, will be higher. In spite of a number of non-idealities, which give rise to certain quantitative deviations of experimental observations from the theoretical relationships given above, the suggested ways of varying experimental conditions in FFF make this method very advantageous.

10.3.4 Zone spreading in focusing FFF

The *H* value for zones focused in channels with modulated cross sections (trapezoidal or parabolic) can be expressed by a relationship which can be transformed to the above-mentioned dimensionless system of coordinates

$$H = 2D / \bar{v}(\Phi_o) + w^2(\Phi_o)\bar{v}(\Phi_o) / 105D \qquad (10.13)$$

where $w(\Phi_o)$ is the channel thickness in a Φ_o coordinate [46,47].

Eqn. 10.13 is applicable to the description of dispersion in channels with modulated cross section in cases where one may assume that the focused zone width in the direction of the Φ axis is far smaller than the channel width in the same direction. If this condition is not fulfilled, the contribution of dispersion resulting from the proper width of the focused zone must also be considered. Details and derivations of the relationships described above can be found in the original papers cited.

10.3.5 Relaxation

Injected into the channel, the sample is homogeneously distributed across the channel. Only under the action of an external physical field is a concentration gradient created, which proceeds until the steady state is reached. The time until a quasi-equilibrium is established is called relaxation time. Relaxation processes contribute to zone broadening. However, this contribution can often be neglected. In such cases where it would considerably increase the total zone spreading, it may be eliminated by stopping the flow through the channel after the injection for the period of time required to reach quasi-equilibrium.

10.4 SCOPE OF VARIOUS TECHNIQUES

10.4.1 Thermal FFF

TFFF ranks among the oldest FFF techniques. It is based on the principle of thermal diffusion. The TFFF channel is relatively simple. It is composed of two metallic blocks with highly polished surfaces between which a spacer foil is clamped. The upper block is heated electrically, while the lower one is cooled by the flow of water. The proper shape of the channel is cut out of the foil, being usually 50 to 250 μm thick. Common temperature gradients between the walls are between 20 and 100°C.

The value of λ is expressed quantitatively by the approximate relationship [6,46]

$$\lambda = [w(\alpha/T)\mathrm{d}T/\mathrm{d}x]^{-1} \qquad (10.14)$$

where α is a dimensionless thermal diffusion factor associated with the thermal diffusion coefficient, D_T, by the relationship

References on p. A477

$$\alpha = D_T \cdot T/D \quad (10.15)$$

From the theoretical viewpoint, TFFF is a rather complicated technique. Regarding the temperature gradient across the channel, the flow is not isoviscous and, as a result of this, neither is the flow velocity profile formed parabolic. With the use of a pressurized system, permitting expansion of the operational temperature region, TFFF can be employed to fractionate effectively low-molecular-mass synthetic polymers (molecular masses of the order of several 10^2 D) and stable polymers of extremely high molecular mass. Temperature gradient programing permits the fractionation of polymers over a wide range of molecular masses within the retention range that is acceptable timewise. Miniaturization of the TFFF channel speeds up the whole analysis time so that it can be reduced to several tens of seconds [66].

An increase in selectivity can be achieved by using a thermogravitational effect, i.e., by using thermal convection in a nonhorizontal channel. This gives rise to a nonparabolic flow velocity profile. A practical application of TFFF to the fractionation of various synthetic polymers soluble in organic solvents was first published by Giddings et al. [67]. A paper by Kirkland and Yau [68], on the other hand, deals with the TFFF of water-soluble polymers. This represents a very important contribution to the FFF literature, because it is a known fact that thermal diffusion coefficients of polymers dissolved in water are generally small. However, these authors showed that a good fractionation of nonionic polymers in water and a reasonable retention of ionic macromolecules can be achieved by adjusting the ionic strength of the mobile phase. Relatively large temperature gradients had to be used for useful separations.

10.4.2 Sedimentation FFF

SFFF, like TFFF, ranks among the oldest FFF techniques. Effective fields in this case are either natural gravitation or the higher centrifugal forces.

For the value of λ of spherical particles it is true that [14]

$$\lambda = 6kT/\Pi d_p^{3} g w \Delta\rho \quad (10.16)$$

where k is the Boltzman constant, g is the gravitational or centrifugal acceleration, and $\Delta\rho$ is the difference between the densities of the particles and the carrier liquid used.

Programing of centrifugal field strength and carrier liquid density considerably extend the range of particle sizes that can be fractionated in a single run. Field strength in SFFF can be programed by gradually decreasing the angular velocity of the centrifuge rotor in the course of the analysis. Carrier liquid density can also be programed by gradually changing the composition, e.g., the concentration of saccharose in water, used as a carrier liquid.

Modern SFFF instruments allow work with very high centrifugal field forces, up to 100 000 x g. The gentle operational conditions of SFFF thus make this technique very attractive for biological applications. In addition to effecting separation, one can also calculate the absolute dimensions or molecular masses of analyzed particles from retention data without the need for calibrating the separation system. The range of published applications cover the fractionation of polymer latex particles, biological cells, viruses, subcellular particles, various inorganic particles, etc. [46].

10.4.3 Electrical FFF

EFFF ranks among the FFF techniques making the highest demands on experimental equipment. In this case, the field is formed by an electrical voltage, which generates an electrical current across the channel. The walls of the channel for EFFF are composed of two semipermeable membranes, permitting the passage of small ions and the separation of the channel space proper from the electrode compartment. This permits creating an undisturbed homogeneous electrical field inside the channel. The dimensionless parameter λ is determined by the electrophoretic mobility, μ, the intensity of the electrical field, E, the diffusion coefficient, D, and the channel thickness, w, by the relationship [14]

$$\lambda = D/\mu E w \qquad (10.17)$$

EFFF separations, similarly to separations by direct electrophoretic methods, may be influenced by changes in experimental conditions, such as pH and ionic strength. These changes affect mobilities and, to a lesser extent, diffusion coefficients. The role of electrophoretic mobility, μ, in electrophoresis corresponds to that of the ratio D/μ in EFFF. For this reason, the fractionated sample components showing only small differences in mobilities but differing considerably in D can be separated successfully by EFFF, in spite of the fact that their electrophoretic resolution in a given case is not particularly good. Hence, EFFF is complementary to direct electrophoretic methods. An experimental advantage of EFFF over electrophoresis is the fact that high voltage gradients per unit of length are achieved at low absolute voltages applied across the channel. To date, most of the application of EFFF have been devoted to fractionating various proteins [14,46].

10.4.4 Flow FFF

FFFF has turned out to be the most versatile FFF technique. In this case, the liquid flow, perpendicular to the flow of the carrier liquid in the channel, creates an external retentive field. In one commonly used configuration the FFFF channel is composed of two parallel semipermeable membranes separated by a thin foil, into which the channel shape is cut. Improved separation speed and efficiency in FFFF has been achieved in a so-called asymmetrical FFFF channel that uses only one semipermeable wall opposite to the accumulation wall [69].

In FFFF, the perpendicular flow with the velocity u acts on all sample components equally. For this reason, FFFF separations are determined solely by differences in the values of the diffusion coefficient, D, of the individual sample components. The retention parameter λ is defined by the relationship [14]

References on p. A477

$$\lambda = \mathcal{R}TV^{\circ}/3\Pi N\eta V_{c}w^{2}d \qquad (10.18)$$

where V_C is the volumetric perpendicular flow (creating the field), η is the viscosity of the medium, V° is the channel volume, d is the effective Stokes diameter, N is Avogadro's number, and the other variables are as defined previously.

FFFF can also be used for continuous separations, such as dialysis or ultrafiltration. For instance, two components, one of which is capable of penetrating a semipermeable membrane while the other is not, can be separated effectively by selecting the proper relations between various experimental parameters. Considerable use of FFFF is confirmed by numerous examples of the effective separation of different samples having particulate character. Viruses and proteins can thus be separated, purified, and characterized by their diffusivities. Colloidal silica gels may be fractionated and analyzed according to their dimensions [14,46].

FFFF of water-soluble polyelectrolytes, sulfonated polystyrene, and the sodium salt of polyacrylic acid has demonstrated that this technique can also be used for the separation of macromolecules dissolved in aqueous carrier liquids [70] as well as in organic solvents [71,72].

10.4.5 Magnetic FFF

MFFF has not been developed sufficiently yet. Two papers [52,73] deal with the fractionation of albumin. However, only weak retentions, if any, were observed. More promising results were obtained by applying MFFF to the separation of suspended iron oxide particles [74,75].

10.4.6 Concentration FFF

CFFF is a technique that was suggested only in theory [76]. It makes use of the concentration gradient of a mixed solvent across the channel, which leads to the formation of a gradient of chemical potential towards the sample components.

10.4.7 Pressure FFF

Pressure FFF and FFFF are very similar techniques. In both instances, it is the transverse flow of fluid that produces an effective field. The principal difference lies in the fact that in FFFF the flow is applied from outside the channel through the semipermeable membranes forming its main walls, whereas in pressure FFF the transverse flow through the wall is caused by a pressure drop inside the channel [54,77].

10.4.8 Shear FFF

Shear FFF has been theoretically designed as a technique in which the shear gradient is used in the capacity of an effective field force perpendicular to the fluid flow [55].

The list of techniques of classical FFF is still incomplete. One may expect further techniques to be designed theoretically and effected experimentally.

10.4.9 Sedimentation/flotation focusing FFF

SFFFFF is a new method in the category of focusing FFF [2]. Separation is based upon sedimentation or flotation of suspended particles in a liquid-phase density gradient. Particles are concentrated by sedimentation/flotation forces in a space where their density is identical with the density of the surrounding medium. Diffusion flux counteracts the focusing processes. After dynamic equilibrium is reached, the solute creates a zone with Gaussian concentration profile.

The density gradient of SFFFFF is formed, for instance, either by the action of a centrifugal force in the liquid medium, consisting of several components with different densities, in a channel which is coiled around the circumference of the rotor of a special centrifuge, or by the gravity in a natural gravitational field. The sample injected into the density gradient is separated according to particle densities into narrow zones, focused at various distances from the channel wall. These focused zones are eluted with different longitudinal velocities of the density-forming carrier liquid according to their positions with respect to the flow velocity profile. The application of the longitudinal flow and, hence, the elution is the main difference between SFFFFF and conventional sedimentation in density gradients (isopycnic sedimentation).

Convenient separation of zones formed in the density gradient can be achieved by using another flow velocity profile than the parabolic one. If the permeability of the channel (as defined in the preceding paragraphs) changes continuously across the channel, i.e., the channel cross section is not rectangular, an asymmetrical shape of the flow velocity profile can be achieved [3,47].

10.4.10 Isoelectric focusing FFF

The electrophoretic mobility of amphoteric macromolecules is a function of pH and is zero at the isoelectric point. When a stable pH gradient is created in the FFF channel (by means of a transverse electric field) amphoteric solutes will be focused at their isoelectric pH. In the steady state, a transverse Gaussian distribution of the focused zone is achieved. The focused zones are separated longitudinally due to the flow velocity profile [57-59].

10.4.11 Elutriation focusing FFF

This technique is based on the counteracting forces of a primary homogeneous external field with constant intensity and of the velocity gradient of the transverse flow across the channel in the direction opposite to the action of a primary field. Combination of these counteracting forces results in focusing of sample components. The longitudinal elution of focused zones is due to the carrier liquid flow velocity profile. Application of the channel with modulated permeability cross section to the combination of an axially asymmetrical

References on p. A477

flow velocity profile and a velocity gradient of the flow in transverse direction has been proposed [60]. The channel has a trapezoidal cross section, two semipermeable walls allowing a transverse flow, and two walls being clamped at a certain angle. The longitudinal flow is applied perpendicular to the primary field. The transverse flow gradient is generated as a result of diverging streamlines in the direction of the field action. Conventional channels with rectangular cross section can be used in this technique. Ratanathanawongs and Giddings [61], who proposed this arrangement, prefer to call this technique flow/hyperlayer field-flow fractionation.

10.4.12 Thermal/hyperlayer FFF

In thermal/hyperlayer FFF a combination is used of a temperature gradient field across the channel with the opposite lift forces, generated at high linear flow velocities [62]. The result is a focusing effect, which, combined with the flow velocity profile inside the channel, produces a longitudinal separation. In principle, this is the focusing FFF technique, but the special name hyperlayer is preferred by the authors [62].

10.5 METHODOLOGY OF POLYMER AND PARTICLE FRACTIONATION AND CHARACTERIZATION

For practical applications of FFF methods and FFF techniques it is important from the methodological viewpoint that both the separation system itself and the actual experimental conditions be optimized. Knowing the particular experimental conditions, one can then evaluate quantitatively the required molecular parameters of the samples being analyzed. These two methodological aspects will be discussed in the following sections.

10.5.1 Optimization of FFF

FFF optimization aims to find the conditions under which interfering effects are suppressed, analysis time is reduced, and either minimum or maximum amounts of sample are separated. Evidently, some requirements will be mutually exclusive. Resolution can be used to evaluate the actual separation by a FFF system. The resolution, R_S, of two sample components is defined by

$$R_S = \frac{2\,(t_{R_1} - t_{R_2})}{t_{w_1} + t_{w_2}} \qquad (10.19)$$

where the $t_{R_{1,2}}$ are the retention times of two different components and the $t_{w_{1,2}}$ are the widths of their elution curves, expressed in time units. An almost complete separation is obtained when $R_S = 1$ (ca. 95% separation for equal concentration Gaussian elution curves). By using Eqn. 10.19 and the equations for retention and zone spreading, R_S can be expressed as

$$R_s = \frac{1/R_1 - 1/R_2}{2\,(1/R_1\sqrt{N_{p_1}} + 1/R_2\sqrt{N_{p_2}})} \tag{10.20}$$

where $N_{p_{1,2}}$ is the number of theoretical plates for sample components 1 and 2.

Eqn. 10.20 is derived by taking into account the definition of the retention ratio, R (Eqn. 10.3), on the basis of retention times for an unretained component, t_{R_o}; and for retained components, t_{R_1} and t_{R_2}, it can be expressed as

$$R = t_{R_o}/t_{R_1} \tag{10.21}$$

for Component 1 and similarly for Component 2. The number of theoretical plates is related to the HETP (as described by Eqn. 10.10) by

$$N_p = L/H \tag{10.22}$$

where L is the length of the FFF channel. When expressed on the basis of $t_{R_{1,2}}$ and $t_{w_{1,2}}$ values, it is true for Gaussian elution curves that

$$N_{p_1} = (4t_{R_1}/t_{w_1})^2 \tag{10.23}$$

for Component 1 and similarly for Component 2. Eqn. 10.20 has been derived on the basis of Eqns. 10.19 and 10.21 - 10.23 [78].

It follows from the above equations and Eqn. 10.11 that the resolution is proportional to the square root of the reduced channel length and decreases with increasing carrier liquid velocity. The efficiency of the FFF separation system increases as the retention increases (i.e., with decreasing retention ratio, R), and thus, the best resolution will be obtained for the sample components that are the most retained. It follows that flow programing (i.e. a gradual increase in the flowrate) would make it possible, on the one hand, to work permanently in the range of the acceptable resolution, and on the other, to decrease the time of analysis. When optimizing the FFF separation system, several important aspects must be taken into account. Attention must be paid to channel construction and assembly of the entire separation system in order to suppress zone spreading to the greatest possible extent. It is necessary to vary experimental conditions to achieve the highest possible resolution. The speed of separation is very important, not only because of time economy, but also because of possible instability of the samples being analyzed. Programing techniques make it possible to achieve the required efficiency and resolution within a reasonable period of time.

10.5.1.1 Channel design

The reduction of deadvolumes in channel, injector, detector, connecting capillaries, etc., is imperative, since all components of the separation system contribute to zone spreading. Some of the following modifications of the separation channel itself allow better

References on p. A477

separating conditions to be achieved. The application of a splitter at the end of the channel permits increasing the concentration of the sample components passing out of the channel into the detector, thus increasing the sensitivity of the detection or decreasing the amount of sample required and thus obviating channel overloading. The degree of enrichment of a sample component at the outlet can be calculated, if its distribution of the concentration across the channel and the outlet stream splitting are known. The use of a splitter on the channel inlet permits sample introduction close to the accumulation wall at the beginning of the channel, thus decreasing the stop-flow time necessary for primary relaxation of the sample.

The most advantageous design in some cases is the annular channel shape. This channel is used not only for shear FFF, but also for other techniques where application of the field in a channel of rectangular cross section is difficult. An advantageous property of the annular channel is the elimination of edge effects. An increase in the retention and in the resolution of the FFF channel can be obtained by modifying the surface of the channel accumulation wall with transversal barriers that form grooves in which the solvent does not move and where the sample components can migrate both in and out by diffusion only.

Except for TFFF and focusing FFF in channels with modulated cross-sectional permeability, all other FFF techniques involve a parabolic flow velocity profile. The retention can be varied by applying liquids exhibiting non-Newtonian behavior. This phenomenon could be utilized to increase the selectivity of strongly retained sample components, due to a changed shape of the flow velocity profile. However, it might be difficult to reach an adequate level of non-Newtonian behavior at the low flow velocities needed for high-resolution FFF separations. Steric FFF might provide a solution, because higher flow velocities can be used. Nonparabolic flow velocity profiles can intentionally be established in thermogravitational FFF. Instead of situating the channel horizontally, its longitudinal axis is either vertical or inclined at a given angle from the vertical axis. Owing to varying densities of the carrier liquid in a temperature gradient, apart from forced flow, flow caused by thermal convection occurs. The resulting flow velocity profile is the sum of the components of the velocity of the external forced flow and the velocity of the internal convection. Thermogravitational FFF has proved to be suitable for the separation of polymers of low molecular mass, which show only small differences in retention in normal TFFF.

10.5.1.2 Programing

If samples to be fractionated are polydisperse, a high field strength must be applied to separate the least-retained macromolecules or particles. As a result, strongly retained macromolecules or particles leave the fractionation channel after a very long period of time. This problem can be solved by a programed decrease in the intensity of the field forces. In principle, two approaches can be used: decreasing the external field strength or changing the sample component property that is decisive for the retention.

Effective properties of the sample components can be changed, e.g., by changing the pH, i.e. by changing the diffusion coefficient and/or electrophoretic mobility in EFFF. The effective size of macromolecules in TFFF can be varied, for example, by changing the solvent quality. Uniform programing means that some parameters (e.g., external field strength) are changed equally and simultaneously throughout the channel. In solvent programing the change is induced by the inflowing solvent and continues gradually in time and distance from the beginning of the channel. An exponential course of uniform programing is particularly advantageous.

In addition to the continuous types of uniform programing, other procedures can also be used; e.g., a discontinuous step-by-step program, where programed properties change stepwise at a time determined in advance. Programing of the carrier liquid velocity, which increases in the course of a FFF experiment, can also be applied. An effect similar to that which occurs with a decrease field strength, i.e. reduction in the analysis time, can be achieved. Yau and Kirkland [79] described a sophisticated time-delayed exponential field decay program, in which constant field strength is maintained for a certain period of time, followed by its exponential decrease. A simple linear relationship between the mass of particles and their retention times in SFFF is the result. Power-law field-decay-programed SFFF can be applied to perform separation while maintaining constant fractionating power [80].

A special case of programing the intensity of the field force was proposed [81,82]. Cyclic changes of the field intensity are used. As a consequence, the mechanism of separation is modified in such a way that a new FFF technique is created. Comparing unprogramed and programed FFF and programing of various types, a number of advantages (e.g., reduced analysis time) can be found, but also some disadvantages. The relationship between efficiency and flowrate, secondary relaxation phenomena occuring with the use of programing, possible solute/solute and solute/accumulation channel wall interactions at high field strengths, and other effects must be taken into account. All these factors can influence decisions in the selection of experimental conditions.

10.5.2 Data treatment

The simplest analytical information obtained from FFF is evidence for the presence of the component investigated in the sample being separated. A peak at the expected interval of retention volume is such evidence, provided that a component with similar retention volume or an artifact is excluded. The area of the fractogram or its height can serve for quantitative evaluation of the components analyzed. The dimensionless retention parameter λ is a function of the size of macromolecules and particles, molecular mass, diffusion coefficient, thermal diffusion coefficient, electrophoretic mobility, electric charge, density, etc., depending on the technique used. With some techniques the relation between the size or molecular mass of the retained sample component and the parameter λ is explicit, but with others this relationship is not straightforward.

The molecular mass distribution is characteristic of most polymers. Weight-average molecular mass (M_w) and number-average molecular mass (M_n) are the most important

References on p. A477

average values easily measured experimentally. These average molecular masses are defined by the relationships

$$M_w = \frac{\int_0^\infty M f_w(M) \mathrm{d}M}{\int_0^\infty f_w(M) \mathrm{d}M} \tag{10.24}$$

and

$$M_n = \frac{\int_0^\infty f_w(M) \mathrm{d}M}{\int_0^\infty M^{-1} f_w(M) \mathrm{d}M} \tag{10.25}$$

where M is a discrete value of the molecular mass, which corresponds to the value of the unnormalized distribution function, $f_w(M)$ being proportional to the detector response. The fractogram can be interpreted to yield an unnormalized differential distribution curve and to calculate the average molecular masses according to Eqns. 10.24 and 10.25.

The evaluation of the particle size distribution is complicated, because the detector response is dependent on the particle size. The most important average values include the number-average particle diameter

$$d_{p,n} = \frac{\sum n_i d_{p,i}}{\sum n_i} \tag{10.26}$$

the weight-average particle diameter

$$d_{p,w} = \frac{\sum n_i d_{p,i}^{\ 4}}{\sum n_i d_{p,i}^{\ 3}} \tag{10.27}$$

the number-average particle mass

$$M_n = \frac{\sum n_i M_i}{\sum n_i} \tag{10.28}$$

and the weight-average particle mass

$$M_w = \frac{\sum n_i M_i^{\ 2}}{\sum n_i M_i} \tag{10.29}$$

where n_i is the number of particles having the diameter $d_{p,i}$ or mass M_i. Bandbroadening processes cause the eluate to exit the channel not as a monodisperse fraction but rather as a mixture of macromolecules with molecular masses both above and below the value corresponding to the particular retention position. The resulting fractogram thus can be described by Tung's integral equation [83]

$$h(V) = \int_{-\infty}^{+\infty} g(Y)G(V,Y)\,dY \quad (10.30)$$

where the function $h(V)$ is the resulting uncorrected fractogram in units of the retention volume, V. The function $g(V)$ is the corrected fractogram, and $G(V,Y)$ denotes the spreading function relative to a unit amount of a monodisperse fraction. Eqn. 10.30 can be solved only under certain conditions. It is the requirement of a uniform spreading function. In such a case, Eqn. 10.30 becomes a convolution integral [84]

$$h(V) = \int_{-\infty}^{+\infty} g(Y)G(V-Y)\,dY \quad (10.31)$$

In a number of cases, especially if the spreading is small, the spreading function can be approximated by the normal Gaussian distribution function, G, in the form [84]

$$G = \left[\frac{1}{2\Pi\sigma^2}\right]^{1/2} \exp\left[-\frac{(V-Y)^2}{2\sigma^2}\right] \quad (10.32)$$

where σ is the standard deviation of the spreading function independent of the retention volume.

A nonuniform spreading function [84]

$$h(V) = \int_{-\infty}^{+\infty} g(Y)\left[\frac{1}{2\Pi\sigma(Y)^2}\right]^{1/2} \exp\left[-\frac{(V-Y)^2}{2\sigma(Y)^2}\right] dY \quad (10.33)$$

complies with samples with a wide distribution. This function can be well approximated by the relationship

$$h(V) = \left[\frac{1}{2\Pi\sigma(Y)^2}\right]^{1/2} \int_{V_i}^{V_f} g(Y)\,\exp\left[-\frac{(V-Y)^2}{2\sigma(Y)^2}\right] dY \quad (10.34)$$

and solved numerically. V_i and V_f are the initial and final retention volumes between which the integration of experimental fractograms is performed. The dependence of $\sigma(Y)$ can be calculated from the equation derived from theoretical relationships for zonebroadening [83]

References on p. A477

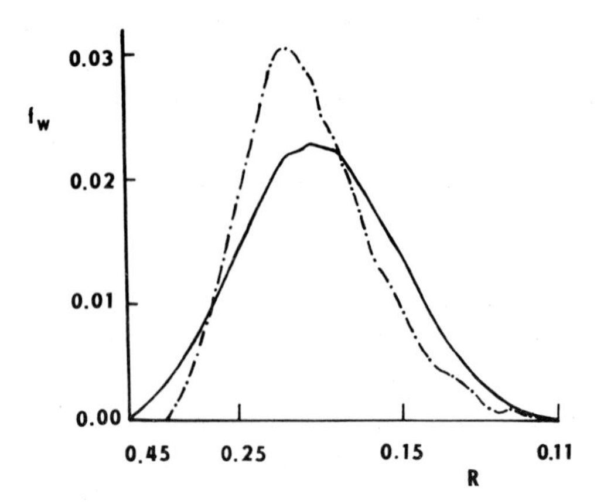

Fig. 10.5. Comparison of uncorrected fractogram (——) and fractogram corrected for zonebroadening (– · – · –). (Reproduced from Ref. 83 with permission.)

$$\sigma(Y) = \left[\frac{\chi(\lambda) w^2 \bar{v}(x) L}{D}\right]^{0.5} \tag{10.35}$$

Fig. 10.5 gives an example of the correction of an experimental fractogram. A considerable difference between the corrected and uncorrected fractograms has been observed for fractionation at high flowrates [83].

10.6 CONCLUDING REMARKS ON THE RELIABILITY OF FFF RESULTS

The invention of FFF has opened new possibilities of high-selectivity and high-resolution fractionation of polymers and particles. Whereas two decades ago the separation of individual polymers according to the degree of polymerization was not a practical reality, and obtaining the polydispersity index of fractions, $M_w/M_n = 1.2$, was considered to be a success, it is now possible to determine analytically and to obtain preparatively fractions with polydispersity values lower than $M_w/M_n = 1.1$ [85].

The application of FFF to the separation and characterization of low-molecular-mass substances does not seem to be very useful. The problem is the high intensity of the physical field that must be applied to achieve measurable retention. Of course, the simple separation of low-molecular-mass compounds from polymers or particles is possible. To separate different small molecules exploitation of the different secondary equilibria between small molecules and large particles has been suggested. The particles can easily be retained under the usual experimental conditions of FFF, and small molecules are consequently retained [86-88].

The fractogram provides information on average molar masses or particle sizes and their distributions, and on peakbroadening [84]. All quantitative results can be obtained by evaluating and interpreting an experimental fractogram in a suitable manner. Each type of information obtained will suffer from the primary error of fractogram recording. The precision of the determination of retention volumes is in direct relation to the precision of the results of FFF analysis. The precision of the detector response to an eluted sample component determines the error of the reading of the physical quantity of the other fractogram coordinate.

The processing of experimental fractograms and the precision of physical quantities that determine the shape of the fractogram are reflected in the precision of the calculated molecular parameters of the polymers and particles analyzed.

Important parameters of the numerical calculation are the total width of the interval at which the reading of the peak heights in the fractogram is taken and the density of the fractogram segmentation. For the evaluation of narrow distributions, neglecting the fractions eluted in the interval outside the limits of detectability has no significant influence. The segmentation density, provided that it is selected within rational limits, has no significant influence on fractogram evaluation [89]. Even systematic errors, the magnitude of which is equal to the precision of the reading of $h(V)$ values, have no substantial influence on the polydispersity values calculated, provided that the distributions are calculated within the limit where the reading of $h(V)$ values is still reliable [89].

REFERENCES

1 J.C. Giddings, *Sep. Sci.,* 1 (1966) 123.
2 J. Janča, *Makromol. Chem. Rapid Commun.,* 3 (1982) 887; 4 (1983) 267.
3 J. Janča and V. Jahnová, *J. Liq. Chromatogr.,* 6 (1983) 1559.
4 J.C. Giddings, *Sep. Sci. Technol.,* 18 (1983) 765.
5 J.C. Giddings, *J. Chem. Educ.,* 50 (1973) 667.
6 E. Grushka, K.D. Caldwell, M.N. Myers and J.C. Giddings, in E.S. Perry, C.J. Van Oss and E. Grushka (Editors), *Separation and Purification Methods,* Dekker, New York, 1973, p.127.
7 J.C. Giddings, *J. Chromatogr.,* 125 (1976) 3.
8 J.C. Giddings, M.N. Myers, F.J.F. Yang and L.K. Smith, in M. Kerker (Editor), *Colloid and Interface Science,* Vol. 4, Academic Press, New York, 1976.
9 J.C. Giddings, M.N. Myers, G.C. Lin and M. Martin, *J. Chromatogr.,* 142 (1977) 23.
10 J.C. Giddings, S.R. Fisher and M.N. Myers, *Am. Lab.,* 10 (1978) 15.
11 J.C. Giddings, M.N. Myers and J.F. Moellmer, *J. Chromatogr.,* 149 (1978) 501.
12 J.C. Giddings, *Pure Appl. Chem.,* 51 (1979) 1459.
13 P.P. Nefedov and P.N. Lavrenko, in *Transport Methods in Analytical Chemistry of Polymers* (in Russian), Khimia, Leningrad, 1979, p. 96.
14 J.C. Giddings, M.N. Myers, K.D. Caldwell and S.R. Fisher, in D. Glick (Editor), *Methods of Biochemical Analysis,* Vol. 26, Wiley, New York, 1980, p. 79.
15 B. Cassatt, *Anal. Chem.,* 52 (1980) 873A.
16 J.C. Giddings, *Anal. Chem.,* 53 (1981) 1170A.
17 J.C. Giddings, M.N. Myers and K.D. Caldwell, *Sep. Sci. Technol.,* 16 (1981(549.
18 E.N. Lightfoot, A.S. Chiang and P.T. Noble, *Annu. Rev. Fluid Mech.,* 13 (1981) 351.
19 H. Hatano, *Kagaku Kyoto,* 36 (1981) 488.
20 M. Martin, *Spectra 2000,* 10 (1982) 51.
21 J. Janča, *Chem. Listy,* 76 (1982) 785.

22 J. Janča, *Trends Anal. Chem.*, 2 (1983) 278.
23 J.C. Giddings, K.A. Graff, K.D. Caldwell and M.N. Myers, *Adv. Chem. Ser.*, 203 (1983) 257.
24 G.S. Karaiskakis, *Chem. Chrom. Genike Ekdosis*, 48 (1983) 75.
25 B.G. Belenkii and L.Z. Vilenchik, *Modern Liquid Chromatography of Macromolecules*, Elsevier, Amsterdam, 1983, p. 415.
26 L.J. Mathias and R. Vaidya, *Polym. News*, 9 (1983) 115.
27 T.H. Maugh, *Science*, 222 (1983) 259.
28 J.C. Giddings, *Sep. Sci. Technol.*, 19 (1984) 831.
29 K.D. Caldwell, in H.G. Barth (Editor), *Modern Methods of Particle Size Analysis*, Wiley, New York, 1984, p. 211.
30 J. Janča, K. Klepárník, V. Jahnová and J. Chmelík, *J. Liq. Chromatogr.*, 7(S-1) (1984) 1.
31 J. Janča, in Z. Deyl (Editor), *New Comprehensive Biochemistry*, Vol. 8, *Separation Methods*, Elsevier, Amsterdam, 1984, p. 497.
32 J. Janča, *Vesmír*, 63 (1984) 80.
33 Y. Gao, *Huexue Tongbao*, 11 (1984) 22.
34 H.G. Barth and S.T. Sun, *Anal. Chem.*, 57 (1985) 151R.
35 L.E. Schallinger and L.A. Kaminski, *Biotechniques*, 3 (1985) 124.
36 V.P. Andreev and L.S. Reifman, *Nauchn. Appar.*, 1 (1986) 3.
37 J. Janča, *Trends Anal. Chem.*, 6 (1987) 147.
38 J. Janča, *Chem. Listy*, 81 (1987) 1034.
39 J.J. Kirkland and R.M. McCormick, *Chromatographia*, 24 (1987) 58.
40 G.B. Levy, *Am. Lab.*, 19 (1987) 84.
41 K.D. Caldwell, *Anal. Chem.*, 60 (1988) 959A.
42 J.C. Giddings, *Polym. Mater. Sci. Eng.*, 59 (1988) 1.
43 J.C. Giddings, *Polym. Mater. Sci. Eng.*, 59 (1988) 156.
44 J.C. Giddings, *Chem. Eng. News*, 66 (1988) 34.
45 G.B. Levy and A. Fox, *Am. Biotechnol. Lab.*, 6 (1988) 14.
46 J. Janča, *Field-Flow Fractionation: Analysis of Macromolecules and Particles*, Chromatogr. Sci. Series, Vol. 39, Dekker, New York, 1988.
47 J. Janča and J. Chmelík, *Anal. Chem.*, 56 (1984) 2481.
48 G.H. Thompson, M.N. Myers and J.C. Giddings, *Sep. Sci.*, 2 (1967) 797.
49 H.C. Berg and E.M. Purcell, *Proc. Natl. Acad. Sci. USA*, 58 (1967) 862.
50 K.D. Caldwell, L.F. Kesner, M.N. Myers and J.C. Giddings, *Science*, 176 (1972) 296.
51 J.C. Giddings, F.J.F. Yang and M.N. Myers, *Anal. Chem.*, 48 (1976) 1126.
52 T.M. Vickrey and J.A. Garcia-Ramirez, *Sep. Sci. Technol.*, 15 (1980) 1297.
53 J.C. Giddings, F.J.F. Yang and M.N. Myers, *Sep. Sci.*, 12 (1977) 499.
54 H.L. Lee, J.F.G. Reis, J. Dohner and N. Lightfoot, *AIChE J.*, 20 (1974) 776.
55 J.C. Giddings and S.L. Brantley, *Sep. Sci. Technol.*, 19 (1984) 631.
56 J.C. Giddings, *Sep. Sci. Technol.*, 13 (1978) 241.
57 J. Chmelík, M. Deml and J. Janča, *Anal. Chem.*, 61 (1989) 912.
58 J. Chmelík and J. Janča, *Chem. Listy*, 83 (1989) 321.
59 W. Thormann, M.A. Firestone, M.L. Dietz, T. Cecconie and R.A. Mosher, *J. Chromatogr.*, 461 (1989) 95.
60 J. Janča, *Makromol. Chem. Rapid Commun.*, 8 (1987) 233.
61 S.K. Ratanathanawongs and J.C. Giddings, *J. Chromatogr.*, 467 (1989) 341.
62 J.C. Giddings, S. Li, P.S. Williams and M.E. Schimpf, *Makromol. Chem. Rapid Commun.*, 9 (1988) 817.
63 J.C. Giddings and K.D. Caldwell, *Anal. Chem.*, 56 (1984) 2093.
64 S.N. Semyonov and K.I. Maslow, *J. Chromatogr.*, 446 (1988) 151.
65 J. Janča, M. Janíček, D. Přibylová and Klesníl, *Anal. Instrum.*, 15 (1986) 149.
66 J.C. Giddings, M. Martin and M.N. Myers, *J. Chromatogr.*, 158 (1978) 419.
67 J.C. Giddings, M.N. Myers and J. Janča, *J. Chromatogr.*, 186 (1979) 37.
68 J.J. Kirkland and W.W. Yau, *J. Chromatogr.*, 353 (1986) 95.
69 A. Litzen and K.G. Wahlund, *J. Chromatogr.*, 476 (1989) 413.
70 J.C. Giddings, G.C. Lin and M.N. Myers, *J. Liq. Chromatogr.*, 1 (1978) 1.

71 S.L. Brimhall, M.N. Myers, K.D. Caldwell and J.C. Giddings, *Polym. Mater. Sci. Eng.*, 50 (1984) 48.
72 S.L. Brimhall, M.N. Myers, K.D. Caldwell and J.C. Giddings, *J. Polym. Sci., Polym. Lett. Ed.*, 22 (1984) 339.
73 S. Mori, *Chromatographia,* 21 (1986) 642.
74 T.C. Schunk, J. Gorse and M.F. Burke, *Sep. Sci. Technol.,* 19 (1984) 653.
75 J. Gorse, T.C. Schunk and M.F. Burke, *Sep. Sci. Technol.,* 19 (1985) 1073.
76 J.C. Giddings, F.J.F. Yang and M.N. Myers, *Sep. Sci.,* 12 (1977) 381.
77 A. Carlshaf and J.A. Jonsson, *J. Chromatogr.*, 461 (1989) 89.
78 M. Martin and A. Jaulmes, *Sep. Sci. Technol.,* 16 (1981) 691.
79 W.W. Yau and J.J. Kirkland, *Sep. Sci. Technol.,* 16 (1981) 577.
80 J.C. Giddings, W.P. Stephen and R. Beckett, *Anal. Chem.*, 59 (1987) 28.
81 J.C. Giddings, *Anal. Chem.*, 58 (1986) 2052.
82 S. Lee, M.N. Myers, R. Beckett and J.C. Giddings, *Anal. Chem.,* 60 (1988) 1129.
83 V. Jahnová, F. Matulík and J. Janča, *Anal. Chem.*, 59 (1987) 1039.
84 A.E. Mamielec, in J. Janča (Editor), *Steric Exclusion Liquid Chromatography of Polymers,* Dekker, New York, 1984, p. 117.
85 M.E. Schimpf, M.N. Myers and J.C. Giddings, *J. Appl. Polym. Sci.*, 33 (1987) 117.
86 A. Berthod and D.W. Armstrong, *Anal. Chem.*, 59 (1987) 2410.
87 A. Berthod, D.W. Armstrong, M.N. Myers and J.C. Giddings, *Anal. Chem.,* 60 (1988) 2138.
88 A. Berthod, D.W. Armstrong, M.N. Myers and J.C. Giddings, *Anal. Chem.,* 61 (1989) 90.
89 J. Janča and K. Klepárník, *J. Liq. Chromatogr.,* 5 (1982) 193.

Chapter 11

Electrophoresis

PIER GIORGIO RIGHETTI

CONTENTS

11.1 INTRODUCTION

The theory of electrophoresis has been adequately covered in the Fourth Edition of this text [1] and in the excellent textbook of Andrews [2] as well as in specific manuals [3,4]. For discussion on electrophoresis in free liquid media, e.g. curtain, free-flow, endless-belt, field-flow-fractionation, particle, and cell electrophoresis the reader is referred to a comprehensive review by Van Oss[5] and to a book largely devoted to continuous-flow electrophoresis [6]. Here the focus will be mostly on electrophoresis in a capillary support, i.e. in gel-stabilized media.

Electrophoresis is based on the differential migration of electrically charged particles in an electric field. As such, the method is applicable only to ionic or ionogenic materials, i.e. substances convertible to ionic species (a classical example: neutral sugars, which form negatively charged complexes with borate). In fact, with the advent of capillary zone electrophoresis (CZE) it has been found that a host of neutral substances can be induced to migrate in an electric field by inclusion in charged micelles, e.g., of anionic (sodium dodecyl sulfate, SDS) or cationic (cetyltrimethylammonium bromide, CTAB) surfactants [7]. Even compounds that are not ionic, ionogenic, or complexable can often be analyzed by CZE as they are transported past the detector by the strong electroosmotic flow on the capillary walls [8].

Basically, electrophoretic techniques can be divided into four main types: zone electrophoresis (ZE), moving-boundary electrophoresis (MBE), isotachophoresis (ITP), and isoelectric focusing (IEF). This classification is based on the following two criteria: (a) initial component distribution (uniform vs. sharply discontinuous) and (b) boundary permeability (to all ions vs. only to H^+ and OH^-). Alternatively, electrophoretic techniques may be enumerated in chronological order, as follows: moving boundary electrophoresis (MBE), zone electrophoresis (ZE), disc electrophoresis, isoelectric focusing (IEF), sodium dodecyl sulfate/polyacrylamide gel electrophoresis (SDS/PAGE), two-dimensional (2-D) maps, isotachophoresis (ITP), DNA sequencing, immobilized pH gradients (IPG), pulsed-field gel electrophoresis, and capillary zone electrophoresis. The latest addition is chromatophoresis, based on the direct coupling of HPLC with SDS/PAGE and thus providing a new type of 2-D map.

ZE became a reality when hydrophilic gelatins (acting as an anticonvective support) were discovered. Grabar and Williams [9] in 1953 first proposed the use of an agar matrix (today abandoned in favor of a highly purified agar fraction, agarose). They also combined, for the first time, electrophoresis on a hydrophilic support with biospecific detection (immunoelectrophoresis). Barely two years after that, Smithies [10] applied another gel, potato starch. The starch blocks were highly concentrated matrices (12-14% solids) and, as it turned out, introduced a new parameter in electrophoretic separations: molecular sieving. Human sera, which in cellulose acetate or paper electrophoresis were resolved in barely 5 bands, now produced a spectrum of 15 zones.

11.2 ELECTROPHORETIC MATRICES

Of all the capillary matrices used as anticonvective media, essentially none has survived in modern electrokinetic methods. Only three matrices are in use today: cellulose acetate, agarose, and polyacrylamide.

11.2.1 Cellulose acetate

Cellulose acetate is cellulose in which each hexose ring of the polysaccharide chain contains two hydroxyl groups esterified by acetate (in general, in the C-3 and C-6 positions). This medium was developed for electrophoresis largely through the work of Kohn

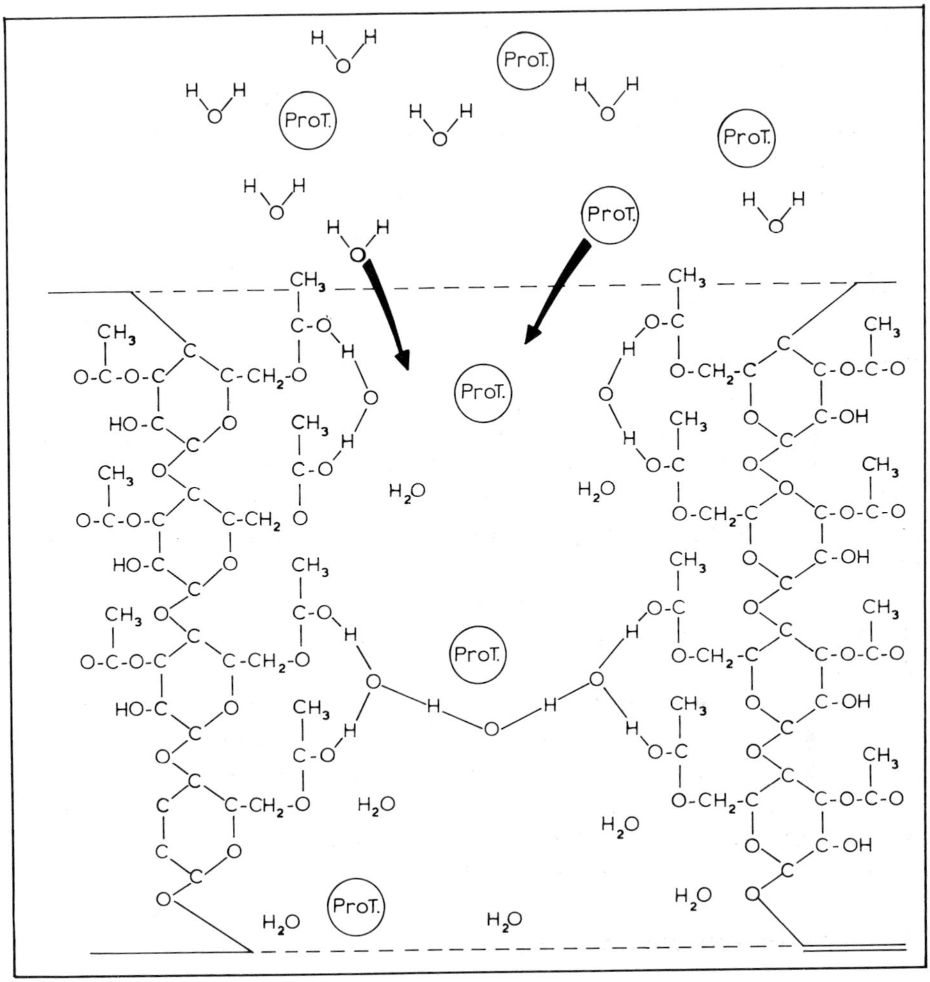

Fig. 11.1. Model of the pore structure in cellulose acetate films. Two polysaccharide pillars are shown delimiting an empty space or pore. Each hexose ring is shown acetylated on C-3 and C-6. The hydration water on the polysaccharide chains is shown being hydrogen bonded to two adjacent carbonyls in the acetyl groups. Additional, freely diffusible molecules of water permeating the pores can be seen. Some H_2O molecules can form hydrogen-bonded chains spanning the length of the pore. The dimensions of these H_2O molecules are grossly exaggerated as compared to the protein (Prot.) ions permeating the gelatin pores (by courtesy of Dr. Del Campo, Chemetron SpA, Milan, Italy).

References on p. A514

[11], who also designed the original tank, produced by Shandon Southern Instruments Ltd. as Kohn's Electrophoresis Apparatus, Model U77, and imitated many times throughout the world.

Cellulose acetate has been (and still is) immensely popular in clinical chemistry, as it offers a ready-made support that can be equilibrated with buffer in a few seconds and produces good separations of proteins from biological fluids in ca. 20 min. Completely automatic systems (for sample loading, staining, destaining, and densitometry) have been built around it, so that this system has become the first example of combined electrophoresis and robotics, an integration which has had its testing ground and development in clinical chemistry laboratories and is now spreading to basic-research laboratories.

Further development of this matrix has resulted in Cellogel, which is a gelatinized film of cellulose acetate. Cellogel, in effect, is a water film: the water is ca. 70% hydrogen-bonded to the polysaccharide chains, and ca. 30% is present as pore-impregnation solvent. A pictorial representation of Cellogel is given in Fig. 11.1. This is a section showing a pore delimited by two polysaccharide pillars, through which proteins and buffer ions can freely permeate. In reality the pores should be conical, since, due to the lamination process, the bottom of the membrane is not permeable to solutes: the pores are produced by solvent evaporation during membrane production, and their size can be controlled by the rate of this evaporation.

Cellogel allows migration of even large serum proteins, like pre-β-lipoprotein ($M_r > 5 \times 10^6$ D) and is amenable to immunofixation in situ [12,13] without any need for prior transfer by blotting, as required, e.g., after SDS/PAGE. In addition, the immunoprecipitate, obtained directly on the cellulose acetate, can be enhanced by use of a second antibody, conjugated with, e.g., alkaline phosphatase [14]. Moreover, due to their large porosity and their availability as thin films, cellulose acetate can be stained with gold micelles, increasing the detection sensitivity to the subnanogram range [15], whereas agarose and polyacrylamide are incompatible with this gilding process [16]. It is true that cellulose acetate is an opaque film, but this does not pose any problems with modern densitometers (e.g., laser scanners). Moreover, the film can be rendered transparent by a short treatment with special solvents (e.g., diacetone alcohol/water/lactic acid, 3:6:1). However, the use of cellulose acetate is limited to clinical electrophoresis and is not much in vogue in basic research for high-resolution runs, even though focusing and two-dimensional techniques on cellulose acetate membranes have been described [17].

11.2.2 Agarose gels

Agarose is a purified linear galactan hydrocolloid, isolated from agar or recovered directly from agar-bearing marine algae, such as the *Rhodophyta*. Genera from which agarose is extracted include *Gelidium, Gracilaria, Acanthopeltis, Ceramium, Pterocladia*, and *Campylaephora*. Agar is extracted from red algae by boiling in water and is then isolated by filtering off the particulates, gelling and freeze-thawing the colloid to remove water-soluble impurities, and precipitation with ethanol [18]. The resultant product is a mixture of polysaccharides, which are alternating copolymers of 1,4-linked 3,6-anhydro-α-

L-galactose and 1,3-linked β-D-galactose. The repeating disaccharide is referred to as agarobiose [19]. Of course, this is an idealized structure, since even purified agarose contains appreciable amounts of the following substituents: sulfate, pyruvate, and methoxyl groups. Sulfate is generally esterified to the hydroxyl at the 4-carbon of β-D-galactopyranose, and the 2- and 6-carbons of 3,6-anhydro-α-L-galactose. Pyruvate is linked to both the 4- and 6-carbons of some residues of β-D-galactopyranose. The resulting compound is referred to as 4,6-*O*-carboxyethylidene. Evidence has been presented that pyruvate is found preferentially on agarose chains with comparatively low amounts of sulfate [20]. In addition, the hydroxyl group on C-6 of some residues of β-D-galactopyranose can also be methylated.

Agarose was first used to form gels for electrophoresis by Hjertén [21] and also for immunodiffusion [22] and chromatography [23]. Agarose can be separated from agaropectin by several procedures, including preferential precipitation of agaropectin by a quaternary ammonium salt, followed by fractionation of agarose with polyethylene glycol.

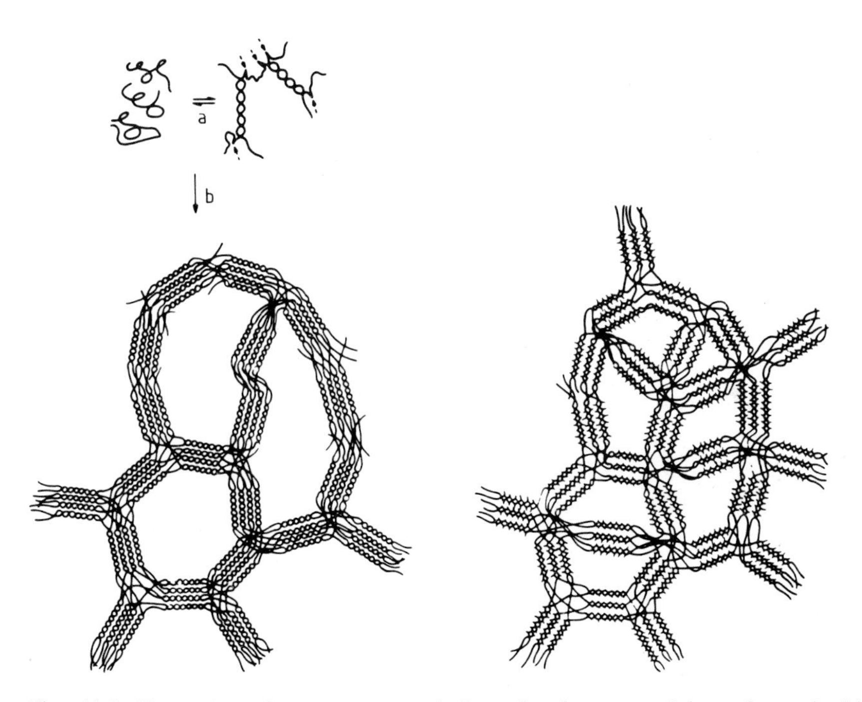

Fig. 11.2. Formation of an agarose gel. Strands of agarose join to form double helices (Transition a). At kinks (arrows) the strands exchange partners. Higher-order aggregation results in suprafibers (Transition b). In hydroxyethylated agarose, it appears that the pillar looses its structure and more of the double helices are dispersed into the surrounding space; this results in a decrement of the average pore size, i.e. higher sieving for the same % matrix as compared with underivatized agarose (by permission from Ref. 31).

Using a combination of polyethylene glycol precipitation and anion-exchange chromatography, an agarose with no detectable pyruvate and 0.02% sulfate has been prepared [24]. A further treatment consists in alkaline desulfation in the presence of $NaBH_4$, followed by reduction with $LiAlH_4$ in dioxane [25]. When an agarose solution is cooled, during gelation single strands of agarose (average molecular mass 120 kD or ca. 392 agarobiose units) [26] dimerize to form a double helix. The double helix has a pitch of 1.9 nm, and each strand has a three-fold helical symmetry [27]. However, formation of the double helix is not sufficient to produce a gel; it is assumed that these helices aggregate laterally during gelation to form suprafibers (Fig. 11.2). In fact, the width of an agarose double helix is no more than 1.4 nm, whereas in gelled agarose fibers at least one order of magnitude wider than this have been observed by electron microscopy and light scattering. In order to explain these data, supercoiling has been invoked, with formation of 'ropes', as shown in Fig. 11.2. The greater the number of aggregated double helices (i.e., the thicker the diameter of the 'pillar'), the larger is the 'pore' defined by the supporting beam [28]. Thus, it is reasonable to assume that it is the presence of these suprafibers that causes agarose gels to have strengths and pore sizes greater than those of gels that do not form suprafibers, like starch and 2.5-5% crosslinked polyacrylamide gels. In electrophoresis, agarose gels are used for fractionation of several types of particles, including serum lipoproteins [29], nucleic acids [30], virus and related particles [31], and subcellular organelles, like microsomes. However, a suspension of 0.14% agarose particles is used for the latter rather than a coherent gel mass [32]. More recently, by exploiting ultra-dilute gels [barely 0.03% agarose, a special high-strength gel, Seakem LE, from FMC BioProducts (formerly Marine Colloids, a division of FMC Corporation)] Serwer et al. [33] have even been able to fractionate *E. coli* cells. Such separations are strongly voltage-dependent, since at 0.5 V/cm *E. coli* penetrates the gel pores, whereas at 2 V/cm not a particle migrated through. This is quite remarkable and could pave the way for separations of intact cells by gel electrophoresis. Previously, this had been thought to be impossible, and in fact Righetti et al. [34] had reported a maximum pore diameter for even the most dilute gelatin (at that time, it had not been possible to gel agarose below 0.16%) of 500-600 nm. Clearly, to force *E. coli* (a rod of roughly 0.5 x 1-3 μm) to migrate into an agarose gel, the average pore diameter must be quite a bit larger than that. According to Serwer et al. [33], the effective pore radius (P_E) of a 0.05% agarose gel is 1.2 μm; thus, the average pore of a 0.03% gel should be several μm wide. The most recent theories on hydrodynamic sieving of spherical particles in dilute agarose gels assume additionally that the pores can be described as cylindrical tubes of radius P_E [33].

Perhaps no company has a greater knowledge of agaroses than FMC BioProducts. It has produced a number of highly sophisticated agarose to meet the diverse requirements of electrophoresis users. However, it is impossible to list all of them and their characteristics here. The reader is therefore referred to a recent, extensive review on this topic [35].

11.2.3 Composite agarose/polyacrylamide matrices

Composite gels can be formed in two different ways: either the warm buffer solution containing agarose and all the ingredients for a polyacrylamide gel can be kept above 35°C until the polyacrylamide has polymerized and then the agarose is allowed to solidify on subsequent cooling, or alternatively the solution is cooled to 20°C so that the agarose solidifies first. Uriel and Berges [36] have used the former procedure in preparing gels containing 0.8% agarose and variable amounts (2.5-9%T) of acrylamide. In reality, composite gels containing >4%T acrylamide seem to be of little interest, since above 4%T polyacrylamide gels alone have perfectly satisfactory mechanical and separation properties and are simpler to prepare. According to Peacock and Dingman [37], above 3%T there was no observable difference between gels in which the agarose had gelled first and those in which acrylamide was the first component to solidify. However, below 3%T, when the polymerized acrylamide remained fluid or formed only a very weak gel, it was important that the agarose solidify first. They used this procedure for making composite gels containing 0.5-2% agarose and 1-3%T acrylamide. The structure of these composite gels is schematically presented in Fig. 11.3. Such gels were quite popular in the Seventies for nucleic acid fractionation, as they allowed fine resolution of DNAs and RNAs with a lower molecular mass (M_r), but they are seldom used in modern electrokinetic methods. However, such gels (under the trade name of Ultrogels) in a spherical form are still popular for chromatographic purposes. Five types of beads exist, with a fractionation in the 1- to 1200-kD range [38]: they are of particular interest also because they can be used as ideal carriers for immobilized ligands by exploiting reactions with both the hydroxyl groups of agarose and the nitrogen of the amido group of polyacrylamide.

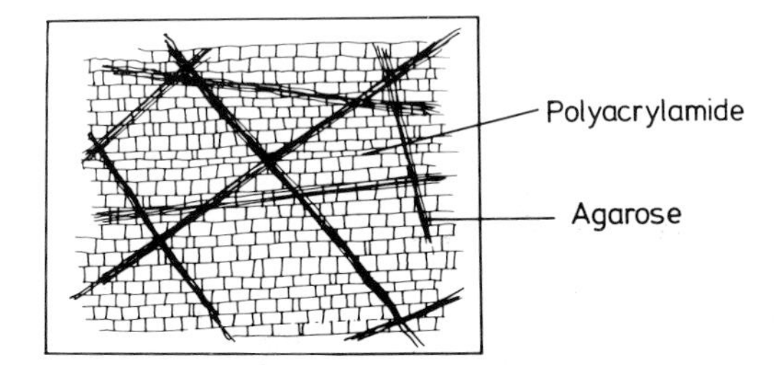

Fig. 11.3. Schematic representation of a composite agarose/polyacrylamide matrix. The thick agarose fibers support the weak structure of highly diluted polyacrylamides. Such a composite matrix could allow fine resolution of smaller fragments of nucleic acids over a large M_r range (by permission from Ref. 38).

11.2.4 Polyacrylamide gels

The most versatile of all matrices are the polyacrylamide gels. Their popularity stems from several fundamental properties: (a) optical clarity, including ultraviolet (280 nm) transparency; (b) electrical neutrality, due to the absence of charged groups; and (c) availability in a wide range of pore sizes. Their chemical formula, as commonly polymerized from acrylamide and *N,N'*-methylene bisacrylamide (Bis), is shown in Fig. 11.4, together with that of the two most widely employed catalysts, persulfate (ammonium or potassium) and *N,N,N',N'*-tetramethylethylenediamine (TEMED) [39]. Four groups simultaneously and independently developed polyacrylamide gel electrophoresis: Raymond and Weintraub [40], Davis [41], Ornstein [42], and Hjertén [43]. In addition, Hjertén [44] introduced polyacrylamide beads also as a matrix for chromatography. Normal polyacrylamide gels (i.e., crosslinked with standard amounts of 3-5% Bis) have porosities that decrease linearly with the monomer content (%T). Their porosities are substantially smaller than those of corresponding agarose gels (cf. Fig. 11.3). However, highly crosslinked gels (%C greater than 10%) can, at very high crosslink levels ($>$30% C), have porosities comparable with or even greater than those of agarose matrices (for a definition of %T and %C, see Hjertén [45]). In fact, the original observation that at progressively higher %C protein mobility would increase concomitantly was already made in 1969 by Hjertén et al. [46], who showed a much increased migration of phycoerythrin in going from 7 to 32%C at constant 5%T. However, high %C polyacrylamide matrices are brittle and opaque, and they tend to collapse and exude water above 30%C [34]. Thus, at present, their use is limited.

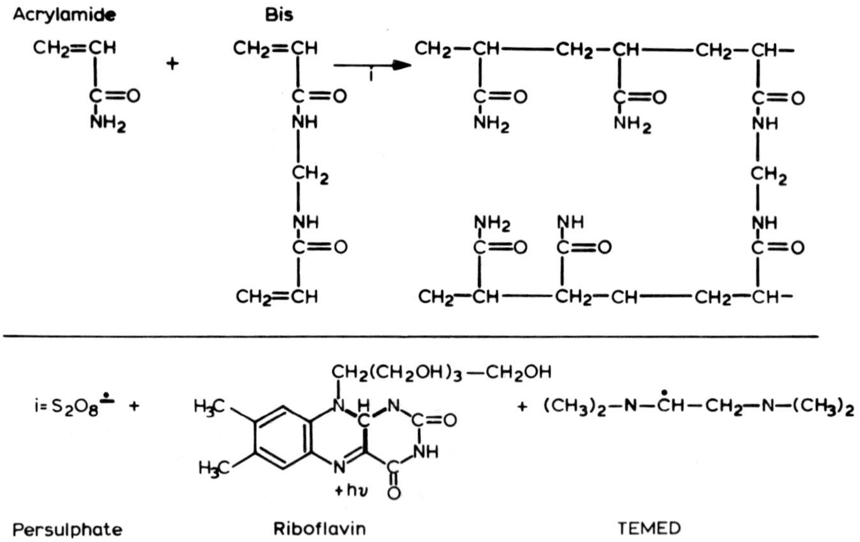

Acrylamide Bis Persulphate Riboflavin TEMED

Fig. 11.4. The polymerization reaction of acrylamide. The structures of acrylamide, *N,N'*-methylenebisacrylamide (Bis), and of a representative segment of crosslinked polyacrylamide are shown. Initiators, designated by *i*, shown are persulfate, riboflavin, and *N,N,N',N'*-tetramethylethylenediamine (TEMED). Light is designated as $h\nu$ (by permission from Ref. 39).

Hydrophilic gels are considered to be a network of flexible polymer chains with interstices into which macromolecules are forced to migrate by an applied potential difference, according to a partition governed by steric factors. Large molecules can only penetrate into regions where the meshes in the net are large, while small molecules find their way into tightly knit regions of the network, even close to the crosslinks. Different models of gel structures have been described: geometric, statistical, and thermodynamic [28,34]. Perhaps the best way to envision a hydrophilic matrix is to consider it as being composed of two interlaced fluid compartments, endowed with different frictional coefficients [47]. No matter how we describe the gel structure, it should be remembered that, in general, agaroses are more porous than polyacrylamides and that these two matrices are used complementary to each other. Several variants of polyacrylamide gels will be described below.

11.2.4.1 Different crosslinkers

The versatility of polyacrylamide gels is also shown by the large number of crosslinkers, besides Bis, that can be used to cast gels with particular properties for different fractionation purposes. *N,N′*-(1,2-dihydroxyethylene)bisacrylamide (DHEBA) can be used for casting reversible gels (i.e. gels that can be liquefied), since the 1,2-diol structure of DHEBA renders them susceptible to cleavage by oxidation with periodic acid [48]. The same principle should also apply to *N,N′*-diallyltartardiamide (DATD) gels [49]. Alternatively, ethylene diacrylate (EDA) gels [50] could be used, since this crosslinker contains ester bonds that undergo base-catalyzed hydrolytic cleavage. The poly(ethylene glycol) diacrylate crosslink belongs to the same series of ester derivatives [34]. As the latest addition to the series, *N,N′*-bisacrylylcystamine (BAC) gels, which contain a disulfide bridge cleavable by thiols, have been described [51]. These gels appear to be particularly useful for fractionation of RNA and are the only ones that can be liquefied under very mild and almost physiological conditions. Practically any crosslinker can be used, but DATD and *N,N′,N′′*-triallylcitrictriamide (TACT) should definitely be avoided, since, being allyl derivatives, they are inhibitors of gel polymerization when mixed with compounds containing acrylic double bonds. Their use at high %C is simply disastrous [52,53].

11.2.4.2 Trisacryl gels

These gels are derived from the polymerization of the unique monomer *N*-acryloyl-2-amino-2-hydroxymethyl-1,3-propanediol. The resulting polymer is shown in the lower part of Fig. 11.5. The trisacryl monomer creates a microenvironment that favors the approach of hydrophilic solutes (proteins) to the gel polymer surface. As shown in Fig. 11.5, the polyethylene backbone is buried underneath a layer of hydroxymethyl groups. Thus, this type of matrix has an obvious advantage over polyacrylamide supports, which, lacking this protective hydrophilic layer, have a more pronounced hydrophobic character. In addition, due to the much larger M_r of the monomer, as compared with acrylamide, gels made with the same nominal matrix content (%T) should be more porous, since they

References on p. A514

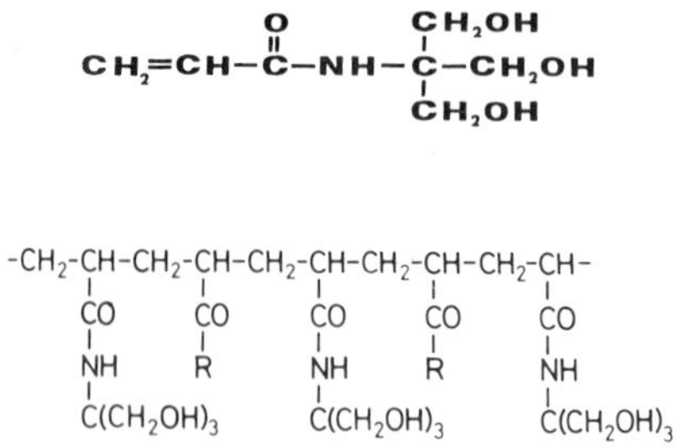

$$CH_2{=}CH{-}\overset{\displaystyle O}{\overset{\|}{C}}{-}NH{-}\underset{\displaystyle CH_2OH}{\overset{\displaystyle CH_2OH}{C}}{-}CH_2OH$$

$$-CH_2-CH-CH_2-CH-CH_2-CH-CH_2-CH-CH_2-CH-$$

CO	CO	CO	CO	CO
NH	R	NH	R	NH
$C(CH_2OH)_3$		$C(CH_2OH)_3$		$C(CH_2OH)_3$

Fig. 11.5. Structure of the trisacryl monomer (upper panel). The lower part depicts a 1:1 copolymer of trisacryl and acrylamide (in the case where R = NH_2). When R is a protolytic group, a series of ion exchangers can be manufactured (by permission from Ref. 38).

would have on the average thicker fibers than the corresponding polyacrylamide matrix. This type of gel is extensively used by IBF Reactifs for preparing gel filtration media or ion exchangers, but its use in electrophoresis had not been reported till 1987, when Righetti et al. [54] tried it for IEF in IPGs. As there did not appear to be an improvement in electrophoretic patterns, this type of matrix was not further investigated. However, more recently, trisacryl gels have been used for DNA separations, with better resolution than with standard gels [55].

11.2.4.3 Hydroxyalkyl methacrylate gels

Hydrophilic hydroxyalkyl methacrylate gels, first introduced by Wichterle and Lim [56], may be prepared by polymerization of a suspension of hydroxyalkyl esters of methacrylic acid and alkylene dimethacrylate by varying the ratio of the concentrations of monomers and inert components. Again, these gels are mostly used for chromatography and are sold as beads under the tradename Spheron (Lachema or Realco Chem. Co.). There are no reports on their use as supports for electrophoresis, although it may be anticipated that, due to their ester groups in both monomers, they might be readily hydrolyzed during electrophoresis at basic pH.

11.2.4.4 Acryloylmorpholine/bisacrylylpiperazine gels

In 1984, Artoni et al. [57] reported a couple of novel monomers, imparting peculiar properties to polyacrylamide-type gels: acryloylmorpholine (ACM) and bisacrylylpiperazine (BAP). These gels exhibited some interesting features: due to the presence of the morpholino ring on the nitrogen involved in the amido bond, the latter was rendered extremely stable against alkaline hydrolysis, which bedevils conventional polyacrylamide matrices. In addition, such matrices are fully compatible with a host of hydro-organic solvents, thus allowing electrophoresis in mixed phases. The formula of the monomers is shown in Fig. 11.6. However, such matrices have found little applications so far, even

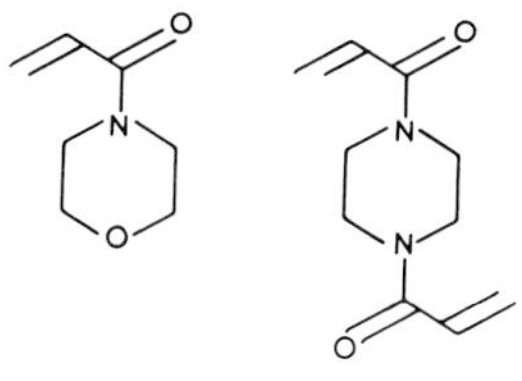

ACM BAP

Fig. 11.6. Chemical structure of acryloylmorpholine (ACM, left) and bisacrylylpiperazine (BAP, right) (by permission from Ref. 57).

though, recently, the use of the BAP crosslinker has been reported to have beneficial effects when silver-staining polyacrylamide matrices [58].

11.2.5 Hydrolink gels

This is a new, mechanically strong, gelatinous matrix, having electrophoretic resolving properties intermediate between those of polyacrylamide and agarose gels [59]. This matrix has the unique property of being amphiphilic (i.e. of swelling in both plain water and protic organic solvents) and seems particularly well suited for electrophoresis of DNA. It is compatible with organic solvents, including 50% dimethylsulfoxide, 50% tetramethylurea, 50% acetonitrile, and 50% tetrahydrofuran. The matrix is believed to consists of brush-like pillars, having a hydrophobic core and a hydrophilic coating (see Fig. 9 in Ref. 59). The latter is formed by short chains, protruding into the surrounding liquid and able to coordinate large amounts of water or polar solvent molecules. Perhaps this could be the birth of a new class of matrices. Even in chromatography, research seems to be moving along this direction. Thus, Gisch et al. [60] have recently described a novel type of stationary phase, which excludes proteins while interacting with small molecules and is called shielded hydrophobic phase (SHP). The SHP material consists of a polymeric bonded phase, containing hydrophobic regions enclaved by a hydrophilic network. The hydrophilic network forms a water-solvated interface through which small analytes, such as drugs, penetrate and interact with the hydrophobic groups, while large, water-solvated molecules, such as proteins, are prevented from such interactions by hydrophilic shielding. Under appropriate chromatographic conditions, the bulk protein will be eluted as an unretained band without affecting the retention of the smaller components. In the SHP packing, the hydrophilic network consists of bonded poly(ethylene oxide) with embedded hydrophobic phenyl groups. It will be of interest to see how both types of matrices will develop in the respective fields.

It might appear excessive to devote so much space to capillary anticonvective media in a general treatment of electrokinetic methodologies. Yet, it should be remembered that, up to the present time, the vast majority of electrophoretic analyses have been performed in gel matrices, so that an overview of such gels encompasses practically most of the

References on p. A514

history of electrophoresis. Of course, this does not apply to the work of research groups such as those of Bier [61], Hannig [62], or Everaerts [1], who have experimented mainly with free liquid phases.

11.3 DISCONTINUOUS ELECTROPHORESIS

In 1959 Raymond and Weintraub [40] described the use of polyacrylamide gels (PAG) in ZE. These gels had substantial advantages over the previously used starch gels: UV and visible transparency (starch gels are opalescent) and the ability to sieve macromolecules over a wide range of sizes. Adding that, in 1964, Ornstein [42] and Davis [43] created discontinuous (disc) electrophoresis by applying to PAG a series of discontinuities (of leading and terminating ions, pH, conductivity, and porosity) and thus further increasing the resolving power of electrophoresis, we can see that electrophoresis in the early Sixties made a great leap forward. In disc electrophoresis (the principle of which is outlined in Fig. 11.7) the proteins are separated on the basis of two parameters: surface charge and molecular mass. (Coincidentally, when disc electrophoresis is performed in tubes, disc-shaped zones are formed.) Ferguson [63] showed that one can distinguish between the two by plotting the results of a series of experiments with polyacrylamide gels of varying porosity. For each protein under analysis, the slope of the curve log m_T

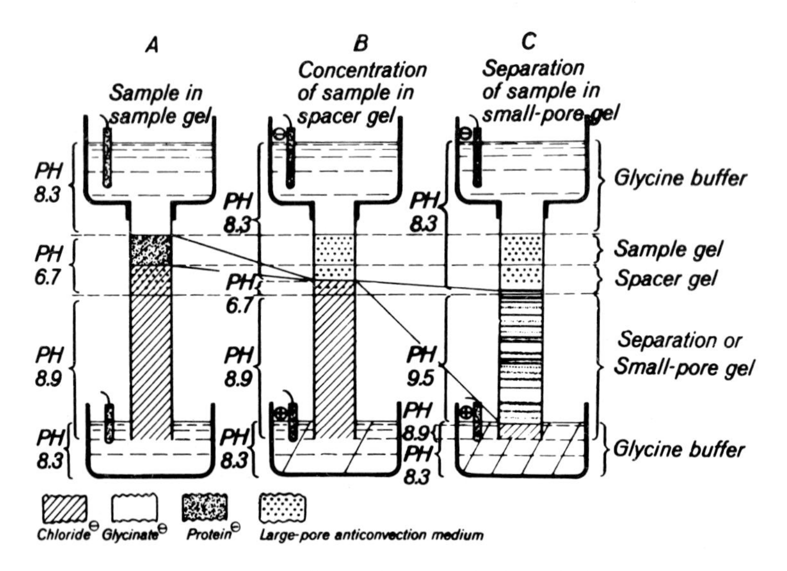

Fig. 11.7. Principle of disc electrophoresis. A: Sample in sample gel; B: sample concentration in stacking gel; C: sample separation in running gel. From top to bottom, the following phases are encountered: glycine buffer at pH 8.3 in the cathodic reservoir; sample gel and spacer gel, both titrated to pH 6.7; small-pore running gel, titrated to pH 8.9 and again glycine buffer in the anodic reservoir at the bottom (reproduced by permission from Ref. 42).

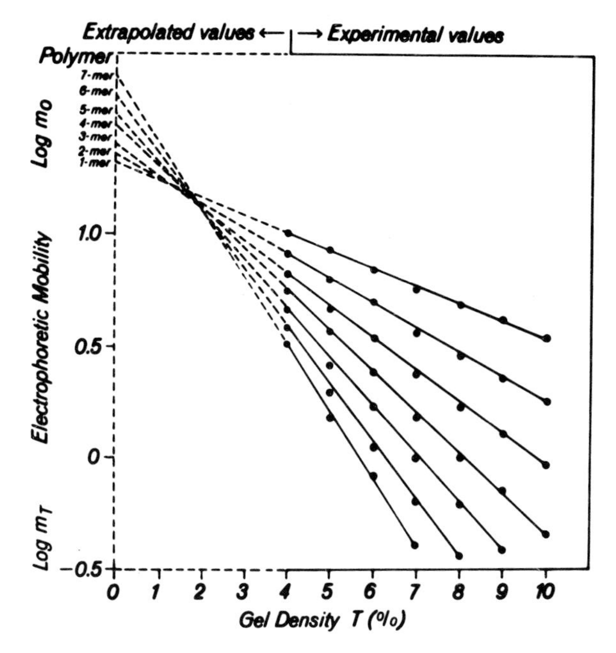

Fig. 11.8. Ferguson plots for polymeric forms of serum albumin. m_T = Relative mobility; %T = total concentration of monomers (acrylamide and Bis) in the gel. The meeting point of the extrapolated curves represents the mobility in the free phase. Parallel lines would indicate charge isomers, whereas lines of different slopes indicate size variants (reproduced by permission from Ref. 64).

(electrophoretic mobility) vs. gel density (%T) is proportional to molecular mass, while the *y*-intercept (Y_0) is a measure of surface charge (see Fig. 11.8) [64]. Being quite tedious, this technique has rarely been used in biochemical work. Moreover, recently, nonlinear Ferguson plots have been reported [65], casting some doubts on the validity of this approach.

11.4 ISOELECTRIC FOCUSING

New routes were thus explored toward a simpler assessment of charge and mass of proteins. As early as 1961 Svensson [66] experimented with IEF. It is an electrophoretic technique by which amphoteric compounds are fractionated according to their isoelectric point (p*I*) values along a continuous pH gradient. Contrary to ZE, where the constant pH of the separation medium establishes a constant charge density at the surface of the molecule and causes it to migrate with constant mobility (in the absence of sieving), the surface charge of an amphoteric compound in IEF keeps changing and decreasing according to its titration curve, as it moves along the pH gradient, approaching its steady-

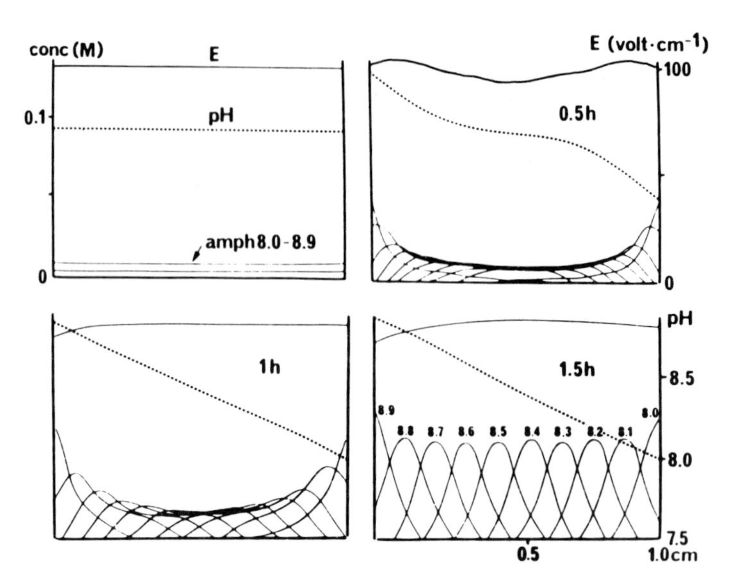

Fig. 11.9. Calculated time development of an isoelectric focusing process involving 10 ampholytes in a closed vessel. The isoionic points of the ampholytes are evenly distributed in the pH 8.0-8.9 range. The initial distribution of the amphoteric buffers is indicated in the upper left diagram. The calculation was performed assuming a constant voltage (100 V/cm) across the system. The anode is positioned to the right in the diagrams. Each *x*-axis represents the distance from the cathode on the same scale as the bottom right figure (reproduced by permission from Ref. 69).

state position, i.e. the region where the pH matches its p*I*. There, its mobility equals zero and the molecule comes to a stop.

The gradient is created and maintained by the passage of current through a solution of amphoteric compounds with closely spaced p*I* values, encompassing a given pH range. The electrophoretic transport causes these buffers (carrier ampholytes, CA) to stack according to their p*I* values, and a pH gradient, increasing from anode to cathode, is established. This process is shown in the computer simulation of Fig. 11.9 [69]. At the beginning of the run, the medium has a uniform pH, which equals the average p*I* of the CAs, and uniform CA concentration throughout (Fig. 11.9, upper left panel). At $t = 0.5$ h, the most acidic (p*I* 8.0) and the most alkaline (p*I* 8.9) species have collected close to the anode and cathode, respectively, so that portions of the pH gradient begin to develop. At $t = 1$ h, the ampholytes have separated further, and at this point an almost linear pH gradient has been established that spans the pH range defined by the p*I* values of the ampholytes. In the simulation, a system of 10 ampholytes, in equimolar ratios, has p*I* values spaced at 0.1 pH unit increments in the pH 8.0-8.9 range. At $t = 1.5$ h, the ampholytes have separated into symmetrical zones having overlapping Gaussian profiles, as predicted by Svensson [66]. At this point, the system has achieved a steady state, maintained by the electric field, and no further mass transport is expected, except from symmetric, to-and-fro micromovements of each species about its p*I*, generated by the action of two opposite forces, diffusion and voltage gradient, acting on each focused component. This pendulum movement, diffusion/electrophoresis, is the primary cause of the residual current under isoelectric steady-state conditions.

The technique only applies to amphoteric compounds and, more precisely, to good ampholytes with small $pI - pK_1$ differences, i.e. with a steep titration curve around their p*I*, conditio sine qua non for any compound to focus in a narrow band. This seldom creates a problem with proteins but this may be so with short peptides, which need to contain at least one acidic or basic amino acid residue, besides the $-NH_2$ and -COOH terminus (which would make them isoelectric between ca. pH 4 and pH 9 and prevent them from focusing). Another limitation with short peptides is encountered at the level of the detection methods: CAs react with most peptide stains. This problem may be circumvented by using specific stains when appropriate or by resorting to immobilized pH gradients (IPG) that show no background reactivity to ninhydrin and other common stains for primary amino groups (e.g., dansyl chloride, fluorescamine).

In practice, notwithstanding the availability of CAs covering the pH 2.5-11 range, the limit of CA/IEF is in the pH 3.5-10 interval. On statistical grounds, this poses a minor problem, as the p*I* values of most proteins cluster between pH 4 and 6, but it might well be a major one for specific applications. When a restrictive support like polyacrylamide (PAA) is used, a limit exists also for the size of the largest molecules exhibiting an acceptable mobility through the gel. A conservative evaluation sets an upper limit around 750 kD for standard techniques. The molecular form in which the proteins are separated strongly depends upon the presence of additives, like urea and/or detergents. Moreover, supramolecular aggregates or complexes with charged ligands can be focused only if their K_d is lower than 1 μM and if the complex is stable at $pH = pI$. Aggregates with higher K_d are easily split by the pulling force of the current, whereas most chromatographic procedures are unable to modify the native molecular form.

The general properties of the carrier ampholytes, i.e. of the amphoteric buffers used to generate and stabilize the pH gradient in IEF, can be summarized as follows:

1. Fundamental 'classical' properties: (a) buffering ion has a mobility of zero at p*I*; (b) good conductance; (c) good buffering capacity.
2. Performance properties: (a) good solubility; (b) no influence on detection system; (c) no influence on sample; (d) separable from sample.
3. 'Phenomena' properties: (a) plateau effect, drift of the pH gradient; (b) chemical change in the sample; (c) complex formation.

The fundamental and performance properties are required for a well-behaved IEF system, whereas the 'phenomena' properties are the drawbacks or failures inherent in the technique. For instance, the 'plateau effect' or 'cathodic drift' is a slow decay of the pH gradient with time, whereby, upon prolonged focusing at high voltages, the pH gradient with the focused proteins is lost in the cathodic compartment. There seems to be no remedy to it (except for abandoning CA/IEF in favor of the IPG technique), since there are complex physicochemical causes underlying it, including a strong electroosmotic flow, generated by the covalently affixed negative charges in the matrix (carboxyls and sulfate in both polyacrylamide and agarose).

References on p. A514

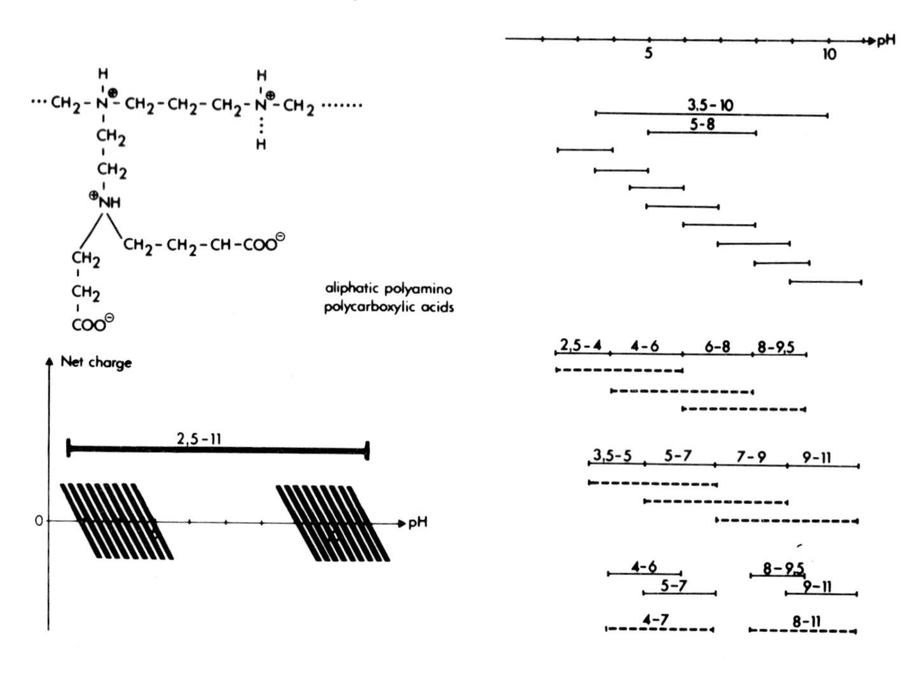

Fig. 11.10. Composition of carrier ampholytes for isoelectric focusing. Upper left: representative chemical formula (oligoamino-oligocarboxylic acids). Lower left: portions of titration curves of carrier ampholytes near the p*I*. Right: different, narrow pH intervals, obtained by subfractionation of the wide pH 3.5-10 gradient components (reproduced by permission from LKB Produkter AB).

In chemical terms, CAs are oligoamino, oligocarboxylic acids, available from different suppliers under different trade names (Ampholine from LKB Produkter AB, Pharmalyte from Pharmacia Fine Chemicals, Biolyte from Bio-Rad, Servalyte from Serva GmbH, Resolyte from BDH). There are two basic synthetic approaches: (a) the Vesterberg approach [67], consisting in allowing different oligoamines (tetra-, penta-, and hexaamines) to react with acrylic acid and (b) the Pharmacia synthetic process [68], which involves the copolymerization of amines, amino acids, and dipeptides with epichlorohydrin. The wide-range synthetic mixture (pH 3-10) seems to contain hundreds, possibly thousands of different amphoteric chemicals having p*I* values evenly distributed along the pH scale (Fig. 11.10).

CAs, from any source, should have a M_r around 750 D (size interval 600-900 D, the higher M_r referring to the more acidic CA species). Thus, they should easily be separable from macromolecules by gel filtration, unless they are hydrophobically complexed to proteins. Dialysis is not recommended, because the CAs tend to aggregate, but salting out of proteins with ammonium sulfate seems to eliminate any contaminant CAs completely. A further complication arises from the chelating effect of acidic CAs, especially towards Cu^{2+} ions, which could inactivate metallo-enzymes. In addition, focused CAs represent a medium of very low ionic strength (< 1 meq/l at steady state) [3]. Since the isoelectric state involves a minimum of solvation and thus of solubility for the protein macroion, there could be a tendency for some proteins, e.g. globulins, to precipitate during IEF in the proximity of the p*I*. This is a severe problem in preparative IEF, but in analytical procedures it can be lessened by reducing the total amount of sample applied.

11.5 SODIUM DODECYL SULFATE ELECTROPHORESIS

SDS electrophoresis was the next logical step after disc electrophoresis. While the latter discriminates among macromolecules on the basis of both size and surface charge, SDS electrophoresis fractionates polypeptide chains essentially on the basis of their size. It is therefore a simple, yet powerful and reliable method for M_r determination. In 1967 Shapiro et al. [70] first reported that electrophoretic migration in SDS is proportional to the effective molecular radius and, thus, to the M_r of the polypeptide chain. This means that SDS must bind to proteins and cancel out differences in molecular charge, so that all components will migrate solely according to size. Surprisingly large amounts of SDS appear to be bound (an average of 1.4 g SDS/g protein). This means that the number of SDS molecules bound is of the order of half the number of amino acid residues in a polypeptide chain. This amount of highly charged surfactant molecules is sufficient to overwhelm effectively the intrinsic charges of the polymer coil, so that their net charge per unit mass becomes approximately constant. If migration in SDS (and disulfide reducing agents, such as 2-mercaptoethanol, in the denaturing step, for a proper unfolding of the proteins) is proportional only to M_r, then, in addition to canceling out charge differences, SDS also equalizes molecular shape differences (e.g., globular vs. rod-shaped molecules). This seems to be the case for protein/SDS mixed micelles. These complexes can be assumed to behave as ellipsoids of constant minor axis (ca. 1.8 nm) and a major axis proportional to the length of the amino acid chain (i.e. to molecular mass) of the protein. The rod length for the 1.4 g SDS/g protein complex is of the order of 0.074 nm per amino acid residue.

A few words should be added about the properties and use of surfactants in the analysis of macromolecules and biomembranes. Four major classes of surfactants are commonly used in biological work: anionic, cationic, zwitterionic, and nonionic. In addition to a polar group, each surfactant possesses a hydrophobic moiety, and the combination of hydrophobic and hydrophilic portions of the amphiphilic molecule provides the basis for detergent action. Nonionic surfactants form a special class, since they are virtually never a single molecular species, but rather a group of structurally related compounds. This is due to a statistical distribution of chain lengths, produced in the manufacture of these surfactants by the polymerization of ethylene oxide. The result, at best, is a single hydrophobic moiety to which polyethylene chains of variable lengths are attached. Due to their amphiphilic properties, all surfactants, when dispersed in water, tend to form aggregates

References on p. A514

in which the hydrophobic portion of the molecules is protected from the solvent. These occur as monolayers at the solvent interface or as micelles in solution. Of these two, the micelle is our primary concern, since this is the functional unit in protein solubilization. In the micelle (containing a variable number of monomers), molecules are arranged in such a way that polar groups point out and are thus exposed to water, while the hydrophobic segments are buried in the interior, shielded from water. Clusters of SDS are ellipsoidal, being some 2.4 nm across the minor axis and composed of ca. 100 SDS molecules (depending on ionic strength and type of solvent). Surfactants in solution are characterized by an important parameter, called critical micelle concentration (CMC), which is defined as the concentration of surfactant at which micelles form. Below the CMC, the monomer concentration naturally increases as a function of total surfactant, as no micelle exists. However, above the CMC value, added surfactant raises the concentration of micelles, while leaving constant the monomer concentration, since the CMC is an effective solubility limit of surfactant monomer. Each surfactant is further characterized by two other functional parameters, in addition to the CMC value: an aggregation number (i.e. the number of monomers in the cluster) and the micellar molecular mass. Two phenomena are generally seen: (a) a single surfactant can exist in different micellar sizes (depending on ionic strength and solvent type) and (b) among the different classes of surfactants, the micellar mass can vary enormously. For further information, the review of Helenius and Simons [71] is recommended.

In SDS electrophoresis, the proteins can be prelabeled with dyes that covalently bind to their $-NH_2$ residues. The dyes can be conventional (like the blue dye Remazol) or fluorescent, like dansyl chloride, fluorescamine, *o*-phthaldialdehyde, and MDPF (2-methoxy-2,4-diphenyl- 3[2*H*]-furanone). Prelabeling is compatible with SDS electrophoresis, as the size increase is minimal, but would be anathema in disc electrophoresis or IEF, as it would generate a series of bands of slightly altered mobility or p*I* from an otherwise homogeneous protein.

For data treatment, the sample and M_r standards are electrophoresed side-by-side in a gel slab. After detection of the polypeptide zones, the migration distance (or R_f) is plotted against log M_r to produce a calibration curve (Fig. 11.11) [72] from which the M_r of the sample can be calculated. It should be noted that, in a gel of constant %T, linearity is obtained only in a certain range of molecular sizes. Outside this limit (in Fig. 11.11, above 60 kD) a new gel matrix of appropriate porosity should be used.

Two classes of proteins show anomalous behavior in SDS electrophoresis: glycoproteins (because their hydrophilic oligosaccharide units prevent hydrophobic binding of SDS micelles) and strongly basic proteins, e.g., histones (because of electrostatic binding of SDS micelles through their sulfate groups). The first can be partially alleviated by using alkaline Tris/borate buffers [73], which will increase the net negative charge on the glycoprotein and thus produce migration rates well correlated with molecular size. The migration of histones can be improved by using pore gradient gels and allowing the polypeptide chains to approach the pore limit [74].

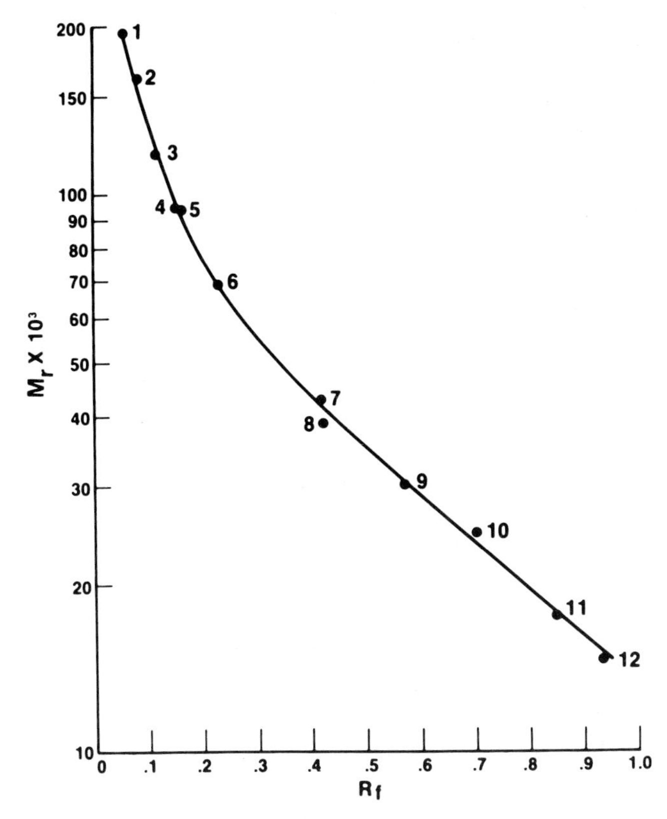

Fig. 11.11. Typical log M_r vs. R_f plot after SDS/PAGE. Note that the plot, under these experimental conditions, is linear only in the M_r range 15 to 60 kD. Markers: 1 = myosin; 2 = RNA polymerase (subunits); 3 = β-galactosidase; 4 = phosphorylase B; 5 = RNA polymerase (α-subunit); 6 = bovine serum albumin; 7 = ovalbumin; 8 = RNA polymerase (β-subunit); 9 = carbonic anhydrase; 10 = trypsinogen; 11 = β-lactoglobulin; 12 = lysozyme (by permission from Ref. 72).

11.6 POROSITY GRADIENT GELS

When macromolecules are electrophoresed in a continuously varying matrix concentration (which results in a porosity gradient) rather than in a gel of uniform concentration, the protein zones are compacted along their track, as the band front is, at any given time, at a gel concentration somewhat higher that the rear of the band, so that the former is decelerated continuously. A progressive band sharpening thus results. There are other reasons for resorting to gels of graded porosity. We have seen that disc electrophoresis separates macromolecules on the basis of both size and charge differences. If the influence of molecular charge could be eliminated, then clearly the method could be used with a suitable calibration for measuring molecular size. This has been accomplished by overcoming charge effects in two main ways. In one, a relatively large amount of charged ligand, such as SDS, is bound to the protein, effectively swamping the initial charges present on the protein molecules and giving a quasi-constant charge-to-mass ratio. However, in SDS electrophoresis proteins are generally dissociated into their constituent polypeptide subunits, and the concomitant loss of functional integrity and antigenic properties cannot be prevented. Therefore, the size of the original, native molecule must be evaluated in the absence of denaturing substances.

In the second method for M_r measurements, this can be done by relying on a mathematical canceling of charge effects, following measurements of the mobility of native proteins in gels of different concentrations. This is the so-called 'Ferguson plot' [63], discussed in Section 11.3. However, this approach to M_r determination has never found widespread application, as it is tedious and expensive in terms of time and sample material.

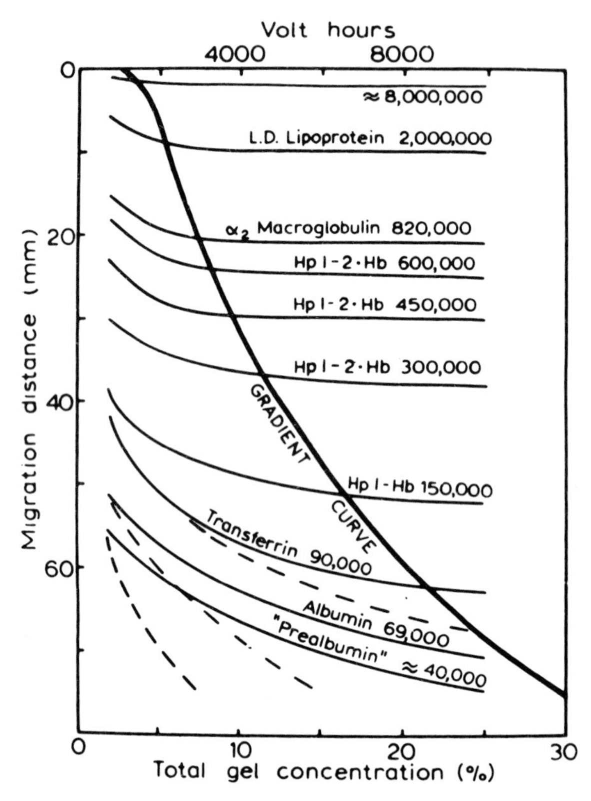

Fig. 11.12. Plots of the mobilities of serum proteins against time for pore gradient gels. Solid lines: uniform crosslinking (5%C); dashed lines: variable crosslinking (by permission from Ref. 75).

This leads us to a third method for molecular size measurements: the use of gels of graded porosity. This represents a very useful addition to the array of gel electrophoretic techniques presently available, and is characterized by high resolving power and relative insensitivity to variability in experimental conditions. The principle of the technique is shown in Fig. 11.12. It can be seen that, under appropriate conditions (at least 10 kV x hours), the mobility of most serum proteins becomes constant and eventually ceases as each constituent reaches a gel density region in which the average pore size approaches the diameter of the protein (pore limit) [75]. Thus, the ratio between the migration distance of one protein to that of any other becomes a constant after the proteins have all entered a gel region in which they are subjected to drastic sieving conditions. This causes the electrophoretic pattern to become constant after prolonged migration in a gel gradient. The gel concentration at which the migration rate for a given protein becomes constant is called the 'pore limit': if this porosity is properly mapped with the aid of a suitable set of marker proteins, it is possible to correlate the migration distance to the molecular mass of any constituent in the mixture.

After electrophoresis is over and the proper experimental data are gathered, they can be handled in two ways: by either two-step or one-step methods. Among the former, the most promising approach appears that of Lambin and Fine [76], who observed that there is a linear relationship between the migration distance of proteins and the square root of electrophoresis time, provided that time is kept between 1 and 8 h. The slopes of the regression lines of proteins in the above graph are clearly an indication of molecular size. When the slopes of the various regression lines thus obtained are plotted against the respective molecular masses, a good linear fit is obtained, which allows M_r measurements of proteins between 2×10^5 and 10^6 D. The shape of the proteins (globular or fibrillar), their carbohydrate content (up to 45%) and their free electrophoretic mobilities (between 2.1 and 5.9×10^{-5} cm^2 V^{-1} sec^{-1}) do not seem to be critical for proper M_r measurements by this procedure. One-step methods have been described by Rothe and Purkhanbaba [77]. These authors have found that, when log M_r is plotted against either D (distance migrated) or %T (acrylamide + Bis), a nonlinear correlation is always obtained. However, when log M_r is plotted vs. $\sqrt{\%T}$ or $\sqrt{D}$, a linear regression line is obtained, which allows the accurate determination of M_r values of proteins. The correlations log $M_r/\sqrt{\%T}$ or log $M_r/\sqrt{D}$ are not significantly altered by the duration of electrophoresis. Therefore, a constant M_r value should be obtained for a stable protein, no matter how long electrophoresis has been going on. The reliability of PAGE with a homogeneous buffer system and a linear porosity gradient is excellent, as indicated by an average standard deviation of $\pm$3.7%. More recently, Rothe and Maurer [78] have demonstrated that the relationship log M_r vs. $\sqrt{D}$ is also applicable to SDS electrophoresis in linear polyacrylamide gel gradients.

11.7 TWO-DIMENSIONAL MAPS

By sequentially coupling pure-charge (IEF) to pure-size (SDS/PAGE; the latter orthogonal to the first) fractionations one can distribute the polypeptide chains two-dimensionally, with charge and mass as coordinates (IEF/SDS or ISO/DALT, according to the Andersons' nomenclature; ISOelectric for charge and DALT for mass separation) [79]. When

References on p. A514

the first dimension is performed in immobilized pH gradients, the technique is called IPG/DALT [4]. The technique was reported for the first time by Barrett and Gould [80] and then described in more detail in 1975 by O'Farrell [81], Klose [82], and Scheele [83]. Large gels (e.g., 30 x 40 cm) [84] and prolonged exposure to radiolabeled material (up to two months) have allowed the resolution of as many as 12 000 labeled peptides in a total mammalian cell lysate. Thus, there is a good chance that, in a properly prepared 2-D map, a spot will represent an individual polypeptide chain, uncontaminated by other material. On this assumption, and provided that enough material is present in an individual spot (about 1 μg), it is possible to blot it on a fiberglass filter and perform sequencing on it [85].

11.8 ISOTACHOPHORESIS

The theoretical foundations of ITP were laid in 1897 by Kohlrausch [86], who published a theoretical treatment of the conditions at a migrating boundary between two salt solutions, showing that the concentrations of ions at the boundary were related to their effective mobilities (Kohlrausch autoregulating function). The term isotachophoresis underlines the most important aspect of this technique, namely the identical velocities of the sample zones at equilibrium [87]. ITP will take place when an electric field is applied to a system of electrolytes, consisting of:

(1) a leading electrolyte, which must contain only one ionic species, the leading ion (L^-), having the same sign as the sample ions to be separated, and an effective mobility higher than that of any sample ions;

(2) a second, terminating electrolyte, which contains one ionic species, the terminating ion (T^-) having the same sign as the sample ions to be separated, and an effective mobility lower than that of any sample ions;

(3) an intermediate zone of sample ions, having the same sign as the leading and terminating ions, and intermediate mobilities.

The three zones are juxtaposed, with the proviso that sharp boundaries must be created at the start of the experiment. The polarity of the electric field must be such that the leading ion migrates in front of the ITP train at all times. When the system has reached the steady state, all ions move with the same speed, individually separated into a number of consecutive zones, in immediate contact with each other, and arranged in order of effective mobilities. Once the ITP train is formed, the ionic concentration in a separated sample zone will adapt itself to the concentration of the preceding zone. The Kohlrausch equation, which is given at the boundary leading/terminating ions, will in fact give the conditions at any boundary between two adjacent ions (A^-, B^-) with one common counterion (R^+) when the boundary migrates in the electric field.

There are two fundamental properties of ITP built into the autoregulating function: (a) the concentrating effect and (b) the zone-sharpening effect. Suppose Component A^- is introduced at very low concentration, even lower than that of the terminating ion, T^-. By virtue of the Kohlrausch equation, since the mobility of A^- is intermediate between that of

L^- and T^-, its concentration will also have to be intermediate between those of L^- and T^-. This will result in A^- being concentrated (decrease in zone length) until it reaches the theoretically defined concentration. Conversely, if the concentration of A^- is too high (even higher than that of L^-), the A^- zone will be diluted (zone lengthening) until the correct equilibrium concentration is reached. Here is an example of Property b: If A^- diffuses into the B^- zone, it is accelerated, since it enters a zone of higher voltage gradient; therefore, it automatically migrates back into its zone. Conversely, if it enters the L^- zone, it finds a region of lower voltage, is decelerated, and thus falls back into its zone. This applies to all ions in the system. It should be noted that the first few minutes of disc electrophoresis, during migration in the sample and stacking gels, represent in fact an ITP migration; this is what produces the spectacular effects of disc electrophoresis: stacking into extremely thin starting zones (barely a few μm thick, from a zone width up to 1 cm in thickness, a concentration factor of 1000 to 10 000) and sharpening of the zone boundaries. An extensive treatise of ITP can be found in the 4th edition of this book [1, p. A345]. A characteristic of ITP is that the peaks, unlike in other separation techniques (except for displacement chromatography), are square-shaped, as a result of the Kohlrausch autoregulating function, i.e. the concentration of the substance within a homogeneous zone is constant from front to rear boundary [88]. In fact, under ideal conditions, the diffusive forces, which cause an eluted peak to spread into a Gaussian shape, are effectively counteracted.

11.9 IMMUNOELECTROPHORESIS

Two different principles – simple and crossed – exist for IEP. Simple IEP consists of a combination of electrophoresis and immunodiffusion in gels. The method is based on the fact that in agarose gels the movement of molecules in an electric field is similar to that in a liquid medium, with the advantage that free diffusion during and after electrophoresis is reduced. The constituents of a mixture are then defined both by their electrophoretic mobilities and by their antigenic specificities.

The basic technique was described for the first time by Grabar and Williams in 1953 [9]. The antigen sample is applied to a hole placed in the middle of a glass plate, coated with a 1-mm layer of agarose gel. Electrophoresis separates the antigen mixture into various zones. A longitudinal trough, parallel to the path of electromigration, is then made in the gel and filled with the antiserum against the antigen mixture. A passive, double diffusion takes place; the antiserum diffuses into the gel (advancing with a linear front) while the antigens diffuse radially in all directions from the electrophoresis zones. The antigen/antibody complexes are formed at equivalence points, the number of precipitates formed corresponding to the number of independent antigens present. The precipitates take the shape of a system of arcs, resulting from the combination of linear and circular fronts. A number of variants have been described over the years. Some of the most popular among them are reported below.

11.9.1 Rocket immumoelectrophoresis

In this technique, the antigen/antibody reaction occurs during the electrophoresis of an antigen in an antibody-containing medium. Both antigen and antibody move according to their electrophoretic mobilities, and they also react with each other, resulting in flame-shaped precipitation zones of antigen/antibody complexes (Laurell's rocket technique) [89]. Under the influence of the electric field, unbound antigen within the peak of the flame-shaped precipitate migrates into the precipitate, which redissolves in the excess antigen. Thus, the leading edge of the flame is gradually displaced in the direction the antigen is migrating. The amount of antigen within the leading boundary edge is successively diminished because of the formation of soluble antigen/antibody complexes. When the antigen is diminished to equivalence with the antibody, the antigen/antibody complexes can no longer be dissolved and a stable precipitate forms at the leading edge, which is thereafter stationary. The distance finally traveled by the peak depends on the relative excess of antigen over antibody and can be used as a measure of the amount of antigen present. Every precipitation band featured as a flame represents an individual antigen [89].

11.9.2 Crossed immunoelectrophoresis

Rocket immunoelectrophoresis actually works best with single antigens or monospecific antibodies. In the case of polydisperse samples, crossed immunoelectrophoresis (CIE) according to Clarke and Freeman [90] is employed. The principle is illustrated in Fig. 11.13A. A mixture of protein antigens is subjected to conventional electrophoresis in agarose. Then the agarose gel is cut into strips, which are transferred to a second glass plate. Melted agarose (1% w/v) containing the antiserum is then cast to form a gel of the same thickness as that of the first-dimension gel and so as to have a secure junction between the two layers by melting them to fuse the edges. During the second-dimension electrophoresis, which is performed perpendicular to the first dimension, each antigen migrates independently, and precipitation zones are formed which resemble the Laurell rockets but have a wider base. A schematic diagram of the overall pattern from human serum is shown in Fig. 11.13B. It is remarkable that two antigens present in a single zone, i.e., possessing equal electrophoretic mobilities, can be distinguished by the second-dimension electroimmune assay as the shape and height of the respective rockets generally do not exactly coincide. In a typical product of the Clarke and Freeman method, ca. 50 rockets can be counted in human serum, and their relative abundance can be assessed by measuring the areas of the respective peaks. Peak assessment and evaluation have recently been highly facilitated by the development of a computer system for the specific analysis of CIE patterns [91].

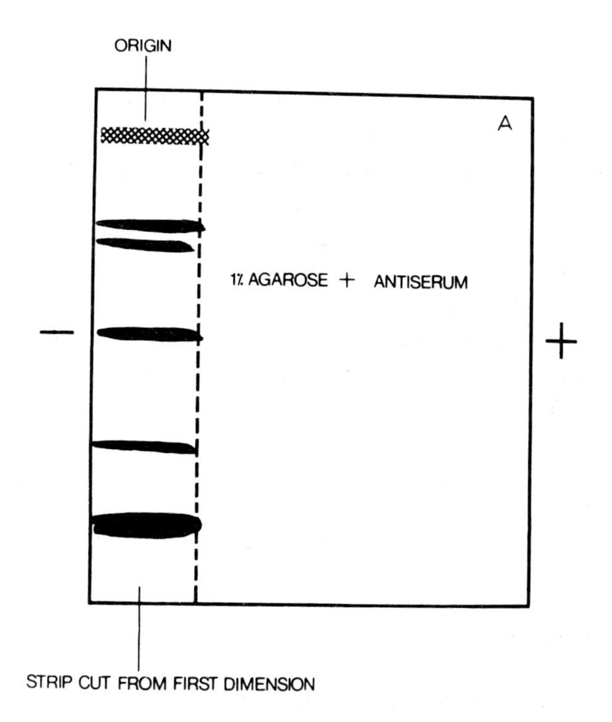

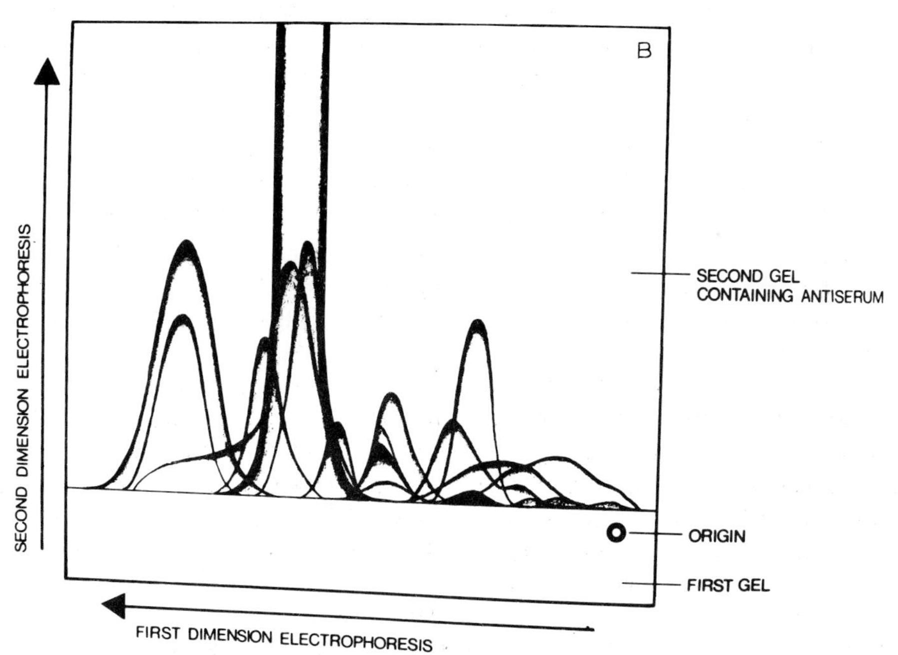

Fig. 11.13. Crossed immunoelectrophoresis. A: First dimension. The strip on the left of the dashed line is transferred to a second plate for electrophoresis in the second dimension (B). Analysis of human serum shown. The albumin peak pierces the plate, as its concentration is much too high compared to that of other serum proteins (by permission from

References on p. A514

11.9.3 Tandem crossed immunoelectrophoresis

It would be quite difficult to identify a single antigen in such a complex pattern as in Fig. 11.13B. One of the proposed methods for complex mixtures is the tandem CIE technique. Before the electrophoresis in the first dimension, two wells rather than one are cut into the gel strip, positioned one after the other in the direction of electrophoretic migration. One well is loaded with a mixture of the antigens to be analyzed, while the pure antigen the peak of which is to be identified in the mixture is introduced in the second well. All the remaining manipulations are then performed according to the classical CIE. In the final pattern there will be two peaks which fuse smoothly and are separated by exactly the distance between the two wells. These two peaks may be of different heights (due to a difference in antigen concentration in the sample and reference wells), whereas the smooth transition between them is indicative of the fact that they are caused by the same protein antigen. It is this canceling of the inner flanks of the two rockets and the fusion process that allows the unknown antigen to be located. By repeating this process with different, purified antigens, it is possible to map, one by one, the different components of a heterogeneous mixture [92].

11.9.4 Intermediate gel crossed immunoelectrophoresis

The tandem CIE, described above, aimed at the identification of a single antigen in a mixture, requires the availability of a pure marker antigen. In an alternative approach, identification can also be made when monospecific antiserum to the antigen is available. This approach is called intermediate gel CIE. After the first-dimension electrophoresis, the slab is prepared as for conventional CIE. Then a 1- to 2-cm strip of polyspecific antiserum-containing agarose gel nearest the first-dimension gel is excised and a monospecific antiserum-containing agarose is cast instead. In the course of the second-dimension electrophoresis, the antigen to be detected will precipitate with the specific antiserum just as it enters the intermediate gel, while other antigens will pass through this gel without being retarded. The bases of their precipitation peaks, formed in the polyspecific antiserum-containing agarose, will be positioned on the borderline between the two latter gels. The antigen under study is thus distinguishable from the others [93].

11.9.5 Fused-rocket crossed immunoelectrophoresis

Contrary to the Laurell rocket, this method is not used for quantitative determinations. Rather, it is applied to detect heterogeneity in seemingly homogeneous protein fractions, obtained by gel chromatography or ion-exchange chromatography. A set of wells is punched in a checker-board pattern into a strip of antibody-free agarose gel. Aliquots of each fraction eluted from the column are placed in the wells, and the gel is left in a humid chamber for 30-45 min to allow the proteins in the wells to migrate and fuse, thus reproducing the continuous elution profile on an extended scale (hence the term fused rockets). On the remainder of the plate a thin layer of agarose is cast, containing a polyvalent antiserum, and electrophoresis is performed as usual. If the collected chromatographic peak is indeed homogeneous, a single fused rocket will appear, while nonhomogeneity will be revealed by an envelope of sub-peaks within the main eluate fraction [94].

In this section only some of the main forms of immunoelectrophoretic techniques have been reviewed. For a more extensive treatise, the reader is referred to certain special issues of *Scand. J. Immunol.* [95-97].

11.10 STAINING TECHNIQUES

Staining methods have been reviewed up to 1975 [98]. In 1979 Merril et al. [99] described a silver-staining procedure in which the sensitivity, which for Coomassie Blue is merely of the order of a few μg/zone, is increased to a few ng of protein/zone, thus approaching the sensitivity of radioisotope labeling. In the gilding technique of Moeremans et al. [100] polypeptide chains are coated with 20-nm particles of colloidal gold and detected with a sensitivity of <1 ng/mm^2. Proteins can also be stained with micelles of Fe^{3+}, although the sensitivity is about one order of magnitude lower than with the gold micelles [101]. These last two staining techniques became a reality only after the discovery of yet another electrophoretic method, the so called Southern [102] and Western [103] blots, in which nucleic acids or proteins are transferred from hydrophilic gels to nitrocellulose or any of a number of other membranes, where they are immobilized by hydrophobic adsorption or covalent bonding. The very large porosity of these membranes makes them accessible to colloidal dyes. In addition, transfer of proteins to thin membranes greatly facilitates detection by immunological methods. This has resulted in new, high sensitivity methods called 'immunoblotting'. After saturation of potential binding sites, the antigens transferred to the membrane are first made to react with a primary antibody, and then the precipitate is detected with a secondary antibody, tagged with horseradish peroxidase, alkaline phosphatase, gold particles, or biotin, which will then be allowed to react with enzyme-linked avidin [104]. In all cases, the sensitivity is greatly augmented. It appears that 'immunoblotting' will pose a serious threat to quite a few of the standard immunoelectrophoretic techniques described above [105]. In terms of colloidal staining, a direct staining method for polyacrylamide gels with colloidal particles of Coomassie Blue G-250 is said to be as sensitive as silver-staining [106].

11.11 IMMOBILIZED pH GRADIENTS

In 1982, immobilized pH gradients (IPG) were introduced, resulting in an increase in resolution by one order of magnitude compared with conventional IEF [107]. By 1980, it was apparent to many IEF users that there were some inherent problems with the technique, which had not been corrected in more than 20 years of use and were not likely to be solved. They can be summarized as follows: (a) very low and unknown ionic strength; (b) uneven buffering capacity; (c) uneven conductivity; (d) unknown chemical environment; (e) inapplicability of pH gradient engineering techniques; (f) cathodic drift (pH

References on p. A514

TABLE 11.1

ACIDIC ACRYLAMIDO BUFFERS

Reprinted by permission from Ref. 108.

pK^*	Formula	Name	M_r	Source
1.0	$CH_2=CH-CO-NH-C(CH_3)(CH_3)-CH_2-SO_3H$	2-Acrylamido-2-methylpropane-sulfonic acid (AMPS)	207	Polysciences Inc.
3.1	$CH_2=CH-CO-NH-CH(OH)-COOH$	2-Acrylamido-glycolic acid	145	P.G. Righetti
3.6	$CH_2=CH-CO-NH-CH_2-COOH$	*N*-Acryloylglycine	129	Pharmacia-LKB Biotechnology
4.6	$CH_2=CH-CO-NH-(CH_2)_3-COOH$	4-Acrylamido-butyric acid	157	Pharmacia-LKB Biotechnology

* The p*K* values for the three Immobilines and for 2-acrylamidoglycolic acid are given at 25°C; for AMPS (p*K* 1.0) the temperature of p*K* measurement is not reported.

gradient instability); (g) low sample capacity. In particular, a most vexing phenomenon was the near-isoelectric precipitation of samples of low solubility at the isoelectric point or of components present in large amounts in heterogeneous samples. The inability to reach stable steady-state conditions (resulting in a slow pH gradient loss at the cathodic gel end) and to obtain narrow and ultranarrow pH gradients, aggravated matters. Perhaps most annoying was the lack of reproducibility and linearity of pH gradients produced by the so-called 'carrier ampholyte' buffers [3]. IPG proved to solve all these problems.

IPG are based on the principle that the pH gradient, which exists prior to IEF itself, is copolymerized, and thus rendered insoluble, within the fibers of a polyacrylamide matrix. This is achieved by using, as buffers, a set of six commercial chemicals (called Immobiline, by analogy with Ampholine, produced by Pharmacia-LKB Biotechnologies) having p*K* values distributed in the pH 3.6-9.3 range. Until now, not much was known about the Immobiline chemicals, except that they are acrylamido derivatives, with the general formula: $CH_2=CH\text{-}CO\text{-}NH\text{-}R$, where R denotes a set of two weak carboxyls, with p*K* values of 3.6 and 4.6, for the acidic compounds, and a set of four tertiary amino groups, with p*K* values of 6.2, 7.0, 8.5, and 9.3, for the basic buffers. These structures have now been

TABLE 11.2

BASIC ACRYLAMIDO BUFFERS

Reprinted by permission from Ref. 108.

pK^*	Formula	Name	M_r	Source
6.2	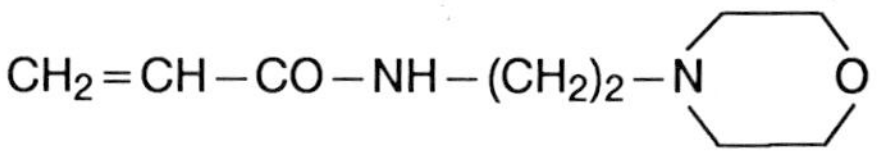 $CH_2=CH-CO-NH-(CH_2)_2-N(CH_2CH_2)_2O$ (morpholine ring)	2-Morpholino-ethylacrylamide	184	Pharmacia-LKB Biotechnology
7.0	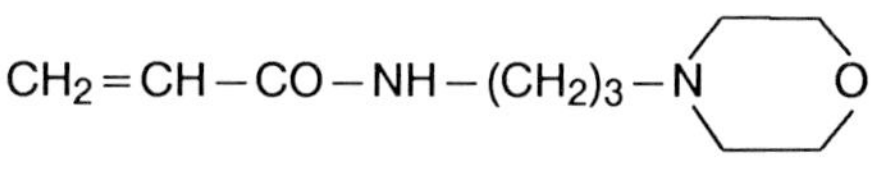 $CH_2=CH-CO-NH-(CH_2)_3-N(CH_2CH_2)_2O$ (morpholine ring)	3-Morpholino-propylacrylamide	198	Pharmacia-LKB Biotechnology
8.5	$CH_2=CH-CO-NH-(CH_2)_2-N(CH_3)-CH_3$	*N,N*-Dimethyl-aminoethyl-acrylamide	142	Pharmacia - LKB Biotechnology
9.3	$CH_2=CH-CO-NH-(CH_2)_3-N(CH_3)-CH_3$	*N,N*-Dimethyl-aminopropyl-acrylamide	156	Pharmacia - LKB Biotechnology
10.3	$CH_2=CH-CO-NH-(CH_2)_3-N(C_2H_5)-C_2H_5$	*N,N*-Diethyl-aminopropyl-acrylamide	184	P.G. Righetti
>12	$CH_2=CH-CO-NH-(CH_2)_2-N^+(C_2H_5)_2-C_2H_5$	*N,N,N*-Triethyl-aminoethyl-acrylamide (QAE)	198	IBF

* All pK values (except for pK 10.3) measured at 25°C. The value of pK 10.3 refers to 10°C.

established and their synthesis and purification processes have been described [108]. Tables 11.1 and 11.2 list the compounds now available, their chemical names, formulae, and physicochemical data. There are quite a few more than the six buffers commercially available from Pharmacia-LKB [107]. Ten of them are listed: Eight are weak acids and bases, with pK values covering the pH range 3.1-10.3, while the other two are a strongly acidic (pK 1.0) and a strongly basic (pK > 12) titrant, introduced in 1984 by Gianazza et al. [109] for producing linear pH gradients covering the entire pH 3-10 range. Computer simulations had shown that, in the absence of these two titrants, extended pH intervals would exhibit strong deviations from linearity at the two extremes, as the most acidic and most basic of the commercial Immobilines would act simultaneously as buffers and titrants [110]. The recently synthesized 2-acrylamidoglycolic acid (p*K* 3.1) [111] is useful for separating strongly acidic proteins, since it extends the pH gradient to as low as pH 2.5. Also, *N,N*-diethylaminopropylacrylamide (p*K* 10.3) was recently utilized for analysis of strongly alkaline proteins [112,113]. Given the fairly evenly spaced p*K* values along the pH scale, it is clear that the set of 10 chemicals proposed here (8 buffers and 2 titrants) is quite adequate to ensure linear pH gradients along the pH 2.5-11 axis (the ideal ΔpK for linearity would be 1 pH unit between two adjacent buffers). The rule $\Delta pK = 1$ is fairly well respected, except for two 'holes': between the p*K* 4.6 and 6.2 and between the p*K* 7.0 and 8.5 species. For a more detailed treatise on how to use an IPG gel and IPG recipes, the reader is referred to an extensive manual [4] and to a recent review [114]. Due to the much increased resolution of IPG, quite a number of so-called 'electrophoretically silent' mutations (bearing amino acid replacements with no ionizable groups in the sidechains) have now been fully resolved.

11.12 PULSED-FIELD GEL ELECTROPHORESIS

In 1984, Schwartz and Cantor [115] introduced another electrophoretic technique in the field of nucleic acids: pulsed-field gel electrophoresis (PFGE). With this technique, it is possible to separate mega fragments of DNA (up to 10 mega base pairs), while only a few

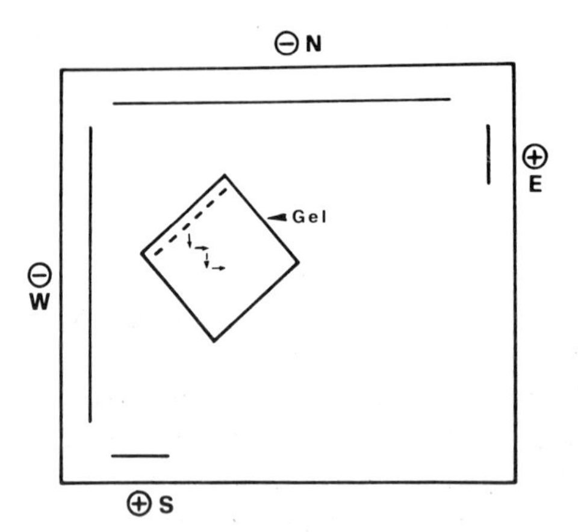

Fig. 11.14. Principle of pulsed-field gel electrophoresis. Series of point electrodes coupled at 120° in the N/S and E/W directions (or in a hexagonal array) are placed along the perimeter of the electrophoretic cell. The N/S and E/W pulses are generated by a microprocessor, coupled to a standard power supply. The DNA fragments migrate in a zig-zag fashion. The slots at the N/W corner are for sample application (DNA inserts). The symbols – and + represent the electrode polarity (by permission from Ref. 117).

years ago the maximum allowable size in agarose gels was barely 20 kilo base pairs. Fig. 11.14 outlines the principle. The electrophoretic chamber is of the 'submarine' type (i.e., the gel is submerged under a thin veil of buffer), but it contains electrodes all along the periphery, coupled at an angle generally >90°, an angle of ca. 120° appearing best for resolution. The electrode couples are called North/South (N/S) and East/West (E/W). The gel is a square with a 20-cm side and is placed at 45° angle at the cell center. The electric field is activated alternately between the N/S and E/W poles, with pulses that can last from milliseconds up to tens of minutes. In the example shown in Fig. 11.14, the DNA molecules, exposed to alternating electric fields, first migrate to the South, than to the East, etc., in a zig-zag path. In the electric field the DNA helices are able to orient themselves and penetrate the pores of the agarose gel via their short axes. In this way, molecular sieving is no longer operating, so that in principle, small and large fragments should migrate together. However, in PFGE they are still separated on the basis of molecular size. This can be explained by assuming that, at each pulse, the molecules spend progressively larger fractions of time in relaxation (i.e. in changing shape and reorienting themselves). Thus, smaller fragments spend more time in net migration while larger DNAs loose extra time in reorienting themselves. According to Olson [116], nucleic acids are assumed to be wedge-shaped, with two limiting configurations. In the passage between these two structures, some molecules could acquire a transient conformation which keeps resonating with pulses of a given frequency and, thus, they never leave the application point. This is in fact what happens in field inversion electrophoresis. A recent review [117] will apprize the readers of progress in that field.

11.13 CAPILLARY ZONE ELECTROPHORESIS

Capillary zone electrophoresis (CZE) appears to be a most powerful technique, perhaps equalling the resolving power of IPG. If one assumes that longitudinal diffusion is the only significant source of zone broadening, then the number of theoretical plates (*N*) in CZE is given by [118]

$$N = \mu V/2D$$

where μ and D are the electrophoretic mobility and diffusion coefficient, respectively, of the analyte, and V is the applied voltage. This equation shows that high voltage gradients are the most direct way to high separation efficiencies. For proteins, it has been calculated that *N* could be as high as 1 million theoretical plates. Fig. 11.15 is a schematic drawing of a CZE system [119]. The fused-silica capillary has a diameter of 50-80 μm and a length up to 1 m. It is suspended between two reservoirs, connected to a power supply that is able to deliver up to 30 kV (typical currents being of the order of 10-100 μA). One of the simplest ways to introduce the sample into the capillary is by electromigration, i.e. by dipping the capillary extremity into the sample reservoir, under voltage, for a few seconds. Detection is usually accomplished by on-column fluorescence and/or UV absorption.

References on p. A514

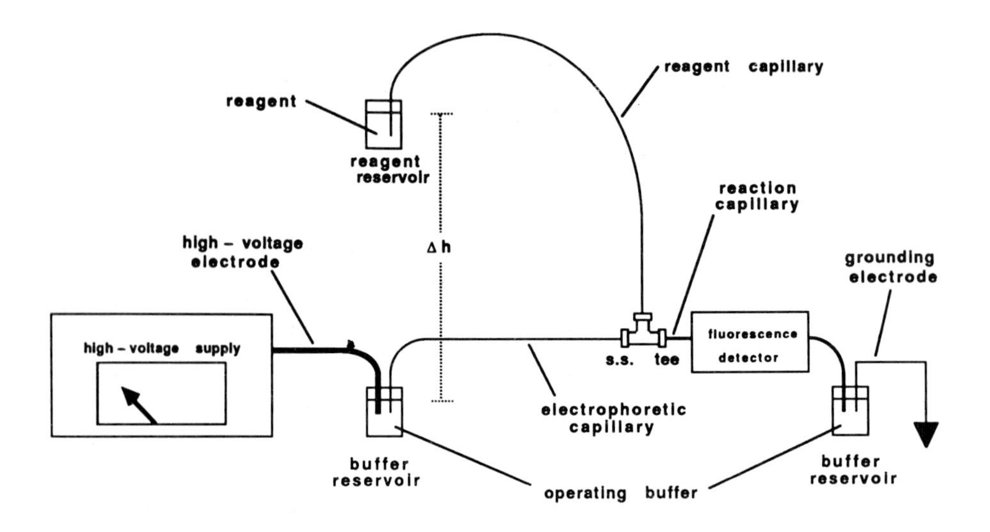

Fig. 11.15. Scheme of a capillary zone electrophoresis apparatus. The high-voltage (HV) power supply can deliver up to 30 kV. The fused-quartz capillary usually has an ID of 50-80 μm. The detector consists of a beam from a high-pressure mercury/xenon arc lamp source, oriented perpendicular to the migration path, at the end of the capillary. The sample signal (generally, emitted fluorescence) is measured with a photomultiplier and a photometer connected to the analog/digital converter of a multifunction interface board, mounted on a microcomputer. In the scheme here reproduced, postcolumn sample derivatization is obtained by allowing the sample zones to react with *o*-phthaldialdehyde (by permission from Ref. 119).

Conductivity and thermal detectors, as usually employed in ITP, exhibit too low a sensitivity in CZE. The reason for this stems from the fact that the flowcell where sample monitoring occurs has a volume of barely 0.5 nl, allowing sensitivities down to the femtomol level. In fact, with the postcolumn derivatization method of Fig. 11.15, a detection limit for amino acids of the order of femtograms is claimed, which means working in the attomol range [118]. By forming a chiral complex with a component of the background electrolyte (Cu aspartame) it is possible to resolve racemates of amino acids [8]. Even neutral organic molecules can be made to migrate in CZE by complexing them with charged ligands, such as SDS. This introduces a new parameter, a hydrophobicity scale, in electrokinetic migrations. For more on CZE, the readers are referred to the Proceedings of the First [120] and Second [121] International Symposium on High-Performance Capillary Electrophoresis.

11.14 CHROMATOPHORESIS

Chromatophoresis is a variant of 2-D techniques, but it is unique in many respects [122]. Classical 2-D maps combine a charge (IEF) with a mass (SDS/PAGE) fractionation, whereas in chromatophoresis reversed-phase HPLC is coupled in a real-time automated

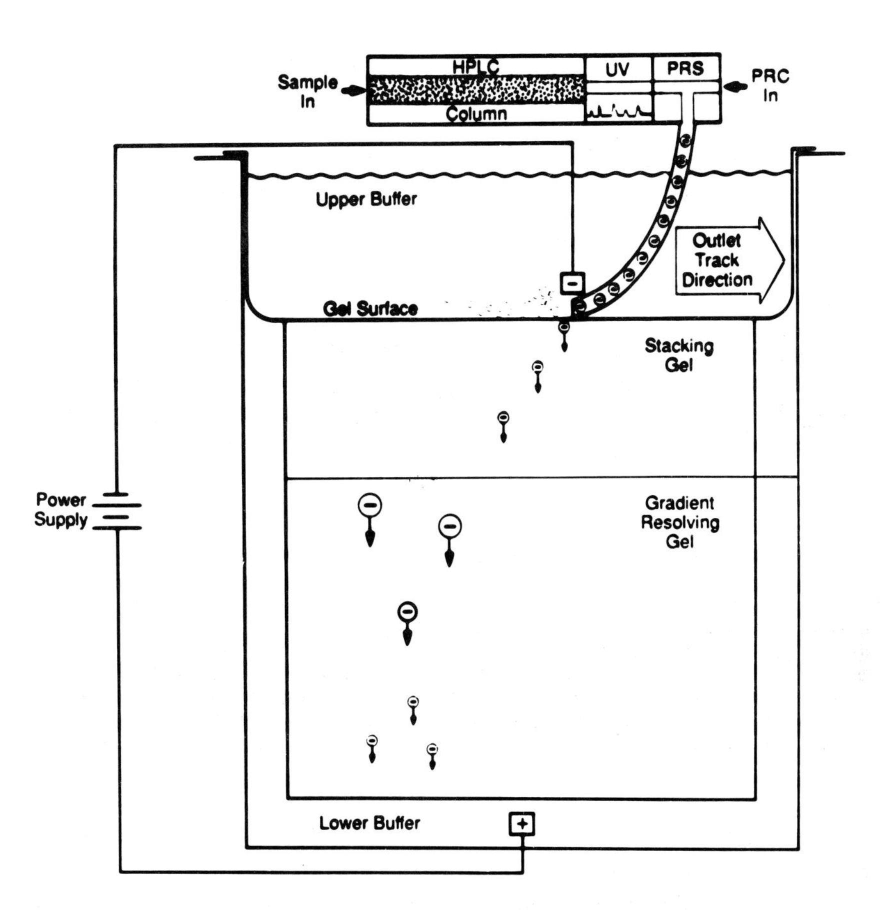

Fig. 11.16. Schematic illustration of the electrophoretic transfer of proteins in the chromatophoresis process. For details, see text (by permission from Ref. 122).

system to SDS/PAGE. Fig. 11.16 shows the principle of how this is achieved. Proteins eluted from the HPLC column pass through a UV detector (UV) to a heated mixing chamber (protein reaction system, PRS). Polypeptides in the eluate are denatured and complexed with SDS upon mixing with the standard protein reaction cocktail (PRC), containing SDS, 2-mercaptoethanol, and buffer. The SDS/protein complexes reach the SDS gel slab through an outlet lying flush on the surface of the stacking gel of a discontinuous polyacrylamide gradient gel. The capillary delivering the column eluate is moved across the gel surface by a computer-controlled tracking system in such a way that the gel width accommodates, by the end of this sweeping process, the entire column eluate. The eluate is delivered to the SDS/gel slab under a moderate voltage gradient, so that each liquid element dispenses immediately its protein content to the stacking gel, thus minimizing side diffusion (there are no sample application troughs). The height of the stacking gel is generally higher than in conventional gels, so that the stacking process can

References on p. A514

be continued up to 1 h, if needed. Thus, by the time the entire column eluate reaches the surface of the running gel, all the protein zones are aligned horizontally.

ACKNOWLEDGEMENTS

Our research quoted here has been supported over the years by grants from Consiglio Nazionale delle Ricerche (CNR) Progetti Finalizzati Biotecnologie e Biostrumentazione and Chimica Fine II and by Ministero della Pubblica Istruzione (MPI), Rome.

REFERENCES

1 F.M. Everaerts, F.E.P. Mikkers, Th.P.E.M. Verheggen and J. Vacik, in E. Heftmann (Editor), *Chromatography,* Elsevier, Amsterdam, 1983, pp. A331-A368.
2 A.T. Andrews, *Electrophoresis: Theory, Techniques and Biochemical and Clinical Applications,*, Clarendon Press, Oxford, 1986.
3 P.G. Righetti, *Isoelectric Focusing: Theory, Methodology and Applications,* Elsevier, Amsterdam, 1983.
4 P.G. Righetti, *Immobilized pH Gradients: Theory and Methodology,* Elsevier, Amsterdam, 1990.
5 C.J. Van Oss, *Separ. Purif. Methods,* 8 (1979) 119.
6 P.G. Righetti, C.J. Van Oss and J.W. Vanderhoff (Editors), *Electrokinetic Separation Methods,* Elsevier, Amsterdam, 1979.
7 B.L. Karger, *Nature (London),* 339 (1989) 641.
8 P. Gozel, E. Gassmann, H. Michelsen and R.N. Zare, *Anal. Chem.,* 59 (1987) 44.
9 P. Grabar and C. A. Williams, *Biochim. Biophys. Acta,* 10 (1953) 193.
10 O. Smithies, *Biochem. J.,* 61 (1955) 629.
11 J. Kohn, *Nature (London),* 180 (1957) 986.
12 L.P. Cawley, B.J. Minard, W.W. Tourtellotte, P.I. Ma and C. Clelle, *Clin. Chem.,* 22 (1976) 1262.
13 A.A. Keshgegian and P. Peiffer, *Clin. Chim. Acta,* 108 (1981) 337.
14 J. Kohn, P. Priches and J.C. Raymond, *J. Immunol. Methods,* 76 (1985) 11.
15 P.G. Righetti, P. Casero and G.B. Del Campo, *Clin. Chim. Acta,* 157 (1986) 167.
16 P. Casero, G.B. Del Campo and P.G. Righetti, *Electrophoresis,* 6 (1985) 373.
17 T. Toda, T. Fujita and M. Ohashi, in R.C. Allen and P. Arnaud (Editors), *Electrophoresis '81,* De Gruyter, Berlin, 1981, pp. 271-280.
18 P. Grabar, *Methods Biochem. Anal.,* 7 (1957) 1.
19 C. Araki, *Bull. Chem. Soc. Jpn.,* 29 (1965) 543.
20 M. Duckworth and W. Yaphe, *Carbohydr. Res.,* 16 (1971) 189.
21 S. Hjertén, *Biochim. Biophys. Acta,* 53 (1961) 514.
22 S. Brishammar, S. Hjertén and B. Van Hofsten, *Biochim. Biophys. Acta,* 53 (1961) 518.
23 S. Hjertén, *Arch. Biochem. Biophys.,* 99 (1962) 466.
24 M. Duckworth and W. Yaphe, *Anal. Biochem.,* 44 (1971) 636.
25 T.J. Låås, *J. Chromatogr.,* 66 (1972) 347.
26 T.G.L. Hickson and A. Polson, *Biochim. Biophys. Acta,* 168 (1965) 43.
27 S. Arnott, A. Fulmer, W.E. Scott, I.C.M. Dea, R. Moorhouse and D.A. Rees, *J. Mol. Biol.,* 90 (1974) 269.
28 P.G. Righetti, in R.C. Allen and P. Arnaud (Editors), *Electrophoresis '81,* De Gruyter, Berlin, 1981, pp. 3-16.
29 J.J. Opplt, in L.A. Lewis and J.J. Opplt (Editors), *Handbook of Electrophoresis,* CRC Press, Boca Raton, FL, 1980, pp. 151-180.
30 N.C. Stellwagen, *Adv. Electr.,* 1 (1987) 177.

31 P. Serwer, *Electrophoresis,* 4 (1983) 375.
32 S. Hjertén, *J. Chromatogr.,* 12 (1963) 510.
33 P. Serwer, E.T. Moreno and G.A. Griess, in C. Schafer-Nielsen (Editor), *Electrophoresis '88,* VCH, Weinheim, 1988, pp. 216-222.
34 P.G. Righetti, B.C.W. Brost and R.S. Snyder, *J. Biochem. Biophys. Methods,* 4 (1981) 347.
35 P.G. Righetti, *J. Biochem. Biophys. Methods,* 19 (1989) 1.
36 J. Uriel and J. Berges, *C. R. Acad. Sci. Paris,* 262 (1966) 164.
37 A.C. Peacock and C.W. Dingman, *Biochemistry,* 7 (1968) 668.
38 E. Boschetti, in P.G.D. Dean, W.S. Johnson and F.A. Middle (Editors), *Affinity Chromatography,* IRL Press, Oxford, 1985, pp. 11-15.
39 A. Chrambach and D. Rodbard, *Science,* 172 (1971) 440.
40 S. Raymond and L. Weintraub, *Science,* 130 (1959) 711.
41 B.J. Davis, *Ann. N.Y. Acad. Sci.,* 121 (1964) 404.
42 L. Ornstein, *Ann. N.Y. Acad. Sci.,* 121 (1964) 321.
43 S. Hjertén, *J. Chromatogr.,* 11 (1963) 66.
44 S. Hjertén, *Anal. Biochem.,* 3 (1962) 109.
45 S. Hjertén, *Arch. Biochem. Biophys.,* Suppl. 1 (1962) 276.
46 S. Hjertén, S. Jersted and A. Tiselius, *Anal. Biochem.,* 27 (1969) 108.
47 J.H. Bode, in B.J. Radola (Editor), *Electrophoresis '79,* De Gruyter, Berlin, 1980, pp. 39-52.
48 P.B.H. O'Connell and C.J. Brady, *Anal. Biochem.,* 76 (1976) 63.
49 H.S. Anker, *FEBS Lett.,* 7 (1970) 293.
50 P.N. Paus, *Anal. Biochem.,* 42 (1971) 327.
51 J.N. Hansen, *Anal. Biochem.,* 76 (1976) 37.
52 C. Gelfi and P.G. Righetti, *Electrophoresis,* 2 (1981) 213.
53 A. Bianchi-Bosisio, C. Loherlein, R.S. Snyder and P.G. Righetti, *J. Chromatogr.,* 189 (1980) 317.
54 P.G. Righetti, C. Gelfi, M.L. Bossi and E. Boschetti, *Electrophoresis,* 8 (1987) 62.
55 B. Kozulic, K. Mosbach and M. Pietrzak, *Anal. Biochem.,* 170 (1988) 478.
56 O. Wichterle and D. Lim, *Nature (London),* 185 (1960) 117.
57 G. Artoni, E. Gianazza, M. Zanoni, C. Gelfi, M.C. Tanzi, C. Barozzi, P. Ferruti and P.G. Righetti, *Anal. Biochem.,* 137 (1984) 420.
58 D. Hochstrasser, M.G. Harrington, A.C. Hochstrasser and C.R. Merril, in C. Schafer-Nielsen (Editor), *Electrophoresis '88,* VCH, Weinheim, 1988, pp. 245-248.
59 P.G. Righetti, M. Chiari, E. Casale, C. Chiesa, T. Jain and R. Schorr, *J. Biochem. Biophys. Methods,* 17 (1989) 21.
60 D.J. Gisch, B.T. Hunter and B. Feibush, *J. Chromatogr.,* 433 (1988) 264.
61 M. Bier, in J. Asenjo and H. Hong (Editors), *Separation, Recovery and Purification in Biotechnology,* ACS Symp. Ser. 314, American Chemical Society, Washington, DC, 1986, pp. 320-330.
62 K. Hannig, in M. Bier (Editor), *Electrophoresis,* Vol. II, Academic Press, New York, 1967, pp. 423-471.
63 K.A. Ferguson, *Metabolism,* 13 (1964) 985.
64 H.R. Maurer, *Disc Electrophoresis,* De Gruyter, Berlin, 1972.
65 A. Chrambach, in C. Schafer-Nielsen (Editor), *Electrophoresis '88,* VCH, Weinheim, 1988, pp. 28-40.
66 H. Svensson, *Acta Chem. Scand.,* 15 (1961) 325, 16 (1962) 456.
67 O. Vesterberg, *Acta Chem. Scand.,* 23 (1969) 2653.
68 L. Söderberg, D. Buckley and G. Hagström, *Protides Biol. Fluids,* 27 (1980) 687.
69 C. Schafer-Nielsen, in C. Schafer-Nielsen (Editor), *Electrophoresis '88,* VCH, Weinheim, 198, pp. 41-48.
70 A.L. Shapiro, E. Vinuela and J.V. Maizel, *Biochem. Biophys. Res. Commun.,* 28 (1967) 815.
71 A. Helenius and K. Simons, *Biochim. Biophys. Acta,* 415 (1975) 29.
72 J.L. Neff, N. Munez, J.L. Colburn and A.F. de Castro, in R.C. Allen and P. Arnauds (Editors), *Electrophoresis '81,* De Gruyter, Berlin, 1981, pp. 49-63.

73 J.F. Poduslo, *Anal. Biochem.,* 114 (1981) 131.
74 P. Lambin, *Anal. Biochem.,* 85 (1978) 114.
75 J. Margolis and K.G. Kenrick, *Anal. Biochem.,* 25 (1968) 347.
76 P. Lambin and J.M. Fine, *Anal. Biochem.,* 98 (1979) 160.
77 G.M. Rothe and M. Purkhanbaba, *Electrophoresis,* 3 (1982) 33.
78 G.M. Rothe and W.D. Maurer, in M.J. Dunn (Editor), *Gel Electrophoresis of Proteins,* Wright, Bristol, 1986, pp. 37-140.
79 N.G. Anderson and N.L. Anderson, *Clin. Chem.,* 28 (1982) 739.
80 T. Barrett and H.J. Gould, *Biochim. Biophys. Acta,* 294 (1973) 165.
81 P. O'Farrell, *J. Biol. Chem.,* 250 (1975) 4007.
82 J. Klose, *Humangenetik,* 26 (1975) 231.
83 G.A. Scheele, *J. Biol. Chem.,* 250 (1975) 5375.
84 R.A. Colbert, J.M. Amatruda and D.S. Young, *Clin. Chem.,* 30 (1984) 2053.
85 R.H. Aebersold, J. Leavitt, L.E. Hood and S.B.H. Kent, in K. Walsh (Editor), *Methods in Protein Sequence Analysis,* Humana Press, Clifton, NJ, 1987, pp. 277-294.
86 F. Kohlrausch, *Ann. Phys. Chem.,* 62 (1987) 209.
87 F.M. Everaerts, J.L. Beckers and Th.P.E.M. Verheggen, *Isotachophoresis. Theory, Instrumentation and Applications,* Elsevier, Amsterdam, 1976.
88 S.G. Hjalmarsson and A. Baldesten, *CRC Critical Reviews Anal. Chem.,* CRC Press, Boca Raton, FL, 1981, pp. 261-352.
89 C.B. Laurell, *Anal. Biochem.,* 15 (1966) 45.
90 H.G.M. Clarke and T. Freeman, *Protides Biol. Fluids,* 14 (1967) 503.
91 I Sondergaard, L.K. Poulsen, M. Hagerup and K. Conradsen, *Anal. Biochem.,* 165 (1987) 384.
92 J. Kroll, in N.H. Axelsen, J. Kroll and B. Weeke (Editors), *A Manual of Quantitative Immunoelectrophoresis,* Universitetsforlaget, Oslo, 1973, pp. 61-67.
93 P.J. Svendsen and N.H. Axelsen, *J. Immunol. Methods,* 2 (1972) 169.
94 P.J. Svendsen and C. Rose, *Sci. Tools,* 17 (1970) 13.
95 N.H. Axelsen, J. Kroll and B. Weeke (Editors), A Manual of Quantitative Immunoelectrophoresis, *Scand J. Immunol.,* Vol. 2, Suppl. 1, 1973.
96 N.H. Axelsen (Editor), Quantitative Immunoelectrophoresis, *Scand. J. Immunol.,* Suppl. 2, 1975.
97 N.H. Axelsen (Editor), *Scand. J. Immunol.,* Vol. 17, Suppl. 10, 1983.
98 J.R. Sargent and S.G. George, *Methods in Zone Electrophoresis,* BDH Chemicals, Poole, 1975.
99 C.R. Merril, R.C. Switzer and M.L. Van Keuren, *Proc. Natl. Acad. Sci. U.S.,* 76 (1979) 4335.
100 M. Moeremans, G. Daneels and J. De Mey, *Anal. Biochem.,* 145 (1985) 315.
101 M. Moeremans, D. De Raeymaeker, G. Daneels and J. De Mey, *Anal. Biochem.,* 153 (1986) 18.
102 E.M. Southern, *J. Mol. Biol.,* 98 (1975) 503.
103 H. Towbin, T. Staehelin and J. Gordon, *Proc. Natl. Acad. Sci. U.S.,* 76 (1979) 4350.
104 H. Towbin and J. Gordon, *J. Immunol. Methods,* 72 (1984) 313.
105 O.J. Bjerrum and N.H.H. Heegaard, *J. Chromatogr.,* 470 (1989) 351.
106 V. Neuhoff, R. Stamm and H. Eibi, *Electrophoresis,* 6 (1985) 427.
107 B. Bjellqvist, K. Ek, P.G. Righetti, E. Gianazza, A. Görg, W. Postel and R. Westermeier, *J. Biochem. Biophys. Methods,* 6 (1982) 317.
108 M. Chiari, E. Casale, E. Santaniello and P.G. Righetti, *Appl. Theor. Electr.,* 1 (1989) 99 and 1 (1989) 103.
109 E. Gianazza, F. Celentano, G. Dossi, B. Bjellqvist and P.G. Righetti, *Electrophoresis,* 5 (1984) 88.
110 G. Dossi, F. Celentano, E. Gianazza and P.G. Righetti, *J. Biochem. Biophys. Methods,* 7 (1983) 123.
111 P.G. Righetti, M. Chiari, P.K. Sinha and E. Santaniello, *J. Biochem. Biophys. Methods,* 16 (1988) 185.
112 C. Gelfi, M.L. Bossi, B. Bjellqvist and P.G. Righetti, *J. Biochem. Biophys. Methods,* 15 (1987) 41.

113 P.K. Sinha and P.G. Righetti, *J. Biochem. Biophys. Methods,* 15 (1987) 199.
114 P.G. Righetti, E. Gianazza, C. Gelfi, M. Chiari and P.K. Sinha, *Anal. Chem.,* 61 (1989) 1602.
115 D.C. Schwartz and C.R. Cantor, *Cell,* 37 (1984) 67.
116 M.V. Olson, *J. Chromatogr.,* 470 (1989) 377.
117 L.H.T. van der Ploeg, *Amer. Biotech. Lab.,* Jan/Febr (1987) 8.
118 J.W. Jorgenson, in J.W. Jorgenson and M. Phillips (Editors), *New Directions in Electrophoretic Methods,* ACS Symp. Ser. 335, American Chemical Society, Washington, DC, 1987, pp. 70-93.
119 J.D. Rose, Jr. and J.W. Jorgenson, *J. Chromatogr.,* 447 (1988) 117.
120 B.L. Karger (Editor), *J. Chromatogr.,* 480 (1989).
121 B.L. Karger (Editor), *J. Chromatogr.,* 516 (1990).
122 W.G. Burton, K.D. Nugent, T.K. Slattery, B.R. Summers and R.L. Snyder, *J. Chromatogr.,* 443 (1988) 363.

Manufacturers and Dealers of Chromatography and Electrophoresis Supplies

(Reproduced in part from the *International Laboratory Buyers' Guide* with permission of International Scientific Information, Inc.)

Aabspec Instrumentation Ltd., 16 Rathmore Ave., Stillorgan, Dublin 18, IRELAND.
ABC Laboratories, P.O. Box 1097, Columbia, MO 65205, USA.
AC Analytical Controls BV, Postbus 374, NL-2600 AJ Delft, THE NETHERLANDS.
Accurate Chemical & Scientific Corp., 300 Shames Drive, Westbury, NY 11590, USA.
Ace Glass Inc., 1430 N.W. Boulevard, Vineland, NJ 08360, USA.
Ace Scientific Supply Co., 40-A Cotters La., East Brunswick, NJ 08816, USA.
Advanced Chemtech Inc., 2500 Seventh Street Rd., P.O. Box 1403, Louisville, KY 40201, USA.
Advanced Chromatographic Technologies, Blücher Strasse 22, D-1000 Berlin 61, GERMANY.
Advanced Separation Technologies (Astec), 37 Leslie Court, Box 297, Whippany, NJ 07981, USA.
AHL Inc., P.O. Box 742, Laurel, MD 20707, USA.
Ahlstrom Filtration Inc., 15 West High Street, Carlisle, PA 17013, USA.
Air Products and Chemicals Inc., P.O. Box 538, Allentown, PA 18105, USA.
Alameda Chemical and Scientific, 922 E. Southern Pacific Drive, Phoenix, AZ 85034, USA.
Alcott Chromatography, 5300 Oakbrook Pkwy., Ste. 100, Norcross, GA 30093, USA.
Aldrich Chemical Co., 940 W. St. Paul Ave., P.O. Box 355, Milwaukee, WI 53201, USA.
Alltech Associates, 2051 Waukegan Road, Deerfield, IL 60015, USA.
ALPHA Applied Research, 2355 McLean Blvd., Eugene, OR 97405, USA.
Alphagaz, Specialty Gases Div., Liquid Air Corporation, 2121 N. California Blvd., Walnut Creek, CA 94596, USA.
Altex Division, Beckman Instruments Inc., 4550 Norris Canyon Rd., San Ramon, CA 94583, USA.
AMBIS Systems Inc., 3939 Ruffin Road, San Diego, CA 92123, USA.
American Bionetics Inc., 21377 Cabot Blvd., Hayward, CA 94545, USA.
American Research Products Co., 30175 Solon Industrial Pkwy., Solon, OH 44139, USA.
American Scientific Products, 1430 Waukegan Rd., McGaw Park, IL 60085, USA.
Amersham Corp., 2636 S. Clearbrook Dr., Arlington Heights, IL 60005, USA.
Amicon Division, W.R. Grace & Co., 17 Cherry Hill Drive, Danvers, MA 01923, USA.
Analabs, 140 Water St., Norwalk, CT 06854, USA.
Analog & Digital Peripherals Inc., P.O. Box 499, Troy, OH 45373, USA.
Analtech Inc., 75 Blue Hen Drive, P.O. Box 7558, Newark, DE 19711, USA.
Analytical Bio-Chemistry Laboratories, P.O. Box 1097, Columbia, MO 65205, USA.
Analytical Chromatography Support, 10606 Brooklet, Ste. 202, Houston, TX 77099, USA.
Analytical Instruments Pty. Ltd., P.O. Box 215, Scarborough, Qld. 4020, AUSTRALIA.
Analytical Measuring Systems, London Rd., Pampisford, Cambridge CB2 4EF, GREAT BRITAIN.
Analytical Products Inc., 511 Taylor Way, Belmont, CA 94002, USA.
Analytichem International, 24201 Frampton Avenue, Harbor City, CA 90710, USA.
Analytic Parameters, P.O. Box 25035, Chicago, IL 60625, USA.
Anamed Instruments (Pvt.) Ltd., New Bombay 400 706, INDIA.
Angar Scientific Co., P.O. Box 538, Florham Park, NJ 07932, USA.
Anotec Separations Ltd., Wildmere Rd., Banbury, Oxon OX16 7JU, GREAT BRITAIN.
Anspec Co. Inc., 50 Enterprise Dr., P.O. Box 7730, Ann Arbor, MI 48107, USA.
Antek Instruments Inc., 6005 N. Freeway, Houston, TX 77076-3998, USA.

Applied Analytical Industries Inc., Rt. 6, P.O. Box 55, Wilmington, NC 28405, USA.
Applied Automation Inc., Pawhuska Rd., Bartlesville, OK 74004, USA.
Applied Biosystems Inc., 850 Lincoln Center Dr., Foster City, CA 94404, USA.
Applied Chromatography Systems Ltd., The Arsenal, Heapy Street, Macclesfield, Cheshire SK11 7JB, GREAT BRITAIN.
Applied Science Labs., P.O. Box 440, State College, PA 16804, USA.
Applied Separations, Box 6032/B, Franklin Tech Center, Bethlehem, PA 18001, USA.
Arnel Inc., 3141 Bordentown Ave., Parlin, NJ 08859, USA.
Asahi Chemical Ind. Co., 1-3-2 Yakoo, Kawasaki-ku, Kawasaki-shi, Kanagawa-ken 210, JAPAN.
Associated Laboratories, 806 N. Batavia, Orange, CA 92668, USA.
Atto Corp., 2-3, Hongo 7-chome, Bunkyo-ku, Tokyo 113, JAPAN.
Aura Industries Inc., P.O. Box 898, Staten Island, NY 10314, USA.
Autochrom Inc., P.O. Box 207, Milford, MA 01757-0207, USA.
Automated Microbiology Systems Inc., 3939 Ruffin Rd., San Diego, CA 92123, USA.
Automatic Switch Co., 60 Hanover Rd., Florham Park, NJ 07932, USA.
Axxiom Chromatography Inc., 23966 Craftsman Rd., Calabasas, CA 91302, USA.
Bacharach Inc., 625 Alpha Dr., Pittsburgh, PA 15238, USA.
Baekon Inc., 18866 Allendale Ave., Saratoga, CA 95070-5239, USA.
J.T. Baker Chemical Company, 222 Red School Lane, Phillipsburg, NJ 08865, USA.
Bal Seal Engineering Co. Inc., 620 W. Warner Ave., Santa Ana, CA 92707, USA.
Balston Inc., 703 Massachusetts Ave., Lexington, MA 02173, USA.
Barspec, P.O. Box 560, Rehovot 76103, ISRAEL.
Baxter Scientific Products, 1430 Waukegan Rd., McGaw Park, IL 60085, USA.
Beckman Instruments Inc., 2500 Harbor Blvd., Fullerton, CA 92634, USA.
Bendix Kansas City Division, Allied Bendix Aerospace, 2000 East 95th Street, Kansas City, MO 64131, USA.
Benson Polymerics, P.O. Box 12812, Reno, NV 89510, USA.
Laboratorium Prof. Dr. Berthold, Calmbacher Str. 22, D-7547 Wildbad, GERMANY.
J. C. Binzer Papierfabrik GmbH, Berleburger Strasse 71, Postfach 44, D-3559 Hatzfeld/ Eder, GERMANY.
Bioanalytical Systems Inc., 2701 Kent Avenue, West Lafayette, IN 47906, USA.
Biocom, B.P. 53, F-91942 Les Ulis, FRANCE.
Bio-Fractionations, 1725 S. State Highway 89-91, Logan, UT 84321, USA.
Biomedical Enterprises Inc., P.O. Box 257, Irvington, NY 10533, USA.
Biomed Instruments Inc., 1020 S. Raymond Ave., Ste. B, Fullerton, CA 92631, USA.
Biometra Biomed. Analytik GmbH, Wagenstieg 5, D-3400 Göttingen, GERMANY.
Bio-Probe International, 14272 Franklin Ave., Tustin, CA 92680, USA.
Bioprocessing Ltd., 1 Industrial Estate, Cosett, Durham DH8 6TJ, GREAT BRITAIN.
Bio-Rad Chemical Div., 1414 Harbour Way South, Richmond, CA 94804, USA.
BIOS Corp., 291 Whitney Avenue, New Haven, CT 06511, USA.
Bioscan Inc., 4590 McArthur Blvd. NW, Washington, DC 20007, USA.
Biotec-Fischer GmbH, Daimlerstrasse 6, D-6301 Reiskirchen, GERMANY.
Biotech Instruments Ltd., 183 Cambord Way, Luton, Beds. LU3 3AN, GREAT BRITAIN.
Biotronik Wissenschaftliche Geräte GmbH, Benzstrasse 28, D-8039 Puchheim-Bahnhof, GERMANY.
Bischoff Analysentechnik und -geräte GmbH, Umler Str. 2, D-7250 Leonberg, GERMANY.
Bodman Chemicals, P.O. Box 2221, Aston, PA 19014, USA.
Boehringer-Mannheim GmbH, Sandhoferstrasse 116, D-6800 Mannheim 31, GERMANY.
Herman Bohlender, Postfach 1145, Bischofsheimer Weg 14, D-6970 Lauda-Königshofen, GERMANY.
Bomem Inc., 625 Marais, Vanier, Que., G1M 2Y2, CANADA.
John Booker & Co., 3825 Bee Cave Rd., Austin, TX 78746, USA.
Brinkmann Instruments Inc., Cantiague Road, Westbury, NY 11590, USA.
B. Brown Diessel Biotech GmbH, Postfach 120, D-3508 Meisungen, GERMANY.
Brownlee Labs. Inc., 2045 Martin Avenue, Santa Clara, CA 95050, USA.
Bruker Analytische Messtechnik GmbH, Silberstreifen, D-7512 Rheinstetten 4, GERMANY.
Büchi Laboratory Techniques Ltd., Meierseggstr. 40, CH-9230 Flawill, SWITZERLAND.
Buchler Instruments, 9900 Pflumm Rd., 17 Lenexa Business Center, Lenexa, KS 66215-1223, USA.
Buck Scientific Inc., 58 Fort Point St., East Norwalk, CT 06855-1097, USA.

Burdick & Jackson, 1935 South Harvey Street, Muskegon, MI 49442, USA.
Burrell Corp., 2223 Fifth Ave., Pittsburgh, PA 15219, USA.
Cal Glass for Research Inc., 3012 Enterprise Ave., Costa Mesa, CA 92626, USA.
CAMAG, CH-4132 Muttenz, SWITZERLAND.
Camlab Ltd., Nuffield Road, Cambridge, Cambs. CB4 1TH, GREAT BRITAIN.
Carnegie Medicin AB, Roslagsvagen 101, S-104 05 Stockholm, SWEDEN.
Carolina Biological Supply Co., 2700 York Road, Burlington, NC 27215, USA.
Cavro Scientific Instruments Inc., 242 Humboldt Court, Sunnyvale, CA 94089, USA.
Cecil Instruments Ltd., Milton Industrial Estate, Cambridge Road, Milton, Cambs. CB4 4AZ, GREAT BRITAIN.
Cera Inc., 14180 Live Oak Ave., Ste. F, Baldwin Park, CA 91706, USA.
C.G.A. Strumenti Scientifici S.p.A., Via del Della Robbia N. 38, I-50132 Firenze, ITALY.
Chemcon Inc., 34 Mann St., South Attleboro, MA 02703, USA.
Chemetron, Via Gustavo Modena 24, I-20129 Milano, ITALY.
Chemical Data Systems, 7000 Limestone Rd., Oxford, PA 19363, USA.
Chemical Dynamics Corp., P.O. Box 395, South Plainfield, NJ 07080, USA.
Chemical Research Supplies, P.O. Box 888, Addison, IL 60101, USA.
Chemtrix Inc., P.O. Box 1329, Hillsboro, OR 97123, USA.
Chromacol Ltd., Glen Ross House, Summers Row, London N12 0LD, GREAT BRITAIN.
Chromapon Inc., P.O. Box 4131, Whittier, CA 90607, USA.
ChromatoChem, Inc., 2837 Fort Missoula Road, Missoula, MT 59801, USA.
Chromatofield, Zila Valampe, F-13220 Châteauneuf-les-Martigues, FRANCE.
Chromatography Sciences Co., 5750 Vanden Abeele, Ville Saint-Laurent, Que. H4S 1R9, CANADA.
Chromatography Services Ltd., Carr Lane Industrial Estate, Hoylake, Merseyside, GREAT BRITAIN.
Chromatography Technology Services, 3301 W.134th St., Burnsville, MN 55337, USA.
Chromatronix, Inc., 2300 Leghorn Street, Mountain View, CA 94043, USA.
Chrompack International BV, Postbus 8033, NL-4330 EA Middelburg, THE NETHERLANDS.
Chrom Tech Inc., P.O. Box 24248, Apple Valley, MN 55124, USA.
Ciba Corning Analytical, Colchester Road, Halstead, Essex C09 2DX, GREAT BRITAIN.
CJB Developments Ltd., Airport Service Road, Portsmouth P03 5PG, GREAT BRITAIN.
Clarkson Chemical Co., P.O. Box 97, Williamsport, PA 17703-0097, USA.
Clontech, 4030 Fabian Way, Palo Alto, CA 94303, USA.
Cluzeau Info-Lab Sarl, B.P. 88, 35 rue Jean-Louis Fauré, F-33220 Ste. Foy-la-Grande, FRANCE.
P. J. Cobert Associates Inc., P.O. Box 12668, St. Louis, MO 63141, USA.
Cole-Parmer Instruments, 7425 N. Oak Park Ave., Chicago, IL 60648, USA.
Combined Sciences Corp., 433 Boston Post Rd., Darien, CT 06820, USA.
Computer Chemical Systems, P.O. Box 683, Rt. 41 & Newark Road, Avondale, PA 19311, USA.
Consort pvba, Parklaan 36, B-2300 Turnhout, BELGIUM.
Core Laboratories Inc., 1300 E. Rochelle Blvd., Irving, TX 7515-2053, USA.
Cortex Biochem Inc., 459 Hester St., San Leandro, CA 94677, USA.
Coulter Electronics Ltd., Northwell Dr., Luton, Beds. LU3 3RH, GREAT BRITAIN.
Crane Co., 175 Titus Ave., Warrington, NY 18976, USA.
Crescent Chemical Co., 1324 Motor Pkwy., Hauppague, NY 11788, USA.
Crown Glass, 990 Evergreen Dr., Somerville, NJ 08876, USA.
Cryogenic Rare Gas, 913 Commerce Circle, Hanahan, SC 29410, USA.
Cuno Inc., Life Science Div., 400 Research Parkway, Meriden, CT 06450, USA.
Curtin Matheson Scientific, 9999 Veterans Memorial Dr., Houston, TX 77038, USA.
Cyborg, 94 Bridge Street, Newton, MA 02158, USA.
Daicel Chemical Industries Ltd., 8-1 Kasumigaseki 3-chome, Chiyoda-ku, Tokyo 100, JAPAN.
Daiichi Pure Chemicals Co., 13-5 Nihombashi 3-chome, Chuo-ku, Tokyo 103, JAPAN.
Dani SpA, V. le Elvezia 42, I-20052 Monza (Milano), ITALY.
Data Translation Inc., 100 Locke Dr., Marlboro, MA 01752, USA.
Del Electronics Corp., 250 E. Sandford Blvd., Mount Vernon, NY 10550, USA.
Delsi Nermag, 15701 W. Hardy, Houston, TX 77060, USA.
Delta Technical Products Co., 7259 W. Devon, Chicago, IL 60631, USA.

DESAGA GmbH, Maass Strasse 26-28, Postfach 101969, D-6900 Heidelberg 1, GERMANY.
Detector Engineering & Technology Inc., 2212 Brampton Rd., Walnut Creek, CA 94598, USA.
Dexsil Chemical Corp., 1 Hamden Park Dr., Hamden, CT 06517, USA.
Digital Equipment Corporation, 4 Results Way, MR04-2/C16, Marlborough, MA 01752-9122, USA.
Dionex Corp., 1228 Titan Way, Sunnyvale, CA 94086, USA.
Diversified Biotech, 46 Marcellus Drive, Newton Centre, MA 02159, USA.
Domnick Hunter Filters Ltd., Durham Rd., Birtley Co., Durham DH3 2SF, GREAT BRITAIN.
Dorr-Oliver, 77 Havemeyer Lane, P.O. Box 9312, Stamford, CT 06904, USA.
Dracard Ltd., Wallis Ave., Park Wood, Maidstone, Kent ME15 9HE, GREAT BRITAIN.
Drew Scientific, 12 Barley Mow Passage, London W4 4PH, GREAT BRITAIN.
E. I. du Pont de Nemours & Co., Barley Mill Plaza, Wilmington, DE 19898, USA.
Duryea Assoc. Inc., 701 Alpha Dr., Pittsburgh, PA 15238, USA.
Dychrom, P.O. Box 70116, Sunnyvale, CA 94086, USA.
Dynamic Solutions Corp., 2355 Portola Rd., Ste. B, Ventura, CA 93003, USA.
Dyson Instruments Ltd., Hetton Lyons Industrial Estate, Hetton, Houghton-le-Spring, Tyne and Ware DH5 0RH, GREAT BRITAIN.
Eastman Kodak Co., P.O. Box 92894, Rochester, NY 14692-9939, USA.
E-C Apparatus Corp., 3831 Tyrone Blvd. North, St. Petersburg, FL 33709, USA.
E-D Scientific Specialties, P.O. Box 369, Carlisle, PA 17013, USA.
EG&G Princeton Applied Research Corp., CN 5206, Princeton, NJ 08543, USA.
Electro Biotransfer Inc., 790 Lucerne Drive, Sunnyvale, CA 94086, USA.
Electronic & Scientific Devices, 100 U.B. Jawahar Nagar, Delhi 110 007, INDIA.
EM Science, 11 Woodcrest Rd., Cherry Hill, NJ 08034-0395, USA.
Carlo Erba Strumentazione, Strada Rivoltana, I-20090 Rodano (Milano), ITALY.
Erma Inc., 2-4-5 Kajicho Chiyoda-ku, Tokyo 101, JAPAN.
ES Industries, 8 S. Maple Ave., Marlton, NJ 08053, USA.
ESA Inc., 45 Wiggins Ave., Bedford, MA 01730, USA.
ETPCORTEC Pty. Ltd., 31 Hope Street, Ermington, Sydney, N.S.W. 2115, AUSTRALIA.
Extrel Corporation, 240 Alpha Drive, P.O. Box 11512, Pittsburgh, PA 15238, USA.
E-Y Laboratories Inc., 105-127 N. Amphlett Blvd., San Mateo, CA 94401, USA.
F.E.R.O.S.A., Fabricación Espanola de Reactivos Organicos SA, C/la Jota 86, E-08016 Barcelona, SPAIN.
FFFractionation Inc., P.O. Box 8718, Salt Lake City, UT 84108, USA.
Fiatron Systems Inc., 510 S. Worthington St., Oconomowoc, WI 53066, USA.
Finnigan MAT, Barkhausenstrasse 2, Postfach 144062, D-2800 Bremen, GERMANY.
Fischer Labor- und Verfahrenstechnik GmbH, Industriepark Kottenforst, D-5309 Meckenheim, GERMANY.
Fisher Scientific Co., 711 Forbes Avenue, P.O. Box 1962, Pittsburgh, PA 15219, USA.
Fisher & Porter, Lab-Crest Scientific Div., East County Line Road, Warminster, PA 18974, USA.
Fisons Scientific Equipment, Bishop Meadow Road, Loughborough, Leicester LE11 0RG, GREAT BRITAIN.
Flow Laboratories SA, Via Campagna, Centro Nord-Sud, Bioggio, CH-6934 Lugano, SWITZERLAND.
Fluid Management Systems, 125 Walnut St., Watertown, MA 02172, USA.
Fluid Metering Inc., 29 Orchard St., P.O. Box 129, Oyster Bay, NY 11771, USA.
Fluka Chemie AG, Industriestrasse 25, CH-9470 Buchs, SWITZERLAND.
FMC Corp., 5 Maple St., Rockland, ME 04841, USA.
Foss Electric Ltd., Sandyford Industrial Estate, Foxrock, Dublin 18, IRELAND.
Fotodyne Inc., 16700 W. Victor Rd., New Berlin, WI 53151-4131, USA.
Foxboro Co., Bristol Park, Foxboro, MA 02035, USA.
Funakoshi, Pharmaceutical Co., 2-3 Surugadai, Kanda, Chiyoda-ku, Tokyo 101, JAPAN.
Galactic Industries Corp., 417 Amherst St., Nashua, NH 03063, USA.
Gallard-Schlesinger Industries Inc., 584 Mineola Ave., Carle Place, NY 11514, USA.
Gallenkamp, Belton Road West, Loughborough LE11 0TR, GREAT BRITAIN.
Gargya Research Inst., A-25, Rajouri Garden, Najafgarh Rd., New Delhi, Delhi 110 027, INDIA.
Gasukuro Kogyo Inc., 6-12-18 Nishi Shinjuku, Shinjuku-ku, Tokyo, JAPAN.

GAT Gamma Analysentechnik GmbH, Dionysiusstr. 6, D-2850 Bremerhaven, GERMANY.
Gelman Sciences Inc., 600 South Wagner Road, Ann Arbor, MI 48106, USA.
Geltech Inc., 934 Salem Pkwy., Salem, OH 44460-313, USA.
Genetic Research Instrumentation Ltd., Gene House, Dunnow Road, Felsted, Dunnow, Essex CM6 3LD, GREAT BRITAIN.
Genex Corporation, 16020 Industrial Drive, Gaithersburg, MD 20877, USA.
Genie Scientific, 17430 Mt. Cliffwood Circle, Unit A, Fountain Valley, CA 92708, USA.
Gibco Ltd., P.O. Box 35, Trident House, Renfrew Rd., Paisley, PA3 4EF, GREAT BRITAIN.
Gilson Medical Electronics Inc., 3000 West Beltline Hwy., Middleton, WI 53562, USA.
Gold Biotech, 10143 Paget, St. Louis, MO 63132, USA.
Gow-Mac Instruments Co., P.O. Box 32, Bound Brook, NJ 08805, USA.
Graphic Controls Corp., P.O. Box 1271, Buffalo, NY 14240, USA.
Griffin & George, Bishop Meadow Rd., Loughborough, Leicestershire LE11 ORG, GREAT BRITAIN.
Grupo Químico Industrial Ltd., Rua Jacurunta, 628 Penha, 21.020 Rio de Janeiro, BRAZIL.
Guelph Chemical Laboratories Ltd., 246 Silvercreek Parkway N., Guelph, Ont. N1H 1E7, CANADA.
Haake Buchler Instruments Inc., 244 Saddle River Road, Saddle Brook, NJ 07662-6001, USA.
Hach Co., P.O. Box 389, Loveland, CO 80539, USA.
Hamilton Company, 4970 Energy Way, Reno, NV 89502, USA.
Harrison Co., Palo Alto, CA 94303, USA.
Haskel Inc., 100-88 E. Graham Place, Burbank, CA 91502, USA.
Helena Laboratories, 1530 Lindbergh Dr., P.O. Box 752, Beaumont, TX 77704-0752, USA.
Hellma Cells Inc., P.O. Box 544, Borough Hall Station, Jamaica, NY 11424, USA.
HETP LC Components, 4 Victoria Road, Wilmslow, Cheshire SK9 5HN, GREAT BRITAIN.
Hewlett-Packard, P.O. Box 10301, Palo Alto, CA 94303-1501, USA.
Hichrom Ltd., 6 Chiltern Enterprise Ctr., Station Road, Theale, Berks. RG7 4AA, GREAT BRITAIN.
Hirschmann Gerätebau GmbH, Lohestrasse 5, D-8025 Unterhaching, GERMANY.
Hitachi Scientific Instruments, 460 E. Middlefield Rd., Mountain View, CA 94043, USA.
HNU Systems Inc., 160 Charlemont St., Newton, MA 02161, USA.
Hoefer Scientific Instruments, 654 Minnesota, P.O. Box 77387, San Francisco, CA 94107, USA.
HP Genenchem, 460 Point San Bruno Blvd., South San Francisco, CA 94080, USA.
HPLC Technology Ltd., Waterloo Street West, Macclesfield, Cheshire SK11 6PJ, GREAT BRITAIN.
HT Chemicals Inc., 4221 Forest Park Ave., St. Louis, MO 63108-2810, USA.
Iatron Laboratories Inc., Tokyo, JAPAN.
IBF Biotechnics Inc., 8510 Corridor Rd., Savage, MD 20763, USA.
IBF Reactifs, 35 Ave. Jean-Jaurès, F-92390 Villeneuve la Garenne, FRANCE.
IBM Instruments Inc., P.O. Box 3020, Wallingford, CT 06492, USA.
ICI Australia Operations Pty. Ltd., 5 Lake Dr., Redwood Gardens, Dingley, Vic. 3172, AUSTRALIA.
ICN Biochemicals Inc., 3300 Hyland Avenue, Costa Mesa, CA 92626, USA.
Idea Scientific Co., P.O. Box 2078, Corvallis, OR 97339, USA.
Ikemoto Scientific Technology Co., P.O. Box 14, Hongo, Bunkyo-ku, Tokyo 113, JAPAN.
Illinois Water Treatment Co., 4669 Shepherd Trail, Rockford, IL 61103, USA.
Immunetics Inc., 380 Green Street, Cambridge, MA 02139, USA.
Infometrix Inc., 2200 Sixth Ave., Ste. 833, Seattle, WA 98121, USA.
Ingold Electrodes Inc., 261 Ballardvale St., Wilmington, MA 01887, USA.
Innovative Chemistry Inc., P.O. Box 90, Marshfield, MA 02050, USA.
Institut für Chromatographie, Postfach 1141, D-6702 Bad Dürkheim 1, GERMANY.
Instrumentation Laboratory, 113 Hartwell Ave., Lexington, MA 02173, USA.
Integrated Separation Systems, 1 Westinghouse Plaza, Hyde Park, MA 02136, USA.
Interaction Chemicals, 1615 Plymouth Street, Mountain View, CA 94043, USA.
Interactive Microware Inc., P.O. Box 139, State College, PA 16804, USA.
Interfacial Dynamics Cop., P.O. Box 279, Portland, OR 97207, USA.
International Biotechnologies Inc., 25 Science Park, P.O. Box 9558, New Haven, CT 06535, USA.
International Equipment Co., 300 Second Avenue, Needham Heights, MA 02194, USA.

Introtek International LP, 120-C Jefryn Blvd., E. Dear Park, NY 11729, USA.
Invicta Biosystems Inc., 2225 Faraday Ave., Carlsbad, CA 92008, USA.
Ion Exchange Products Inc., 4834 S. Halsted Street, Chicago, IL 60609, USA.
Ionics Inc., 65 Grove St., Watertown, MA 02172, USA.
Isco Inc., P.O. Box 5374, Lincoln, NE 68505, USA.
Isolab Inc., P.O. Box 4350, Akron, OH 44321, USA.
Ithaca Laboratory Equipment Co., 305 W. Green St., Ithaca, NY 14850, USA.
Janus Laboratories Inc., 9307 Rock Canyon Way, P.O. Box 1406, Orangevale, CA 95662, USA.
Japan Analytical Industry Co., 208 Musashi, Mizuho, Nashitama, Tokyo 190-12, JAPAN.
JASCO, Japan Spectroscopic Co., 2967-5 Ishikawa-cho, Hachioji City, Tokyo 192, JAPAN.
Jaytee Biosciences Ltd., Kent Research & Development Centre, University of Kent, Canterbury, Kent CT2 7PD, GREAT BRITAIN.
JEOL Ltd., 1-3 Musashino 3-chome, Akishima City, Tokyo 196, JAPAN.
JJ's (Chromatography) Ltd., Hardwick Industrial Estate, Kings Lynn PE30 4JG, GREAT BRITAIN.
JM Science Inc., 5820 Main Street, Ste. 300, Buffalo, NY 14221, USA.
Jobin Yvon, 16-18 Rue du Canal, F-91160 Longjumeau, FRANCE.
Jones Chromatography Ltd., New Road, Hengoed, Mid Glamorgan CF8 8AU, GREAT BRITAIN.
Jookoo Co Ltd., 3-19-4 Hongo, Bunkyo-ku, Tokyo 113, JAPAN.
Jordan Scientific Co., 4315 S. State Rd. 446, Bloomington, IN 47401, USA.
Jordi Associates Inc., 26 Pearl St., Bellingham, MA 02019, USA.
Joyce-Loebl, Marquisway, Team Valley, Gateshead NE11 0QW, GREAT BRITAIN.
JPS Chimie, B.P. 343, CH-2022 Bevaix, SWITZERLAND.
Jule Biotechnologies Inc., 25 Science Park, Hew Haven, CT 06511, USA.
J & W Scientific Inc., 91 Blue Ravine Rd., Folsom, CA 95630, USA.
Kalex Scientific Co., 7 Mora Court, Manhasset, NY 11030, USA.
Keystone Scientific Inc., Penn Eagle Industrial Park, 320 Rolling Ridge Dr., Bellefonte, PA 16823, USA.
Keystone Valve, 9700 W. Gulf Bank Dr., Houston, TX 77040, USA.
Kinetek Systems Inc., 11802 Borman Dr., St. Louis, MO 63146, USA.
Kipp & Zonen BV, Mercuriusweg 1, NL-2624 Delft, THE NETHERLANDS.
Dr. Herbert Knauer KG, Heuchelheimer Strasse 9, D-6380 Bad Homburg, GERMANY.
Koch-Light Ltd., Edison House, 163 Dixons Hill Rd., North Mymms, Hatfield, Herts., GREAT BRITAIN.
Koken Co., 14-18, 3-chome Mejiro, Toshima-ku, Tokyo 171, JAPAN.
Konik Instruments S.A., Ctra. Cerdanyola 65-7, P.O. Box 136, Sant Cugat del Valles, E-08024 Barcelona, SPAIN.
Kontes Biotechnology, P.O. Box 729, Vineland, NJ 08360-2899, USA.
Kontron Instruments AG, Bernerstrasse Süd 169, CH-8010 Zürich, SWITZERLAND.
Kratos Analytical, Barton Dock Rd., Urmston, Lancas. M31 2LD, GREAT BRITAIN.
Kronus Inc., P.O. Box 1075, Ste. 312, Dana Point, CA 92629, USA.
Kupper & Co., Montalinweg 12, Postfach 55, CH-7402 Bonaduz, SWITZERLAND.
Labclear, 508 - 29th Ave., Oakland, CA 94601, USA.
Lab-Crest Scientific, E. County Line Rd., Warminster, PA 18974, USA.
Lab Glass Inc., 1172 NW Blvd., Vineland, NJ 08360, USA.
Labindustries Inc., 620 Hearst Ave., Berkeley, CA 94710-0128, USA.
Labomatic AG, Im Kirschgarten 30, CH-4124 Schönenbuch, SWITZERLAND.
Labotron Instruments AG, Forrlibuckstrasse 66, CH-8005 Zürich, SWITZERLAND.
Lachema, Research Institute for Pure Chemicals, Karásek 28, CS-621 33 Brno, CZECHOSLOVAKIA.
LaSalle Scientific Inc., 103 Elmslie St., Lasalle, Que. HBR 1V4, CANADA.
Laser Precision Analytical, 17819 Gillette Ave., Irvine, CA 92714, USA.
LC Packings, Wilhelminastraat 118, NL-1054 WP Amsterdam, THE NETHERLANDS.
LC Resources, 3182 C Old Tunnel Road, Lafayette, CA 94549, USA.
LDC Analytical Inc., 3661 Interstate Industrial Park Road North, P.O. Box 10235, Riviera Beach, FL 33404, USA.
Lee Scientific Corp., 4426 S. Century Drive, Salt Lake City, UT 84123, USA.
Lida Manufacturing Corp., 9115 26th Ave., Kenosha, WI 53140, USA.

Life Science Laboratories Ltd., Sedgewick Rd., Luton LU4 9DT, GREAT BRITAIN.
Linear Instruments Corp., 500 Edison Way, P.O. Box 12610, Reno, NV 89510, USA.
Liquid Carbonic, 135 S. LaSalle, Chicago, IL 60603, USA.
LKB-Produkter AB, P.O. Box 305, S-161 26 Bromma, SWEDEN.
Logos Scientific Inc., 700 Sunset Rd., Henderson, NV 89015, USA.
Lurex Manufacturing Co., 1298 North West Blvd., Vineland, NJ 08360, USA.
MAC-MOD Analytical Inc., 127 Commons Court, Chadds Ford, PA 19317, USA.
Macherey-Nagel, Neumann-Neander Strasse, Postfach 307, D-5160 Düren, GERMANY.
Mallinckrodt Inc., 675 McDonell Blvd., St. Louis, MO 63134, USA.
M.A.L.T.A., srl, Via Gustavo Modena 24, I-20129 Milano, ITALY.
Malvern Instruments Ltd., Spring Lane South, Malvern, Worcester WR14 1AQ, GREAT BRITAIN.
Manville Filtration & Minerals, Ken-Caryl Ranch, Denver, CO 80217-5108, USA.
Markson Science, 10201 S. 51st Street, Ste. 100, Phoenix, AZ 85044, USA.
Marsh Biomedical Products, 274 N. Goodman St., Rochester, NY 14607, USA.
Mattson Instruments Inc., 1001 Fourier Court, Madison, WI 53717, USA.
May & Baker Ltd., Liverpool Rd., Eccles, Manchester M30 7RT, GREAT BRITAIN.
MCRA Applied Technologies Inc., P.O. Box 377, Rockaway, NJ 07866, USA.
Medatronics Corp., 3901 Clark St., Seaford, NY 11783, USA.
Medical Air Technology Ltd., Canto House, Wilton St., Denton, Manchester M34 3LZ, GREAT BRITAIN.
Memtek Corp., 28 Cook St., Billerica, MA 01821, USA.
E. Merck, P.O. Box 4119, D-6100 Darmstadt, GERMANY.
Metrohm Ltd., CH-9101 Herisau, SWITZERLAND.
Michrom BioResources, 4193 Sundown Road, Livermore, CA 94550, USA.
Micro Filtration Systems, 6800 Sierra Ct., Dublin, CA 94568, USA.
Micromeritrics, 1 Micromeritrics Drive, Norcross, GA 30093, USA.
Micron Separations, Inc., 135 Flanders Rd., Westborough, MA 01581, USA.
Microphoretic Systems Inc., 750 N. Pastoria Avenue, Sunnyvale, CA 94086, USA.
Microsensor Technology Inc., 41762 Christy St., Fremont, CA 94538, USA.
Midwest Scientific Inc., 228 Meramec Station Rd., Valley Park, MO 63088, USA.
Mikrolab Aarhus A/S, Axel Kiers Vej 34, DK-8270 Hojbjerk, DENMARK.
Miles Scientific, 30 W 475 N. Aurora Rd., Naperville, IL 60566, USA.
Milevac Scientific Glass Ltd., 38/40 Broton Dr. Trading Estate, Halstead, Essex C09 1HB, GREAT BRITAIN.
Millipore Corporation, 75C Wiggins Avenue, Bedford, MA 01730, USA.
Milton Roy LDC Div., P.O. Box 10235, Riviera Beach, FL 33404, USA.
Mitsui Toatsu Chemicals Inc., 2-5 Kasumigaseki, 3-chome, Chiyoda-ku, Tokyo 100, JAPAN.
Modchrom Inc., 8666 Tyler Blvd., Ste. 3, Mentor, OH 44060, USA.
Molecular Dynamics, 230 Santa Ana Circle, Sunnyvale, CA 94086, USA.
Molecular Instruments Co., P.O. Box 1652, Evanston, IL 60201, USA.
Mott Metalurgical Corp., Spring Lane, Farmington, CT 06032, USA.
The Munhall Company, 5655 N. High St., Worthington, OH 43085, USA.
Muromachi Kagaku Kogyo Kaisha Ltd., No. 3, 4-chome, Muromachi, Nihonbashi, Chuo-ku, Tokyo 103, JAPAN.
National Diagnostics Inc., 1013-7 Kennedy Blvd., Manville, NJ 08835, USA.
PE Nelson Systems, 10061 Bubb Road, Cupertino, CA 95014, USA.
Neslab Instruments Inc., P.O. Box 1178, Portsmouth, NH 03801, USA.
The Nest Group, 45 Valley Rd., Southboro, MA 01771, USA.
New Brunswick Scientific Co., 44 Talmadge Rd., P.O. Box 4005, Edison, NJ 08818-4005, USA.
Newman-Howells Assoc. Ltd., Wolvesey Palace, Winchester, Hamps. S023 9NB, GREAT BRITAIN.
Nicolet Instrument Corp., 5225-1 Verona Rd., Madison, WI 53711, USA.
Nihon Seimitsu Kagaku Co, 25-10 Futaba-cho, Itabashi-ku, Tokyo 173, JAPAN.
Nimbuchem SA, 115 Ch. de Charleroi, B-5900 Jodoigne, BELGIUM.
Nissei Sangyo America, 1701 Golf Rd., Ste. 401, Rolling Meadows, IL 60008, USA.
Noah Technologies Corp., Tool Fairgrounds Parkway, San Antonio, TX 78238, USA.
Nordion Instruments Oy Ltd., P.O. Box 1, SF-00371 Helsinki, FINLAND.
Nuclear Sources & Services Inc., P.O. Box 34042, Houston, TX 77234, USA.

Nuclide Corp., 1155 Zion Rd., Bellefonte, PA 16823, USA.
Ohio Valley Specialty Chemical Inc., 115 Industry Rd., Marietta, OH 45750, USA.
O.I. Analytical, P.O. Box 2980, College Station, TX 77841-2980, USA.
Olympus Clinical Instruments Div., 4 Nevada Dr., Lake Success, NY 11042, USA.
Omnifit Ltd., 51 Norfolk St., Cambridge, Cambs. CB1 2LE, GREAT BRITAIN.
Oncor, P.O. Box 870, Gaithersburg, MD 20884, USA.
On-Line Instruments Systems Inc., Route 2, P.O. Box 111, Jefferson, GA 30549, USA.
Owl Scientific Plastics Inc., P.O. Box 566, Cambridge, MA 02139, USA.
Oyster Bay Pump Works Inc., No. 1 Bay Ave., P.O. Box 96, Oyster Bay, NY 11771, USA.
Pall Biosupport Co., 77 Crescent Beach Rd., Glen Cove, NY 11542, USA.
Pallflex Inc., Kennedy Dr., Putnam, CT 06260, USA.
Parker Hannifin Corp., 9400 S. Memorial Pkwy., P.O. Box 4288, Huntsville, AL 35802, USA.
Particle Data Inc., 11 Hahn St., P.O. Box 265, Elmhurst, IL 60126, USA.
P.C. Inc., 11805 Kim Pl., Potomac, MD 20854, USA.
PCP Inc., 2155 Indian Rd., West Palm Beach, FL 33409-3287, USA.
Pen Kem Inc., 341 Adam St., Bedford Hills, NY 10507, USA.
Peris Industries Inc., P.O. Box 1008, State College, PA 16804-1008, USA.
Perkin-Elmer Corp., 761 Main Avenue, Norwalk, CT 06859-0090, USA.
Peptides International Inc., 10101 Linn Station Rd., Ste. 445, Louisville, KY 40223, USA.
PerSeptive Biosystems Inc., 222 Third Street, Ste. 0300, Cambridge, MA 02142, USA.
Petazon Co., Zug, SWITZERLAND.
Pharmacia LKB Biotechnology AB, S-751 82 Uppsala, SWEDEN.
Pharma-Tech Research Corp., 8800 Kelso Dr., Baltimore, MD 21221, USA.
Phase Separations Inc., 140 Water Street, Norwalk, CT 06854, USA.
Phenomenex, 2320 W 205 St., Torrance, CA 90501, USA.
Philips Analytical, NL-5600 MD Eidhoven, THE NETHERLANDS.
Phortran Inc., 344 Lekeside Drive, Foster City, CA 94404, USA.
Photometrics, 2010 N. Forbes Blvd., Tucson, AZ 85745, USA.
Photovac Inc., 134 Doncaster Ave., Thornhill, Ont. L3T 1L3, CANADA.
PI Technologies, 3182 C Old Tunnel Road, Lafayette, CA 94549, USA.
Pickering Laboratories Inc., 1951 Colony St., Ste. S, Mountain View, CA 94043, USA.
Pierce Chemical Co., P.O. Box 117, Rockford, IL 61105, USA.
Poly Labo Paul Block & Cie, 305 Route de Colmar, F-67100 Strasbourg, FRANCE.
Poly LC, 9052 Belwart Way, Columbia, MD 21045, USA.
Polymer Laboratories Inc., 160 Old Farm Road, Amherst, MA 01002, USA.
Polymicro Technologies Inc., 3035 N 33rd Dr., Phoenix, AZ 85017, USA.
Polysciences Inc., 400 Valley Rd., Warrington, PA 18976-2590, USA.
PQ Corporation, Conshohocken, PA 19428, USA.
PreComp Inc., 17 Barstow Rd., P.O. Box 461, Great Neck, NY 11021, USA.
Preiser Scientific Inc., 94 Oliver Street, St. Albans, WV 25177, USA.
Princeton Separations Inc., P.O. Box 300, Adelphia, NJ 07710, USA.
Princeton Testing Laboratory, P.O. Box 3108, Princeton, NJ 08543, USA.
Process Analyzers Inc., 8 Headley Pl., Fallsington, PA 19054, USA.
Prochrom SA, Chemin de Blanches Terres, B.P. 9, F-54250 Champigneulles, FRANCE.
Puregas/General Cable Co., P.O. Box 666, 5600 W. 88 Ave., Westminster, CO 80030, USA.
Philips Analytical Pye Unicam Ltd., York Street, Cambridge CB1 2PX, GREAT BRITAIN.
Quadrant Scientific Ltd., 36 Brunswick Rd., Gloucester GL1 1JJ, GREAT BRITAIN.
Quadrex Corp., P.O. Box 3881, New Haven, CT 06525, USA.
Radiomatic Instruments & Chemical Co., 5102 S. Westshore Blvd., Tampa, FL 33611, USA.
Radnoti Glass Technology Inc., 227 West Maple Ave., Monrovia, CA 91016, USA.
Rainin Instrument Co., Mack Road, Woburn, MA 01801, USA.
Realco Chemical Co., New Brunswick, NJ 08903, USA.
Reanal, P.O. Box 54, H-1441 Budapest 70, HUNGARY.
Regis Chemical Company, 8210 Austin Ave., P.O. Box 519, Morton Grove, IL 60053, USA.
Reichelt Chemietechnik GmbH, Englerstr. 18, D-6900 Heidelberg 1, GERMANY.
Reliable Scientific Inc., 881 Richland Dr., Memphis, TN 38116, USA.
Reliance Glass Works, 220 Gateway Rd., P.O. Box 825, Bensenville, IL 60106, USA.
Repligen Corp., 1 Kendall Square, Bldg. 700, Cambridge, MA 02139, USA.

Restek Corp., 110 Benner Circle, Bellefonte, PA 16823, USA.
Rheodyne Inc., P.O. Box 996, Cotati, CA 94931, USA.
Rhône-Poulenc Ltd., Liverpool Rd., Eccles, Manchester M30 7RT, GREAT BRITAIN.
Richard Scientific Inc., 250 Bel Marin Keys Blvd., Ste. D3, Novato, CA 94949, USA.
Riedel de Haen AG, Wunstorfer Str. 40, D-3016 Seelze, GERMANY.
Rockland Inc., P.O. Box 316, Gilbertsville, PA 19380, USA.
Rohm & Haas Company, 727 Norristown Road, Spring House, PA 19477, USA.
Ruska Laboratories Inc., P.O. Box 630009, Houston, TX 77263-0009, USA.
Russell pH Ltd., Station Rd., Auchtermuchty, Fife KY14 7DP, GREAT BRITAIN.
SAC Chromatography Ltd., Summerhouse Hill, Cardington, Bedford MK44 3SD, GREAT BRITAIN.
Sadtler Research Labs., 3316 Spring Garden Street, Philadelphia, PA 19104, USA.
Saitron SpA, 39, Via del Crocifisso, Ponte a Ema, I-50126 Firenze, ITALY.
Sanki Engineering Ltd., 2-16-10 Imazato, Nagaokakyo, Kyoto 617, JAPAN.
Serasep Inc., 1600 Wyatt Drive, Ste. 10, Santa Clara, CA 95054, USA.
Sargent-Welch Scientific Co., 7300 N. Linden Ave., Skokie, IL 60077, USA.
Sarstedt Inc., P.O. Box 4090, Princeton, NJ 08543, USA.
Sartorius GmbH, Weender Landstr. 94/108, D-3400 Göttingen, GERMANY.
Sartorius Filters Inc., 30940 San Clemente St., Hayward, CA 94544, USA.
Säulentechnik, Dr. Ing. H. Knauer GmbH, Am Schlangengraben 16, D-1000 Berlin 20, GERMANY.
Savant Instruments Inc., 110 Bi-County Blvd., Farmingdale, NY 11735, USA.
Scanivalve Corp., 10222 San Diego Mission Rd., San Diego, CA 92108, USA.
Schleicher & Schuell GmbH, Postfach 4, D-3354 Dassel, GERMANY.
Scientific Equipment & Instrument Co., 265 Houret Dr., Milpitas, CA 95035, USA.
Scientific Glass Engineering Co., 2007 Kramer La., Austin, TX 78758, USA.
Scientific Instrument Services Inc., Rt. 179, P.O. Box 593, Ringoes, NJ 08551, USA.
Scientific Logics Inc., 21910 Alcazar Ave., Cupertino, CA 95014, USA.
Scientific Systems Inc., 349 N. Science Park Road, State College, PA 16803, USA.
Scientific Technologies Inc., 3121 Glen Royal Rd., Raleigh, NC 27612, USA.
Scott Specialty Gases, Route 611, Plumsteadville, PA 18949, USA.
S-Cubed, P.O. Box 1620, La Jolla, CA 92038, USA.
SEAC srl, Via Carlo del Prete 139, I-50127 Firenze, ITALY.
Seamark Corp., 11618 Busy St., Richmond, VA 23236, USA.
Dr. R. Seitner Mess- und Regeltechnik GmbH, Mühlbachstrasse 20, D-8031 Seefeld/Obb., GERMANY.
Separation Industries, P.O. Box 4338, 4 Leonard Street, Metuchen, NJ 08840, USA.
Separations Group Inc., 17434 Mojave Street, P.O. Box 867, Hesperia, CA 92345, USA.
Separations Technology, P.O. Box 352, Wakefield, RI 02880-0352, USA.
Sepragen Corp., 2126 Edison Ave., San Leandro, CA 94577, USA.
Serapine Corp., 821 Franklin Ave., Garden City, NY 11530, USA.
Serva Fine Biochemicals Inc., 200 Shames Dr., Westbury, NY 11590, USA.
Servomex Ltd., Crowborough, Sussex TN6 3DU, GREAT BRITAIN.
Severn Analytical Ltd., Unit 2B, St. Francis' Way, Shefford Industrial Park, Shefford, Beds. SG17 5DZ, GREAT BRITAIN.
SGE International Pty. Ltd., 2/76 Charles St., Ryde, NSW 2112, AUSTRALIA.
Shandon Southern Products Ltd., Chadwich Rd., Astmoor, Runcorn, Ches. WA7 1PR, GREAT BRITAIN.
Shimadzu Corporation, 1 Nishinokyo-Kuwabaracho, Nakagyo-ku, Kyoto 604, JAPAN.
Showa Denko K.K., 2-24-25 Tamagawa, Ohta-ku, Tokyo 146, JAPAN.
Siemens AG, Mess- u. Prozesstechnik, E 687 Postfach 21 1262, D-7500 Karlsruhe, GERMANY.
Sievers Research Inc., 1930 Central Ave., Ste. C, Boulder, CO 80301, USA.
Sigma Chemical Co., 3050 Spruce St., P.O. Box 14508, St. Louis, MO 63178, USA.
SKALAR Analytical BV, Spinveld 2, NL Breda, THE NETHERLANDS.
Small Parts Inc., P.O. Box 381966, Miami, FL 33238-1966, USA.
Sonntek/Knauer, P.O. Box 8589, Woodcliff Lake, NJ 07675, USA.
Sopar-Biochem, 124 rue J. Besme, B-1080 Bruxelles, BELGIUM.
Sota Chromatography, P.O. Box 693, Crompond, NY 10517, USA.
Southland Cryogenics, 2424 Lacy Lane, P.O. Box 110627, Carrollton, TX 75011, USA.
Spark Holland BV, P.O. Box 388, NL-7800 AJ Emmen, THE NETHERLANDS.

Spectra Gases Inc., 277 Coit St., Irvington, NJ 07111, USA.
Spectramass Ltd., Radnor Park Industrial Estate, Congleton, Cheshire, CW12 4XR, GREAT BRITAIN.
SpecTran Corp., 50 Hall Rd., Sturbridge, MA 01566, USA.
Spectra-Physics, 3333 N. First St., San Jose, CA 95134, USA.
Spectra-Tech Europe Ltd., Genesis Centre, Science Park South, Birchwood, Warrington WA3 7BH, GREAT BRITAIN.
Spectron Instrument Corp., 1342 W. Cedar Ave., Denver, CO 80223, USA.
Spectronics Corp., 956 Brush Hollow Rd., P.O. Box 483, Westbury, NY 11590, USA.
SpectroVision Inc., 25 Industrial Ave., Chelmsford, MA 01824, USA.
Spectrum Medical Industries Inc., 8430 Santa Monica Blvd., Los Angeles, CA 90069, USA.
Spectrum Scientific Inc., 15413 Vantage Pkwy, East Houston, TX 77032, USA.
Spellman High Voltage Electronics Corp., 7 Fairchild Ave., Plainview, NY 11803, USA.
Spinco Division, Beckman Instruments Inc., Palo Alto, CA 94304, USA.
SPIRAL Sarl, 3, rue des Mardors, Z.A. Couternon, F-21560 Arc-sur-Tille, FRANCE.
SRI Instruments, 548 South Gertruda Ave., Redondo Beach, CA 90277, USA.
Sterling Organics Ltd., Hadrian House, E. Gefield Ave., Fawdon, Newcastle-on-Tyne NE3 3TT, GREAT BRITAIN.
Sterogene Bioseparations, 140 E. Santa Clara Street, Arcadia, CA 91006, USA.
St. John Associates Inc., 4805 Prince George's Ave., Beltsville, MD 20705, USA.
Stratagene Inc., 11099 N. Torrey Pines Rd., La Jolla, CA 92037, USA.
Strawberry Tree Inc., 160 S. Wolfe Rd., Sunnyvale, CA 94086, USA.
Sugiyama Shoji Co., Ato Bldg., 1-2-7 Shibadaimon, Minato-ku, Tokyo 105, JAPAN.
Sun Brokers Inc., P.O. Box 2230, Wilmington, NC 28405, USA.
Supelco Inc., Supelco Park, Bellefonte, PA 16823, USA.
Suprex Corporation, 125 William Pitt Way, Pittsburgh, PA 15238, USA.
Swagelok Co., 31400 Aurora Rd., Solon, OH 44139, USA.
Sycopel Scientific Ltd., The Laboratory, Station Road, E. Bolden, Tyne and Ware NE36 OEB, GREAT BRITAIN.
Synchrom Inc., P.O. Box 310, Lafayette, IN 47902, USA.
Synthetic Peptides Inc., Dep. of Biochemistry, University of Alberta, Edmonton, Alta. T6G 2P5, CANADA.
Systems Instrument Corp., America/SICA, 106 Centre St., Dover, MA 02030, USA.
Tecan U.S., P.O. Box 8101, Hillsborough, NC 27278, USA.
Techlab GmbH, Evessener Str. 2, D-3305 Erkerode, GERMANY.
Technical & Analytical Solutions, 49 Pelham Street, Ashton under Lyne, Lancashire OL7 ODT, GREAT BRITAIN.
Technical Assoc., 7051 Eaton Ave., Canoga Park, CA 91303, USA.
Technicon Industrial Systems, 511 Benedict Ave., Tarrytown, NY 10591, USA.
Technimed Corp., 4987 NW 23rd Ave., Ft. Lauderdale, FL 33309, USA.
Technology Applications Inc., 26 West Martin Luther King Drive, Cincinnati, OH 45219, USA.
Tegal Scientific Inc., P.O. Box 5905, Concord, CA 94524, USA.
Tekmar Co., P.O. Box 371856, Cincinnati, OH 45222-1856, USA.
Teknivent Corp., 11684 Lilburn Park Rd., St. Louis, MO 63146, USA.
Tescom Corp., 12616 Industrial Blvd., Elk River, MN 55330, USA.
Tessek, 1070 Catleton Way, Sunnyvale, CA 94087, USA.
Texas Instruments, P.O. Box 655012, Dallas, TX 75240, USA.
Thermedics Inc., 470 Wildwood Street, Woburn, MA 01888-1799, USA.
Thermo Environmental Instruments Inc., 8 West Forge Parkway, Franklin, MA 02038, USA.
TM Analytic Inc., 303 E. Robertson, Brandon, FL 33511, USA.
Tohoku Electronic Industrial Co., 6-6-6 Shirakashidai, Rifu-cho, Miyagi 981-01, JAPAN.
Tokyo Kasei Kogyo Co., 3-9-4 Nihonbashi-Honcho, Chuo-ku, Tokyo 103, JAPAN.
Tokyo Rikakikai Co., Toei Bldg. 4-3, 4-chome, Muromachi, Nihombashi, Chuo-ku, Tokyo, JAPAN.
Toso Haas, Independence Mall West, Philadelphia, PA 19105, USA.
Tosoh Corporation, 4560 Tonda, Shinnanyo, Yamaguchi 746, JAPAN.
Toyo Soda Mfg. Co., 1-14-15 Akasak, Minato-ku, Tokyo 107, JAPAN.
Trace Analytical, P.O. Box 2523, Stanford, CA 94305, USA.
Tracor Instruments, 6500 Tracor La., Austin, TX 78725, USA.

Triangle Laboratories Inc., P.O. Box 13485, Research Triangle Park, NC 27709, USA.
Trivector Systems International Ltd., Sunderland Rd., Sandy, Beds. SG19 1RB, GREAT BRITAIN.
Tudor Scientific Glass Co., 555 Edgefield Rd., Belvedere, SC 29841, USA.
Tyler Research Corp., 6128 103 St., Edmonton, Alta. T6H 2H8, CANADA.
Ultra-Lum Inc., 217 E. Star of India La., Carson, CA 90746, USA.
Ultra Scientific Inc., 1 Main St., Hope, RI 02831, USA.
Ultra-Violet Products Ltd., Science Park, Milton Rd., Cambridge CB4 4BN, GREAT BRITAIN.
U-MicroComputers Ltd., Winstanley Industrial Estate, Long Lane, Warrington WA2 8PR, GREAT BRITAIN.
Unimetrics Corp., 501 Earl Rd., Shorewood, IL 60436, USA.
Union Carbide Industrial Gases, 200 Cottontail La., Somerset, NJ 08873, USA.
United States Biochemical Corp., P.O. Box 22400, Cleveland, OH 44122, USA.
Universal Absorbents Inc., 2801 Bankers Industrial Dr., Atlanta, GA 30360, USA.
Universal Biochemicals, 6 Sathya Sayee Bagar, Madurai, Tamil Nadu 625 003, INDIA.
Universal Scientific Inc., 2801 Bankers Industrial Dr., Atlanta, GA 30306, USA.
Upchurch Scientific Inc., 2969 N. Goldie Rd., Oak Harbor, WA 98277, USA.
USA Scientific Plastics, P.O. Box 3565, Ocala, FL 32678, USA.
UTI Instruments Co., 325 N. Mathilda Ave., Sunnyvale, CA 94063, USA.
UTI-tect Inc., 2233-F Northwestern Ave., Waukegan, IL 60087, USA.
UVP Inc., 5100 Walnut Grove Ave., San Gabriel, CA 91778, USA.
Vangard International, 1111-A Green Grove Rd., P.O. Box 308, Neptune, NJ 07754-0308, USA.
Varex Corporation, Burtonsville Commerce Center, 4000 Blackburn La., Burtonsville, MD 20866, USA.
Varian Instrument Group, 2700 Mitchell Drive, Walnut Creek, CA 94598, USA.
Verifio Corp., 250 Canal Blvd., Richmond, CA 94804, USA.
Vested Corp., 9299 Kirby Dr., Houston, TX 77054, USA.
Vetter Laborgeräte GmbH, Postfach 1348, Rudolf-Diesel-Str. 21, Bad Wurt/BRD, D-6908 Wiesloch 4 (Baiertal), GERMANY.
VG Analytical Ltd., Floats Road, Wythenshawe, Manchester M23 9LE, GREAT BRITAIN.
VG Masslab Ltd., Tudor Rd., Altrincham WA14 5RZ, GREAT BRITAIN.
VICI Valco Instruments Co., P.O. Box 55603, Houston, TX 77255, USA.
Vickers Instruments Inc., P.O. Box 99, 300 Commercial St., Malden, MA 02148, USA.
Violet, Via Giovanni Giorgi No. 22, I-00149 Roma, ITALY.
VWR Scientific, P.O. Box 7900, San Francisco, CA 94120, USA.
Waitaki International Biosciences, 55 Glen Scarlett Rd., Toronto, Ont. M6N 1P5, CANADA.
Wako Chemicals USA Inc., 12300 Ford Rd., Ste. 130, Dallas, TX 75234, USA.
Wakunaga Pharmaceutical Co., Shimokotachi 1624, Koda-cho, Takata-gun, Hiroshima 729-64, JAPAN.
Walden Precision Apparatus Ltd., The Old Station, Linton, Cambridge, CB1 6NW, GREAT BRITAIN.
Wale Apparatus Co., 400 Front St., P.O. Box D, Hellertown, PA 18055, USA.
Waters Chromatography, 34 Maple Street, Milford, MA 01757, USA.
Watson Products Inc., 1068 1/2 N. Allen Ave., Passadena, CA 91104, USA.
Wescan Instruments Inc., 3018 Scott Blvd., Santa Clara, CA 95054-0984, USA.
Whatman Ltd., Springfield Mill, Maidstone, Kent ME14 2LE, GREAT BRITAIN.
Wheaton Scientific, 1301 N. Tenth St., Millville, NJ 08332, USA.
Wilmad Glass Co., Route 40 & Oak Rd., Buena, NJ 08310, USA.
Wyatt Technology, P.O. Box 3003, Santa Barbara, CA 93130, USA.
Xavier Industries, 3627 W. Warner Ave., Santa Ana, CA 92704, USA.
Xerox Analytical Laboratories, 0114-42D Joseph C. Wilson Ctr. for Technology, Rochester, NY 14644, USA.
Xydex Corporation, 4 Alfred Cr., Bedford, MA 01730, USA.
YMC Inc., 51 Gibraltar Dr., Ste. 2D2, Morris Plains, NJ 07950, USA.
Yokagawa Corp., 200 W. Park Dr., Peachtree City, GA 30269, USA.
J. Young (Scientific Glassware) Ltd., 11 Colville Rd., London W3 8B3, GREAT BRITAIN.
Zeta-Meter Inc., 50-17 5th Street, Long Island City, NY 11101, USA.
Zinsser Analytic GmbH, Eschborner Landstr. 135, D-6000 Frankfurt 94, GERMANY.
Zymed Laboratories Inc., 52 S. Linden Ave., Ste. 4, South San Francisco, CA 94080, USA.

Subject Index

B

C

D

E

N

O

P

Q

R

S

T

U

V

W

JOURNAL OF CHROMATOGRAPHY LIBRARY

A Series of Books Devoted to Chromatographic and Electrophoretic Techniques and their Applications

Although complementary to the *Journal of Chromatography*, each volume in the Library Series is an important and independent contribution in the field of chromatography and electrophoresis. The Library contains no material reprinted from the journal itself.

Other volumes in this series

Volume 1	**Chromatography of Antibiotics** (*see also* Volume 26) by G.H. Wagman and M.J. Weinstein
Volume 2	**Extraction Chromatography** edited by T. Braun and G. Ghersini
Volume 3	**Liquid Column Chromatography**. A Survey of Modern Techniques and Applications edited by Z. Deyl, K. Macek and J. Janák
Volume 4	**Detectors in Gas Chromatography** by J. Ševčík
Volume 5	**Instrumental Liquid Chromatography**. A Practical Manual on High-Performance Liquid Chromatographic Methods (*see also* Volume 27) by N.A. Parris
Volume 6	**Isotachophoresis**. Theory, Instrumentation and Applications by F.M. Everaerts, J.L. Beckers and Th.P.E.M. Verheggen
Volume 7	**Chemical Derivatization in Liquid Chromatography** by J.F. Lawrence and R.W. Frei
Volume 8	**Chromatography of Steroids** by E. Heftmann
Volume 9	**HPTLC — High Performance Thin-Layer Chromatography** edited by A. Zlatkis and R.E. Kaiser
Volume 10	**Gas Chromatography of Polymers** by V.G. Berezkin, V.R. Alishoyev and I.B. Nemirovskaya
Volume 11	**Liquid Chromatography Detectors** (*see also* Volume 33) by R.P.W. Scott
Volume 12	**Affinity Chromatography** by J. Turková
Volume 13	**Instrumentation for High-Performance Liquid Chromatography** edited by J.F.K. Huber
Volume 14	**Radiochromatography**. The Chromatography and Electrophoresis of Radiolabelled Compounds by T.R. Roberts
Volume 15	**Antibiotics**. Isolation, Separation and Purification edited by M.J. Weinstein and G.H. Wagman

Volume 16 **Porous Silica**. Its Properties and Use as Support in Column Liquid Chromatography
by K.K. Unger

Volume 17 **75 Years of Chromatography — A Historical Dialogue**
edited by L.S. Ettre and A. Zlatkis

Volume 18A **Electrophoresis**. A Survey of Techniques and Applications. **Part A: Techniques**
edited by Z. Deyl

Volume 18B **Electrophoresis**. A Survey of Techniques and Applications. **Part B: Applications**
edited by Z. Deyl

Volume 19 **Chemical Derivatization in Gas Chromatography**
by J. Drozd

Volume 20 **Electron Capture**. Theory and Practice in Chromatography
edited by A. Zlatkis and C.F. Poole

Volume 21 **Environmental Problem Solving using Gas and Liquid Chromatography**
by R.L. Grob and M.A. Kaiser

Volume 22A **Chromatography**. Fundamentals and Applications of Chromatographic and Electrophoretic Methods. **Part A: Fundamentals and Techniques**
edited by E. Heftmann

Volume 22B **Chromatography**. Fundamentals and Applications of Chromatographic and Electrophoretic Methods. **Part B: Applications**
edited by E. Heftmann

Volume 23A **Chromatography of Alkaloids. Part A: Thin-Layer Chromatography**
by A. Baerheim Svendsen and R. Verpoorte

Volume 23B **Chromatography of Alkaloids. Part B: Gas–Liquid Chromatography and High-Performance Liquid Chromatography**
by R. Verpoorte and A. Baerheim Svendsen

Volume 24 **Chemical Methods in Gas Chromatography**
by V.G. Berezkin

Volume 25 **Modern Liquid Chromatography of Macromolecules**
by B.G. Belenkii and L.Z. Vilenchik

Volume 26 **Chromatography of Antibiotics. Second, Completely Revised Edition**
by G.H. Wagman and M.J. Weinstein

Volume 27 **Instrumental Liquid Chromatography**. A Practical Manual on High-Performance Liquid Chromatographic Methods. **Second, Completely Revised Edition**
by N.A. Parris

Volume 28 **Microcolumn High-Performance Liquid Chromatography**
by P. Kucera

Volume 29 **Quantitative Column Liquid Chromatography**. A Survey of Chemometric Methods
by S.T. Balke

Volume 45A **Chromatography and Modification of Nucleosides. Part A: Analytical Methods for Major and Modified Nucleosides — HPLC, GC, MS, NMR, UV and FT-IR**
edited by C.W. Gehrke and K.C.T. Kuo

Volume 45B **Chromatography and Modification of Nucleosides. Part B: Biological Roles and Function of Modification**
edited by C.W. Gehrke and K.C.T. Kuo

Volume 45C **Chromatography and Modification of Nucleosides. Part C: Modified Nucleosides in Cancer and Normal Metabolism — Methods and Applications**
edited by C.W. Gehrke and K.C.T. Kuo

Volume 46 **Ion Chromatography.** Principles and Applications
by P.R. Haddad and P.E. Jackson

Volume 47 **Trace Metal Analysis and Speciation**
edited by I.S. Krull

Volume 48 **Stationary Phases in Gas Chromatography**
by H. Rotzsche

Volume 49 **Gas Chromatography in Air Pollution Analysis**
by V.G. Berezkin and Yu. S. Drugov

Volume 50 **Liquid Chromatography in Biomedical Analysis**
edited by T. Hanai

Volume 51A **Chromatography, 5th edition.** Fundamentals and Applications of Chromatography and Related Differential Migration Methods. **Part A: Fundamentals and Techniques**
edited by E. Heftmann

Volume 51B **Chromatography, 5th edition.** Fundamentals and Applications of Chromatograpy and Related Differential Migration Methods. **Part B: Applications**
edited by E. Heftmann